AF694791

Modern Project Management

Holger Timinger

Modern Project Management

Successful projects with plan-based, agile and hybrid approaches

WILEY

WILEY-VCH GmbH

The Author

Prof. Dr. Holger Timinger
Mörikeweg 16
84036 Landshut

Cover Picture: Leigh Prather - stock.adobe.com

Copy editing: Sparks Publishing Services Ltd

All books published by **Wiley-VCH** are carefully produced. Nevertheless, authors, editors, and publisher do not warrant the information contained in these books, including this book, to be free of errors. Readers are advised to keep in mind that statements, data, illustrations, procedural details or other items may inadvertently be inaccurate.

Bibliographic information published by the Deutsche Nationalbibliothek

The Deutsche Nationalbibliothek lists this publication in the Deutsche Nationalbibliografie; detailed bibliographic data are available on the Internet at <http://dnb.d-nb.de>.

© 2025 Wiley-VCH GmbH, Boschstr. 12, 69469 Weinheim, Germany

All rights reserved (including those of translation into other languages). No part of this book may be reproduced in any form – by photoprinting, microfilm, or any other means – nor transmitted or translated into a machine language without written permission from the publishers. Registered names, trademarks, etc. used in this book, even when not specifically marked as such, are not to be considered unprotected by law.

Print ISBN: 978-3-527-53058-8
ePub ISBN: 978-3-527-85063-1

Typesetting: Straive, Chennai, India

Contents

Introduction

About this book

This is a book about modern project management, which should enable you to use suitable processes, methods, tools and roles for successfully manage your projects. You will learn what characterizes traditional, plan-based project management and in which situations it has strengths and weaknesses. Together, we will work through important fundamentals of agile project management and examine when and where we use them in whole or in part. We will then combine traditional, plan-based and agile approaches, *when it makes sense.*

In this book, the terms "traditional project management" and "plan-based project management" are used synonymously. "Traditional" is intended to express that the corresponding processes, methods and roles have been known and successfully used for many years or even decades. Sometimes such projects are also referred to as predictive. Predictability is based on plans for what the future of the project will look like. The great importance of plans is an expression of such an approach, which is why the term "plan-based project management" is becoming increasingly popular. We want to adopt this term in this book.

To help you choose the right process model, I will go into detail about the HyProMM structuring framework. HyProMM stands for **Hybrid Project Management Model**. The model supports you in choosing the right processes, methods, tools and roles for your projects. Ideological barriers between traditional, plan-based and agile perspectives are overcome. Instead, you can select suitable components according to the situation and combine them into a holistically optimized project management process model.

Digital businesses, **agile transformation** and technology and innovation management are buzzwords that concern many companies. The COVID-19 pandemic, collapsing supply chains and other crises have shown how drastically and quickly the business and market environment can change. Companies must cope with such challenges by shortening their innovation cycles and increasing their competitiveness. There were and are a number of prominent companies that dominated their respective markets not so long ago and no longer exist today. The latter must of course be prevented from the perspective of the respective companies. Efficient and effective project management can make an important contribution to this by supporting change and enabling a rapid response to a changing environment. This book will help you with your agile transformation and show you how to organize your project management accordingly.

You will learn the basics of leadership so that you can support and manage projects. I will go into detail about the special characteristics of leadership in a project environment. And since you are usually not alone in the company with your project, we will also look at the interfaces to cross-project organizational units such as programs and portfolios.

Aim of this book

This book teaches you the **basics of plan-based and agile project management**. It shows you how you can combine the best of these two worlds to create what is known as **hybrid project management** - and tailor it to your specific project. Although the book uses current standards such as the Individual Competence Baseline (ICB) 4.0 of the International Project Management Association (IPMA) and the PMBOK of the Project Management Institute (PMI), it goes beyond them. The aim is to select and adapt the project management to the needs of your project rather than to strictly follow one of the given standards.
In order to provide you with a broad range of tools for the successful management of your projects, we will also systematically and comprehensively cover hybrid project management using the innovative HyProMM framework. The framework structures project management processes, methods, tools and roles. You can think of it like a construction kit. You will learn when to use which components of the kit. So that you can apply your newly acquired knowledge as quickly as possible in your respective environment, there are many **practical examples** and **case studies** with solutions in each chapter.
This book contains a toolbox with many plan-based and agile methods. This provides you with a comprehensive **reference book** and enables you to expand your knowledge and select the right methods and tools depending on the situation.

Target audience

This book has been written for readers who want to acquire in-depth knowledge in the areas of plan-based, agile and hybrid project management in a short period of time, for example in order to accompany the agile transformation in their companies. With its clearly structured layout, the presentation of different process models, methods and tools and the discussion of practical examples and case studies, it is aimed at both **students** and **participants in further education seminars** as well as **practitioners** who are looking for suitable tools for specific challenges or want to modernize their company's individual project management.

Prerequisites

The book does not require any special prior knowledge. However, in order to benefit from the content, you should have an interest in questions of project organization, management and leadership.

Structure of the book

In **Chapter 1** we start with **Project management standards and norms**. This ensures that we are guided by internationally established processes and methods. However, we also discuss the limitations of these standards, some of which we have to overcome in order to carry out our own projects as efficiently and effectively as possible.

In **Chapter 2**, you will learn about **plan-based (traditional) process models** of project management. These include, for example, the waterfall and the V-model. I go into typical methods such as project structuring and discuss the strengths and weaknesses of plan-based approaches.

Plan-based project management itself is presented in **Chapter 3**. **Chapter 4** is dedicated to **agile project management**. You will learn about its origins and current representatives such as Scrum and Kanban. We will discuss the advantages and disadvantages compared to plan-based process models and start to think about possible combinations of plan-based and agile approaches.

We will then expand on this in **Chapter 5** on **hybrid project management**. The aforementioned HyProMM framework for combining plan-based and agile processes, methods, tools and roles is presented and explained in detail. This framework should enable you to derive and implement hybrid process models that suit you and your projects.

In **Chapter 6** you will learn the basics of **leadership**. We focus in particular on typical project management constellations and address plan-based as well as agile and hybrid projects and their special features in terms of leadership. Modern project management does not end at the project boundaries, but links the projects with the corporate strategy, other projects, programs and portfolios. We look at these **interfaces of hybrid projects to programs and portfolios** in **Chapter 7**.

Chapter 8 presents the **toolbox** of plan-based, agile and hybrid processes and methods. The aforementioned HyProMM framework now enables you to combine these processes and methods into individual, customized process models. This gives you the tools to successfully define, plan, manage and complete projects in different constellations.

In addition to the toolbox, the **glossary** in **Chapter 9** will help you if you want to look up terms and definitions. The **Certificate in Modern Project Management** is presented in **Chapter 10**. Finally, in **Chapter 11** you will find the **solutions to all practical examples and case studies** of this book.

Use of colors, symbols and formats

The symbols and markings in the book are largely self-explanatory: important terms that you should not overlook when reading quickly are printed in **bold**. *Italics* is used for emphasis in context.

The book uses symbols and special formats to help you find your way around the book.

> ***Definitions** in this book are used to create a standardized terminology in order to be able to communicate clearly and professionally with other people in the project. Definitions are in italics; the term to be defined is additionally printed in bold. All definitions can also be found in the book's glossary.*

Tip

Tips are presented in this book like this box. There are similar formats for warnings and to highlight places where caution is required.

At the end of each chapter you will find practical examples and case studies. Key questions have been formulated for many of the case studies, the answers to which can be found in Chapter 11. Some of the chapters contain short examples, which are marked as follows:

Example

Examples of projects include the organization of a vacation trip or company party, the improvement of company processes, the construction of buildings and the development of new products and services.

For better orientation in the book, color codes are used that immediately indicate whether we are talking about plan-based, agile or hybrid project management. Separate colors are also used for the topic of leadership and the interfaces between hybrid projects and programs and portfolios. In many chapters, these color codes are printed in the outer margins to make it easier for you to navigate through the book. The color codes used can be taken from Figure 1.

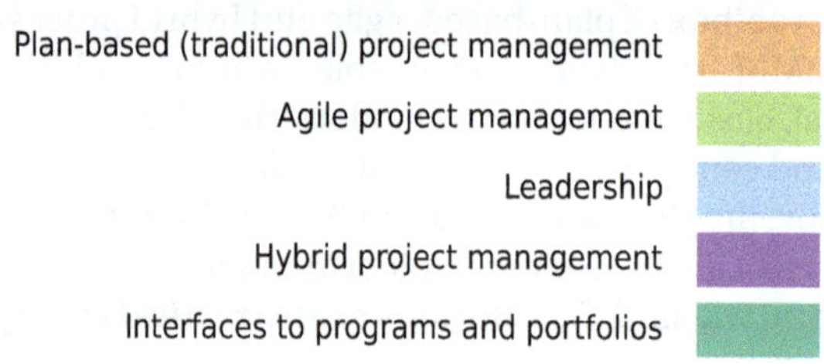

Figure 1: Color codes for different topics and chapters of this book.

A final note: The book contains many practical examples and case studies based on the author's experience. To make the book easier to read, people and companies are sometimes given specific names. However, these have been changed to protect real people and companies. The names used are purely fictitious. Any similarities with real persons or companies are purely coincidental.

1 Standards and norms in project management

In this Chapter ...

- you will learn about common project management standards and norms and
- you will acquire the skills to evaluate these standards and to identify and apply the standard that is right for you and your project.

Overview

Before we begin to take a closer look at projects, project management and standards, we define important terms in order to develop a common understanding:

> *A **project** is usually a unique endeavor that can be distinguished from other tasks. It has limited time, financial, personnel and material resources. Projects pursue defined goals and have a project-specific organization.*
>
> ***Project management** refers to the entirety of tasks, methods and means from the areas of definition, planning, control, closure and management for the successful completion of projects.*

Anyone managing a small project in the private sector, in a company or other organization probably does not think about whether the underlying procedures and methods used are standardized or not. Small and medium-sized companies are reluctant to use the major project management standards because they fear that the project management process will be too bureaucratic and not tailored to their situation [Turner et al., 2010] or because they are not familiar with the established standards in detail. For projects with a complex project scope or many different stakeholders such as

- project owner or project sponsor
- project team members
- other companies, for example suppliers
- other projects with at least partially shared resources

a lack of common understanding of the procedure makes collaboration more difficult. Typical collaboration interfaces are illustrated in Figure 1.1. If the approach to project definition, planning and management is not clear to the people involved or is communicated in a misleading way, plans, key performance indicators for project control and expectations do not match. This prevents or makes it more difficult to achieve the project's objectives.

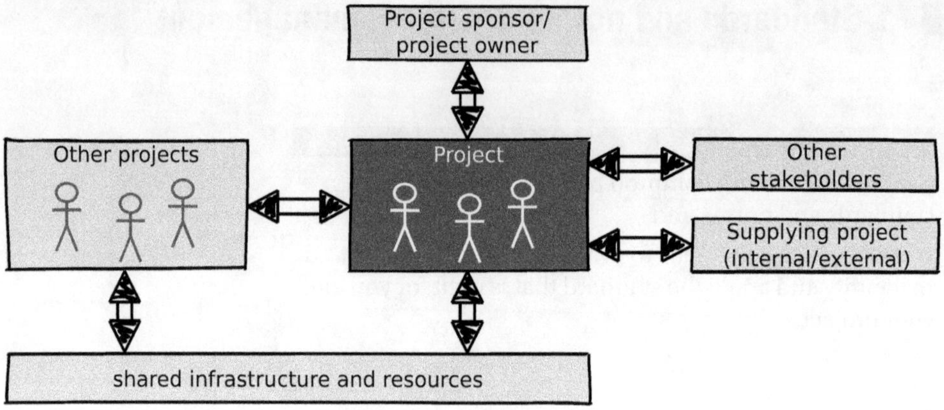

Figure 1.1: A project at the center with interfaces to the project owner or project sponsor, other projects, suppliers, other stakeholders such as a steering committee and infrastructure shared with others, such as software and reporting and shared resources.

Project management helps us on different levels:

- With project management, we can develop the **strategic objectives** of a company, formulate a vision and derive a suitable mission.
- The **measures to achieve** the strategic goals can also be identified, initiated and controlled with project management.
- In general, good project management helps us to implement the strategy **operationally** and to implement many initiatives effectively and efficiently.

In order to develop a common understanding of the aforementioned terms, I propose the following definitions:

> *The **vision** of an organization makes statements about its long-term goal (what are we doing?), motivation (why are we doing this?) and target group (who are we doing this for?)*

> *The **mission** of an organization describes the way to implement the vision. It defines how the vision is to be achieved (daily actions), which offers (products, services, etc.) characterize the organization and which values are important in the process.*

Vision and mission are the basis of the strategy

> *A **strategy** represents a fundamental, long-term behavior of an organizational unit towards its environment to achieve long-term goals.*

The strategy must be scrutinized from time to time and adapted to the environment. New technologies, markets and behavioral habits of the target group can remove the basis (business model) of an organization. Here, too, good project management helps to take appropriate measures at an early stage and when the need for change is identified.

Example

In the 20th century, a number of mail order companies emerged that offered an extremely wide range of goods in thick, printed catalogs. Orders were placed by letter, fax or telephone. Around the turn of the millennium, the internet became increasingly efficient and online providers attracted customers with simple ordering processes, generous return policies and fast deliveries. It turned out that many of the previously successful companies were unable to make the transition from analog mail order to online shops and have effectively ceased to exist today. Others, on the other hand, recognized the signs of the times early on, invested in their own expertise to develop online-based retail and are still successful on the market today.

There is one more term we should look at, as it is often mentioned in the context of strategy and project management: VUCA.

> *__VUCA__ stands for __V__olatility, __U__ncertainty, __C__omplexity and __A__mbiguity. VUCA describes a constantly changing complex environment (business world).*

Many organizations perceive their current environment as very complex, volatile (in the sense of unstable) and characterized by uncertainty. For a long time, the real and concrete dangers of pandemics, wars and totalitarian countries were ignored. However, the impact on your own organization can be considerable. Added to this are the aforementioned technological changes brought about by digitalization, social change and changing habits and values.

This makes it all the more important for organizations to focus on their project management. This is one of the basic prerequisites for recognizing the need for change at an early stage and implementing the necessary measures quickly. Established standards and norms can provide initial guidance on the right project management approach.

Many companies are starting to assign all employees who

- lead projects
- work on them or
- work in line functions with interfaces to the projects,

according to one of the major project management standards.

The top priority when selecting and adapting a project management standard should be its suitability for the project and the company's working methods. We will discuss criteria and the procedure for selecting and adapting project management standards later on.

Example

If a company does not have a standardized approach to projects, a common set of methods or a standardized use of software, project employees must first be trained in the processes, methods and software to be used for each project. A project handover or the spontaneous involvement of additional or other employees for support is hardly possible. This makes project implementation inefficient and control options in the form of flexible resource deployment more difficult. In addition, a lack of a uniform approach makes project learning more difficult, as there is no common basis for improvements.

However, standardization also entails risks: If a standard is chosen that does not suit the company and its projects, the problems multiply. It is not a single project that is then inefficient or ineffective, but all projects that adhere to this unsuitable standard.
In this chapter, we briefly summarize some common standards. These include

- the Project Management Body of Knowledge (PMBOK) of the Project Management Institute
- the PRINCE2 project management system of the British company Axelos Ltd.
- the Individual Competence Baseline ICB 4.0 of the International Project Management Association,
- the international ISO 21502 series of standards and
- the series of German standards DIN 69901.

There are also standardization efforts in agile project management. I will therefore briefly introduce the requirements of the Scrum Alliance and Scrum.org.
Efforts are also being made in the academic sector to make qualifications in project management comparable. For example, several universities have created a competence framework "Modern Project Management" that introduces students to a modern understanding of project management that takes plan-based, agile and hybrid process models and methods into account.
All of the standards presented here can largely be understood as a framework that needs to be further refined with specific processes, methods and roles. The standards leave more or less room for individual adaptation. Even if most of the standards mentioned are updated from time to time, you will notice that some of the standards date back to a time when agile project management approaches did not yet play a significant role. Current versions now attempt to incorporate agile principles, which is not always successful without contradictions.
The following sections are intended to give you an initial impression of the characteristics of the standards mentioned. This should help you in later decisions for or against the selection and adaptation of a standard.
In the following Chapters 3 and 4, you will learn about plan-based, traditional and agile process models. With this knowledge, we will develop criteria for the selection of one or more suitable process models and their adaptation to company-specific requirements.

PMI Project Management Body of Knowledge

Introduction

The Guide to the Project Management Body of Knowledge, or PMBOK® Guide [PMI, 2021] for short, published by the Project Management Institute (PMI) is also a standard of the American National Standards Institute (ANSI). The current version of the PMBOK® Guide dates from 2021 and has the version number 7.
This latest version brings with it many structural and content changes. Until the previous version 6, the PMBOK® was very process-oriented. Instead of the process orientation, the latest edition now focuses on a value and principle-oriented perspective. However, the PMI

emphasizes that process orientation is still relevant in practice and that there will be many organizations that will continue to follow the predecessor standard.

The PMBOK® is divided into two main parts:

- *Standard for project management*: The standard contains a system of value delivery and 12 principles for project management instead of the process groups initiation, planning, execution, monitoring and control as well as closure.
- *Guide to the Project Management Body of Knowledge*: The previous ten knowledge domains (integration, content and scope, schedule, cost, quality, resources, communication, risk, procurement and stakeholders) have been abandoned. Instead, eight performance domains are now used and the topics of tailoring, models, methods and artifacts are included.

The new structure of the PMBOK® is illustrated in Figure 1.2.

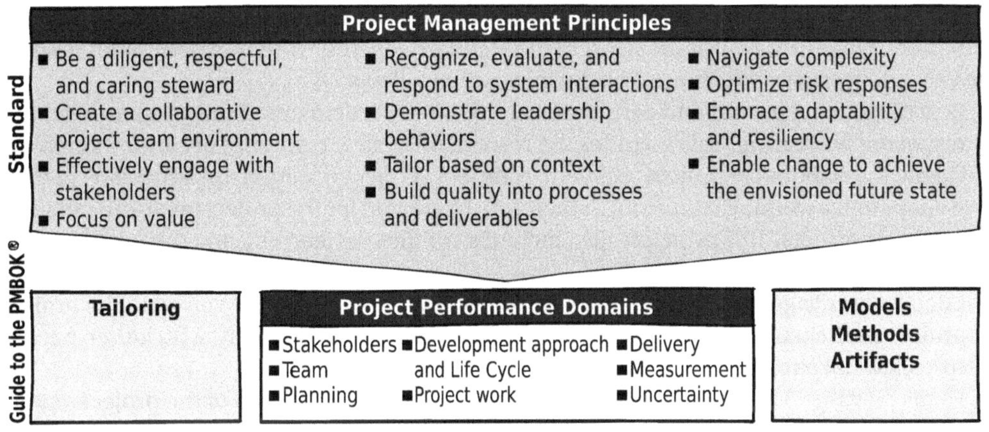

Figure 1.2: Structure of the standard for project management and the performance domains. Own illustration based on [PMI, 2021].

With these changes, agile and hybrid project management approaches have also been integrated into the PMBOK®. The PMI offers various certification options for people working in the project business based on the PMBOK® Guide:

- CAPM® – Certified Associate in Project Management: Basic certification for people working on projects
- PMP® – Project Management Professional: Certification for project managers who can demonstrate experience in managing projects
- PgMP® – Program Management Professional: Certification for managers who lead several complex projects and are responsible for their strategic and economic objectives
- PfMP® – Portfolio Management Professional: Certification for managers who manage and are responsible for entire portfolios.

There are other certificates, such as the PMI Agile Certified Practitioner or the PMI Professional in Business Analysis. The aforementioned certifications do not build on each other hierarchically. For example, the PMP® certificate does not require prior CAPM® certification.

For students there is the "PMI Project Management Ready" certificate, which does not require any practical experience.

Standard for project management according to PMBOK®

The standard for project management is based on a code of ethics and professional conduct with four fundamental values and 12 principles based on them. The values and principles form the basis for the people working on the project. The four values are as follows:

- Responsibility
- Respect
- Fairness
- Honesty

According to the PMI, these values form a moral basis on which we should base our actions. Together with people from the field, 12 principles were developed for more concrete implementation, which are now at the heart of the standard together with the orientation towards the value delivery system. These 12 principles are as follows:

- Be a diligent, respectful and caring steward: Project participants should be hardworking, respectful and caring. This includes the responsible use of resources of various kinds.
- Create a collaborative project team environment: Project results are produced by people. A cooperative working culture must therefore be created for the project team and potential external partners. This principle also includes continuous learning and the further development of team skills.
- Effectively engage with stakeholders: In addition to those directly involved in the project, project participants are also responsible for dealing appropriately with other people, so-called stakeholders.
- Focus on value: Value can be created during the project, at the end of the project or after the project has been completed. At the same time, value creation is an important indicator of project success. The evaluation of value creation should therefore be part of regular project management.
- Recognize, evaluate and respond to system interactions: The standard understands projects as a system and requires project participants to think and act in terms of systems. This also includes addressing the dynamics of systems.
- Demonstrate leadership behaviors: Effective leadership requires appropriate leadership behavior that takes into account the values of the standard.
- Tailor based on context: The so-called continuous tailoring (adaptation) of project management to the (possibly changing) environment of the project is an important success factor.
- Build quality into processes and deliverables: Compliance with the agreed acceptance criteria must be ensured. Effective processes must be established for this purpose.
- Navigate complexity: Complexity goes hand in hand with many interfaces and interactions. The degree of complexity can change during the project life cycle. The project participants are aware of this and act accordingly.

- Optimize risk responses: Risks and opportunities are identified and taken into account appropriately. Coping strategies should be planned and implemented appropriately. They are coordinated with relevant stakeholders and are the responsibility of one person.
- Embrace adaptability and resiliency: This principle is about strengthening the ability to adapt to changing conditions or a changing project environment. It also aims to strengthen resilience. This should enable the project team to cope well with setbacks and mistakes.
- Enable change to achieve the envisioned future state: The ability to deal with change and to shape change should be strengthened. Changes may become necessary due to internal and external project influences. A sound stakeholder management and motivational approaches should accompany the change.

Performance domains of a project

The performance domains of a project are not part of the standard, but describe the principles stated in the standard as part of the *Guide to the PMBOK®*. In the current edition, significantly more considerations regarding agile or hybrid project management are now taken into account, so each domain provides suggestions for either more plan-based or agile implementations.

Figure 1.2 summarizes these performance domains in the lower part of the illustration.

Stakeholders represent the first performance domain for which appropriate stakeholder management must be applied. Stakeholder management, in turn, consists of identifying and analyzing stakeholders, prioritizing stakeholders according to their importance, defining measures for stakeholder communication and monitoring stakeholders and measures.

The second performance domain consists of the **team**. Among other things, this involves creating suitable team structures and choosing a suitable management style.

In **planning** domain, all planning measures for addressing the project scope are summarized. In particular, this includes task, schedule, resource and cost-related plans. Emphasis is also placed on questions of estimation accuracy as part of the planning process.

In the domain **development approach and life cycle**, it must be ensured that the appropriate project management is well selected. The selection should take into account the scope of the project and the project environment. Degrees of freedom here are, for example, the choice of predictive (plan-based) or adaptive (iterative) process models and methods.

The **project work** is about creating an effective and efficient working environment for the development of the project object.

The **delivery** of the project object is an important event for evaluating the success of the project. Procedures are used to determine and check the acceptance criteria.

The **measurement** domain contains tasks for recording and analyzing the status of the project.

Dealing with **uncertainty** is characteristic of projects. Uncertainty results from the degree of novelty inherent in projects. Appropriate time or cost-related buffers and risk management, for example, are used to deal with uncertainty.

PRINCE2®

Introduction

PRINCE2® stands for Projects In Controlled Environments and is a project management system consisting of principles, roles and practices. PRINCE originated in the UK, where it became the government standard for IT projects in 1989. In 1996, it was further developed into PRINCE2® and presented as a comprehensive, cross-industry project management system. PRINCE2® is now published and distributed by the British company Axelos Ltd.
The last major revisions were made in 2009, 2017 and 2023. However, to make it clear that PRINCE2® has remained true to its principles, the number 2 has not been changed. Changes from the 6th to the current 7th edition include a stronger focus on people and sustainability context. In addition, the previous themes have been renamed to practices.
For PRINCE2® there is a staged certification system [Axelos, 2022]:
- PRINCE2®-Foundation: basic certification
- PRINCE2®-Practitioner: certification that builds on the foundation level and includes the processing of practical case studies.

As with the PMI, there are also other certificates, such as the PRINCE2-Agile® certificate, which accentuates agile skills.
PRINCE2® is a flexible, customizable project management system. It is able to integrate agile working methods and process models and there are numerous practical examples of this. Figure 1.3 shows an overview of PRINCE2®.

The seven principles of PRINCE2®

The following principles represent the core values of a PRINCE2® project:
- **Ensure continued business justification**
  A PRINCE2® project needs a justified reason to be carried out. This reason can be, for example, a specific strategic contribution or can be proven by a profitability analysis. Projects that lose this justification in the course of the project must be terminated.
- **Learn from experience**
  In order to ensure the transfer of knowledge despite a time-limited project duration and corresponding time-limited project organization PRINCE2® puts great emphasis on knowledge management. Knowledge must be safeguarded and transferred to new projects.
- **Define roles, responsibilities and relationships**
  Roles and responsibilities must be defined, structured and communicated to all relevant stakeholders.
- **Manage by stages**
  PRINCE2® PRINCE2 requires further management phases in addition to the initiation phase. Phase transitions are approved by a project board (steering committee). The standard emphasizes that this principle can also be applied to agile projects.

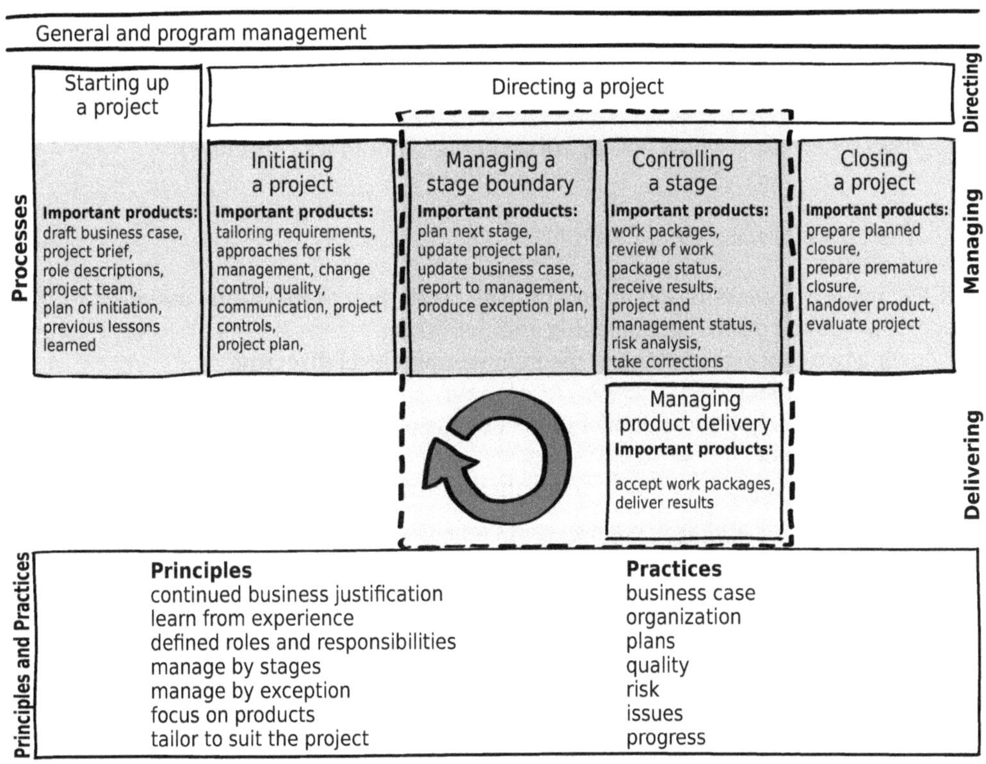

Figure 1.3: Overview of PRINCE2® based on [Murray et al., 2023].

- **Manage by exception**
  Tolerances are defined for important project objectives. These define the scope of action for those responsible. If the tolerances are exceeded, escalation is required.
- **Focus on products**
  The project scope is understood as a product or result. The project must be directed towards this and optimized.
- **Tailor to suit the project**
  PRINCE2® includes a collection of best practices that need to be tailored, i.e. adapted, to the project context (e.g. environment, size, complexity, team).

The seven practices of PRINCE2®

The seven practices of PRINCE2® can be understood as knowledge areas and represent typical project management disciplines. They are intended to answer the questions on project justification, content, structure and status.
The practices and associated questions are listed below:
- business case (Why?)
- organization (Who?)

- plans (How? How much? When?)
- quality (What?)
- risks (What if?)
- issues (What now?)
- progress (Where are we now, where are we going? Should we continue?)

The seven processes of PRINCE2®

PRINCE2® takes a process-based approach. The seven processes can be assigned to the 3 management – levels directing, managing and delivering.
The following two processes belong to the management level **directing**:

- **Starting up a project**
  Examination of the justification of a project by the project board
- **Directing a project**
  Activities of the project board to manage the project

The management level **managing** includes the following five processes:

- **Starting up a project**
  Elaboration of the business case and preparation of the project start by the project manager
- **Starting up a project**
  Definition and planning of the project by the project manager
- **Managing a stage boundary**
  Preparing the next phase, updating the business case, preparing the phase report, obtaining approval for the phase; if there are several phases, this process is run through several times accordingly
- **Controlling a stage**
  Release of new work packages, status review, risk management, initiation and monitoring of control measures; if there are several phases, this process is run through several times accordingly
- **Closing a project**
  Prepare project completion, handover products, evaluate project and recommend completion

The following process belongs to the management level **delivering**:

- **Managing product delivery**
  The team ensures the delivery of deliverables to the project manager. Partial deliveries are also possible with iterative or incremental working methods.

ISO 21502

The ISO 21502 is a rather comprehensive standard, which was released in 2020. It takes account of the project environment and activities before and after conducting the project.

It also contains role descriptions and a lot of definitions, which can help to gain a common understanding.
The central part is about integrated project management practices, which are outlined in Figure 1.4. The sponsoring organization evaluates whether it is worth starting a project as part of the pre-project activities. The organization oversees the project during project execution. One of the tasks in the post-project activities is to ensure that the project results are sustainable for the organization.

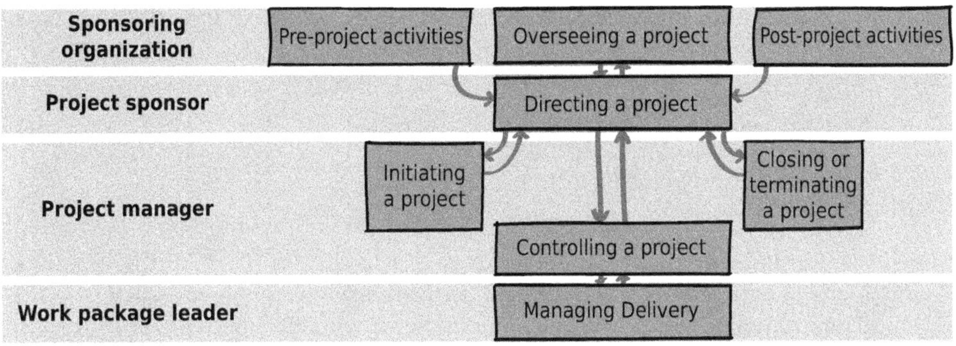

Figure 1.4: Integrated project management practices and associated roles according to ISO 21502 [ISO21502, 2020].

The project sponsor serves as an interface to the project manager, who initiates, manages and ultimately completes the project. Each work package leader ensures the contribution of work package deliverables to the overall project scope.
While the *integrated* project management practices serve as a framework for the overall project management organization, *individual* practices facilitate project execution at the project level. These individual practices are planning, benefits management, scope, resources, schedule, cost, risk, issues, change control, quality, stakeholders, communication, change, reporting, information, procurement and lessons learned. Each of these practices is briefly described in the standard.
Overall, ISO 21502 is a standard that relies heavily on plan-based project management with traditional hierarchies. As such, it does not really do justice to a modern understanding of project management. However, the defined terminology within the standard can help to promote a common global understanding.

DIN 69901

The predecessor of the aforementioned ISO 21502:2020 was ISO 21500:2012, which in turn was heavily influenced by the German standard DIN 69900 and the DIN 69901 series. These are structured as follows:

- DIN 69900: Network technology, description and terms
- DIN 69901-1: Project management systems – Fundamentals
- DIN 69901-2: Project management systems – Processes and process model
- DIN 69901-3: Project management systems – Methods

- DIN 69901-4: Project management systems – Data and data model
- DIN 69901-5: Project management systems – Terms and definitions

There are additional DIN standards related to project management, such as the DIN 69909 series on multi-project management or DIN ISO 10007 on configuration management. The DIN series was published in 2009 and does not include the term "agile project management" or specific process models such as the waterfall or V-model. The DIN series rather represents the plan-based, traditional view of project management with sequential phases, a predictability of the future course of the project that can be achieved through good plans.

IPMA Individual Competence Baseline

Introduction

Version 4.0 of the Individual Competence Baseline (ICB) [IPMA, 2017] of the International Project Management Association (IPMA) has been available since fall 2015. The ICB is widely used internationally with a certain focus on Europe. The ICB does not propose project management processes or methods, but rather defines competencies that are important for successful project management.
In earlier versions, the abbreviation ICB stood for IPMA Competence Baseline. To emphasize more strongly that the standard focuses on individuals working in project management, the new term "Individual Competence Baseline" was chosen. This also makes it easier to classify the standard within other IPMA standards, namely the Organizational Competence Baseline (OCB) and the Project Excellence Baseline (PEB):

- The **OCB** explains how project-oriented organizations function and how their project management can be improved.
- The **PEB** shows how outstanding project and program management skills can be achieved. It helps organizations assess their own capabilities and develop improvements.

In the following explanations, we will limit ourselves to the ICB, i.e. the Individual Competence Baseline [IPMA, 2017]. ICB 4.0 focuses on people and their competencies for processing projects, programs and portfolios. It is internationally harmonized and has been translated into various languages.
The IPMA offers a four-stage certification system via its respective national representatives:

- IPMA® Level D – Certified Project Management Associate: Basic certification for people working on projects
- IPMA® Level C – Certified Project Manager: Certification for project managers with significant professional experience
- IPMA® Level B – Certified Senior Project Manager: Certification for project managers of complex projects and programs
- IPMA® Level A – Certified Project Director: Certification for managers of project-oriented organizations

The individual levels build on each other. However, combination courses are offered so that higher levels can be certified at the same time with the appropriate qualification. The certification usually consists of several examinations (oral and written) as well as proof of experience in the form of written transfer reports or project experience reports. For levels C, B and A, certain practical experience in project management must be demonstrated.
The IPMA/GPM certifications focus on plan-based, traditional project management. Agile or hybrid skills are only examined in peripheral fields. In order to be able to offer something in this area too, some local country representations of IPMA has created the hybrid+ certificate, which focuses on agile and hybrid project management.

Individual competencies

The project management standard Individual Competence Baseline 4.0 (ICB) from the International Project Management Association (IPMA) differs significantly from other standards in terms of its structure. The ICB does not define processes or organizational structures, but focuses on the individual competencies of people involved in the project business.
The ICB defines individual competence as a person's ability to apply knowledge, skills and abilities in such a way that a desired result is achieved. Knowledge, skills and abilities are defined in hierarchical order:

- **knowledge** means knowing something, for example being able to read and understand a certain plan
- **skills** allows us to use the knowledge, for example to create our own plans
- **abilities** build on knowledge and skills and enable their application in a specific context, for example interpreting a plan and deriving control measures in the event of deviations from the plan.

The ICB 4.0 describes a total of 29 different competencies, which are assigned to the following three competence areas:

- **People competencies** must be demonstrated by people in order to successfully participate in or shape projects, programs or portfolios. Examples include self-reflection, communication, project team leadership, teamwork and negotiation skills.
- **Practice competencies** include methods, tools and techniques that are used to successfully implement projects, programs and portfolios, such as competencies for formulating and analyzing goals and requirements, planning, controlling and designing changes.
- **Perspective competencies** help with interaction with the environment. Examples include strategic, structural and procedural thinking as well as dealing with power in the context of working with internal and external stakeholders of a project, program or portfolio.

The three areas of competence are also referred to as the Eye of Competence. The use of the 29 competencies is explained for projects, programs and portfolios.

Green Project Management P5

Introduction

The Green Project Management (GPM) organization was founded in the USA in 2009. It considers itself a global leader in the practice of sustainable project management.
GPM offers training courses on sustainable project management in accordance with its own P5 standard, either online or with accredited training partners. In addition, a certification system is available that offers three levels, some of which build on each other:

- **GPM-b** uses a multiple-choice exam to demonstrate basic knowledge of sustainable project management. The knowledge can be acquired either through GPM training courses or through self-study.
- **GPM-s** focuses on in-depth skills and abilities. Candidates must submit a documented case study. Possession of the GPM-b certificate is not required.
- **GPM-m** is the master level of the certificate system. Candidates must already hold a GPM-s certificate and demonstrate their competencies through a documented case study and by providing references.

The GPM P5 standard

The GPM standard for sustainable project management is called P5 [GPM, 2023]. In 2023, it was published in its 3rd edition and licensed under a creative commons license. P5 stands for the five dimensions:

- products
- process
- people
- planet
- prosperity.

The GPM considers these dimensions to be a logical extension of the iron triangle, which consists of the dimensions scope, time and cost. The P5 encompasses the iron triangle and puts it into a broader context, see Figure 1.5.
The standard contains a so-called ontology, which can be regarded as a structure of 49 elements, i.e. topics relating to sustainability in project management. An excerpt of these elements is shown in Figure 1.5. The P5 standard lists all elements, briefly explains them and relates them to the 17 Sustainable Development Goals (SDGs) of the United Nations [UN, 2015].
The GPM P5 standard is strongly aligned to sustainability. Although other standards such as PRINCE2® have started to integrate sustainability topics as well, the P5 standard goes far beyond that. Anyone interested in creating a project management framework with a strong focus on sustainability will find a lot of relevant and helpful information. In contrast, the standard does not provide enough information on general project management to derive a general project management framework.

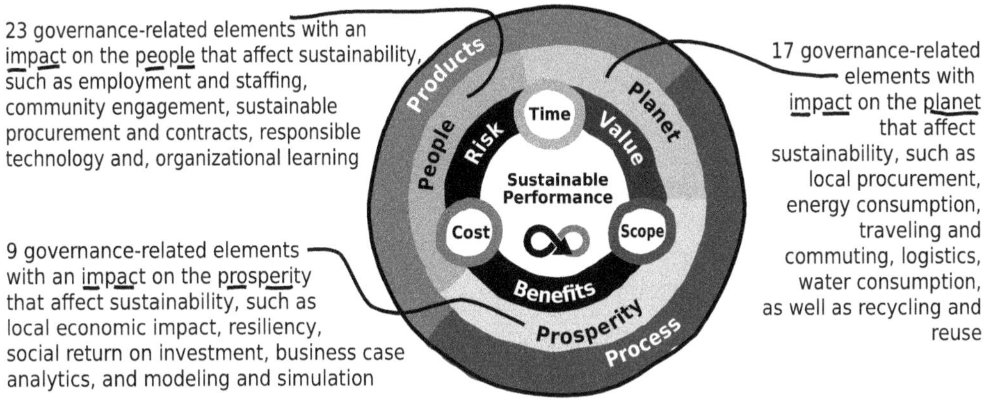

Figure 1.5: Project management dimensions as defined in the GPM P5 standard in accordance with [GPM, 2023].

Agile standards

Introduction

You may now be wondering whether agility and standardization are not mutually exclusive. If we look at the Agile Manifesto [Beck et al., 2001], which you will learn more about in Chapter 4, there are two central values, among others:

- **Individuals and interactions** are valued more than processes and tools.
- **Responding to change** is valued more than following a plan.

Today, there are several standards and certification options for agile process models that consider the values of the agile manifesto. This offers both opportunities and risks: standardization and certification promote visibility and increase the number of people who adopt and disseminate the ideas. It also facilitates collaboration between teams that follow the same standard. Certified experts will write their certification on business cards, publish it on social media and make others curious. They know the underlying considerations that led to agile management and can apply and further develop them. In addition, a common standard can be a good basis for identifying and implementing improvements.
Caution is advised with regard to the ideals of the Agile Manifesto if standardization leads to

- building new hierarchies,
- proceeding like a recipe instead of really reacting to changes, or to
- valuing processes and tools more than individuals and interaction.

Scrum is a widely used process model for agile project management. Scrum is characterized by close, recurring involvement of the customer and an iterative approach. An important document that can be described as the Scrum standard is the Scrum Guide by Ken Schwaber and Jeff Sutherland, which can be obtained free of charge in many languages from www.scrumguides.org.

Important organizations around agile project management

Scrum.org

Scrum.org was founded in 2009 by Ken Schwaber, one of the authors of the Agile Manifesto, to provide an organizational platform for the further development of Scrum.
Scrum.org offers several certifications, including:
- Professional Scrum Master™
- Professional Scrum Product Owner™
- Professional Scrum Developer™

The examinations on which the certificates are based can be taken online. Scrum.org provides extensive literature and videos as well as practice questions for preparation.

Scrum Alliance®

The Scrum Alliance® is an organization with over 400,000 members currently. Its aim is to further establish Scrum. The Scrum Alliance® offers some of the best-known certifications for Scrum. These are structured hierarchically. Depending on the area of application, certification starts with one of the following three:
- Certified Scrum Master®
- Certified Scrum Product Owner®
- Certified Scrum Developer®

Building on this, the Certified Scrum Professional certification can be acquired. The acquisition of the certificates is usually linked to participation in a seminar with an authorized provider.

Other organizations

Today, all major project management organizations, such as the Project Management Institute, the International Project Management Association and Axelos, have also integrated agile working methods and process models into their standards. However, the scope varies and agile concepts sometimes seem like foreign objects in otherwise plan-based project management.
The shift towards an increasing scope of agile content in the project management standards is also reflected in the certifications. For example, agile topics are being added to existing certificates and additional certificates with an agile focus are being offered.

Modern project management

Introduction

Nowadays, project participants are expected to choose the project management method that best suits the respective project task and project environment. Depending on the constellation, this can be plan-based or agile working methods and process models.

Hybrid combinations are also conceivable and are often a very good solution. Hybrid project management refers to the combination of different process models or methods, for example the integration of the agile Scrum process model into a plan-based, phase-oriented project process.
Comprehensive project management skills are required in order to select the right process models for the situation and tailor them accordingly. An initiative has therefore been formed in the academic sector to promote such skills for modern project management. This initiative has developed a system of certificates consisting of three levels: Foundation, Professional and Excellence, see Figure 1.6.

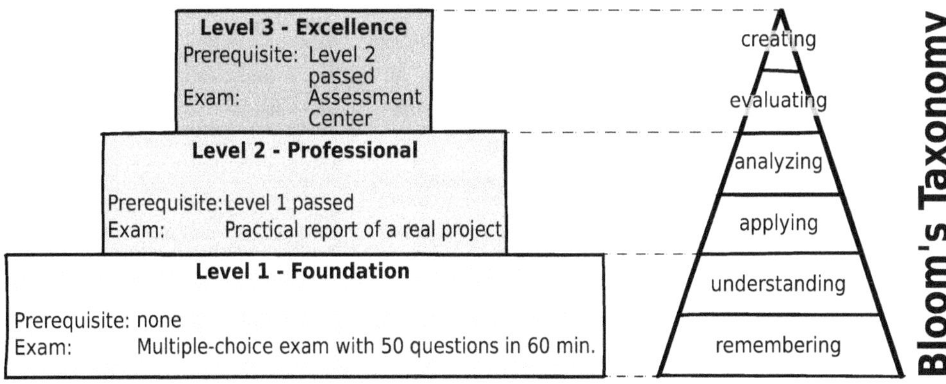

Figure 1.6: Competence levels of the Modern Project Management certificate and mapping to the Bloom taxonomy [Bloom et al., 1956].

The competence levels are based on the Bloom Taxonomy [Bloom et al., 1956], from which the examination formats can also be derived:

- **Level 1 – Foundation:** This level assesses comprehensive knowledge and understanding as well as simple use cases of plan-based, agile and hybrid project management. The exam takes place online as a multiple-choice exam.
- **Level 2 – Professional:** This stage tests practical application and reflection skills. This involves documenting and critically reflecting on the project management of a real (student) project.
- **Level 3 – Excellence:** This level tests the highest competence levels by requiring participants to work on and solve a complex case study as part of an assessment center.

This book covers the essential topics of this certificate, which are summarized in Figure 1.7. There is also a sample exam for Level 1 – Foundation in Chapter 10.

Figure 1.7: Topics of the Modern Project Management certificate, which must be demonstrated in varying degrees of detail depending on the competence level.

At a glance

- Known project management standards are the
  - PMBOK® of the Project Management Institutes (PMI)
  - PRINCE2® from Axelos
  - ISO 21502
  - ICB 4.0 of the International Project Management Association (IPMA)
  - Scrum from Scrum.org or the Scrum Alliance®
- Project managers and employees can be certified in accordance with most standards and thus document their relevant skills.
- Most standards and certificates relate to plan-based project management, which has been known for decades. So far, many standards have only partly included any agile or hybrid approaches.
- There are now increasing efforts to develop a holistic understanding of project management that allows project participants to select and tailor the most suitable process models depending on the situation.

2 Process models for plan-based project management

In this Chapter ...

- you will learn what distinguishes plan-based organized projects
- you will learn why plan-based project management is often also called traditional or predictive project management and
- you will acquire skills for defining, planning and controlling projects that follow sequential, concurrent or repetitive process models.

What does traditional mean here?

Traditional project management refers to a plan-based project process, sometimes also called a predictive process model. At the beginning, scope, time and cost-related targets are defined and plans are drawn up to achieve them. During the execution phase, the plans are implemented and deviations from the plan are minimized.
You may be wondering how modern and traditional project management fit together. At first glance, the two terms "modern" and "traditional" seem contradictory.

> ***Modern project management*** *refers to situationally appropriate project management that facilitates efficient and effective project implementation. It can include both plan-based (traditional, predictive) and agile components.*

Anyone reading current literature on project management will notice that so-called agile project management, such as Scrum and Kanban, is becoming increasingly important. Many companies expect shorter innovation cycles and faster reactions to the market from the use of agile project management.
In contrast to agile project management, non-agile process models, which are still widely used and form the basis for many common certifications, are often referred to as classic, conventional or traditional project management.
These process models are based much more on plans than agile process models, which is why they are usually referred to as plan-based process models in this book.

Tip

The term "traditional project management" should not be understood as a negative label. Traditionally implemented projects are still relevant today. If they are applied correctly and in the right situations, they can lead to excellent project results. In Chapter 5, you will learn to differentiate between situations in which you should use plan-based or agile process models. You will also learn how to combine the two to create so-called hybrid process models.

Structure of process models

Before we start with the presentation of the process models, let's define the term:

> *A **process model** combines methods and elements, processes and phases for a standardized project process. It consists of structural, functional and process-related building blocks.*

Figure 2.1 shows an overview of possible elements of a process model. In practice, not all elements are always present in a process model.

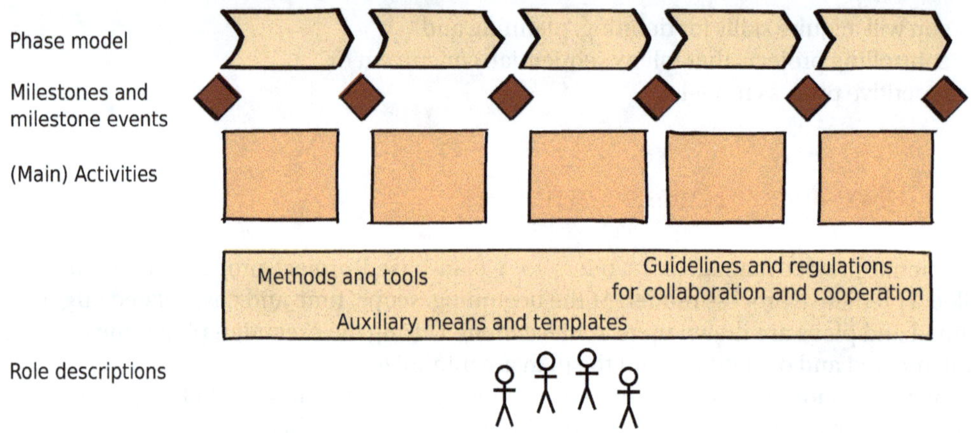

Figure 2.1: Process models can contain the following components: a phase model, milestones with results and main activities to be carried out in the respective phases. They may also integrate methods and tools, rules for processing as well as auxiliary means and templates, such as documentation templates, checklists or templates for planning programs. The roles of the people involved in the project should also be described.

Tip

In the following chapters, you will get to know some common process models in detail. These are so-called **general process models**, also known as universal or cross-industry process models. These are not adapted to a company or a specific industry. In contrast, there are **specific process models**, which – derived from the general process models – have been individually adapted to the needs of a company or industry.

A related term is **project governance**.

> *The **project governance** refers to the management framework in which decisions are made. It comprises rules, processes and structures for the organizational structuring of the project and for ensuring the ability to act.*

General process models often only contain a definition of phases and their chronological order.

> *A **phase** is a time-limited section that differs in content from other sections.*

Specific process models usually also define (at least some of) the other elements explained in Figure 2.1.

In addition to the elements described above, process models can consist of building blocks as described in the FELICS framework (Framework for the Construction and Tailoring of Engineering Development Process Models) [Timinger et al., 2022]. Such building blocks fulfill either structural, functional or process-related requirements:

- **Structural building blocks of process models** structure the course of the project, for example through defined phases and milestones. Role definitions also help to bring order to the project.
- **Functional building blocks of process models** are used to fulfill certain functions. Examples include elements for quality assurance, synchronization of parallel work, communication and collaboration, measurement and knowledge management.
- **Processual building blocks of process models** ensure that the project workflows are interlinked and that a satisfactory project result can be achieved at the end. Examples include process and method definitions and the use of suitable tools.

A detailed description of FELICS and all elements can be found in [Timinger, 2021].

Tip

Analyze the project management in your company based on the elements and building blocks mentioned. You will probably notice some of them, while others may be missing. This is not necessarily a shortcoming. However, it is definitely worth questioning your own process model and checking whether the elements and building blocks that are really relevant to success are integrated to an appropriate extent.

Example

Many companies have defined a product realization process as part of their quality management system. This represents a process model that describes how the company develops products and brings them to market. In addition to the product realization phases, it usually also defines delivery items, individual milestones, methods to be used, rules of cooperation and the roles involved in product realization. For example, verification and validation measures can be defined as quality assurance components in certain project phases. Communication is promoted through defined regular meetings. A recurring retrospective may be defined for knowledge management.

Short review of the history of project management

To understand some of the characteristics of plan-based project management, it is worth taking a look back to its origins. Until the beginning of the 20th century, knowledge of how complex projects should be carried out was heavily dependent on the knowledge of individuals. There were no project management standards or established methods.

In the 1910s, Henry Laurence Gantt developed the bar charts that are still popular today, also known as Gantt charts, see Figure 2.2. When the race for supremacy in space began in the mid-20th century, new process models were needed that would enable unique projects to be completed quickly while guaranteeing high-quality and safe results. At the same time, the American Project Management Institute (PMI) and the International Project Management

Association (IPMA) established organizations that began to standardize process models and methods for successful project management.

Figure 2.2: Overview from the time of the emergence of Henry Gantt's beam plans to important space travel projects and the founding of leading project management organizations to the agile manifesto in the 21st century.

Many of the process models and methods from this period are still part of current certifications and are used successfully in day-to-day project management. Towards the end of the 20th century, the first agile process models spread with Extreme Programming and Scrum. In 2001, the agile manifesto was published, which documents agile values and principles and is still the basis of many agile process models today. In the following sections, however, we will first focus on the plan-based view of project management. I will then discuss agile process models in the fourth chapter.

Common characteristics of plan-based process models

Processes and elements of project planning and control

There is no one and only plan-based project management. Rather, there are various plan-based process models that have certain characteristics in common and differ in others. One of the common characteristics of plan-based process models is that they attempt to predict the future through plans and then follow these plans.

> *A **plan** is a (usually simplified) representation of an expected (future) target state. The plan can contain (work) steps to achieve this target state, premises and other information to be considered.*

Example

Schedules show when which tasks should be completed in the future in order to achieve the project goals.

The assumption of plan-based project management is that the project will be successful if the plan is fulfilled. This imposes high demands on planning. Anything that is forgotten or incorrectly predicted in the plans has an impact on the success of the project. According to Wysocki, plan-based process models can be purely sequential (continuous) or incremental (building on each other), see Figure 2.3 [Wysocki, 2019]. In both cases, the project is first defined, i.e. objectives and the organization are established and then the project is planned in detail until the end of the project. This is followed by the implementation of what has been planned - either in one go or in several increments. The latter allow the interim evaluation of the results.

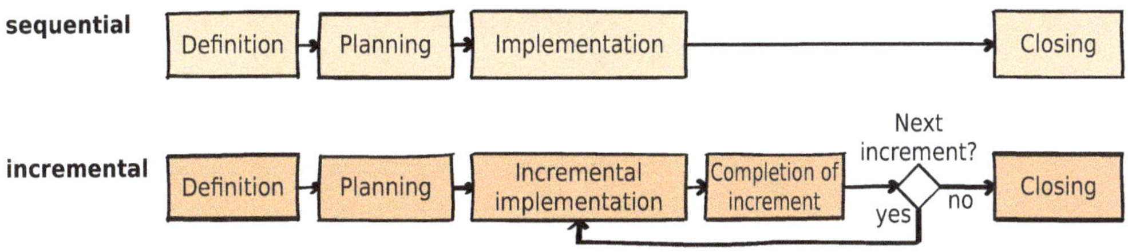

Figure 2.3: Plan-based approach for projects according to Wysocki [Wysocki, 2019]. Plan-based projects run either purely sequentially or incrementally. In both cases, however, the project is completely planned at the beginning.

Tip

Plan-based process models rely on plans that attempt to predict the future course of the project.

Most plan-based process models structure a project into **project management phases** and **project phases.** The former refer to the chronological sequence of project management tasks and can be standardized. For example, in the following project management phases:

- initialization phase
- definition phase
- planning phase
- control phase and
- closing phase.

In fact, many industries or companies have individual phase models for the implementation of projects. For example, there are recommendations from the German Association of the Automotive Industry on process models to be used [VDA, 2020] [Hab and Wagner, 2013]. The US Food and Drug Administration also publishes guidelines for the development of medical devices [FDA, 1997].

Warning

In contrast to project management phases, which can be standardized at least to some extent, project phases are project-specific. For example, a development project can be divided into an analysis phase, a concept phase, an implementation phase, a test phase, a documentation phase and an deployment phase, while a project to improve a production process could be divided into an analysis phase, an optimization phase, a pilot phase and a documentation phase.

Project management tasks and methods have been defined for the individual project management phases, some of which are summarized in Figure 2.4.

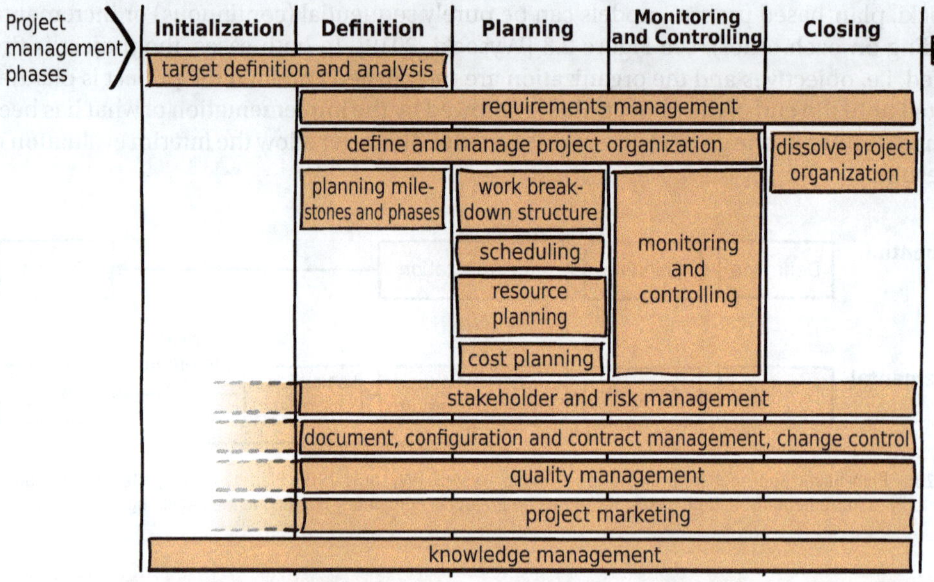

Figure 2.4: Typical timeline of plan-based projects with project management phases and selected elements of project planning and control.

Once a project has been approved on the basis of initial targets and profitability analyses and the **project order** has been issued, the **target definition** and **target analysis** (checking for target conflicts and target prioritization) begin. **Requirements** are specified and kept up to date over the course of the project; their fulfillment is documented at the end. The **project organization** must be defined. This includes deriving the organization chart, clarifying roles including specifying responsibilities, and agreeing which (sub)teams exist and who consults with whom and how often. Escalation paths and reporting boards should also be defined. Initial rough planning based on the targets is carried out using simple **milestone or phase plans.** Both are simplified plans that are often created on a whiteboard by the team. They help to check the plausibility of the project targets in terms of the performance and quality to be achieved, the resources required, the funding needed and the resulting project duration.
Detailed planning only begins with the work breakdown structure. If the **work breakdown structure** defines what needs to be done in the project, the schedule then provides a precise statement of the deadlines to be met. The **schedule** is coordinated with the **resource plan**, as deadlines can only be met if sufficient resources are available. Finally, the **cost plan** provides the costs incurred by the project. For medium-sized and larger projects, the cost plan should not only provide information about the planned total costs, but also show the distribution of costs over time. This is a prerequisite for the earned value analysis, which is a powerful tool for **controlling** during project execution. With the help of the earned value analysis, conclusions can be drawn about the project status and, if necessary, **controlling measures** can be intervened.

In addition, there are a number of activities that need to be carried out across several project management phases, such as stakeholder and risk management. The methods outlined here are explained in more detail later in this chapter and in the toolbox in Chapter 8.

Roles and interfaces

Plan-based process models are generally based on a traditional, hierarchical understanding of leadership and roles, see Figure 2.5.

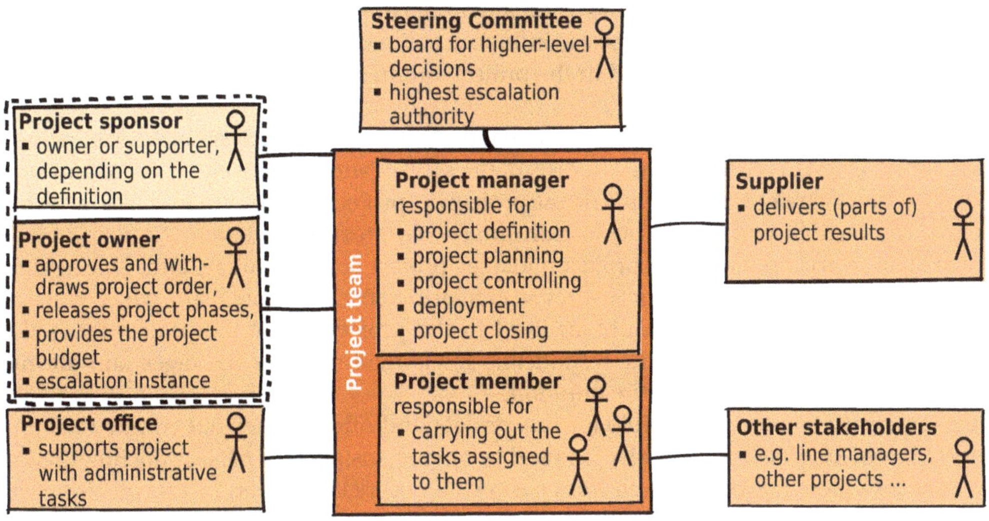

Figure 2.5: Roles and interfaces in project management.

The **project owner** of a project can come from your own or another company. External project owners are often also referred to as **clients**. The project owner provides the financing, is an escalation authority and grants and withdraws the project order. At the end of the project, the project owner approves the project result. Depending on the process model, the project owner can also approve individual project phases.

The role of the **project sponsor** is not clearly defined either in the literature or in practice. In some organizations, it is equivalent to the project owner. In some cases, a distinction is made between the project owner, who is responsible for the scope of the project, and the (financially supporting) project sponsor. The project sponsor usually comes from higher management, for example the executive board. They can provide financial, personnel or organizational support. In some cases, the role also acts more as a mentor and link between the project and management.

Irrespective of the role designations of project sponsor and project owner, it is advisable to ensure that the aforementioned tasks are performed. This means that both the more formal aspects of project assignment and financing are clarified and that a good, supportive relationship with senior management is ensured.

The **steering committee** is a decision-making and escalation committee that is superordinate to the project. It is usually made up of high-ranking representatives of the (external) client and the (internal) project owner. The steering committee always makes a decision if it exceeds the authority of the project manager or if the client and project manager cannot agree on a decision. The steering committee is usually the highest escalation authority, as the appropriate composition of this committee enables a balance to be struck between the interests of the client and the project. Regular reports are submitted to the steering committee.
The **project manager** is appointed by the steering committee (sometimes also by the client or project owner) and is responsible for project planning, control, management and project completion. Depending on the organizational form, the project manager coordinates the tasks of the project team members or is their functional or even disciplinary superior. Project managers should have a certain degree of influence in the selection of project staff. They have the authority to distribute tasks within the project and to structure the project, for example into several sub-projects.
The **project team members** work on the actual project scope in the form of work packages. They are expected to represent the interests of the project and proactively report to and support the project manager. Depending on the project the project manager may delegate further areas of responsibility and authority to individual project team members.
The **project team** consists of the project manager, project team members and, if necessary, sub-project managers. In larger projects, you can form a so-called core team, which consists of employees who are permanently involved in the project, while the project team also includes employees who are only responsible for individual work packages and are not involved in the project for the entire duration.
A **project office** is only responsible for one project and often carries out commercial and controlling-related tasks. It can also be used to ensure sustainable project documentation so that subsequent projects benefit from cost estimates, risks, success factors etc. from the project.
A **project management office** is engaged across all projects and is responsible for the entire project portfolio. It supports the portfolio manager in the preparation of key figures, the further development of project management within the company and generally in the performance of his or her tasks.
Suppliers can be internal (e.g. other departments of the same company) or external. They provide results for individual work packages or entire sub-projects. To avoid misunderstandings caused by contradictory statements, communication with suppliers should be streamlined by designated employees. This can either be the project manager directly or an appointed supplier manager.
Depending on the project, **other stakeholders** include the portfolio and program managers, the development managers, the marketing and sales department, the quality management department and representatives from production. Potential disciplinary supervisors of the project team members are also important stakeholders. Project managers should get a comprehensive overview of the stakeholders relevant to their project as early as possible, for example in the initiation or definition phase, and find out about their responsibilities and expectations of the project. Suitable methods for this can be found in the toolbox in Chapter 8.

Tip

The roles presented here are common in projects. In practice, however, there are sometimes significant deviations. It is therefore always necessary to check how the roles are structured in the respective organization. In particular, the roles of project owner, project sponsor, client and steering committee are not always clearly distinguished from one another. This can lead to conflicts of responsibility during the course of the project. You should therefore clarify the relevant responsibilities and authorities of the existing roles in your organization right at the start of your projects.

Types of organization

Projects can be organized or linked to the organization of a company in different ways.
In the **functional project organization**, project managers report directly to the executive management. They have a staff position without any functional or disciplinary authority. The project manager can only coordinate the project team with members from different departments of the company without having the authority to issue instructions. This type of organization can usually be set up quickly and easily and enables good capacity management of employees across several departments. This means that the workload can be well balanced between different projects and departments. By keeping employees in the permanent line management, the exchange of knowledge with other employees is also guaranteed. Figure 2.6 provides an overview of these and two other organizational forms.
The **pure project organization**, sometimes also referred to as autonomous project organization, assigns the employees of a project to the project manager in terms of functional and disciplinary matters. Conflicts between the line manager and the project manager therefore do not occur. The project manager has a strong position. Due to the clear assignment to a project, employees generally identify strongly with it, which is usually motivating and therefore facilitates success. However, this form of organization is complex to set up, as formal supervisor assignments have to be reorganized for each project. This form of organization is therefore only worthwhile for large, long-term projects. Workload management is also not easy, as each employee must be assigned meaningful tasks in the project that correspond to their qualifications. In addition to functional and project-related management tasks, project managers also have to take on disciplinary leadership tasks. This, in turn, can be at the expense of the actual project work in terms of the time required for this.
The **matrix project organization** attempts to combine the strengths of the two forms already mentioned. The employees are functionally subordinate to the project manager. However, the line manager remains the disciplinary supervisor. This means that employees can easily be assigned to several projects or to both project and line tasks. As a result, workload management is generally good. By remaining in the line organization, the exchange of knowledge across project boundaries is also relatively easy to organize. The sharing of authorities and responsibilities between project managers and line managers can lead to conflicts, which may be carried out on the backs of the employees and lead to an overload of work.
In addition to the internal forms of organization mentioned above, there are also **external forms of project organization**. These are set up exclusively for the management of a project. The executive management is therefore also directly in charge of the project. This form of organization is similar to a pure project organization. It is only appropriate for very large and long-term projects. It is, for example, found when several companies join forces for a specific project, in the form of a joint venture or a consortium.

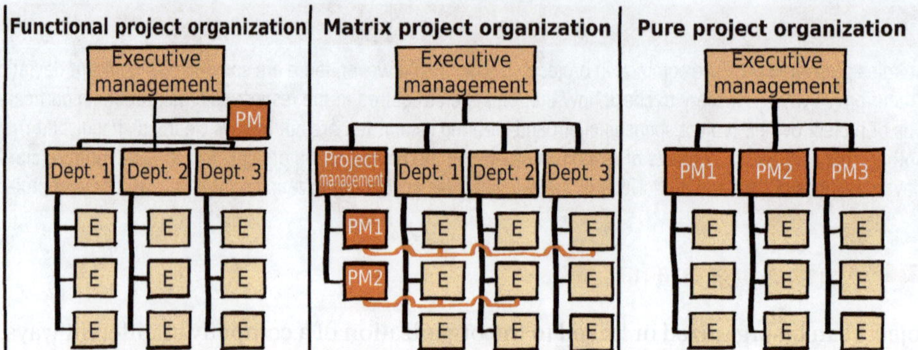

	Functional project organization	Matrix project organization	Pure project organization
Characteristics	employees remain in the line, project manager has only weak authority for coordination	line manager remains disciplinary, project manager becomes functional manager	employees report to the project manager in functional and disciplinary terms
Typical project types	small projects, organizational projects	medium to large projects, research and development projects	large projects, investment projects
Strengths	simple structure, no organizational changes required, good workload balance, good exchange of knowledge	good workload balance, good exchange of knowledge	clear reporting lines and responsibilities, low conflict potential, high level of identification with the project
Weaknesses	no managerial authority, coordination difficult, increase in workload in line departments	potential for conflict due to division of responsibilities between project and line, risk of employees being overloaded by project and line tasks	complex setup and closure of the project organization, difficult workload management, additional disciplinary tasks for the project manager, difficult knowledge sharing

PM: project manager Dept.: Department E: Employees

Figure 2.6: Overview of the organizational forms of functional project organization, matrix project organization and pure project organization.

Types of process models

In the following sections, you will become familiar with various classes of **plan-based process models**:

- **Sequential process models** structure projects in phases that are executed one after the other, i.e. sequentially. The most well-known representatives of this class are the waterfall model and the V-model. Strictly speaking, one phase must be completed successfully before the next phase can begin.
- **Concurrent process models**, also known as parallel process models, use the overlap of phases to accelerate project execution through parallel processing and to identify risks earlier. Concurrent engineering is an important representative of this class of process models.

- **Repetitive process models** run through project phases several times. Accordingly, the project result is developed step by step. Each step is also known as an iteration or an increment. It allows the partial results achieved so far to be evaluated and the experience gained to be taken into account in the next step. Representatives of this class of process models are incremental, iterative, recursive and evolutionary models.

In addition to the classes mentioned, there are others, such as prototypical or reuse-oriented process models [Bunse and von Knethen, 2008], which we will not discuss in detail here. While prototypical process models are based on sequential process models and extend these with repetitive elements to generate early development prototypes, reuse-oriented process models are based on repetitive process models. They rely on the reuse of results from previous projects and deliberately incorporate this reuse into the project management process.

Sequential process models

Waterfall model

The waterfall model is probably the most intuitive process model of all and follows the principle of one step at a time. It is one of the best-known process models and its beginnings date back to the middle of the 20th century. The name comes from the sequential and cascading presentation of the phases, which is reminiscent of a waterfall cascading down into the valley. For each phase, it is necessary to define exactly what needs to be achieved and which documents need to be prepared, for example a specification document in the planning phase or certain design drawings in the design phase. Once a phase is complete - and only then - the transition to the next phase is approved. Phases cannot be skipped or started at overlapping times. The model is shown in Figure 2.7.
Ideally, the project should progress as planned. However, if a phase cannot be successfully completed because, for example, it turns out in the implementation phase that some

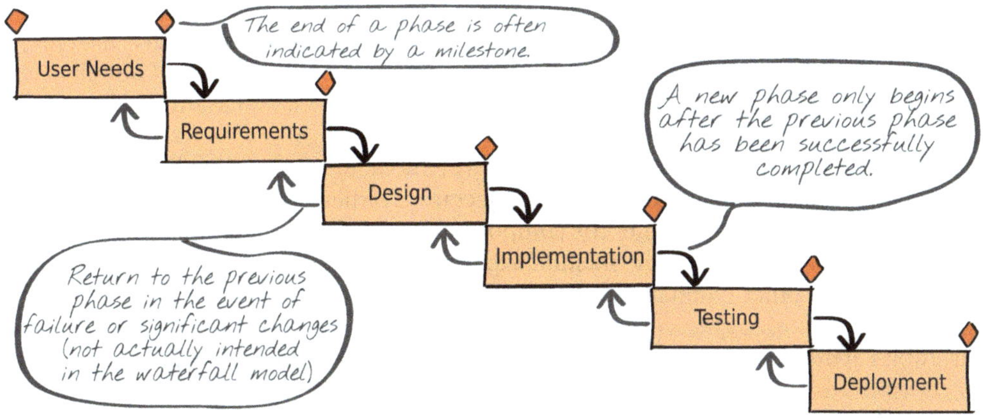

Figure 2.7: In the waterfall model, the individual phases are run through sequentially. The next phase can only begin once the current phase has been fully completed. Phase repetitions and steps backwards are not actually intended. However, they may be necessary if a phase cannot be successfully completed as planned.

specifications cannot be realized as planned, it is necessary to return to one of the previous phases. As this means a deviation from the actual plan, such backward steps cause additional work in the project, for example in the form of escalation meetings or replanning.
Thanks to its clear structure and simple definition of project progress from phase to phase, the waterfall model is still very popular today. It reduces the risk of "getting lost" due to too much parallel work in the wrong order. The hard condition that one phase must be completed before the next phase can begin is intended to ensure the quality of the project deliverables. However, it is important to clearly define and review what is to be achieved at the end of each phase. It is common practice to schedule a milestone at the end of each phase at which this review takes place.

> ***Milestones** are events of particular importance. They define important points in time in the schedule at which certain deliverables must be available.*

According to this definition, the project start is also a milestone at which the project must be approved. The review of the results at the end of the phase is also known as the stage gate or quality gate principle.

> ***Quality gates** are milestones with a review character. A quality gate defines in advance which quality criteria must be met in order to be allowed to enter the next project phase. The quality criteria are checked by a defined committee.*

This means that for each phase (= stage), the activities to be performed and the deliverables (quality) to be achieved are precisely defined. At the end of the phase, i.e. at the gate to the next phase, these activities and results are reviewed. The gate may only be passed and the next phase started if the result of the review is positive.
If a certain product quality is to be achieved with the help of quality gates, it is important that the quality requirements are clearly specified. Once a quality gate has been passed, it is not repeated in the waterfall model. If quality problems are overlooked, they are likely to remain in the finished product. In many companies, quality gates are checked as part of so-called review meetings. In these, a review team, in which the client is often also represented, checks whether a phase transition can be taken responsibility for. If not all activities have been completed or not all results have been achieved, a phase transition can be refused.
In practice, in contrast to the sequential core idea of the waterfall model, a phase transition is sometimes permitted with conditions. The project team is then given conditions for completion within a certain period of time. The composition of the review teams differs significantly from company to company. Some include experts who evaluate the actual project deliverables in detail. In other companies, management representatives predominate, who focus more on the economic aspects of the project. For a meaningful evaluation, the entire project deliverables at the given phase must be examined. This includes technical or content-related aspects as well as the economic prospects of success.

Tip

The scope and importance of the reviews should depend on the significance of the phase transitions in the quality assurance of the project deliverables. If there are further quality assurance measures, the reviews are less important than if there are no other quality assurance activities. If the reviews are an important quality assurance instrument for you, you should staff the review team with the appropriate expertise. For example, it has proven to be a good idea to involve independent reviewers in the evaluation of the project results at the end of the phase. They must not have any responsibility of their own in the project.

In addition to reviewing quality aspects, the quality gates can also be used for the staged release of project budgets. In this case, the project only ever receives its budget for the next phase. The budget for the next phase is only allocated once this has been positively assessed.

V-model

Another sequential process model is the V-model. It is named after the arrangement of the individual project phases, see Figure 2.8. On the left branch of the V, the project scope to be realized is increasingly specified from the rough draft to the detailed design. Once the specified project deliverables have been implemented, the right-hand branch of the V is completed bottom-up: First all components are tested, then they are integrated into a subsystem or directly into the overall system and finally tested in this new context. Finally, the customer requirements from the specifications are validated and accepted.

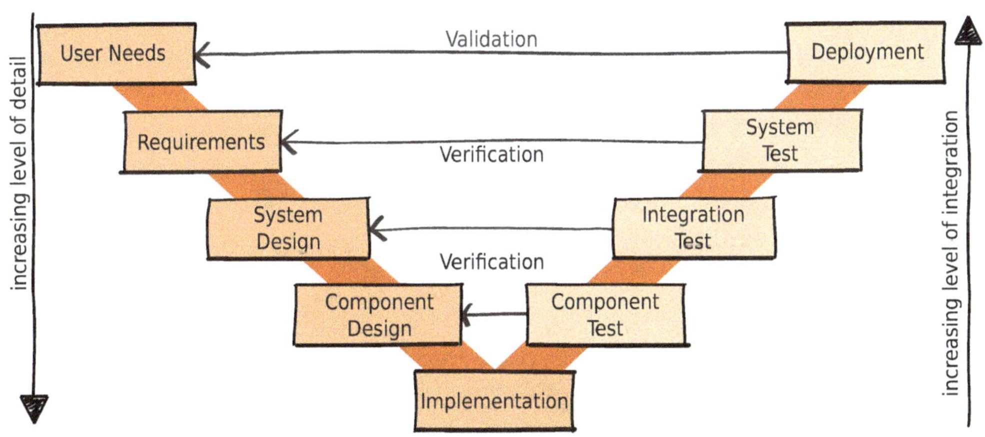

Figure 2.8: In the V-model, the individual phases are run through sequentially. The process takes the form of a V. From left to right and from top to bottom, the project scope to be realized becomes more and more detailed. Each phase on the left-hand branch of the V corresponds to a verification or validation phase on the right-hand branch of the V.

In the V-model, each specification and design phase of the left branch has a counterpart in the form of corresponding tests on the right branch of the V. This structure promotes the idea of quality and integrates it firmly into the model: everything that has been specified and for which a design draft has been created must be assessed later as part of the verification and validation process.

As the latter two terms are important in project management and are sometimes confused, they are briefly defined here:

> ***Verification** means the provision of objective evidence (usually by measurement) that a specific requirement has been fulfilled. Verification answers the question of whether it has been implemented correctly (in terms of the requirement).*
>
> ***Validation** refers to providing objective evidence that the user needs have been met. Validation answers the question of whether the right thing has been developed for the customer.*

Example

A manufacturer of high-quality watches commissions a supplier to produce a case. One of the manufacturer's user needs is for the watch to be water-resistant to a depth of 100 yards. This is WHAT the watch must be able to perform. In the specification sheet, the supplier specifies the specific measures to achieve the water resistance, i.e. HOW this performance can be achieved. This includes appropriate material thicknesses and seals.
As part of the verification process, precise measurements are taken to check whether the correct material thicknesses have been achieved and whether the seals have been selected and used correctly. The final validation checks whether the watch is actually water-resistant to a depth of 100 yards. In the event of a specification or design error, it may happen that the requirements of the specifications are met (verification successful), but the product still does not meet the corresponding user needs.

In sectors where product safety is of particular importance, such as the aerospace industry or medical technology, the V-model is frequently used, as the verification and validation integrated into the model can provide complete proof of compliance with requirements.

Tip

Both the waterfall model and the V-model show that they were developed at a time when the safety of project deliverables was of particular importance: when people were to be transported into space as part of the space programs in the middle of the 20th century, the technology used had to be safe. However, quality assurance is implemented differently in the waterfall and V-model. While in the waterfall model, reviews at the phase transitions are intended to ensure that the project has been processed correctly, the V-model relies on verification and validation phases that correspond to the corresponding requirements and user needs instead.

Concurrent process models

Concurrent engineering

Concurrent engineering is one of the best-known representatives of concurrent process models. It deliberately dispenses with the sequential processing of phases. Instead, parallel processing of the next phase is always started as soon as sufficient information (e.g. preliminary deliverables) is available, see Figure 2.9. Phases can be processed partially or completely overlapping in concurrent engineering.

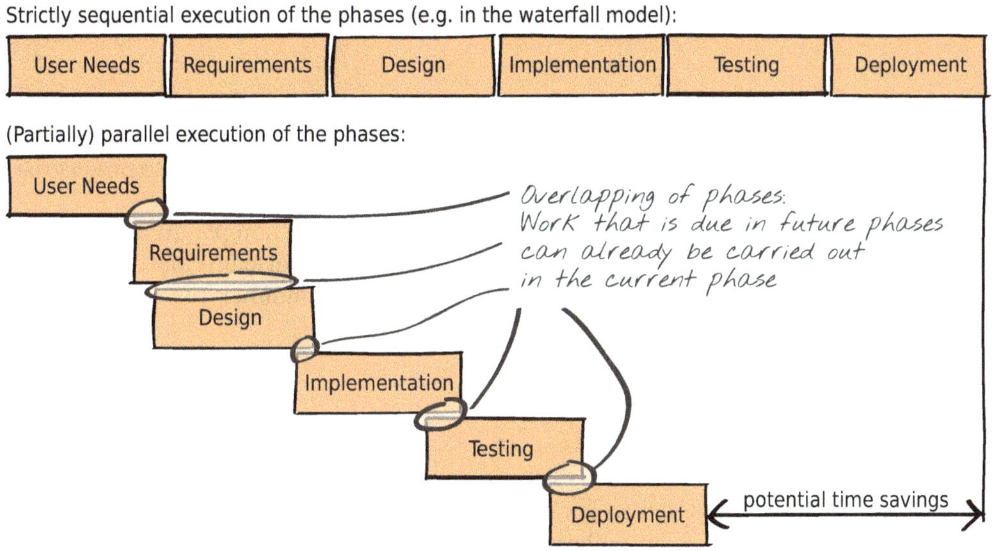

Figure 2.9: Advantages of parallel or partially parallel processing of the phases in concurrent engineering: By starting the work of subsequent phases earlier, the project can possibly be completed more quickly than with strictly sequential processing. In addition, potential problems are revealed sooner and can then be solved.

Example

The development of new software, which is to be offered as software as a service via the internet, is organized as a project that uses concurrent engineering. The first architecture-related software components are already developed in parallel to the specification phase. Architecture-independent requirements are detailed in parallel. A commercialization phase, in which the business model for the software is refined, is carried out completely in parallel. The first beta tests with selected users are already underway during the implementation phase.

The risk that changes to the design are discovered during implementation and that parts of the architecture need to be changed is offset by the opportunity of an earlier market entry.

The example shows the opportunities and risks of parallel processing. But even if the opportunities outweigh the risks, you should plan the parallel start of the subsequent phase carefully. There are two questions that you might want to clarify in advance:

- Is sufficient information available so that the work of the subsequent phase can already be started reasonably and with an acceptable risk of later correction due to changes?
- Are sufficient resources available so that real parallelization can be achieved?

New

The second question is particularly important for success: if resources have to be taken from the current phase for parallel processing of the subsequent phase, parallelization does not lead to any time savings, but can conversely cause delays in the project process, as the employees involved have to switch back and forth between several tasks.

In general, you should carefully analyze the anticipation of work in subsequent phases. This can result in risks that you should include in your project risk management and monitor with due attention. Conversely, the anticipation of work can also be used to reduce project risks at an early stage. Assessing this is always a case-by-case decision.
Concurrent engineering is often used for the early start of production planning and setting up the production of a product. In this way, it is examined in parallel to product development whether the product can be manufactured later in a reasonable manner (in terms of quality, time requirements and costs).

Repetitive process models

In the process models discussed so far, repetitions always occur when changes in earlier phases mean that the work in other phases also has to be changed. These repetitions in the form of corrections lead to additional work and are therefore not desirable.
In contrast, repetitive process models are designed from the very beginning in such a way that repetitions are taken into account. First, a rough draft of the entire project scope is created. The project is then planned iteratively. The phases that immediately follow are planned in more detail than phases that are further in the future. Knowledge gained during the course of the project is incorporated into the detailed planning of future project phases.

Incremental process model

An important representative of the repetitive process models is the **incremental process model**, which is outlined in Figure 2.10.
In each increment, the requirements for the intended partial or intermediate deliverables are formulated, a design draft is created and implementation and testing are conducted. This is repeated until the final deliverables are achieved. It is not necessarily required to repeatedly go through all phases for each increment. If, for example, all requirements can be formulated and analyzed at the beginning, but the overall design has not yet been created, it is sufficient to go through the design, implementation and test phases in each increment.
Repetitive process models are particularly useful when

- parts of the project are subject to very high uncertainties
- relevant experience for processing is missing or
- the requirements are subject to change.

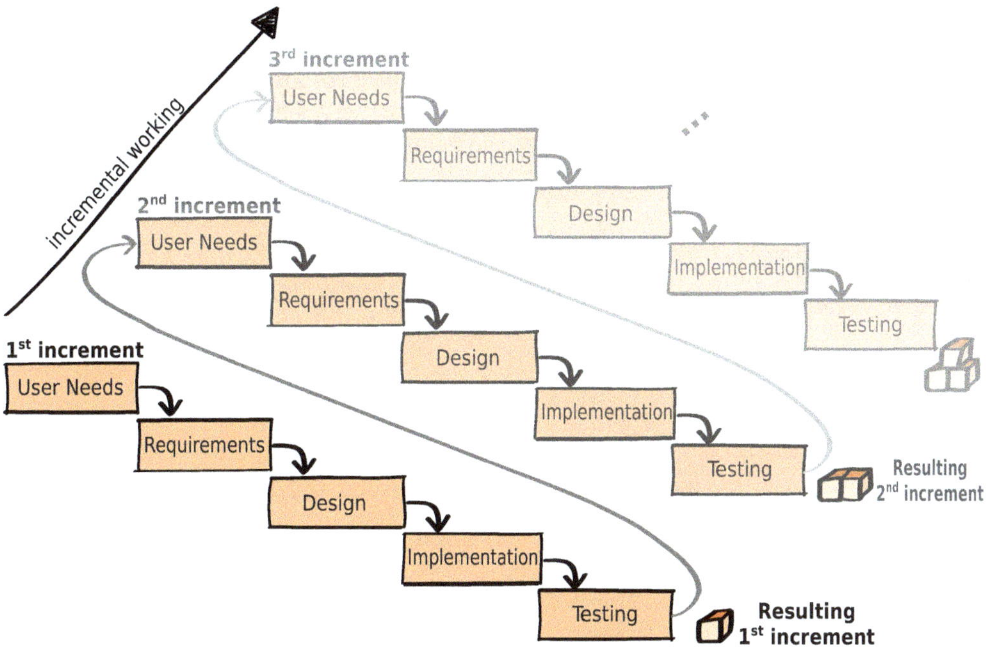

Figure 2.10: Incremental process model: The individual phases from the requirement to the first partial result (resulting 1st increment) are passed through sequentially. Further additions are then made to the requirements and the design and finally a new partial result (resulting 2nd increment). This is repeated until the overall result has been achieved.

In these cases, such process models help to gradually reduce uncertainties from increment to increment and stabilize the requirements. With incremental process models, the project deliverables are supplemented in each phase repetition. This can be done, for example, by adding further functions to a product development project.

Iterative process model

In iterative process models, the project deliverables are also elaborated in recurring phases. In contrast to the incremental process model, however, the deliverables are not gradually supplemented with additional content from phase to phase, but are worked on iteratively from the rough draft to the detailed result. The distinction between incremental and iterative processes may sound academic and in fact an iterative process model would look very similar to the incremental process model shown in Figure 2.10. In terms of content, however, it makes a difference whether something is added incrementally to the project from phase to phase or whether the project results are simply iteratively revised, elaborated in detail and improved.

Spiral model

Probably the best-known representative of repetitive process models is the spiral model, which is shown in Figure 2.11. Each iteration (repetition) serves to detail the requirements, refine the goals and improve the design from the requirements to the rough draft and finally to the detailed design.

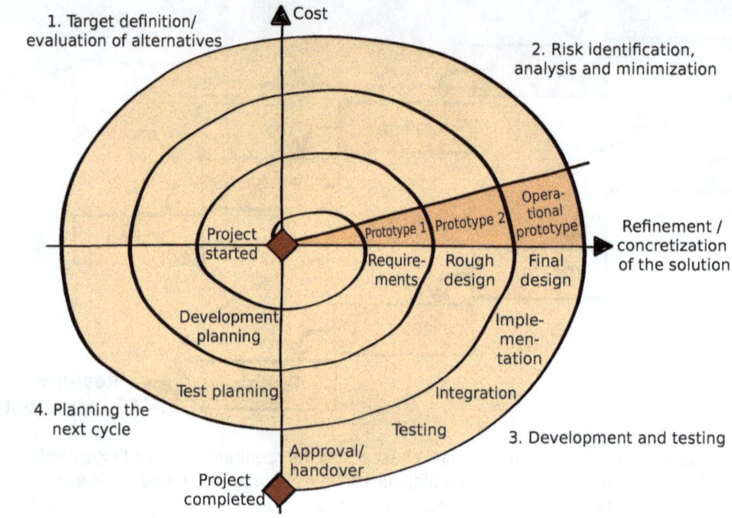

Figure 2.11: The project start in the spiral model is shown in the center. In several recurring iterations, goals and requirements are determined, alternative solutions are evaluated, risks are identified and minimized, the requirements are implemented and tested and finally the next iteration is planned. From iteration to iteration, the project deliverables become more concrete until, after a planned number of iterations, the final project result is available and the project can be closed.

Tip

When planning the recurring loops, care should be taken to ensure that there is something presentable and analyzable at the end of each repetition. The deliverables must be suitable for providing insights for the planning and implementation of the next recurring loop. Otherwise, the overall project result cannot be detailed and stabilized.

The spiral model has similarities with the well-known agile process model Scrum, which we will discuss in Chapter 4, due to its repetitive nature and emphasis on the step-by-step detailing of the project result. However, you should note that agility is not achieved through specific processes, but through a new mindset that differs from traditional ways of thinking in many areas of management.

Other repetitive process models are recursive or evolutionary models, which we will not consider any further here, as the basic idea of recurring loops is comparable.

At a glance

- Plan-based project management refers to plan-driven project execution in which deviations from the plan are minimized during the controlling phase.
- Instead of plan-based project management, the terms "traditional project management" or "predictive project management" are sometimes used.
- A process model combines methods and elements, processes and phases for a standardized project process.
- Well-known process models for plan-based project management include the waterfall model, the V-model, concurrent engineering and the spiral model.
- All plan-based process models have in common that they use plans to model the future of the project.
- Many plan-based process models structure the project into an initialization phase, a definition phase, a planning phase, a control phase and a closing phase.
- The individual process models have different strengths and weaknesses, which are outlined in the following overview:

	Strengths	Weaknesses
Waterfall model	▪ easy to understand ▪ clearly structured ▪ attention to quality through reviews (stage/quality gates) ▪ strict separation of identification of requirements, design draft and implementation facilitates configuration management	▪ quite rigid, only limited flexibility ▪ strictly interpreted, no parallelization possible ▪ project deliverables are only evaluated at the end, early feedback to developers and customers is not provided ▪ documentation-oriented
V-model	▪ easy to understand ▪ clearly structured ▪ assurance of quality through mandatory verification and validation ▪ strict separation of identification of requirements, design draft and implementation facilitates configuration management	▪ project deliverables are only evaluated at the end, early feedback to developers and customers is not provided ▪ documentation oriented and possibly too bureaucratic for small projects
Concurrent Engineering	▪ opportunity to save time through (partially) parallel processing ▪ by anticipating work in later phases, risks are addressed earlier and feedback is provided	▪ risk of additional work in the case of necessary corrections to work already carried out
Spiral model	▪ structured handling of unstable or unclear requirements ▪ consciously developing the subject of the project step by step	▪ risk of additional work in the event of necessary corrections to later increments due to new findings ▪ incremental approach can invite change requests and delay project completion

Practical examples and case studies

Case study 2.1: Definition of quality gates

In the Omikron project, a new paint shop for a customer's industrial plants is to be developed. The company conducts all such projects according to a company-specific adaptation of the waterfall model. The phases of the strictly sequential model can be seen in Figure 2.12.

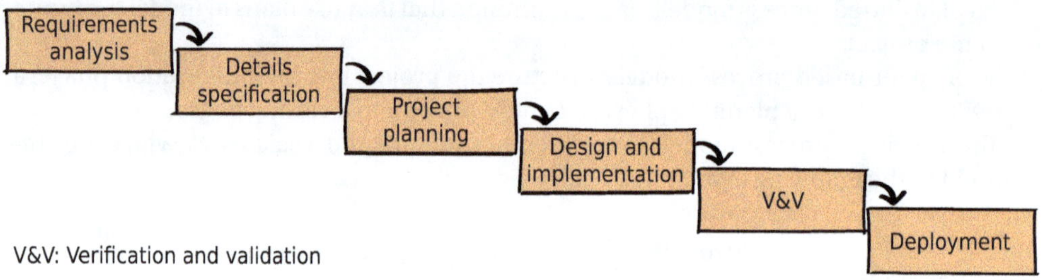

Figure 2.12: Process model in the Omikron project.

The quality of the product to be developed should be ensured by appropriate quality gates. At the end of each phase of the waterfall model (gate), the previous phase is to be evaluated. Only if the evaluation is positive can the project start working on the next phase. The company has failed so far to define the content and process of the evaluations at the gates. As the project manager, it is now up to you to make a proposal to the customer about which gates you want to define and how exactly the evaluation should be conducted.

Guiding questions

- Which gates would you define?
- Who takes part in the evaluations at the gates?
- What would you evaluate in each of the defined gates?

Case study 2.2: Traceability in the V-model

Starting point is the Omikron project outlined in case study 2.1 for the development of a paint shop for industrial plants on behalf of a customer. Originally, the project was to be processed according to the company's own waterfall model. However, the client insisted on processing the project according to the V-model including complete documentation of the fulfillment of all customer requirements.

Guiding questions

- How does the waterfall model from case study 2.1 look in V-model illustration?
- How do you document the complete fulfillment of customer requirements from their definition at project start to handover to the customer? (It may be worth taking a look at the toolbox in Chapter 8 on traceability.)

Case study 2.3: Prototype planning with the spiral model

Moving Forward is an organizational project that aims to optimize sales processes by surveying sales employees, customers and representatives of departments that interface with sales. The requirements are to be used to create a recommendation for a new organizational structure and process organization. Once the requirements have been approved, this new organization is to be implemented. The company does not have a standardized process model for handling organizational projects.

In order to structure the many requirements step by step and derive a prototype in the form of a new organizational structure and process organization, you would like to use the spiral model. As the company's internal client, the management expects you to provide a rough project plan showing how you will proceed.

Guiding questions

- What does your project-specific process model look like?
- How many iterations are you planning?
- What are possible characteristics of the individual iterations?

3 Plan-based project management – tools, methods and ongoing tasks

In this Chapter ...

- you will acquire the basics for the initialization, definition, planning, control and closure of plan-based projects and
- you will learn which ongoing tasks, such as stakeholder and risk management, project marketing, quality management, document and configuration management, change control and contract management, are part of plan-based projects.

Introduction

To begin with, I will introduce you to typical processes, methods and approaches of plan-based project management. As a reminder, I would like to mention again that we use the term "plan-based project management" as a synonym for traditional or predictive project management. Many of the methods can also be found briefly and concisely in the toolbox in Chapter 8. I will refer to specific profiles throughout the following sections. The methods are presented alongside the common project management phases:

- initialization and definition
- planning
- control
- closing.

Tip

Many of the methods presented in the remainder of this chapter can also be used reasonably and successfully in agile and hybrid projects, for example methods for role clarification, stakeholder and risk management or project closure. However, as the methods have their origins in plan-based project management, I present them in this chapter.

Project initialization and definition

Overview – where are we in the project management process?

A project begins with the **initialization phase** and the subsequent **definition phase**, see Figure 3.1, which shows the entire project lifecycle.

> *The **project lifecycle** describes the course of a project from the very start of the project to its end.*

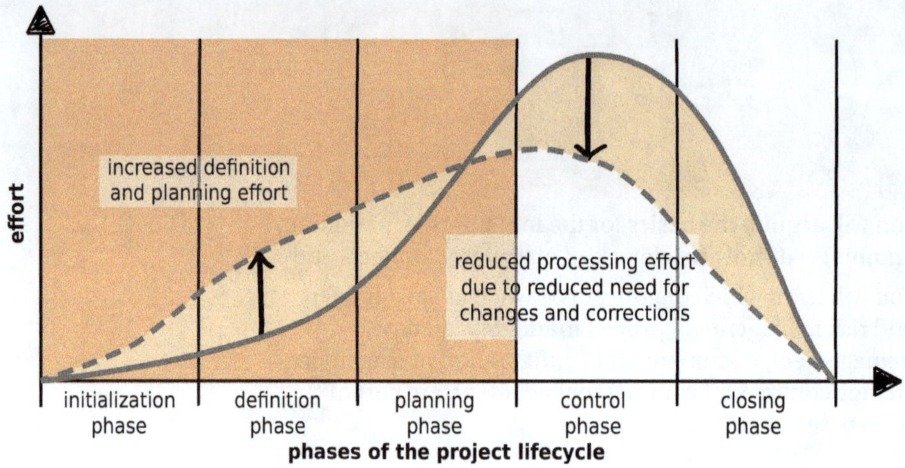

Figure 3.1: The project begins with the initialization and definition phase. This is followed by (detailed) planning and the actual work on the project deliverables, which is accompanied by appropriate project control. Finally, the results are handed over to the customer and the project is completed. A good definition of the project objectives can reduce the effort required for changes and corrections in the later control phase (dotted line).

Experience shows that the initialization and definition phase can make an important contribution to the subsequent success of a project. Experts also like to say: "Tell me how your project starts, and I'll tell you how it will end." If objectives and requirements are not formulated with sufficient precision and analyzed for their achievability, there will be increased effort in later phases due to changes and corrections. Strictly speaking, it is assumed that a project is successful if it achieves its objectives. However, this does not guarantee that all key stakeholders are satisfied with the project. We therefore define project management success in a broader sense:

> ***Project management success*** *refers to the recognition of the project results by the relevant interested parties (= stakeholders).*

Tip

Plan-based project management is characterized by the idea that the future course of the project can be mapped by good plans. This is why it is also referred to as plan-driven project management. Deviations from plans are considered critically and attempts are made to avoid them.

The aim of this early phase is therefore to
- a clear formulation of the objectives and their analysis,
- complete requirements and
- to set up a project organization that matches the project objectives to reduce the effort in later phases.

Project start

"When does a project begin?" – the question sounds simple, but in practice the answer is not always clear. Often

- a new idea for a product,
- a process improvement,
- an upcoming investment or
- a customer order

is the trigger for new work outside the line organization. A formal project assignment does not always exist at this point. Sometimes there is a vague request to start with certain preliminary investigations on a specific issue or an order to start implementing a solution straight away. The latter case might be risky if the problem to be solved has not been sufficiently analyzed beforehand. In such cases, a solution may have been developed, but it does not fit the actual problem and resources are wasted unnecessarily.
Another pitfall of an unstructured project start is the lack of an organizational structure. The most important project roles should be clarified in order to create effective communication channels and derive the project objectives. Otherwise, roles and responsibilities are randomly defined and might not be suitable for the actual project.
These initial considerations should show that often there is no such thing as a unambiguous project start. This seemingly insignificant observation often leads to difficulties, as opportunities to refine and improve the project idea are missed and unnecessary time elapses before the actual work can begin.

Example

The development manager was put under pressure by the marketing department to develop a successor to the previous "cash cow", a CNC milling machine. There are some unstructured ideas for product improvements in the development department. The development manager instructs an employee to collect these and then start implementing them. He wants to see a prototype as soon as possible in order to reduce the pressure on his department and to be seen as a problem solver himself.

It is tempting to jump straight into implementation as this gives the impression of making rapid progress. Later there is a need for change or the project is even terminated. Experienced project managers therefore first analyze and structure the task before starting implementation.

Project idea and project approval

As soon as ideas for a new project emerge, you should clarify the following topics:

- Responsibilities:
  - Who later issues the formal project approval and thus the project order? Who is the project sponsor?

  - Who will be responsible for project management until then (and beyond)?
  - What resources (human, financial, material) can be used?
  - What are the authorities and responsibilities of the project team members and the project manager?
- Project management processes:
  - Which processes must be adhered to?
  - What tailoring options exist (i.e. options to adapt the processes to the current task) in order to be able to work efficiently?
- Objectives:
  - What rough objectives (scope/quality, cost and time-related objectives) can be derived from the project idea?
- Project approval:
  - What information is required for the approval of the project?

It is also important to ensure the economic justification of the project, for example by means of a **business case**:

> *A **business case** is an economic justification for a project and contains a quantitative comparison of the expected income and the corresponding expenses for a project.*

Typically, a business case contains monetary and non-monetary considerations. The typical contents include ...

- a justification of the relevance of the project,
- the presentation of compliance with the organization's overall strategy,
- an initial overview of possible risks,
- the expected expenses/costs,
- the achievable income and
- a rough estimate of the project duration.

The business case may change during the course of the project due to a changing environment or new insights. It should therefore be reviewed from time to time and updated if necessary.

Warning

Do not confuse the business case with a business plan. The latter is common for start-ups. A business plan is much more comprehensive. It is a planning document for the development of an entire enterprise. Together with other information, one or more business cases could be part of a business plan.

Often a **preliminary project approval** is initially granted, which only includes the definition phase. The aim of this preliminary approval is to give the project team sufficient freedom to clarify the project scope and develop the project objectives. This phase can then also be used to further elaborate the business case and support it with sound figures. Once the project has been clearly defined a decision is made as to whether it should actually be conducted. In this case, the budget is then made available for the subsequent phases.

The project approval can be documented in a **project brief**, sometimes also referred to as a project one-pager.
You can find the project brief in the toolbox in Chapter 8 under *I.1.1 Create project brief.* An example of a project profile can be seen in Figure 3.2. This can also be used to issue a gradual project approval from phase to phase.

Project brief		
Name of the project and logo		
Brief description:		
Objectives:		
Start date of project:	End date of project:	
Milestones:	Resources:	Costs:
Risks and mitigation measures:		
Project sponsor / owner:	Client(s):	
Project manager:	Project team:	
Approval for phase: o Initialization	o Definition	o Planning o Control o Closing
Signatures:		

Figure 3.2: Example of a project brief that summarizes important information about the project and is used to approve the project or to release the a certain project phase.

You may also find the *project canvas* helpful when preparing the start of the project. We will discuss this later in the context of the project start workshop.

Success factors

Before you start a project, you should take a look at the factors that promote project success. There are countless studies and field reports on this, some of which come to different conclusions. The reasons for these deviations are different approaches to the studies and the fact that success factors are favorable factors and not guarantees. This means that a project can fail even though many success factors are fulfilled. The reverse is also possible.
Nevertheless, it is worth familiarizing yourself with typical success factors and analyzing their degree of fulfillment. The fulfillment of some factors can be easily controlled and thus increase the probability of project success. Factors that are not fulfilled can be dealt with in stakeholder and risk management, which we will discuss in detail later.
The following list provides an overview of important success factors for orientation. The order is based on the order in which the topics are discussed in this book:

- high degree of project management maturity of the project participants and the project environment (all chapters on modern project management)

- choosing and tailoring a suitable process model (chapters *process models for plan-based project management, agile project management* and *hybrid project management*)
- clear objectives and requirements (chapters *definition and analysis of project objectives* and *from project objectives to requirements*)
- appropriate project governance including a well-defined project organization with assigned responsibilities, authorities and competencies (chapter on *project organization*)
- fast and transparent decision-making processes (chapter on *project organization*)
- support of the (top) management (chapter on *project organization*)
- qualification of the project team (chapter on *project organization*)
- effective communication within the project team (chapters on *project organization* and *leadership*)
- sufficient resources (chapter on *resource and cost planning*)
- good planning and control (depending on process model, see chapters on *plan-based project management, agile project management* or *hybrid project management*)
- good communication with people in the project environment/good stakeholder management (chapter on *stakeholder and risk management*)
- effective risk management (chapter on *stakeholder and risk management*)
- good feedback culture, ability to learn from failures (chapter on *knowledge management*).

Project types

It can be beneficial to characterize the project. This helps us to select the appropriate process model depending on the characteristics. Possible ways of characterizing a project are briefly outlined below.

Dependence on the client

For **external** clients, we usually need certain roles, methods and tools that we do not need for **internal** projects. Examples include supplier managers, contract management methods and some formal specifications.

Dependence of the project scope

Another typical differentiation of projects is the classification into **investment projects, organizational projects** and **research and development projects**. Depending on the type, different aspects of project management come into play here too. For example, financial and legal issues play a major role in many investment projects. In the case of organizational projects, there is often a great need for accompanying change management in order to take the various interest groups into account. In research and development projects, the technical challenge requires special measures for requirements management and the promotion of innovation.

Dependence of the project organization

In the **pure project organization**, the project manager is both the technical and disciplinary supervisor. In the **functional project organization**, the project manager has solely coordinating tasks in a staff function without a supervisor role. In the **matrix project organization**, the project manager shares the supervisor role. While the project manager takes on the functional leadership, the line manager is responsible for disciplinary aspects.

Dependence on complexity and the environment

The characterization of the scope and the social complexity can help to focus the project management, see Figure 3.3. The scope is classified into the categories simple/clear and complex/open. The former stands for a clear, well-communicated task that is not particularly difficult. A complex or open task exists if the project scope contains many interfaces, is not clearly defined or involves considerable challenges in finding a solution. Low social complexity means that the project team is relatively homogeneous and can work largely autonomously. High social complexity means that there are many stakeholders and diverse interests.

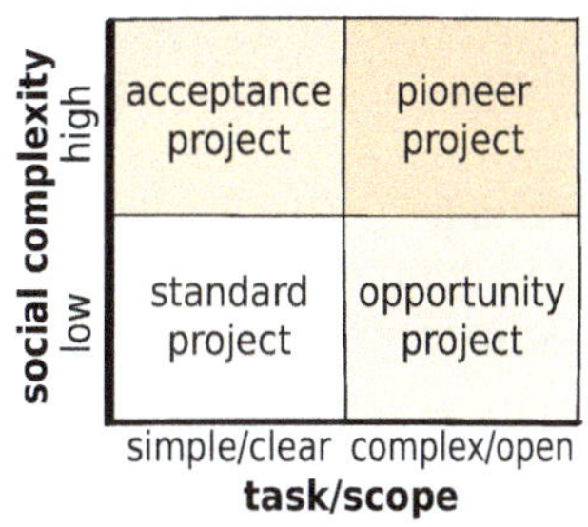

Figure 3.3: Project characterization based on the complexity of the project scope and the environment according to [Boos and Heitger, 1996].

Standard projects are considered easy to plan. In opportunity projects, the project participants should concentrate on defining the task and finding a suitable solution. In acceptance projects, the tasks and solutions are not the biggest challenge. Instead, they require careful stakeholder management with an effective communication strategy to manage the project's various stakeholders. In pioneer projects, both stakeholder management and the project scope itself are a challenge.
We look at further methods for characterizing projects and the corresponding project management, including the effects on the process model, in Chapter 5 on hybrid project management.

Kick-off meeting and project startup workshop

One of the biggest challenges for the project manager in the early phases of the project is to establish a structure and direction for the emerging project given the large number of stakeholders, expectations and information.
Important stakeholders (i.e. people interested in or affected by the project) should definitely be involved here. In many enterprises, it is therefore common to start the project with a **kick-off meeting.** This meeting serves to *inform the stakeholders.* A typical *agenda* contains the following topics:

- welcome and introduction round

- presentation of the rough project objectives and - if already known - important milestones
- introduction of the project organization (project manager, team, client ...)
- next steps in the project.

The kick-off meeting is not actually a working meeting, but an information event for important stakeholders.
Instead, the so-called **project startup workshop** has proven itself as a *working meeting*. This workshop is about preparing an initial project plan together as a team and with the client. *Participants* are:

- project manager and project team
- project sponsor and/or client
- other important stakeholders.

The *agenda* should consider the following topics [Schelle et al., 2006]:

- welcome, introduction and identification of expectations for the meeting
- collection of information
- overview of the given project management
- identification of stakeholders and definition of rough project objectives
- creation of an initial work breakdown structure
- initial effort and cost estimate
- structuring the project into project phases and defining important milestones
- outline of the project organization
- identification and assessment of project risks
- next steps, detailed planning of the next phase if necessary
- feedback session.

Further explanations can be found in the toolbox in Chapter 8 under *I.1.2 Project startup workshop* and *I.1.3 Kick-off meeting*.
At the latest at the end of the project kick-off workshop, it should be possible to complete the project brief for granting the project approval, see Figure 3.2. As an alternative, a **project canvas** can be developed.
A project canvas is a comprehensive overview to describe, visualize and evaluate a project in a language that is understandable for all relevant stakeholders, see Figure 3.4. The project canvas is developed in a team together with key stakeholders, for example as part of the project startup workshop. The predefined structure of the canvas helps to provide a complete and structured description of the project. During project execution, the project canvas can be used for continuous status assessment and reporting.
The exact structure depends on the requirements of the project and the project environment. Frequently used entries are objectives, budget, team, boundary conditions, milestones, project scope and quality assurance, resources, risks and opportunities, deliverables and important stakeholders. The project canvas should be completed collaboratively within the team. A brief explanation of all the fields of the project canvas can be found in the toolbox in Chapter 8 under *I.1.5 Project canvas*.
The *duration of the project startup workshop* depends on the size and complexity of the project. The desired level of planning also influences the duration of the workshop. In order

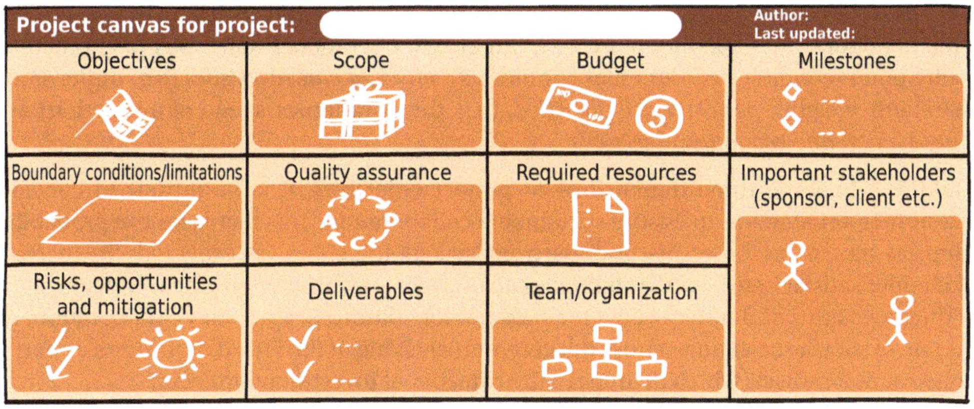

Figure 3.4: Example of a project canvas. The individual fields help you not to forget anything and to proceed in a structured manner. A project canvas is particularly suitable for the initial project definition and approval as well as for subsequent reporting and communication to important stakeholders.

to achieve reasonable and thorough results, you should allow at least half a day, or two to three days for larger projects.

Definition and analysis of project objectives

I have already talked about project objectives and emphasized their importance. However, I have not yet systematically addressed the functions that objectives perform and how objectives can be defined and analyzed.
The **dimensions of the objectives**

- scope
- time
- cost

form the so-called iron triangle of project management, see also toolbox in Chapter 8 *I.2.2 Iron triangle*. If you change one of the objectives, this affects at least one of the other objectives. For example, if an earlier deadline is to be met, more resources will probably have to be used for this, which will increase costs.

- **Cost-related objectives** must be defined comprehensively. They must not only relate to the *project costs* incurred during the project, for example in the context of product development. Instead, the objectives must be formulated for the entire *product lifecycle*, i.e. with regard to production, maintenance, operating and disposal costs. There are many methods for defining cost objectives that help to optimize costs and performance. These include the design-to-cost approach, target costing, value analysis and others.
- When defining the **time-related objectives**, many people first think of the project milestones. But here, too, it is important to consider any deadlines and time periods beyond the actual project. Examples include the definition of maintenance intervals or the minimum product service lifetime.

- **Scope-related objectives** describe what the project must be able to deliver in the end. This refers not only to *functional objectives*, but also to operational objectives (e.g. functionality under certain environmental conditions), *physical characteristics* (e.g. dimensions, strength, weight), *performance objectives* (e.g. the duration or speed of a function) and *production and economic objectives.*

In addition to these three dimensions of project objectives, it is important to consider **stakeholder satisfaction** to ensure sustainable collaboration. This increases the probability of *project success,* as fewer disruptions from stakeholders can be expected. To increase stakeholder satisfaction, **social objectives** can therefore be formulated. These include team-building measures and other events for stakeholder communication and collaboration. An example of such a social objective is the ceremonial laying of the foundation stone as part of a construction project (which belongs to the category of investment projects).
Objectives can either focus on the *result* to be achieved or on the *approach,* i.e. the way in which the objective is achieved.

Example

For example, a result-oriented objective is to create a design drawing for a specific component.
An approach-oriented goal in this context is the use of a specific design software (e.g. to ensure the compatibility of the file with other project partners).

Finally, let's look at the **exclusion of objectives.** Exclusions are used to set boundaries and avoid confusion. Stakeholders often hear what they want to hear and expect more from the project than intended. Exclusions can be defined to avoid later disappointment, costly renegotiations or change requests.
The **meaning of project objectives** go beyond the bare definition of the project objectives:

- Objectives provide a **control function**. It helps to evaluate the project by comparing objectives with what has been achieved.
- Objectives provide **orientation** which helps to categorize the project and understand what it is about. Instead of a long project description, the project is identified via the objectives.
- Objectives can fulfill a **connection function** and create a "feeling of togetherness". The common objective brings the team together. The members support each other in their joint work and are prepared to go the extra mile.
- The **coordination function** of objectives helps us to structure the project and create interfaces with other units of our organization or external partners. This can be achieved by clearly defining who is responsible for which objective.
- The **selection function** is always important when there are things to prioritize. If there are several options to choose from, the objectives can be used to evaluate and then prioritize them.

You should be aware of these functions when defining the project objectives. If the objectives are formulated in such a way that they do not fulfill the functions mentioned, the corresponding function must be fulfilled by other aspects of project management.

Writing project objectives

"Develop a new version of our control system firmware as quickly as possible" - such assignments by the client can be made to project teams. As long as a project has not yet been defined and only exists as a vague idea, it is obvious that it is not easy to make precise statements about project objectives. At the beginning of the project, objectives, efforts and therefore also costs and deadlines are still unknown. The NASA Software Engineering Labs and others have shown in empirical studies that these initial uncertainties lead to deviations between the actual effort and the estimated effort. At the start of the project, this ratio on average has a value of 4 or $\frac{1}{4}$. This means that the actual effort (at the end of the project) can be between a quarter and four times the estimated effort (at the beginning), see also Figure 3.5.

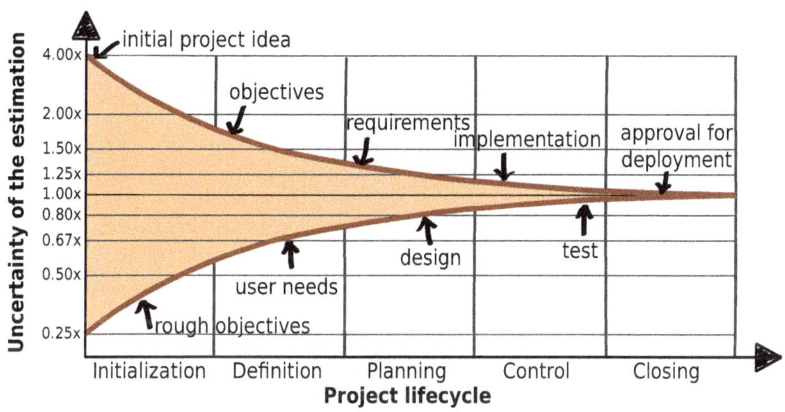

Figure 3.5: The Cone of Uncertainty represents the uncertainty in knowing the effort in the project. This decreases over time if appropriate measures (defining objectives, specifying requirements, drafting design) are taken.

The uncertainties about the project's progress can be significantly reduced by taking suitable measures to specify the objectives and reduce the uncertainties associated with the initially unclear solution. In the example of Figure 3.5, this is achieved through the further definition of objectives, detailed requirements specification, the design draft, implementation and testing of the project object.

These considerations are intended to illustrate the need for a good definition of objectives. The initial request to simply develop something as quickly as possible does not in itself reduce uncertainty.

The SMART principle helps with precise definition of objectives, see Figure 3.6 and the toolbox in Chapter 8 under *I.2.1 Defining SMART objectives.*

S	specific	simple, comprehensible, precise
M	measurable	clear criteria for checking fulfillment
A	achievable	achievable under realistic conditions (but quite demanding)
R	relevant	relevant to the project and related to the overall objective
T	time-bound	can be scheduled and planned

Figure 3.6: The letters SMART stand for **S**pecific, **M**easurable, **A**chievable, **R**elevant and **T**ime-bound. They indicate criteria that well-formulated objectives should fulfill.

Example

The aim outlined at the beginning "Develop a new version of our control firmware as quickly as possible" does not meet the SMART criteria. It is neither specific, measurable nor scheduled. Whether it is achievable is unclear due to a lack of precise specification. A better formulation is: "By November 30, 2026, version 2.0 of the control firmware for the 'VarioChart XL' product must be available in series production." Not every individual objective must meet all SMART criteria, but the totality of all individual objectives should be formulated SMART.

In addition to a SMART formulation, well-defined project objectives have further characteristics:

- Objectives are formulated in such a way that they do not yet define the solution **solution-neutral**. Otherwise, they restrict the freedom in the process of finding an optimal solution, resulting in increased costs or an increasing demand on time.
- Objectives must be **known to all relevant stakeholders**, i.e. in particular the project team, the sponsor and the client.
- Objectives should be defined in such a way that they remain **stable** throughout the entire project lifecycle.
- Objectives must be **consistent with corporate objectives**.
- There must be **no** serious **conflicts between objectives.**

The last two aspects require a precise analysis of the objectives, which we will now look at in more detail.

Hierarchical objectives analysis

The hierarchical objective analysis is an effective method for

- structuring objectives, presenting them clearly and thus checking their completeness
- the identification of interrelationships and dependencies between objectives
- monitoring the measurability of individual objectives and

- further detailing objectives or, in the other direction, for combining several detailed objectives into one overall goal.

Figure 3.7 shows the example of objectives in a hierarchical representation. The corresponding method profile can be found in the toolbox in Chapter 8, *D.3.1 Create objectives hierarchy*. The hierarchical representation helps to refine the overall objective of the project. Based on this, scope-, cost- and time-related objectives are defined. If necessary, a distinction can be made between objectives that focus on the results to be achieved or on the approach, the team should choose. Social objectives for stakeholder satisfaction and exclusions for differentiation can also be taken into account. From top to bottom, the objectives are increasingly detailed until concrete evaluation criteria for measuring the fulfilment of objectives are defined in the second to last line and metrics for their review are defined in the bottom line.

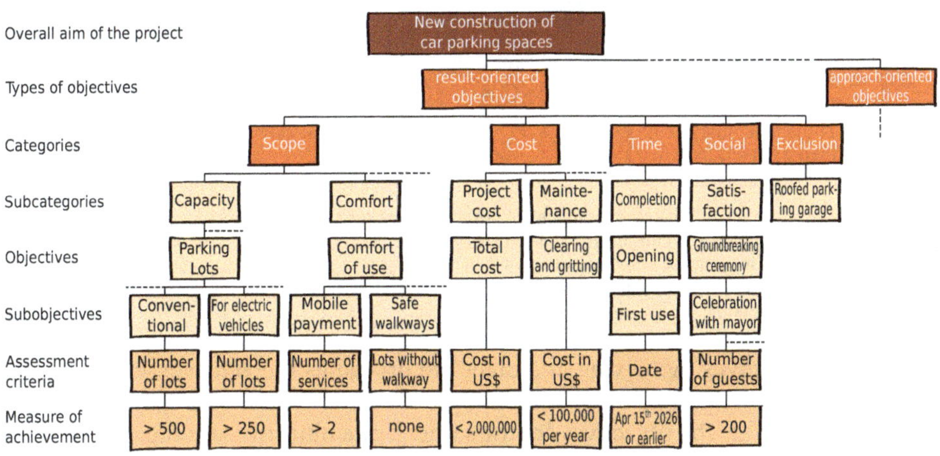

Figure 3.7: Example of a hierarchy of objectives based on Nagel [Nagel, 1992]. The dotted lines indicate that the hierarchy of objectives can be continued and that only an excerpt is shown here. It is important to note that the bottom line contains concrete measures that enable the fulfilment of objectives to be monitored.

Matrix of objectives

The objectives of a project must not be mutually exclusive. However, even competing objectives can jeopardize the success of a project. You should therefore analyze the goals you have set in terms of their relationship to each other.

An objective matrix illustrates such relationships between the project objectives and assesses their compatibility. An example is shown in Figure 3.8 using some roughly formulated objectives for a project to improve public bus transport. The objectives are evaluated as follows:

- Antinomy of objectives: The objectives are mutually exclusive. If one objective is achieved, the other can no longer be reached.
- Conflicting objectives: Achieving one objective makes it more difficult to meet the other objective, but does not make it impossible.
- Unrelated objectives: The objectives do not influence each other.

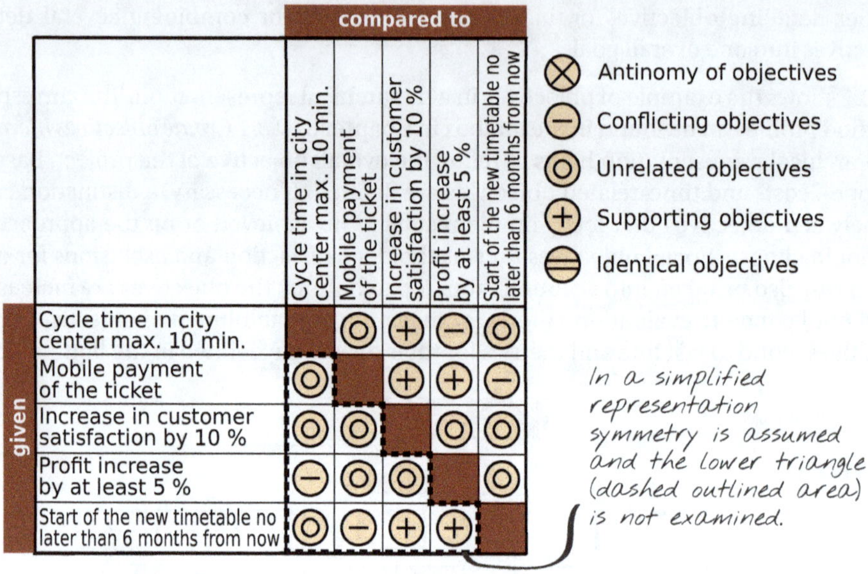

Figure 3.8: Example of a matrix of objectives. The objectives are compared horizontally and vertically. Assuming that the objective "cycle times" is achieved, it can be assumed that the objective "increase in customer satisfaction" is also supported (supporting objectives). In contrast, achieving the objective of "increasing customer satisfaction" does not necessarily indicate whether this also supports the objective of "cycle times". In a simplified version, the dashed area can be omitted and a symmetrical representation is assumed.

- Supporting objectives: Achieving one objective promotes the achievement of the other objective.
- Identical objectives: The objectives are the same. If one objective is achieved, the other is also fulfilled.

In the example shown in Figure 3.8, the matrix is read as follows: Assuming the achievement of the vertically listed objectives (given), it is examined whether this makes it easier or more difficult to achieve the horizontally listed objectives (analyzed). This may result in an asymmetrical evaluation matrix:

- If the cycle time in the city center is no longer than ten minutes, it can be assumed that this is also beneficial to customer satisfaction. The objectives are assessed as supporting.
- On the contrary, this conclusion is not correct, as the increase in customer satisfaction can also be achieved in completely different ways and is therefore not necessarily linked to the cycle time. The objectives are assessed as neutral.

The illustration can be simplified by ignoring the causal relationship and only examining whether the objectives tend to support or compete with each other in general.

Warning

If antinomies of objectives arise during the evaluation, the project will not be able to achieve all the targets. You should then definitely check how this incompatibility of several objectives can be resolved. Otherwise, the success of the project is at risk. The occurrence of competing objectives can also disrupt the subsequent course of the project, as compromises have to be negotiated or the fulfillment of all objectives requires very complex solutions. In the case of conflicting goals (and even more so in the case of antinomies), the project objectives must be prioritized as early as possible in order to be able to decide which objective is more important in the event of a conflict. Conflicting objectives should also be examined and monitored more closely as part of risk management.

The corresponding method profile can be found in Chapter 8 *D.3.2 Create matrix of objectives.*

Prioritization

Objectives should be prioritized. In most cases, this should be done together with the project sponsor or client. In the event of a conflict, a decision can then be made as to which objective is given more importance and which objective may even be dropped altogether.
Prioritization helps, for example, to focus on the really important objectives. If it is then clear which objectives are more important than others, the necessary measures can be initiated with the appropriate emphasis.
It is usually easier to prioritize right at the beginning of the project. The prioritization can then be made by mutual agreement on the basis of facts. The prioritization can be documented as follows:

- Establish a ranking from the most important (priority one) to the least important objective (priority corresponds to the number of objectives), for example from one to ten.
- Prioritization in categories: Must-have-, Should-have-, Could-have- und Won't-have-objectives.

The prioritization should be documented and approved by the project team, the sponsor and the client. This possibly cumbersome procedure helps to reduce the risk of later misunderstandings, see also Chapter 8 *D.3.3 Prioritize objectives.*

From project objectives to requirements

We have learned about objectives and their different functions and characteristics. The objectives must be further specified for project execution. This is done by means of **requirements management.**
First, have a look at the definition of a requirement:

*A **requirement** is the documented representation of a*

- *condition or capability needed to solve a problem or achieve an objective, or a*
- *condition or capability of a system to fulfill a contract, standard, specification or other document.*

Requirements

- can be derived from the project objectives
- form the basis for the design of the project scope and
- are used to evaluate the achievement of project deliverables.

Example

One of the requirements for a new audio system is that the audio system to be developed must be able to play the music tracks stored on smartphones with an Android operating system via a Bluetooth connection.
Based on this requirement, the technical design is laid out during the course of the project, i.e. the transmission protocol and other functions are implemented. Test cases are also specified in order to check that the requirement is fulfilled. In such a test, for example, the compatibility of a representative selection of common smartphone devices with the audio system could be verified.
If the requirement is changed during the course of the project, this has an impact on the design of the system and the test specification. For example, if the system is to be able to support devices with a different operating system, the transmission protocol must be adapted in the design and the test extended to include the corresponding devices.

Like objectives, requirements must be *SMART*, i.e. specific, measurable, achievable, relevant and time-bound. Well-formulated requirements correspond to the syntax shown in Figure 3.9.

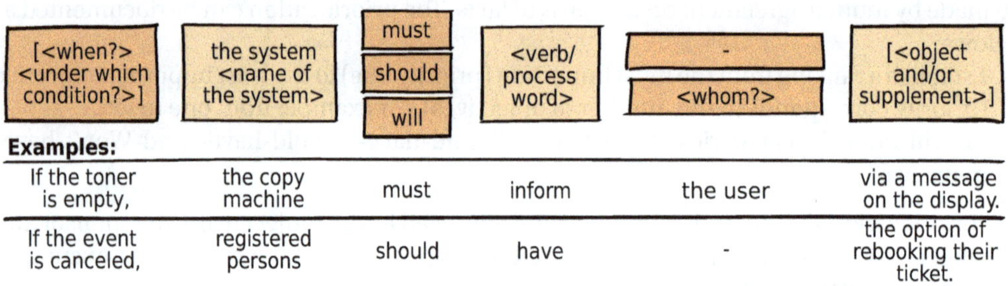

Figure 3.9: Syntax of a well-formulated requirement with two examples derived from Rupp [Rupp, 2021].

You should consider the following aspects when formulating your requirements:

- **Write down requirements separately**, i.e. do not combine several requirements into a single one. This reduces complexity and facilitates subsequent changes and testability.
- Use of **consistent terminology**, i.e. the same terms should be used in the same way throughout. This reduces the risk of misunderstandings and makes it easier to recognize correlations.
- **Full sentences:** Do not formulate requirements in bullet points. This reduces the risk of misunderstandings.
- **Ensure measurability and testability:** Requirements that cannot be verified must be critically reviewed if there is any need at all.

The example of the requirement for an audio system presented at the beginning shows that requirements are relevant throughout the entire project lifecycle. Requirements must therefore be **traceable**.

*The **traceability** of requirements means the ability to trace requirements and the design drafts, test cases or other documents derived from them over the entire project lifecycle.*

Traceability must be ensured for the following aspects:

- Source of a requirement (where does the requirement come from?)
- Relationships and dependencies between individual requirements.
- Deliverables, e.g. design drafts, test cases and test results, derived from requirements.

Warning

Requirements run through the entire project. They form the basis for the design of the project scope and, at the end, its verification and validation. Requirements must therefore be kept up to date throughout the entire project. Every time a requirement is changed, it must be checked whether deliverables and documents based on it, such as design drafts or test specifications, also need to be changed.

The requirements of the project scope, often a product, a service, an investment or a process improvement, are documented in the user needs and the requirements specification.

*The **user needs**, also known as the customer requirements specification, are the client's requirements for the project. They summarize what the client wants. With the **requirements specification**, the project documents how the user needs are to be implemented.*

The client collects his expectations of the project in the form of user needs. These can also be used to obtain offers from several potential suppliers. In the case of in-house research and development projects, user needs are often collected by marketing or sales departments and handed over to the development department for implementation.
The project shall address all user needs in the requirements specification and document how these are to be fulfilled.

- The *user needs* summarize WHAT the customer wants. It should be written by the client.
- The *requirements specification* explains HOW the needs are to be realized. It is usually written by the contracting project.
- It is advisable to stick to the separation of user needs (or customer requirements specification) and requirements specification, which defines the realization by the project in practice. It is advisable not to start further project planning and implementation until the requirements specification has been approved by the client.

Therefore, when compiling the specifications, systematically go through the entire product lifecycle and collect the requirements necessary for each phase. Think about the expectations that the project deliverables must fulfill as well as the environmental conditions, maintenance, reliability, etc. A possible structure of the requirements and functional specification is summarized in Figure 3.10.
Several requirements in the requirements specification can correspond to a single user need (and vice versa). For example, the user need U101.3 from the customer requirements specification,

User needs/customer requirements specification	Requirements specification
1. Introduction *What is it all about?* *Why is there a need for a new project?* *Description of the current and desired target status*	1. Introduction *How is it implemented?* *Description of the product structure?* *Description of the actual and planned status*
2. Objectives from the customer's point of view	2. Objectives
3. Product use *What purpose is the product used for?* *How is the product usually used?* *Other aspects, such as quantity, sales concept ...*	3. Product description *What components does the product consist of?*
4. Functional requirements	4. Functional requirements
5. Non-functional requirements *Requirements for performance, efficiency, scalability, reliability, operating conditions, safety ...*	5. Non-functional requirements *Implementation in terms of performance, efficiency, expandability, reliability, operating conditions, safety ...*
6. Interfaces *What interfaces to the user, the environment and to other products are required?*	6. Interfaces *Which interfaces to the user, to the environment and to other products are implemented?*
7. Testing and acceptance *What tests does the customer expect?* *What are the acceptance criteria?*	7. Testing and acceptance *How is the product tested?* *How is acceptance carried out?*
8. Documentation *Which documents (design drawings, protocols, instructions for use ...) are required?*	8. Documentation *Which documents (design drawings, protocols, instructions for use ...) are being prepared?*
9. Boundary conditions *Laws, standards, guidelines to be complied with ...*	9. Boundary conditions *Laws, standards, guidelines to be complied with ...*
10. Miscellaneous Topics	10. Miscellaneous Topics

Figure 3.10: Illustration of a typical structure of a list of user needs and the corresponding requirements specification. Guiding questions and hints for orientation are written in italics.

> *U101.30: The clock radio must be able to bridge short-term power failures of less than five minutes and guarantee all the functions listed under user need U200.*

can result in several requirements for the product design, which are documented in the requirements specification.

The relationship between the user needs and the requirements specification can be established using standardized identification codes. A requirement of the requirements specification belonging to the user need listed above would then read:

> *S101.30-1: All settings made by the user are stored immediately afterwards in a non-volatile data memory.*
>
> *S101.30-2: In the event of a brief power failure, the clock radio is supplied with power via an internal battery for at least five minutes so that all functions (time display, alarm function, etc.) of the L200 can be maintained.*

The House of Quality can help with the development of ideas for the realization of user needs. This is part of quality function deployment and therefore of quality management. The House of Quality is presented in the toolbox in Chapter 8.

Phase and milestone planning

You already know the importance of objectives and the requirements derived from them. Now you will learn how you can create a **first rough project plan** with simple means in order to check the feasibility of the specified objectives. As we will see later, it is important to set ambitious goals. However, for the project to be successful, the goals must also be realistic. It is often the case that figures for deadlines and costs are announced overhastily, only to turn out to be unachievable once the initial planning steps have been completed.

Tip

Some project tasks seem overwhelmingly challenging at the beginning, especially for new project managers. The complexity of the upcoming project and the pressure from the client to specify costs and deadlines lead to project managers making hasty commitments to certain costs and deadlines. If it turns out later in the planning process that these figures are unrealistic, the client will blame this on the project manager's lack of competence and may make additional claims or demand compensation. Therefore: Don't be discouraged by complex tasks! In this chapter, you will learn how to create an initial, very rough project plan using simple means. You should only commit to time and cost objectives once this initial plan has been prepared.

The objectives of phase and milestone planning are:

- initial rough structuring of the project by dividing the project into sequential and/or parallel phases
- providing an overview of the overall course of the project by assigning tasks and resources to the phases
- first estimates of costs and deadlines.

Important definitions of phase and milestone planning are:

> ***Project phases*** *structure the project into separate time periods and thus define the rough course of the project.*
>
> ***Milestones*** *are events of particular importance. They define important points in time in the project at which certain deliverables must be available.*
>
> ***Phase models*** *are part of a process model and structure the project in phases. Phases can be run through sequentially (one after the other), partially in parallel, completely in parallel or recurrently (multiple times).*
>
> ***Phase plans*** *are rough project plans that structure the project into phases. For each phase, tasks to be completed, required resources, costs and deadlines are roughly defined. Phasing is usually based on a phase model. Its phases are used as the starting point for phase planning.*

The two different terms *phase planning* and *milestone planning* emphasize either

- what has to be accomplished in the individual phases, how much it costs and how long it takes or
- which results have to be achieved at certain milestones, how much they cost and at which points in time the milestones are planned.

In practice, the terms "phase planning" and "milestone planning" are often used interchangeably. However, the milestone plan is sometimes also understood as a straightforward schedule in which the most important milestones are listed separately from the tasks to be

performed for them. Such a plan bears the risk of adopting desired targets as milestones without checking their plausibility through rough planning.
Figure 3.11 illustrates the structure of a phase and milestone plan. It consists of

- a phase-oriented structure of the project lifecycle,
- important milestones (at least one at the end of each phase),
- a rough list of the tasks to be completed for each phase,
- an estimate of the resources required and the resulting deadlines and costs for these tasks,
- the total costs and the total duration of the project.

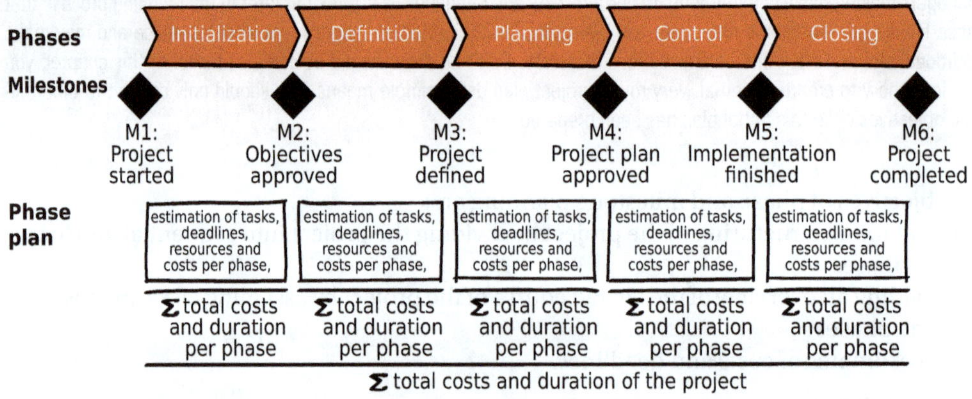

Figure 3.11: Structure of a phase and milestone plan. Costs and deadlines are estimated for each phase and used to determine the total costs and duration of the project.

In the context of phase and milestone plans, we deliberately do not speak of work packages, but of tasks or activities to be carried out in the individual phases. Work packages are part of the work breakdown structure, which we will get to know later.
Phase and milestone plans are often created by just a few people, for example the project manager and core team on a whiteboard. The steps for creating such plans are explained in Figure 3.12 using the example of the construction of a detached house. The numbers in the following list correspond to the numbers in the figure:

1. Development of a common understanding of the **project objectives.**
2. **Structuring the project into phases.** If a company-specific process model exists, the phases defined in its phase model should be adopted in the phase and milestone plan. Alternatively, individual phases that fit the upcoming project can be defined.
3. **Determination of important milestones** at the end of each phase and, if required, within phases.
4. **Identification of important tasks** for each phase and their breakdown under the respective phase.
5. **Identification of the necessary resources and costs as well as deadlines.** You should at least roughly estimate the costs and deadlines. For past projects, a rough estimate or expert opinions are good sources. More precise methods for estimating costs are presented below. However, these can only be used to a limited extent in the current phase, as little is known about the project apart from the project objectives.
6. **Determination of costs and the duration for each phase.**
7. **Determination of costs and the duration for the entire project.**

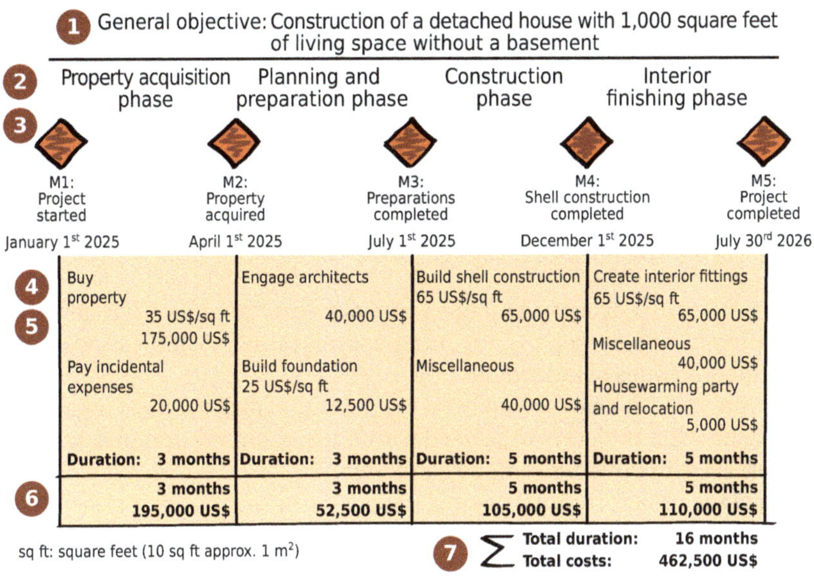

Figure 3.12: Example of a phase and milestone plan. The numbers in the black circles refer to the steps in the preparation of a phase and milestone plan and are explained in the body text.

Tip

Become accustomed to using clear terminology in order to avoid misunderstandings and to recognize what the term refers to based on the chosen wording:

- Express *milestones* as events and not as ongoing activities or simple keywords, for example: "project completed" instead of "project completion" or "project end".
- If possible, you should set a specific date for each milestone.
- Formulate *tasks* (and later work packages and processes) with a verb to emphasize the temporal dimension: "create construction" instead of just "construction".
- Add the suffix "phase" to *phases* to make it clear that you are actually talking about a phase, not an activity, for example "supplier acquisition phase" instead of "supplier acquisition".

A phase and milestone plan is an initial rough plan – no more, but also no less. This plan still contains many uncertainties due to the rather imprecise task descriptions and estimates. Nevertheless, it is a good basis for comparing initial planning results with the objectives. If objectives and planning results differ considerably, you have two options:

- Question the plan and the assumptions you made when preparing it. Check whether you see potential for acceleration or cost savings.
- Adjust the objectives to the planning results.

Warning

In practice, a third, dangerous alternative is often chosen: Although the plan and the objectives do not match, the objectives are maintained unchanged without sufficient examination of whether there really is potential for acceleration or cost savings. The motto is "We'll get there somehow." However, practice also shows that this is often a fallacy.

Project organization

Objectives and an initial rough plan help the project to take shape. Before you can move on to detailed planning, you need a team and have to set up the project organization.

*The **project organization** is a group of people and the related infrastructure for which agreements are made regarding authority, relationships and responsibilities in alignment with the business and functional processes.*

The project organization covers several hierarchically structured levels: It governs the integration of the project in the company, the internal project organization, as well as the cooperation and responsibilities of individuals.
The following objectives can be derived from the definition of the project organization:
- determining how the project is organized within the company and through which interfaces it interacts with other units of the company
- structuring of the project internal organization
- definition of the individual roles, authorities and responsibilities in the project
- selection and use of tools for daily project work.

The objectives are structured in a hierarchical top-down manner, from a company-wide view of the project organization to the organization of daily tasks. Individuals have different responsibilities and authorities and need different skills to fulfill them.

*A **responsibility** is an obligation associated with a specific task to ensure that the right things are done to fulfill the task or achieve a goal and to take ownership of it.*

*An **authority** is a permission and power to perform certain actions and make decisions.*

*A **competence** is the proven ability to apply knowledge and skills appropriately to the situation.*

Important stakeholders and their roles are briefly outlined below.

*A **role** describes a position in the project that is associated with a certain responsibility and the corresponding authority. Usually, a role is tied to certain expectations from other stakeholders.*

An overview of the responsibilities and authorities of important roles can be found in Figure 3.13. You have already learned the basic meaning of the roles in Chapter 2.

Project internal organization

In the project initialization and definition phase, a complete resource plan is not yet available. Nevertheless, the project manager is usually supported by other people and experts. The internal project organization can be defined in an organization chart.

	Project sponsor	Project owner/ client	Steering committee	Project manager	Team member	Project office	Supplier
Responsibility	can be identical with project owner/client supports project financially, organization-ally or ideally	secures funding and provides the project budget escalation authority in the event of problems	reports to executive management monitoring the progress of the project escalation instance	reports to steering committee planning, execution and control of the project project closure	working on the assigned tasks reporting to project manager	usually administrative support for the project manager	delivery of the agreed deliverables in accordance with the contract
Authority	depending on the arrangement, identical to the client or authority as coach or mentor	issues and withdraws the project order releases the subsequent project phase	decides if the limit of the project manager's authority has been reached decides on major changes balancing the interests of the client and the assigned project	participates in project definition and resource allocation depending on organization, coordinating, functional or disciplinary leadership decides on tasks in the project	within the scope of the assigned tasks preparing and, if necessary, participating in project decisions	requesting data in coordination with the project manager	decisions within the agreed framework in accordance with the contract

Figure 3.13: Responsibilities and authorities of important project participants in the project.

*An **organizational chart** visualizes the organizational hierarchy, structures organizational units and depicts relationships between them.*

Figure 3.14 shows the organization chart of a project with several sub-projects.

*A **sub-project** is an organizational sub-unit of the project that is separate from other units in terms of its tasks or duties.*

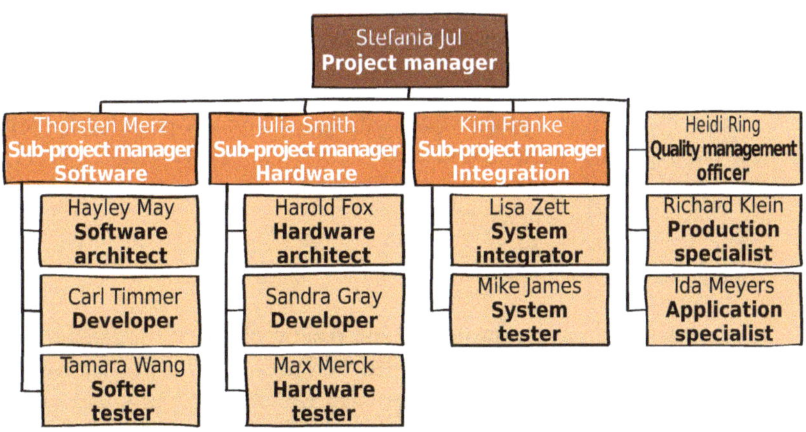

Figure 3.14: Example of a project organization chart with the sub-projects software, hardware and integration as well as employees who directly report to the project manager (right column).

In the example, these are the software, hardware and integration sub-projects. The employees working on these sub-projects are assigned to the respective sub-project managers. Like the sub-project managers, other project employees are assigned directly to the project manager and report to them.
You should consider the following aspects when organizing the project internally:
- promoting efficient and at the same time sufficient communication
- establishing clear responsibilities and reporting channels
- manageable scope of tasks for the project manager.

Example

A project is to develop a new product for surface treatment in a plasma. The project team consists of 15 employees, all of whom report directly to the project manager. The weekly **project meetings** usually last several hours. While one part of the team is bored because irrelevant points are discussed, another part lacks the time for in-depth content. However, this is necessary to solve problems.
Based on this observation, the project manager structures the project into the sub-projects of plasma generation, control panel and system. From now on, the project meeting will only deal with aspects that affect all employees and sub-projects. The technical discussions, on the other hand, will take place in the respective sub-projects. This reorganization will reduce the weekly meeting time from 45 person-hours to just under 30 person-hours. Despite this reduction in workload, there is now enough time available to solve subject-specific problems.

Proceed carefully when filling the sub-project manager positions. The questions you need to clarify are as follows:
- How should the role of sub-project manager be organized? Should the focus be on the holistic management of the sub-project or purely on functional coordination? The answer to this question has an impact on the skills required of the sub-project manager.
- Are there employees in the team who may consider project management as one of their next career steps and who could qualify for this role as a sub-project manager?
- Do the employees have the required skills in organization, project management and leadership in addition to their specialist expertise?

It is often the case that when filling project manager and sub-project manager positions, only purely specialist skills are taken into account. However, a specialist employee is not necessarily a good (sub)project manager with leadership skills.

Escalation

In the event of a conflict, objective, solution-oriented action is often superseded by personal attacks and blockades. This is why you should develop escalation paths when setting up the project organization and coordinate them with the project team and other affected stakeholders.

*An **escalation** is the structured delegation of a problem or decision to the next higher hierarchy level if it cannot be solved at the current hierarchy level.*

Inexperienced project managers often hesitate to escalate at an early stage and try to somehow solve a problem that exceeds their authority. They are worried that escalation could be perceived as weakness or incompetence.
For this reason, and because it is usually easier to agree on who is responsible for decisions in the event of a conflict at the start of the project, appropriate decisions should be made when setting up the project organization.
The escalation path should include project team members, sub-project managers and the project manager and may also involve external stakeholders such as management, the client or the steering committee if the project manager's authority is exceeded.
Escalation paths can be defined as flowcharts, tables or in text form. A simple example is shown in Figure 3.15. The documentation should indicate in which cases escalation applies, to whom escalation is made, and what authority this instance has or to whom it can in turn escalate further.

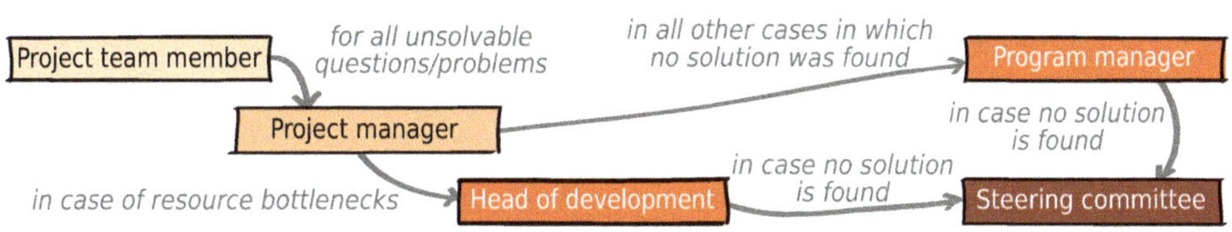

Figure 3.15: Exemplary illustration of an escalation path.

You should take the following aspects into account when escalating:

- Check whether you have exhausted your own scope for action.
- Choose the right addressee for the escalation. The addressee should have sufficient authority to resolve the escalated issue.
- Involve important stakeholders and do not ignore any hierarchical levels.
- Always escalate from a lower hierarchy to the next higher hierarchy.
- Announce the escalation to the affected stakeholders. Otherwise, you will weaken your own position, as others could interpret this as your inability to solve the problem by yourself.
- If you announce an escalation, you must also carry it out. Otherwise, you will also weaken your own position as you will lose credibility.
- Look for supporters for your cause, preferably among influential stakeholders.
- Prepare the escalation thoroughly and request a clear and binding decision from the addressee.

Practical tools for daily project work

In day-to-day project work, a lot has to be organized, tracked, checked, analyzed and documented. The methods, tools and software that support this are too numerous to present them all here. We will take a closer look at a few selected tools.

We have already learned that projects are unique undertakings. Consequently, we have to deal with an issue for the first time and solve it. Structured problem-solving cycles help us to find a solution. These help us to understand the underlying problem and develop a tailor-made solution. Project participants should therefore have good problem-solving skills and be able to apply appropriate methods. Two methods are briefly outlined below and explained in the toolbox in Chapter 8:

- In **structured problem solving**, the problem is first analyzed comprehensively and a target for the solution is derived from this. Several alternative solutions are then developed, analyzed and compared to each other in the synthesis. The best evaluated solution is then presented for a decision and implemented. If it turns out during the problem-solving process that the problem has not been properly understood or that no reasonable solution has been found, the cycle starts all over again, see Figure 3.16.

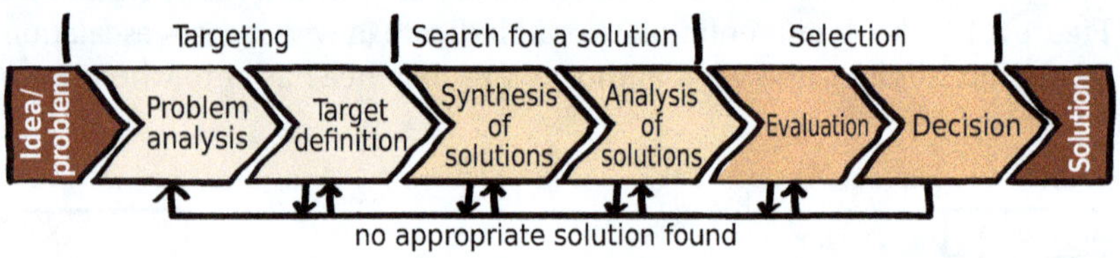

Figure 3.16: Sequence of structured problem solving.

- **Design thinking** is often associated with agile projects. However, the process can actually be used in any project when it comes to developing creative new solutions. An interdisciplinary team participates in order to develop new solutions that go beyond the conventional approach. The participants should consciously develop new ideas without first thinking about their realization and feasibility. This would only unnecessarily limit creativity at this stage. Design thinking is usually moderated or accompanied by a design thinking coach, often also referred to as a facilitator.
  The design thinking process is divided into a *macro cycle*, in which different prototypes are designed, and a *micro cycle*. The latter is about understanding the underlying problem and developing solutions step by step. You can find out more about design thinking in Chapter 4 on agile process models. Macro and micro cycles are also explained in the toolbox in Chapter 8 under *D.4.9 Performing a design thinking macro cycle* and *D.4.10 Performing a design thinking micro cycle.*

In the project business, you will always have to come up with new ideas. Make conscious use of **creativity techniques**, for example:

- **Brainstorming and -writing:** This method runs in two separate stages:
  - During the actual brainstorming, ideas are collected from the participants and either written directly on a whiteboard by the moderator or written down on cards by the participants and then hung up on a pinboard. In order not to limit creativity, evaluations are taboo in this phase.

- In the evaluation phase, the ideas are first structured and summarized in meaningful groups. Duplicates are removed. The remaining ideas can then be evaluated and prioritized.

- **6-3-5 method:** The name describes the procedure of this incremental method. Six participants write three ideas per increment on a sheet of paper, which is then passed clockwise to the next person. This person adds another three ideas. After 5 increments, the game ends and the ideas are collected, evaluated and prioritized. As you can write down your ideas without being disturbed, there are fewer interferences. At the same time, passing on the cards facilitates inspiration. If you read your neighbor's ideas, you may come up with new ideas based on them. The 6-3-5 method involves all team members, as everyone is required to write something down. The numbers should be seen as a rough guideline. Depending on the task and the team, more or fewer people can be involved.
- **Nominal group technique:** The nominal group technique combines idea collection and evaluation. First, ideas are collected in writing (brainwriting). Each idea is then prioritized by each participant and an overall prioritization is determined by summing up the results. This produces a prioritized list of ideas.
- **Morphological box:** For a problem to be solved, features and corresponding characteristics are collected. Both, features and characteristics, are then arranged as in the example in Figure 3.17. Features and characteristics should be selected independently of each other. They can be combined arbitrarily. Interesting combinations can now be discussed, evaluated and prioritized. By presenting the various possibilities, the aim is to encourage thinking beyond known patterns.

<table>
<tr><th>Feature/dimension</th><th colspan="4">Characteristic</th></tr>
<tr><td>Material of the casing</td><td colspan="2">Plastic</td><td colspan="2">Aluminum</td></tr>
<tr><td>Color of the casing</td><td>Gray</td><td>Black</td><td>Red</td><td>Blue</td></tr>
<tr><td>Type of display</td><td>Segment/single line</td><td>Segment/double line</td><td colspan="2">Matrix</td></tr>
<tr><td>Number of colors that can be displayed</td><td>1</td><td>2</td><td>4</td><td>8</td></tr>
<tr><td>...</td><td colspan="4">...</td></tr>
</table>

Figure 3.17: Example of a morphological box for a pocket calculator. The characteristics should be able to be combined as required. For example, the pocket calculator can be developed out of gray plastic with a double line display and four displayable colors. Alternatively, a model made of red anodized aluminum with a two-color matrix display can be implemented.

The following methods, among others, are ideal for structuring and analyzing information and for supporting creativity techniques:

- **Mindmap**: In a mindmap, ideas are collected based on a central topic and structured in a hierarchical tree diagram. An example can be seen in Figure 3.18.
- **5-Why method** (also 5-W method): This method is used to find the root cause of a problem or error. Five "why" questions are asked, which build on each other to lead to the actual cause. In fact, you don't limit yourself to five questions, but keep asking them until the cause is found.

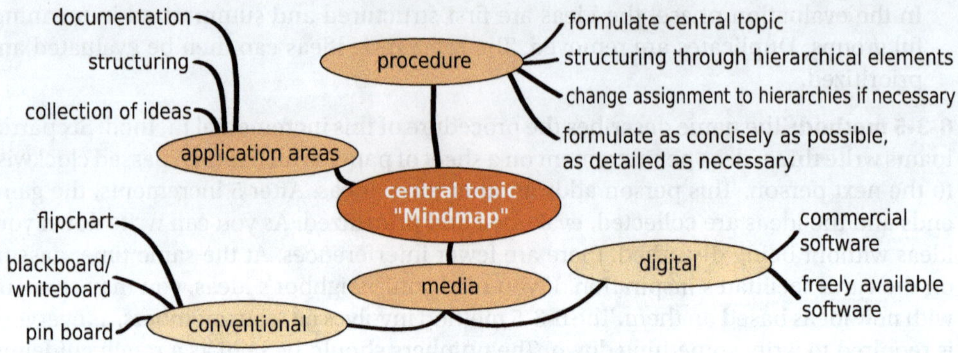

Figure 3.18: Example of a mindmap.

Example

The following consecutive chain of questions illustrates the application of the 5-Why method:

Why does the part not fit with the rest of the component?

Because the wrong component was delivered.

Why was the wrong component delivered?

Because the wrong component was ordered.

Why was the wrong component ordered?

Because the bill of materials (BOM) lacked precise information on the component.

Why did the BOM lack precise information on the component?

Because the BOM had not yet been finalized and released.

Why was the BOM not yet finalized and released?

Because this was not clearly required by the development process.

The root cause is therefore that the development process allows orders based on non-approved bill of materials (BOM). To avoid the error permanently, the process should be changed so that the BOMs must be released before the order is released.

- **Ishikawa diagram** (also cause-effect or fishbone diagram): Based on defined criteria, potential causes are identified for a problem to be solved. For each cause, it is then checked whether there are other, underlying causes. The criteria used to search for the causes of the problem are often referred to as 6M. An example is shown in Figure 3.19 on the left.
- **ABC analysis**: In the ABC analysis, elements to be categorized are lined up in descending order. In the example of Figure 3.19 on the right, this is the share of product costs for the components of the product. The most important elements are in category A. The ABC analysis is often used in conjunction with the Pareto principle. This states that 80% of the

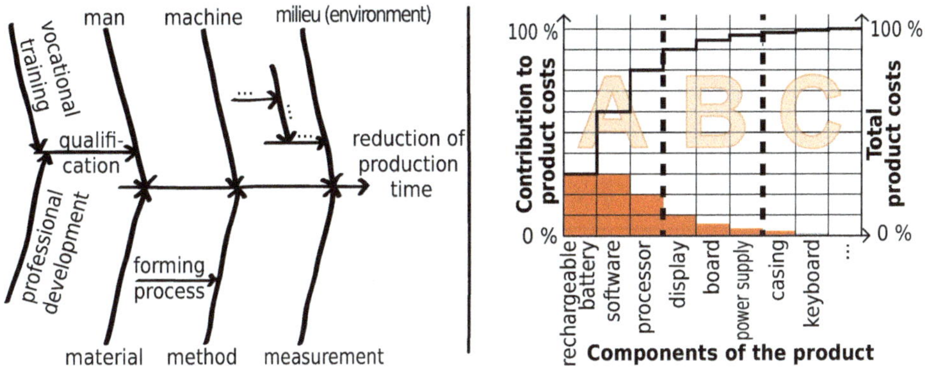

Figure 3.19: Left: Example of an Ishikawa diagram with the 6M: man, machine, milieu (environment), material, method, measurement. Right: Example of an ABC analysis.

results (i.e. category A, e.g. product costs) are achieved with 20% of the input (number of product components).

The presented methods can either be used after an initial creativity workshop in order to structure the collected information, or they can already be used during the workshop. Then, they can be used for a comprehensive presentation of the information.

The following aspects of **self-organization**, including **document management**, should be implemented at the very beginning of the project:

- The **Eisenhower principle** helps you to categorize upcoming tasks, see Figure 3.20. We differentiate between important and urgent tasks. In practice, these two categories are often mixed up and not managed appropriately. This can lead to overload and mistakes. Even if the recommendations of the Eisenhower principle do not always have to be strictly followed, they do provide assistance in assigning tasks.

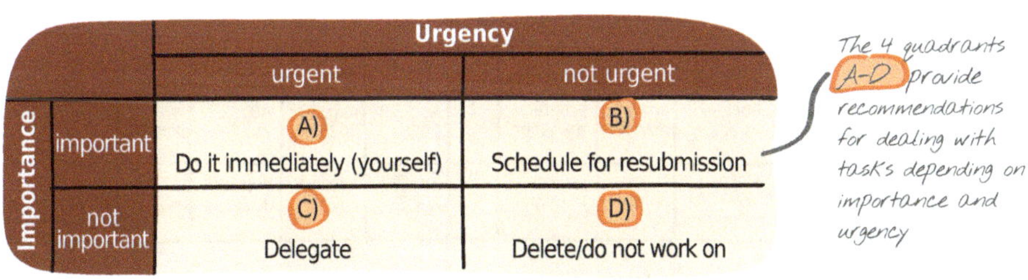

Figure 3.20: Sketch of the Eisenhower principle: Tasks are ordered according to their importance and urgency. The resulting quadrant indicates whether the tasks are to be done immediately (quadrant A), should be done later (B), can be delegated (C) or are not carried out at all (D).

- Determine how documents are named and where they are archived.
- Document responsibilities and obligations to communicate information. You can do this with a **RACI matrix**, for example. RACI stands for **R**esponsible, **A**ccountable, **C**onsulted and **I**nformed. The matrix defines who is responsible for implementation (responsible), who is responsible for approval or has budget authority (accountable), who must be

consulted for advice (consulted) and who must be informed (informed). An example of a RACI matrix can be seen in Figure 3.21.

	Program manager	Project manager	Requirements engineer	System architect	Test manager	Developer	...
Project meeting	I	A/R	C	C	C	C	
Project brief	A	R	C	C	C	C	
Requirements Specifications	C	A	R	I			
System design		A	I	R		C	
Test planning		A	C	I	R	I	
...							

Figure 3.21: Example of a RACI matrix (R: responsible, A: accountable, C: consulted, I: informed).

- Instead of or in addition to a RACI matrix, an TCR matrix can also be created. This defines who has to complete certain **T**asks in the project, which **C**ompetencies are required for this and which **R**esponsibilities exist. The TCR matrix is also explained in the toolbox in Chapter 8 under *D.2.1 Define roles.*
- Create and maintain an **list of action item**, sometimes also called a list of open points or action point list. The complexity of projects increases rapidly and the risk of forgetting things that need to be done increases. The list of action items should be discussed in every project meeting and be accessible to the team members. A typical structure is shown in Figure 3.22. A well-managed list of action items also serves to document completed tasks and any information collected. It is a useful addition to meeting minutes and can also replace them in smaller projects. Today, it is quite common to manage such lists with appropriate software, which then sends automatic reminders at defined times and facilitates the documentation of tasks.

#	action item	description	respon-sible	approved by	prio	started	due	expected	status	comments
1	update project brief	updating the project brief to release the planning phase	project manager	program manager	1	Aug. 10th 2025	Aug. 31st 2025	Aug. 31st 2025	work in progress	-
2	finalize contract with supplier	a contract for work and services is to be established with company XY for the development of the ball bearings	supplier manager	project manager	2	Aug. 17th 2025	Aug. 31st 2025	Sep. 18th 2025	rests	delay due to plant vacations at the supplier

Figure 3.22: Example of a list of action items with consecutively numbered (#) action items.

Project planning

Overview – where are we in the project management process?

The initialization phase and the definition phase have been successfully completed; now the actual **planning phase** begins, see Figure 3.23.

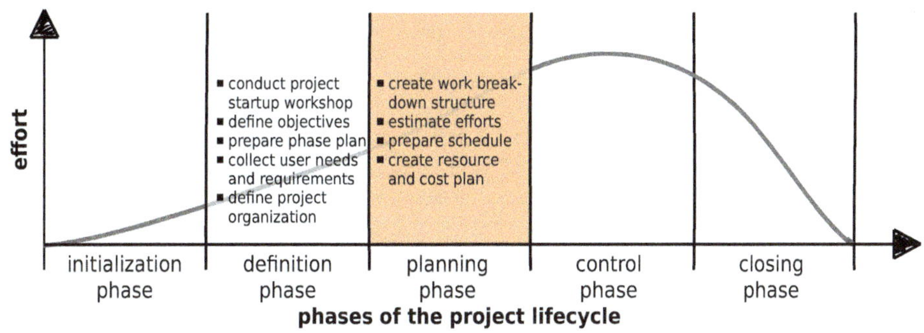

Figure 3.23: After a successfully completed initialization and definition phase, the project objectives have been determined, analyzed and transferred into specific requirements. The project organization has been defined. On this basis, detailed plans can now be prepared for the WHAT (work breakdown structure), the WHEN ("schedule"), the WHOM (resource plan) and the HOW MUCH (cost plan).

In practice, the term "project plan" is used in different ways: Some people understand it to mean only the schedule, others the schedule and cost plan. In fact:

> *The **project plan** includes all plans that exist in a project.*

It is therefore misleading to speak of a single plan. The most important plans include

- the work breakdown structure
- the schedule
- the resource plan and
- the cost plan.

If you change a plan during the course of the project, you must check whether this change affects the other plans and change them accordingly. Therefore, from a certain level of complexity in the project, it is advisable to use integrated project management software that ensures the consistency of the individual plans.

Example

A small development project has been planned in terms of what needs to be done (work breakdown structure) and when the tasks are to be completed (schedule). When preparing the resource plan, the project manager is informed by the head of the development department that the system architect will only be available for the project one month later. Although he would be urgently needed, he has to complete other tasks first. Alternative resources are not available.
The project manager now has to reschedule. The missing system architect resource also affects the schedule, which must be revised as a result. The work breakdown structure is not affected by this change.

Tip

Plans are by no means an end in themselves. Since their preparation and subsequent maintenance takes time:
- you should review each plan to determine whether it is necessary, i.e. whether you will derive a benefit from the information of the plan that justifies the effort it takes to create the plan, and
- you should check each plan to see how detailed the plan should be, for example whether individual project hours, days, weeks or months should be planned and how detailed individual resource and cost items should be considered.

Unfortunately, the second point of the tip is often ignored. A plan that is based on only roughly estimated information will not be more accurate just because you plan every action in great detail.

Planning the work breakdown structure

The work breakdown structure defines **what** is to be done in the project. It does not make any statements about when something will be done, what resources are required or how much it will cost to complete the project.
The timeline, required resources and resulting costs are subsequently planned based on the work breakdown structure. It is therefore essential that the work breakdown structure is *complete*, i.e. that it takes into account all upcoming tasks of the project. In other words: What is forgotten in the work breakdown structure is also very likely to be missing in the schedule, resource plan and cost plan.
As the name suggests, the work breakdown structure is intended to create structures. It provides an important basis for the further course of the project. This is achieved through the following aspects:
- The work breakdown structure *contains all tasks of the project* and thus provides an overview of everything that needs to be done.
- Due to its completeness, the work breakdown structure is the *basis for all further plans* and also for the *controlling* of the project.
- Tasks are clustered and structured in a meaningful way to *prevent forgetting anything or duplicating tasks.*
- A *responsible person is appointed* for each task so that there is a "caretaker" who is responsible for completing it on time and within budget.
- Once responsibilities have been defined, it may be necessary to optimize the *organizational chart* and adapt communication paths.
- Depending on the project and its organization, work breakdown structures can be standardized to a certain degree; they therefore also serve the purpose of *knowledge management* and the transfer of project experience for future projects.

The work breakdown structure itself consists of various elements that are uniformly named in almost all standards. In order to avoid misunderstandings, we would like to follow this nomenclature:

*The **project structure** refers to the entirety of tasks, work packages and activities that together represent the project. Structuring can be based on different aspects, such as time, function or object-related. In principle, the project structure comprises both the process structure and the organizational structure of a project, whereby only the organizational structure is defined in the work breakdown structure.*

*The **work breakdown structure** documents the project structure with its work packages as a hierarchical diagram, table or list.*

*The **work breakdown code** is a coding scheme that uniquely identifies all elements of the work breakdown structure.*

*The **root element** refers to the top level of the work breakdown structure, i.e. the entire project task.*

A work breakdown structure consists of tasks and work packages that can be grouped together. A group of work packages can represent a phase or a sub-project, for example.

*A **task** is a work assignment under defined conditions and using certain resources. Depending on the scope and hierarchical level, a task can be an entire program, a project, a sub-project, a work package, an activity or an even smaller unit. Tasks can be subdivided into sub-tasks.*

*A **sub-task** is obviously a part of a task and can itself be subdivided into further sub-tasks and/or work packages. A sub-task can be used for structuring in the work breakdown structure. However, it can neither represent the overall project nor be at the lowest level of the work breakdown structure, which is reserved for work packages.*

The terms "sub-project" and "sub-task" are often used interchangeably in the context of work breakdown structures. I will only make the following distinction here:

*A **sub-project** is an organizational sub-unit of the project that is separate from other units in terms of its work packages or duties. It is made up of several work packages. A sub-project manager is responsible for it. Sub-projects can refer to other work breakdown structures and be detailed there.*

*A **work package** is a self-contained assignment that is not further subdivided in the work breakdown structure. The sum of all work packages in the work breakdown structure is everything that needs to be accomplished in the project. If required, work packages can be further broken down into so-called activities in the project schedule.*

The last term we consider is activity. Activities do *not* appear in the work breakdown structure, but are part of the project schedule, instead. To differentiate the term, we define it here:

*An **activity** is a temporal process element that represents a self-contained assignment. A work package can be converted into one or more activities. While a work package defines what is to be completed, activities define the (temporal) sequences of the assignments. Activities are not part of the work breakdown structure.*

In the previous definitions, you learned that work breakdown structures be represented by a hierarchical diagram, table or list. Figure 3.24 shows the composition of a work breakdown structure as a hierarchical diagram and as a list.

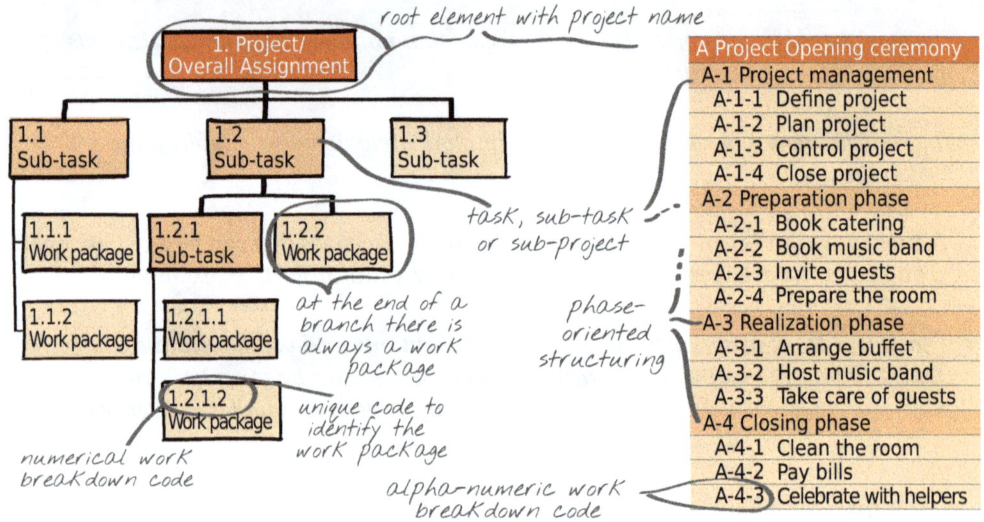

Figure 3.24: Structure of a work breakdown structure as a hierarchical diagram (left) and as a list (right) for the project "Organization of an opening ceremony".

The example in the list emphasizes an aspect already mentioned: the work breakdown structure must be complete! In other words, it also contains the project management duties. These are usually entered in the left-hand column as the first group of work packages, regardless of the chosen structuring scheme.

When structuring, it is helpful to think about a consistent **structuring principle**. You are probably familiar with this problem from your email inbox: you can sort emails by date of receipt, by sender or by topic, for example. For work breakdown structures, it also makes sense to think about a structuring principle for the work packages in order to create a consistent structure.

Common structuring principles are:

- **Phase-oriented structure:** Similar to the phase plan, the project is divided into phases and the sub-tasks and work packages to be completed are assigned to these phases. It is important to emphasize that even if the phases run one after the other, you should not be tempted to turn the work breakdown structure into a schedule. The work breakdown structure deliberately does not contain any time sequences. Whether a work package is at the top or bottom, left or right of the work breakdown structure does not tell you whether it is to be completed earlier or later than the neighboring work packages.

- **Object-oriented structure:** With this structure, the project is divided into objects. Objects can be deliverables like assemblies, individual parts or simply related sub-tasks.
- **Function-oriented structure:** This structure is sometimes also called an activity-oriented structure. The project is structured according to the functions to be performed. The functions are often based on the organizational units of a company that perform certain functions.

Other structuring principles include location-based structure (structured according to the location where the work packages are to be performed) and responsibility-based structure (structured according to persons who are responsible for the assigned work packages).
Of course, so-called **mixed-oriented structuring principles** are also possible. For example, one part of the work breakdown structure is structured phase-oriented and another object-oriented. This can be useful, for instance, if the first hierarchical level is structured according to phases and the second hierarchical level is subdivided into object-oriented sub-tasks.

Warning

Caution is advised with mixed-oriented structuring principles: The use of different principles makes it more difficult to clearly assign work packages. This can result in duplicates or forgotten work packages.

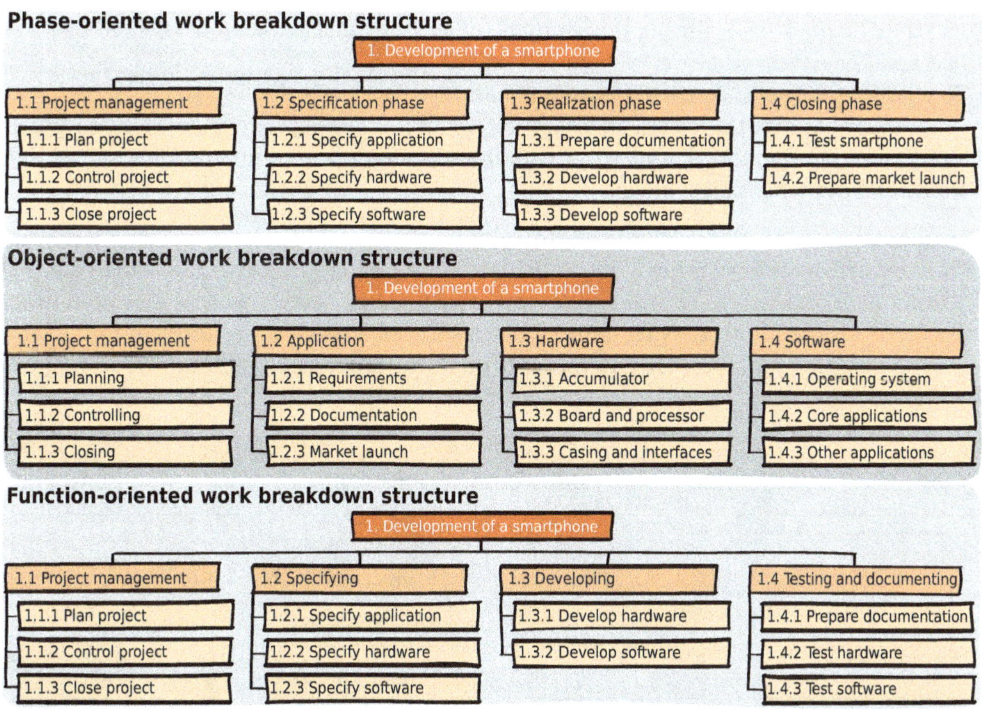

Figure 3.25: Exemplary work breakdown structures with different structuring principles for one and the same project.

Figure 3.25 shows three common structuring principles based on a development project for a smartphone. Even if the different structuring principles lead to different work breakdown

structures, the same things are still done in the project. One aspect should be emphasized again: In the left-hand column, the project management tasks were planned in the aforementioned diagram, regardless of the structuring principle.
Work breakdown structures can be created either top-down or bottom-up. With the top-down approach, you consider sub-tasks and the work packages required for each sub-task based on the overall task of the project. With the bottom-up approach, you first collect all the work packages and then try to group them into sub-tasks.
We now know the aims of project structure planning and how such plans are structured. To do so, we have assigned work packages to specific sub-tasks. The work packages in the work breakdown structure are identified by their code and name. However, behind each work package there is not only a task to be performed, but also additional information that is important for further project planning. Before we consider which of this information should be defined for each work package, let's make some basic considerations about work packages:

- On the one hand, work packages should be sufficiently detailed to provide information for further plans that build on the work breakdown structure. On the other hand, planning a high level of detail is time-consuming and can lead to a loss of clarity. A good guideline for dividing the project into work packages is to divide the overall task of the project into between 20 and 100 work packages. This means that one work package corresponds to approximately 1 to 5% of the project scope. A further subdivision into smaller work units can then be made later during scheduling. There, work packages can be divided into several smaller activities.
- The work packages must be self-contained and have a clearly defined task.
- Work packages may be processed by several people. However, they must be the responsibility of a single person. Shared responsibilities often mean that no one feels responsible in the end or that conflicts arise.

In smaller projects, work packages can be documented in tabular form as a list, for example in a spreadsheet. For larger projects with several people involved, it is worth carefully defining work packages. For each work package, you can specify what is expected of the person responsible for processing it and what resources are available. You can find a large number of templates for this in the literature or on the internet. Figure 3.26 shows an example of a work package form.

Work package form			
WBS Code: 1.3.1	Name: *Create documentation*	Version: *1.0*	
Sub-task/sub-project:	*1.3 Implementation phase*	Status: *approved*	
Responsible:	*Karla Kolumna*		
Description:	*The technical data sheet, the production documents and the user manual must be prepared in accordance with process Doc.235-Y.*		
Deliverable(s):	*1) Technical data sheet 2) Production documents 3) User manual*		
Dependencies on other work packages:	*All work packages of the specification phase 1.2 must be completed.*		
Milestones:	*M1.3.1.1: Preparation of documentation started* *M1.3.1.2: Approval of documentation started* *M1.3.1.3: Documentation approved and released*		
Start date:	*April 1st 2025*	Ende date:	*September 30th 2025*
Effort in staff hours:	*120*	Required resources:	*Content management system* *Technical authors: Byrton, Smith*
Costs:	*14,500 US$*		

Figure 3.26: Example of a form for describing a work package.

Like all plans, work package descriptions must also be kept up to date. Some information such as dates and resources can only be entered later, after scheduling and resource planning has been completed.

In the previous examples in this chapter, you have already become familiar with two types of **work breakdown codes**. In general, coding schemes can be divided into purely identifying and classifying schemes:

- **Identifying coding schemes** assign unique identification codes according to a defined scheme. Typical representatives are:
  - *numeric coding schemes* in which digits are separated by dots or other characters. A digit is added at each hierarchical level, for example 1, 1.1, 1.1.1 etc.
  - *alpha-numeric coding schemes*, in which digits are replaced by letters in one or more places, for example A, A.1, A.1.1 etc.
  - *decadal coding schemes*, in which the number ranges used increase by a factor of ten at each hierarchical level, for example 1, 10, 100 etc.
- In **classifying coding schemes**, the identifier is selected in such a way that further conclusions can be drawn from it. For example, the identifier can directly indicate which suppliers are involved or who the responsible person is. An example of classifying coding is the following coding system: P1.1.3.1AAcc. In this system, P1 stands for the project number, 1.3.1 for the numerical code of the work package, AA for the responsible department and cc for a specific supplier involved.

Of course, other coding systems are also possible. However, to ensure traceability and the unique assignment of codes, these should follow a defined scheme.

Tip

In practice, numerical codes are often used, whereby the root element is assigned a unique company-wide project number.

You should consider the following aspects when planning the work breakdown structure:

- Ensure **completeness** - everything that is worked on or later invoiced must be part of a work package in the work breakdown structure.
- **Don't forget** to plan for the tasks of **project management** - its tasks also take time and must be budgeted for.
- Define exactly **one responsible person** for each work package. This person does not necessarily have to work on the work package on his or her own. However, the person responsible for the work package is responsible for its completion and must take care of it.
- Select a **structuring principle** that is appropriate for the project.
- Don't plan too roughly or in too much detail. Rule of thumb: **circa 20-100 work packages**, depending on the size and complexity of the project.
- Another advice that helps with the dimensioning of work packages is the **8/80 guideline**: It says that the work packages should require no less than 8 hours and no more than 80 hours, or 10 days, to complete.
- Identify the work packages with a **unique work breakdown code**. This prevents confusion and misunderstandings.
- **Keep the plan up to date** during the project lifetime.

With the work breakdown structure, you now have a planning document at your hands that defines what needs to be done in the project. However, in most cases, the work breakdown structure will not be sufficient as the only plan. You will need corresponding schedules and cost plans to control the schedule and cost targets. To create these, you need to know the time required and the resources needed.
As you probably remember, projects are unique endeavors. Consequently, you can only make limited use of past data when determining the effort involved. Instead, you have to make estimates about the effort and resources required in the future based on assumptions.
As the term *effort* is understood differently in different disciplines, it is defined here together with other terms:

> ***Effort*** *refers to the need for or consumption of time, money or human and material resources to complete a task.*

Typical effort figures are given in staff days, man days or machine hours.

> *The* ***duration*** *or* ***lead time*** *is the pure time required to complete a work package or activity, taking into account the resources available.*

Example

The effort required to build a wooden pavilion was estimated at four staff days. This means that one person needs four days. Alternatively, two people can finish in two days or four people in just one day.

The duration can therefore be determined from the time and resource-related effort, knowing the available resources.
The relationship between effort and duration described in the example is only linear to a limited extent. For example, it is unlikely that 16 people can complete the pavilion in just a quarter of a day. Above a certain number of people, there is a risk that the workers will hinder each other and the linear relationship will be broken. Common estimation methods for effort and costs are:

- expert estimation including single and multiple interviews and the Delphi method
- estimation by analogy
- parametric estimation, such as COCOMO II and the function point method
- three-point estimation, also called range estimation or beta method.

These and other methods are explained in the toolbox in Chapter 8. I also refer to the explanations on effort estimation in Chapter 4, where I present critical chain project management.

Scheduling

You are now ready to plan the scope of the project in the work breakdown structure and estimate the effort involved. One of the next questions to be clarified is when and in what order the planned tasks should be completed and when the project deliverables can be delivered.

*The **flow chart** defines the logical sequence of tasks in the project.*

*The **schedule** completes the flow chart with specific deadlines by taking into account realistic lead and waiting times.*

Schedules are typically associated with the following expectations:

- **Sequence** and **deadlines** of the tasks to be completed: You need to know which resources you need and when. To do this, you need the due dates for the upcoming tasks. The client also wants to know when you will provide the deliverables.
- **Visualization of dependencies** between the tasks: The dependencies of the tasks can indicate risks and can be used to optimize the temporal sequence of the tasks.
- **Prerequisite for controlling the project**: As part of the controlling discussed later, you will see that it is important to set the current project progress in relation to the planned progress. The latter results from the schedule.

You have already become familiar with a simple form of scheduling: the milestone plan. A **milestone plan** assigns deadlines to important project events such as the start of the project, the completion of the deliverables and the end of the project. In small projects with few tasks, this simple form of scheduling may be sufficient. Figure 3.27 illustrates the schedule of a small renovation project as a milestone plan. For comparison, the same project is also shown as a Gantt chart and as a network diagram.

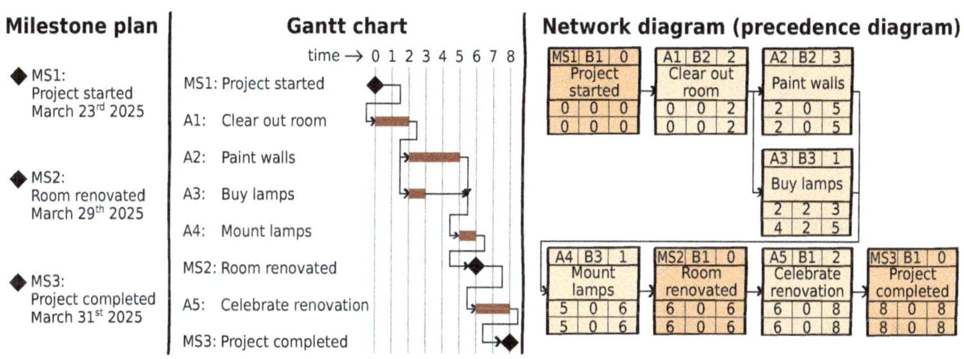

Figure 3.27: Three types of scheduling – left: milestone plan without specifying the tasks to be carried out between the milestones; middle: Gantt chart (bar chart) showing activities as bars and milestones as diamonds along a time scale. Dependencies are shown as arrows; right: network diagram illustrating activities and milestones as nodes (boxes). The abbreviations and contents of the nodes are explained later in the text.

Schedules are often described as **bar charts**, also known as **Gantt chart** after its founder, Henry L. Gantt. A Gantt chart not only lists the milestones of the project, but also the activities to be carried out. The time periods required for working on them are shown as bars along a time axis. Dependencies between the activities are modeled by arrows, so-called relationships.

Network diagrams are another way of visualizing schedules. Two different variants of network diagrams are the arrow diagram and the precedence diagram. In the arrow diagram, the arrow represents the activity to be carried out. In the precedence diagram, the activities are represented by nodes (rectangles). Arrows link the activities and thus model the logical

dependencies. In the rest of the book, we will use the precedence diagram. This offers more possibilities for modeling the temporal course of the project.
Before we delve deeper into scheduling, let's familiarize ourselves with a few terms. While in work breakdown structures we only plan work packages and sub-tasks, in schedules we use activities, milestones and relations, see also Figure 3.28:

> ***Process elements** of a schedule describe activities, milestones and relationships.*

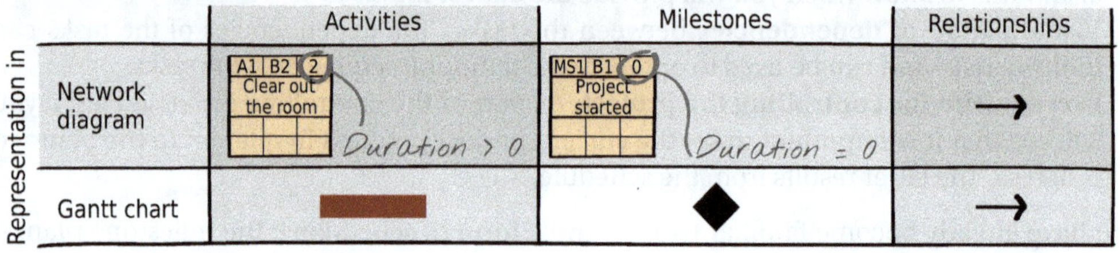

Figure 3.28: Representation of elements in the network diagram and Gantt chart. In the network diagram, activities and milestones differ only in that activities have a duration (top right corner of the node), while milestones represent an event without a duration. Relationships are represented by arrows that connect the individual activities and milestones.

You have already been introduced to the definition of an activity earlier to differentiate it from a work package. Here is a repetition of the term:

> *An **activity** is a temporal process element that represents a self-contained assignment. A work package can be converted into one or more activities. While a work package defines what is to be completed, activities define the (temporal) sequences of the assignments. Activities are not part of the work breakdown structure.*

> ***Milestones** are events of particular importance. They define important points in time in the schedule at which certain deliverables must be available.*

> ***Relationships** link activities and milestones. They establish a quantifiable and logical relationship between the elements of the schedule.*

A first step in converting a work breakdown structure into a schedule is to detail the work packages by breaking them down into tasks and milestones, see Figure 3.29.
With these basics, we can now start scheduling and create network diagrams.

> ***Network diagrams** represent the schedule graphically or in the form of tables with the help of activities, milestones and relationships.*

You have already seen in the previous illustrations how activities and milestones are represented as nodes in the network. Not all of the information contained in the diagram has been explained so far. We will now do this with the help of Figure 3.30.

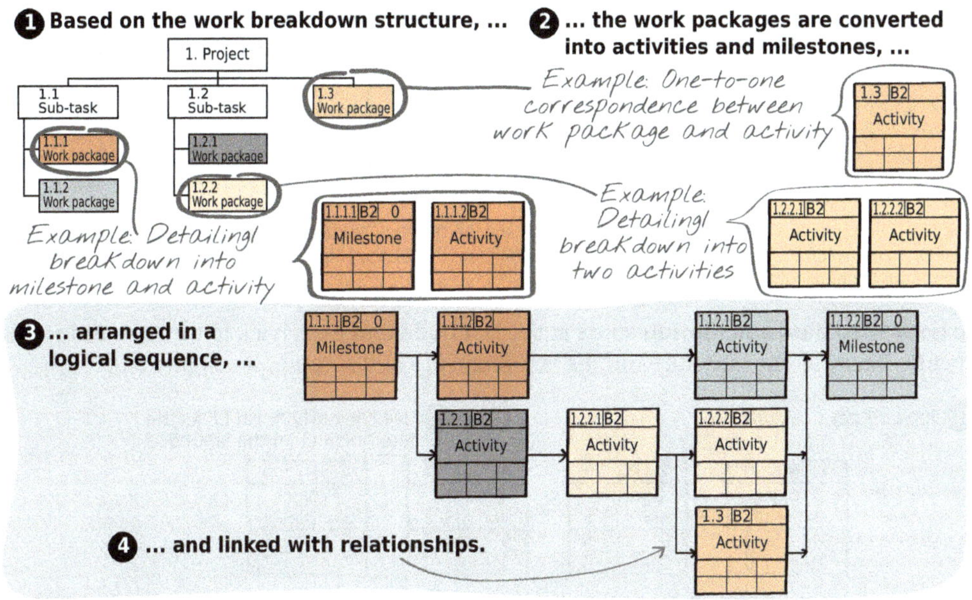

Figure 3.29: Detailing the work packages by breaking them down. The work packages can be transferred into one or more activities or into activities and milestones. Only activities and milestones are then used in the schedule. The numbers one to four highlighted with black circles indicate the processing sequence for detailing and arrangement.

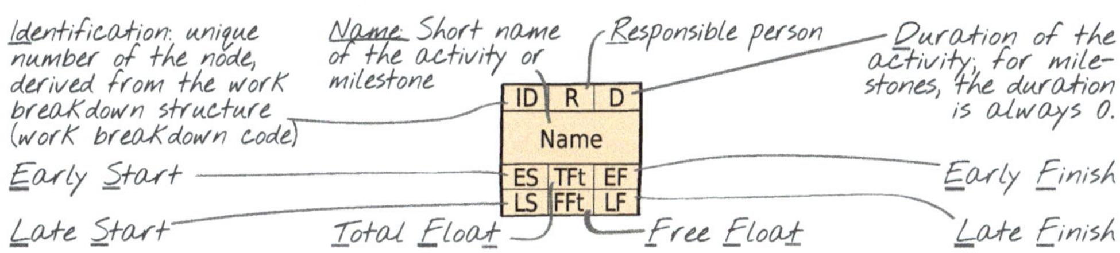

Figure 3.30: Representation of activities and milestones as nodes in the network diagram.

The ID and name are used to identify and describe the activities and milestones. Identification takes place via a unique number, which is usually derived from the work breakdown code of the work breakdown structure. The duration is based on the effort estimation. The early and late times and the buffers (floating times) are calculated. To do this, the activities and milestones must first be linked to each other via relationships.

The simplest form of relationship between the elements of a network diagram is probably the **finish to start relationship**, see Figure 3.31. The determination of the early start and the early finish shown in the figure is done by means of the forward pass calculation. The **forward pass calculation** for the finish to start relationship is calculated as follows:

$$EF_n = ES_n + D_n$$
$$ES_{n+1} = EF_n$$

The early finish EF of a node with ID n is the sum of the early start ES and the duration D of node n. The early start ES of the subsequent node n+1 is equal to the early finish EF of node n.

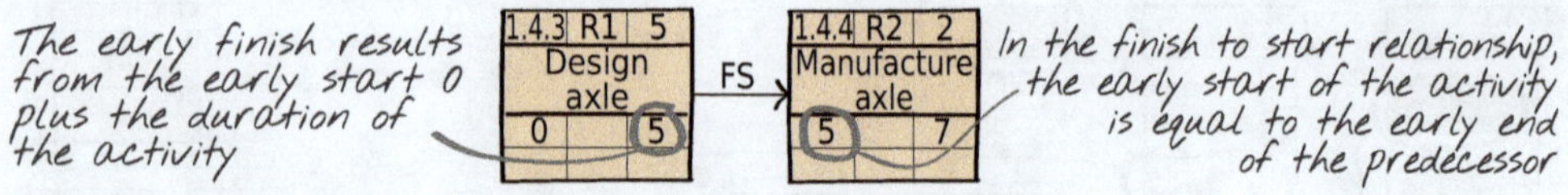

Figure 3.31: Example of a finish to start relationship (FS) and the resulting calculation of the early start of the successor node. The abbreviation FS in the network diagram can be omitted. If a relationship is not defined in more detail, it is always a finish to start dependency.

The **backward pass calculation** starts at the last node and goes back to the first node of the network diagram. The late start and the late finish are determined, see Figure 3.32.

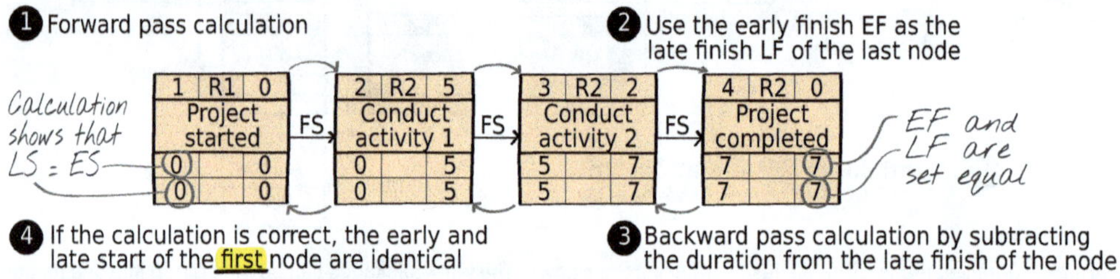

Figure 3.32: Forward and backward pass calculation: 1) In forward pass calculation, the early dates for the start and end of the subsequent nodes are determined based on the early start of the first node. 2) The backward pass calculation begins by setting the late finish equal to the early finish, in this case by transferring the value 7 to the second line of the last node (here: node with number 4). 3) Subtracting the duration from the late finish results in the late start of the node. This is taken as the late finish in the predecessor node. 4) If the calculation was correct, the early and late start of the first node are identical.

The following applies to the backward pass calculation of the normal finish to start relationship:

$$LS_n = LF_n - D_n$$
$$LF_n = LS_{n+1}$$

The late start LS of a node n is equal to the late finish LF the node subtracting its duration D. The late finish LF of a node is equal to the late start of its successor n+1.
In the simple example of Figure 3.32, there are no differences between the early and late dates. Differences arise when activities of different durations are planned in parallel, see Figure 3.33. Relationships also allow the **definition of minimum and maximum distances**. The minimum distances MINZ between two nodes specify the minimum distance. During actual project execution, the distance can be greater, but not smaller. The maximum distances MAXD between two nodes indicate the maximum distance. The successor may start earlier, but not later. Minimum and maximum distances can be used, for example, to plan waiting or delivery times in the schedule. Negative time intervals can be used for modeling parallel activities. An example is shown in Figure 3.34.

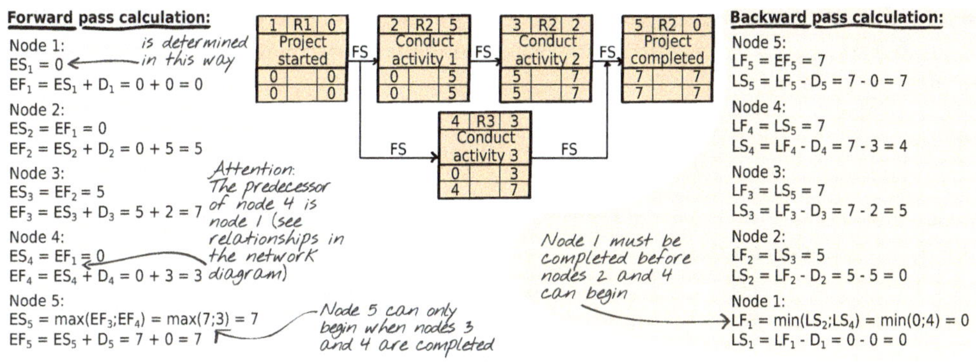

Figure 3.33: Example with activities planned in parallel and illustration of the forward and backward pass calculation.

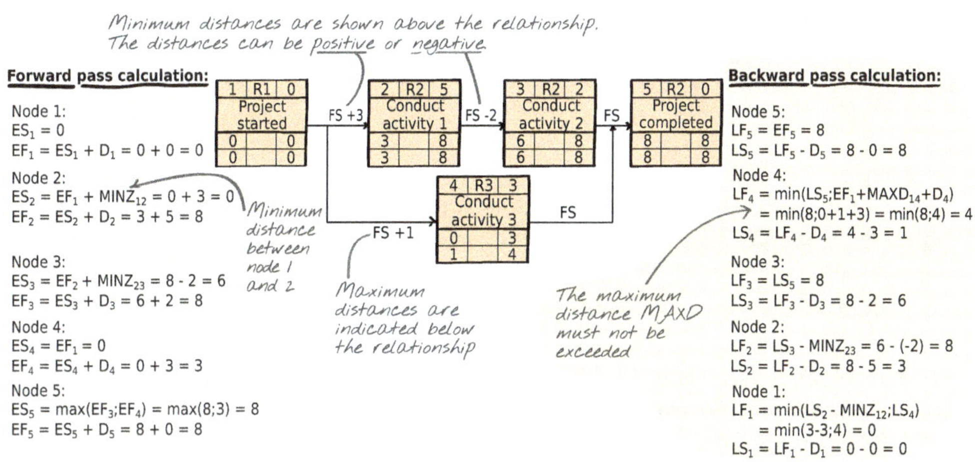

Figure 3.34: Minimum distances are indicated above the relationship, maximum distances below.

The last node elements that we have not yet calculated are the total float TFt and the free float FFt.

> *The **total float** of an activity is the maximum time by which this activity can be delayed without delaying the completion of the project. The total float results from the difference between the late and early time of the activity.*

The total float is either the difference between the finish times or the start times:

$$TFt = LF - EF = LS - ES$$

> *The **free float** is the maximum time of an activity that this activity can be delayed from its early position without delaying the subsequent activity from its early position.*

In the case of a finish to start relationship with a minimum distance MINZ (MINZ = 0 if no minimum distance has been defined), the calculation of the free float (FFt) is based on a node n and its successor n+1:

$$FFt_n = ES_{n+1} - MINZ_{n,n+1} - EF_n$$

There are other floats, but these are rarely taken into account in practice and are not indicated in the network nodes, for example:

*The **independent float** is the time span by which an activity or milestone can be postponed if the predecessor is in the late position and the successor is in the early position.*

Warning

The floats discussed here (total float, free float, independent float) are buffers that result from the different durations of parallel activities. In addition to such buffers, there also exist scheduled buffers. These buffers are also called safety buffers and are hidden in the activities themselves. There, the planned duration is extended by the scheduled safety buffer.

*A **safety buffer**, also known as a risk reserve, is a temporal, financial or qualitative buffer for unplanned situations. This buffer is used to reserve additional time, additional money or additional resources for challenging project phases or work packages.*

We can distinguish between safety buffers for foreseeable risks and unforeseen risks:

*The **contingency reserve** is a buffer for efforts and costs from identified, i.e. known, risks. The amount of the contingency reserve is determined from the assessment of the risks. If a risk occurs, the associated efforts and costs can be taken from the contingency reserve.*

*The **management reserve** is a buffer for efforts and costs from unknown risks. Its amount is determined from the experience of past projects and the company's risk policy. If an unexpected risk occurs, the resulting efforts and costs are compensated from the management reserve.*

You will learn more about the contingency and management reserve later in this chapter when we look at risk management.
Floats as calculated buffers in the schedule and safety buffers should not be mixed up.
The critical path determines the total duration of the project and is defined as follows:

*The **critical path** is the sequence of activities and milestones in which there are the fewest (usually no) floats. The critical path therefore determines the duration of the overall project.*

In other words, any delay of an activity or milestone that is on the critical path leads to a delay of the project (without the initiation of countermeasures).

So far, we have connected two nodes of the network diagram exclusively with the finish to start relationship FS. In addition to this, there are other relationships:

- **Start to start relationship SS:** If two nodes are linked with the start to start relationship, they start at the same time or with a specified offset.
- **Finish to finish relationship FF:** If two nodes are linked with the finish to finish relationship, they end at the same time or with a specified offset.
- **Start to finish relationship SF:** If two nodes are linked with the start to finish relationship, the successor ends at the start of the predecessor.

Minimum and maximum distances can be defined for all relationships. An overview is shown in Figure 3.35.

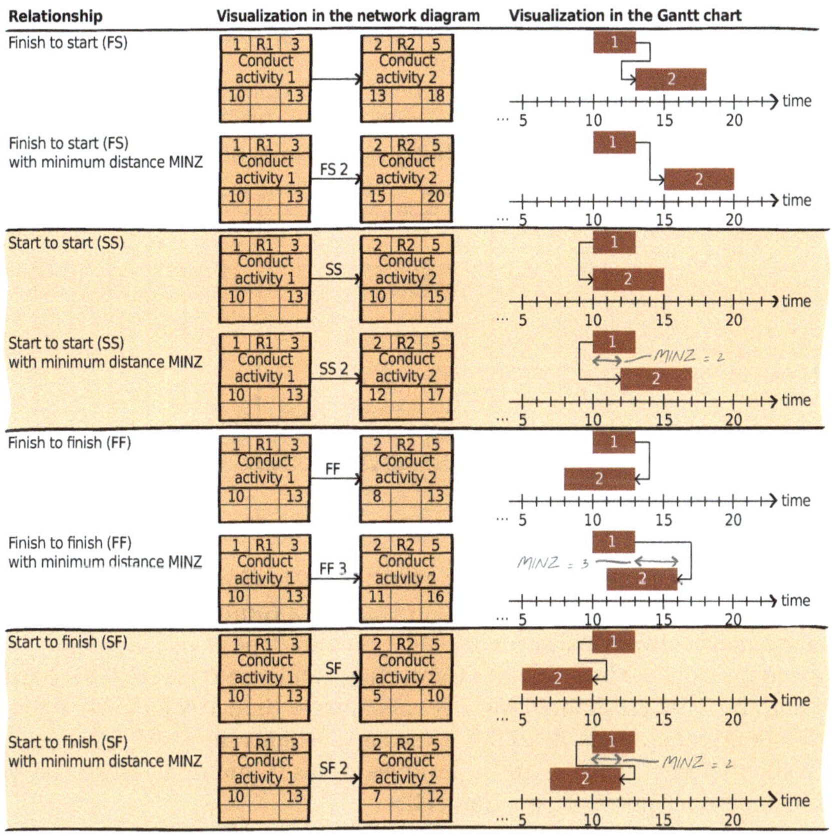

Figure 3.35: Overview of the relationships finish to start, start to start, finish to finish and start to finish.

The network diagram in Figure 3.36 illustrates the application of the different relationships, the buffer calculation and the determination of the critical path. In the exercises at the end of the chapter, you will find further network diagrams to calculate yourself.

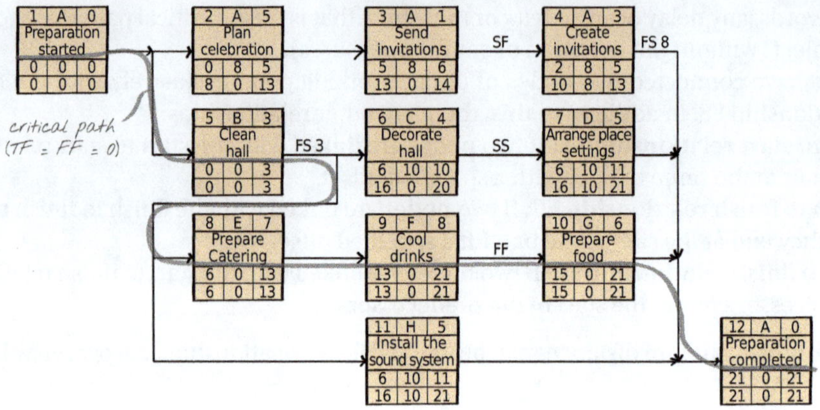

Figure 3.36: Example of a network diagram with various relationships, calculated buffers and critical path.

Alternatively, network diagrams can also be represented in tabular form. The tabular version of the network shown in Figure 3.36 can be seen in Figure 3.37. The information is the same, but due to the compact presentation, the tabular form is sometimes perceived as confusing.

ID	Name	Resp.	Duration	Predecessor and (relationship)	ES	EF	LS	LF	TFt	FFt	C.p.
1	MS: Preparation started	A	0	-	0	0	0	0	0	0	yes
2	Plan celebration	A	5	1 (FS)	0	5	8	13	8	0	no
3	Send invitations	A	1	2 (FS)	5	6	13	14	8	0	no
4	Create invitations	A	3	3 (SF)	2	5	10	13	8	8	no
5	Clean hall	B	3	1 (FS)	0	3	0	3	0	0	yes
6	Decorate hall	C	4	5 (FS+3)	6	10	16	20	10	0	no
7	Arrange place settings	D	5	6 (SS)	6	11	16	21	10	10	no
8	Prepare catering	E	7	5 (FS+3)	6	13	6	13	0	0	yes
9	Cool drinks	F	8	8 (FS)	13	21	13	21	0	0	yes
10	Prepare food	G	6	9 (FF)	15	21	15	21	0	0	yes
11	Install the sound system	H	5	5 (FS+3)	6	11	16	21	10	10	no
12	MS: Preparation completed	A	0	4 (FS+8), 7 (FS), 10 (FS), 11(FS)	21	21	21	21	0	0	yes

Resp. = Responsible person C.p. = Critical path

Figure 3.37: Example of a network diagram in tabular form.

However, many project managers prefer a linear representation of the schedule based on a timeline anyway. This type of representation is called a **Gantt chart** (or bar chart). Most project management software packages are capable of this type of visualization.
Network diagrams can be converted into Gantt charts and vice versa. For this purpose, the early start and the early finish are drawn as bars along a time axis for each activity. Milestones are marked as diamonds. In addition, the late positions can also be shown. Typically, the relationships between activities and milestones are also visualized. Figure 3.38 shows the network diagram from Figure 3.36 as a Gantt chart.

Tip

Many project management software packages for creating Gantt charts offer the option of displaying the underlying network data in tabular form. The table then supplements the graphical representation of the bars as a chronological progression with additional information.

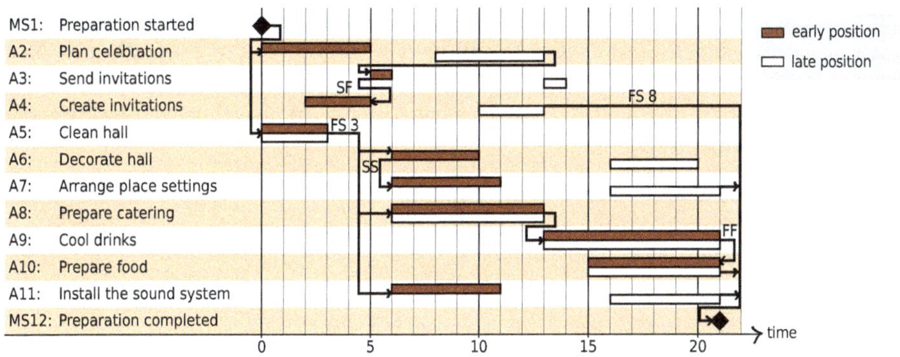

Figure 3.38: Example of a Gantt chart showing the earliest and latest positions and the relationships between activities and milestones.

An overview of the strengths of the two representations:

- Network diagrams ...
  - can be calculated manually, which facilitates transparency
  - help with control and planning
  - have a very high information density (relationships, earliest and latest positions as well as total and free float) and
  - illustrate dependencies between activities and milestones very well.
- Gantt charts ...
  - show processes and milestones along a time axis
  - are well suited for presentations
  - are familiar to many managers even without project management knowledge and
  - are supported by many software packages.

Resource and cost planning

In previous planning, it was initially assumed that all the necessary resources will be available. The demand and availability of the required resources can now be determined on the basis of the existing schedule.

> ***Resources** are human, financial and material resources in quantifiable quantities that are required for the successful completion of the project.*
>
> ***Material resources** include materials (e.g. for the construction of prototypes), consumables, operating resources, infrastructure (offices, laboratories), equipment and energy.*

The **challenges in resource planning** are manifold:

- Dependence on the schedule and the effort estimate: It must be known which and how many resources are needed at what time and for how long.
- Interactions within the overall company: The same resources can be required by several projects or units of the line organization at the same time.

- Flexibility: Effort estimates are first and foremost estimates with uncertainties. In addition, there are spontaneous influences such as illness, job changes, new projects and tasks that affect the availability of the required resources. The scope of the project can also change during the course of the project. This can have an impact on the quantity, quality and skills of the resources required.

In principle, resources also include financial resources. However, we do not take these into account as part of resource planning. Instead, we plan them separately in the subsequent cost plan.
The **goals of resource planning** are
- to determine the demand for resources (in terms of time and type/scope/quantity),
- to clarify the availability of resources and then
- to create a realistic resource allocation for the required resources in line with the schedule.

Cost planning uses the knowledge of the required resources and the costs associated with their use to determine the financial resources required for the project. The minimum goal is the **determination of the total costs of the project**. For advanced project controlling methods, it is not sufficient to know the total costs. Instead, the costs per time period (e.g. per week, month or quarter) must be known. This is also known as **time-resolved cost planning**.

Example

The 50th anniversary of a sports club is to be planned as a small project. In this case, it is usually sufficient to determine the total costs of the project.
A biogas plant is to be built as part of a major project. Knowing the total costs is not enough for a sound project controlling using the earned value analysis. Instead, it is necessary to determine which costs are incurred in which month. This information is important in order to be able to draw conclusions about delays or lead times later on by comparing the actual costs incurred over time.

The scheduling of employees is usually the most difficult part of resource planning. Employees have specific skills that are not available in unlimited quantities and cannot be built up on short notice. Each project creates a resource plan that is aligned with the schedule. The required resources are requested from portfolio management (or generally: from the line organization) and allocated if available. In the event of resource bottlenecks (more resources requested than available), they are allocated depending on the project priority.
In matrix and functional project organizations, in which several projects share resources, this sharing can easily lead to conflicts:
- If fewer resources are available than are required, the resources are distributed according to the priority of the projects.
- Changes in the schedule, e.g. due to delays, mean that resources are needed at different times than originally planned. One way of dealing with such changes is to redistribute resources between several projects running in parallel. However, this leads to undesirable side effects with other projects. Shifting resources in this way can affect the execution of projects that originally had no problem at all.

Cross-project resource planning is based on a comparison of demand and capacity:

> *The* ***capacity*** *indicates how many resources of a certain type (people or material resources) or qualification (e.g. requirements engineers, software developers, integrators) are available per period of time.*

The capacity is compared to the demand of the projects. Both together can be represented in a resource histogram, see Figure 3.39. As resources with different qualifications are not interchangeable, capacity and demand must be analyzed separately for each type of resource and qualification. In Figure 3.39, for example, the capacity and demand for requirements engineers and software developers were examined.

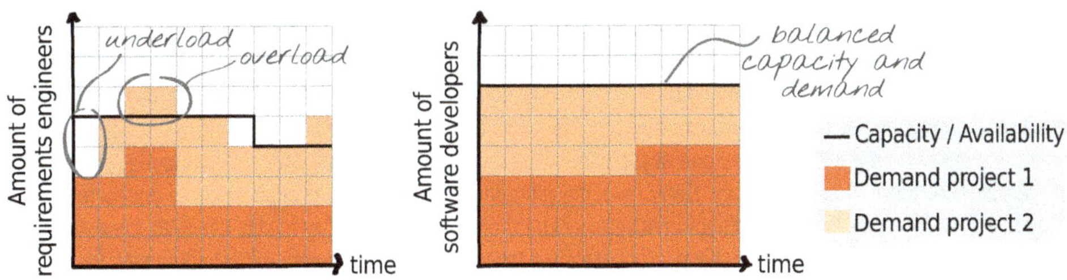

Figure 3.39: Resource histogram for the exemplary resource types requirements engineer and software developer. They are both required in projects 1 and 2. The capacity (availability of resources) can fluctuate due to vacations, new employees and job cuts. The demand in the projects is also subject to temporal changes. There are periods of underload and overload. By shifting the schedule of activities, overload phases can be reduced if necessary. In the case of software developers, capacity and demand are already balanced.

Warning

If the resources are completely utilized, i.e. if the demand corresponds exactly to the capacity, overload situations quickly occur in the event of later changes. The originally balanced allocation can later lead to time phases with overload situations as a result of the change. We take a closer look at the causes, consequences and countermeasures in the chapter on critical chain project management in the context of agile project management, see Chapter 4.

The static comparison between the capacity, i.e. the available resources, and the demand, i.e. the resources required by the projects, is already quite complex. Things get really challenging when the plan changes dynamically during project execution, i.e. when it deviates from the original plan, as explained in Figure 3.40.
This can quickly lead to interactions with other projects. Cross-project portfolio management must clarify whether, in the example in Figure 3.40, resource A may remain in project 1 for a longer period of time, resulting in a resource shortfall in project 2. Alternatively, resource A can be withdrawn from project 1 as planned so that it can work in project 2, but this leads to a further delay in project 1.
The example chosen is quite obvious, but gives an idea of the complexity of real situations.

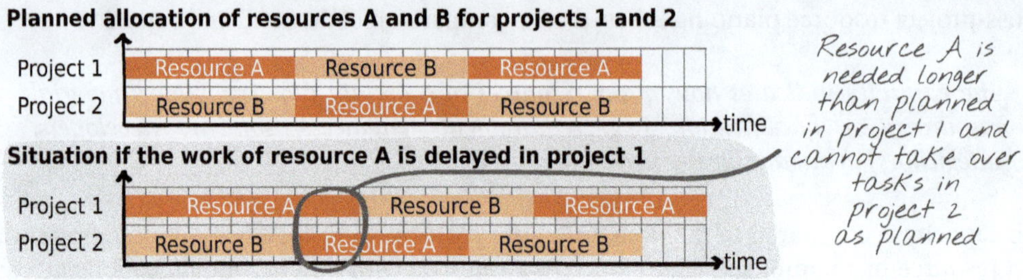

Figure 3.40: Interaction between otherwise independent projects due to resource sharing: If the demand for resource A in project 1 changes compared to the original plan, this affects the availability of the resource for other projects. The delay of one project can result in the delay of several projects.

Tip

Despite the challenges described above, it makes sense to plan resources on a timeline. Every project manager, and even more so every portfolio manager, should bear in mind that dependencies arise when allocating resources across projects. These dependencies should be reduced as much as possible.
Avoid planning under permanent overload and at the same time a high degree of dependency between projects.
Do not plan in more detail than necessary. Daily resource planning for a two-year project leads to poorly readable, confusing plans. At the same time, the planning accuracy can hardly be used on a day-to-day basis, as the estimation accuracy is not sufficient. As a result, you have to constantly adapt the plan and request and schedule new resources.

The starting point for the **creation of a resource plan** is the schedule:
- Arrange the activities of the schedule along a timeline.
- The scaling of the timeline must be appropriate to the duration and complexity of the project. In a five-year project, it is usually not necessary or possible to plan the demand of resources on a daily basis. Conversely, planning on an annual basis would be too vague.
- Determine the required resources or the required resource category (e.g. four software developers, two requirement engineers) for each activity. Whenever the demand for resources is not obvious, the same methods can be used for estimating the demand as for estimating the effort.
- Assign the resources to the corresponding time periods.
- Sum up the resources across all activities per time period.

By doing so, you create a resource histogram for the respective project, which you can later use for cost planning and project controlling. The resource planning as described above is illustrated in Figure 3.41.
The comparison of demand and capacity illustrated there can alternatively also be represented as a histogram as in Figure 3.39.

Warning

You should resolve overload situations before proceeding with the project. Overload means shortage and indicates that your plan cannot be executed as planned.

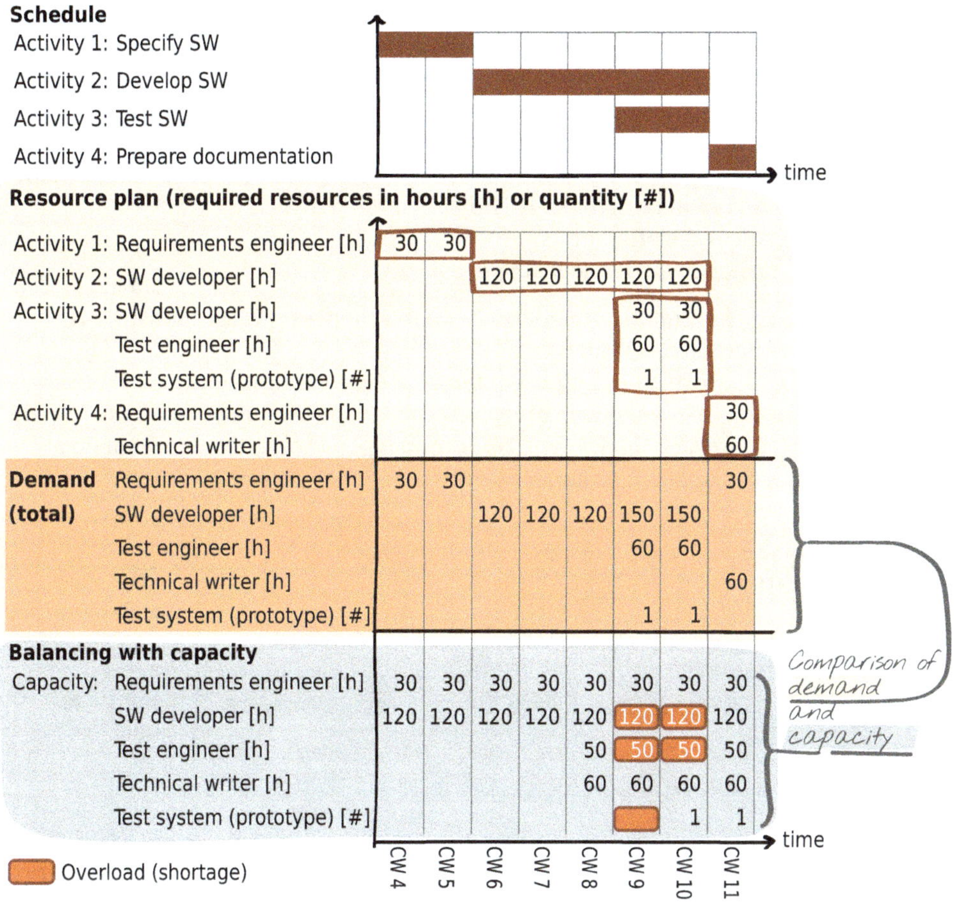

Figure 3.41: Resource plan based on the schedule. The exemplary project was planned on a weekly basis. The resources are determined for each activity and time unit, summed up and compared with the available capacity. In the example, software (SW) developers and testers are missing in calendar weeks (CW) 9 and 10 as well as a test system in CW 9.

If required resources are not available, it is essential to check whether overload can be reduced through optimization, for example by changing the timing of activities or stretching them out. The optimization must then be incorporated back into the schedule so that the schedule and resource plan are aligned.
The aim of optimization is to change the schedule and resource plan so that the required resources are actually available at the planned time. Optimization begins with a review of whether activities can be

- shifted or postponed
- stretched
- compressed or
- split up.

The aim of the optimization is to ensure that despite the changes, the project duration is not delayed.

Overload situations can also be avoided by **increasing capacities**. For this purpose, material resources can be purchased, employees can be hired or temporarily available employees can be brought in via personnel service providers. Increasing capacity is not always possible and is typically associated with costs. The training of new or temporary employees by existing employees can even increase the overload situation in the short term.
If it is not possible to balance the demand of the resource plan with the available capacity using the aforementioned means, the only choice is often between

- extending the **project duration** or
- a reduction or change in the **project scope**.

However, you should coordinate both options with the client or project sponsor, as they usually prevent the fulfillment of certain project objectives.
You can now create the corresponding **cost plan** based on the resource plan. To do this, the allocated resources are calculated using the corresponding cost rates, other costs are added and the total project costs are determined. Important types of costs are shown in Figure 3.42. However, not all types mentioned there are relevant for every project.

Personnel costs	Material costs	Infrastructure and service costs		Capital and other costs
▪ salaries ▪ wages ▪ social costs ▪ training costs	▪ raw materials and supplies ▪ tools ▪ warehousing ▪ packaging	▪ rooms ▪ rents, leases ▪ energy ▪ IT, telephone, internet ▪ postage, courier	▪ travel ▪ consulting ▪ hospitality ▪ licenses/ commissions	▪ depreciation and amortization ▪ capital commitment ▪ risk costs ▪ taxes and fees ▪ insurance

Figure 3.42: Overview of the different kinds of costs in projects based on Schnell [Schnell, 2005].

In contrast to the categorization of cost types shown in Figure 3.42, there are other possibilities. The Project Management Institute, for example, refers to a subdivision into direct costs, indirect costs, fixed costs, variable costs and sunk costs [Epstein and Maltzman, 2013]. The categorization helps you not to forget anything. However, it is even more important to clarify individually which costs need to be included in the project at hand.
Typically the most important of these costs are **personnel and material costs**. Infrastructure and service costs as well as capital costs and other costs can also be taken into account. In many companies, however, these are allocated to the projects as a fixed amount or as a percentage of overhead costs, see Figure 3.43.

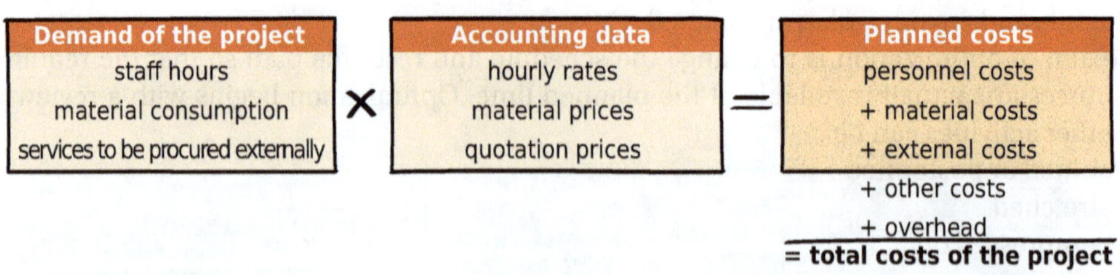

Figure 3.43: Cost calculation based on Fiedler [Fiedler, 2020], taking into account the usual personnel, material and external costs for projects. Other costs may include travel expenses. The overhead includes the use of internal services (rooms, IT, etc.). It can be allocated as a fixed amount or as a percentage, for example depending on the project budget or the number of employees in the project.

In principle, the procedure outlined in Figure 3.43 can be used to determine the total project costs and thus complete the cost planning.
However, we will learn that *knowledge of the cost progression per time period enables improved possibilities for controlling and analyzing the progress of the project.*
Therefore, we now combine our data from the resource plan with the cost calculation and thus obtain a time-resolved cost plan. The procedure is explained in Figure 3.44.

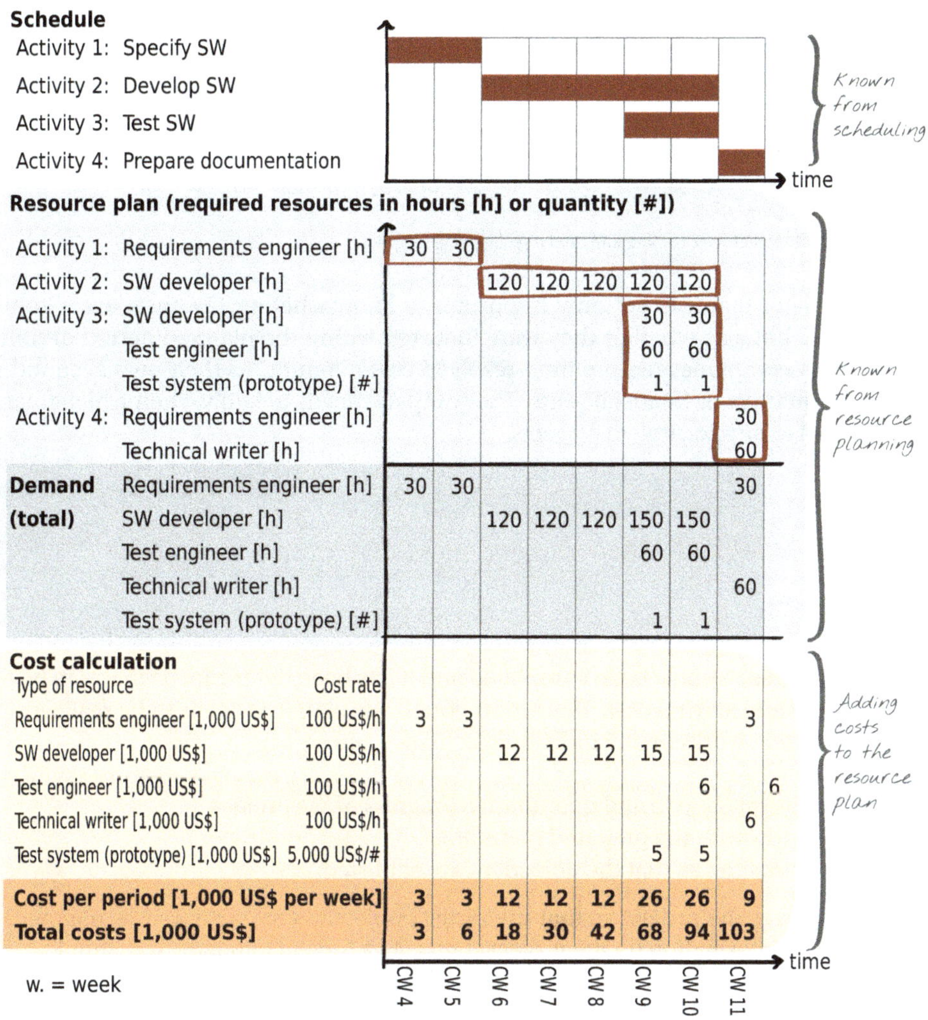

Figure 3.44: Preparation of the cost plan: The resource demand (highlighted in gray) is multiplied by the company-specific cost rates. This results in corresponding costs per period of time (in this case weeks). These can be summed up over time (total costs in the last line of the table). In the example, the total costs are 103,000 US$.

The **costs per time period** are also referred to as **cost line** and the **total costs** as cumulated cost or planned value PV. The latter term is mainly used in earned value analysis, which I will explain in the next chapter. As Figure 3.45 shows, the cost line (per period) and total cost line can also be displayed as a diagram.

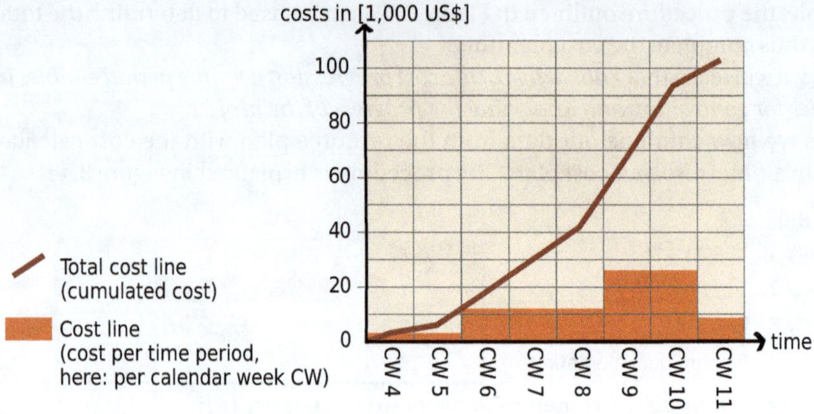

Figure 3.45: Visualization of the cost line and total cost line as a diagram.

I will explain later that it is not only important to know whether the costs are within the planned limits, but also whether they were incurred within the planned period of time. In addition, many companies require time-resolved cost planning, as their own accounting is structured in quarters or financial years. The company needs to know when which financial resources need to be allocated to projects.

Project controlling

Overview – where are we in the project management process?

The planning phase has now been completed and it has been determined what is to be done when, by whom and at what cost. This is now your basic plan that needs to be realized:

- The *project team* works on the individual activities and
- the *project manager*
  - *controls* the qualitative and quantitative progress of the project
  - *compares* this with the prepared plans and
  - *initiates measures* so that the objectives are achieved.

In order to manage the project so that the project objectives are achieved as planned, you need up-to-date information on the project status. As a project manager, you should

- you should objectively and reliably **determine** the qualitative and quantitative **progress**
- **interpret** determined **figures** correctly and
- **analyze the determined project status in the context of the originally established plans.**

Many project managers create elaborate plans but are unable to use them later for progress monitoring and control. This is either because the plans do not provide the required data despite the effort invested or because they have not mastered the methods of progress monitoring and control.

Control options are particularly effective if they are implemented at an early stage. The earlier intervention takes place, the smaller the intervention can usually remain in order to achieve the objectives. However, the prerequisite is that the need for intervention is recognized. The classification of these activities in the project lifecycle is illustrated in Figure 3.46.

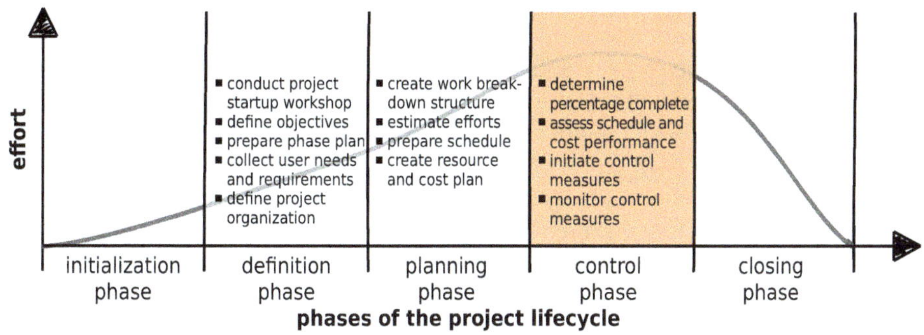

Figure 3.46: The realization of the project plans begins in the control phase. The project team works on the assigned tasks and the project manager monitors the progress of the project, analyzes it and intervenes in the event of deviations from the plan.

Warning

Those who want to control and manage the project purely on the basis of a schedule often only recognize delays when entire activities or work packages have been delayed and these effects also extend to subsequent activities. The aim of progress control is to detect delays or cost overruns at an early stage and take appropriate countermeasures.

The tasks of the project manager in the control phase are summarized in Figure 3.47.

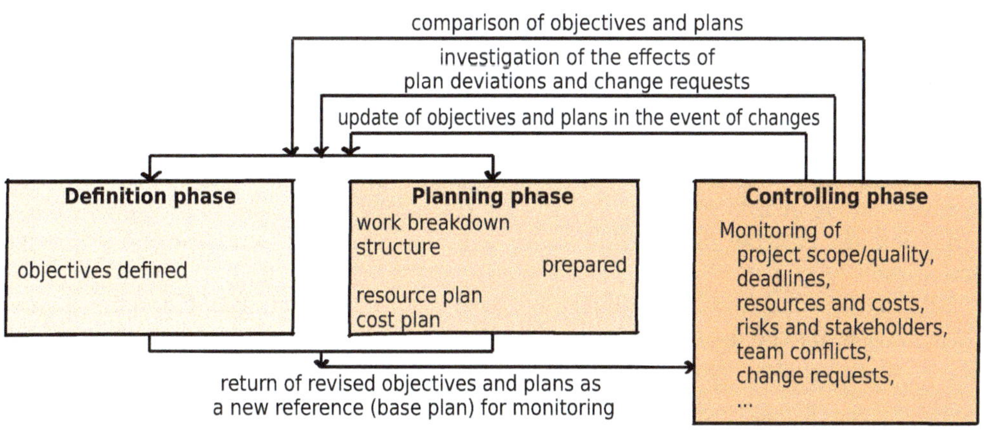

Figure 3.47: Overview of the tasks in the controlling phase of a project. The project manager monitors and compares the target and actual figures with plans and objectives. In the event of deviations (delays, cost overruns or approved change requests), the targets and plans are updated and taken into account for further monitoring and control.

Warning

Some project managers carry out comparisons of target and actual figures and try to control the project. However, they forget to update the plans. But with outdated plans, it is impossible to monitor and control the project properly. The situation can be compared to a car journey using a navigation system with outdated maps. This makes it difficult or even impossible to reach the destination.

Progress monitoring and analysis should be carried out regularly and throughout the entire duration of the project. You will see below that this involves some effort, which immediately raises the question of what is meant by "regularly" and whether it is worth the effort. The overview shown in Figure 3.48 gives a rough indication of how often checks and analyses should be carried out. However, it neglects factors such as particular project complexity or special risks. In practice, this and internal company guidelines may lead to completely different intervals for status assessments.

Project duration	Frequency of status assessment
▪ few months	▪ weekly
▪ 1 to 2 years	▪ bi-weekly
▪ 3 to 5 years	▪ monthly to bi-monthly
▪ more than 5 years	▪ bimonthly to quarterly

Figure 3.48: Relationship between project duration and frequency of progress monitoring and analysis for rough orientation.

Before the individual methods of monitoring and control are discussed, some important terms are introduced. We start with a few terms which we have so far only been used from the perspective of project planning:

> *The **budget at completion BAC** is the total cost of the project at the scheduled completion date, according to the cost plan.*

We therefore receive this as a total cost from the prepared cost plan.

> *The **planned value PV** are the planned costs at a chosen date. At the end of the project, the planned value corresponds to the budget at completion.*

The planned value is the total cost line (cumulated cost) in the cost plan, remember the illustration in Figure 3.45. This also makes it clear why I have attached so much importance to planning the cost per time period so far: We now need this total cost line for project controlling.

In addition to the data from the planning, i.e. the budget at completion BAC and the planned value PV, we also need data on the actual costs in order to analyze the status of the project. This includes the following variables:

> *The **actual cost AC** is the cost actually spent up to a specific reporting date.*

> *The **percent complete PC** is the ratio of work already completed to the total workload on a chosen date.*

As the name suggests, the percent complete is given as a percentage. If half of the planned work has been completed by a certain date, the percent complete PC is 50%.
The earned value EV can be determined from the budget at completion BAC and the percent complete PC:

> *The **earned value EV** is the (planned) value of the (actual) work performed on a chosen date. It is calculated as the product of the percent complete PC and the budget at completion BAC.*

All terms can refer either to an activity, a work package, a sub-task/sub-project or to the overall project. The following convention applies throughout the rest of this book:
If the variable has a lower index (e.g. $EV_{1.4.2}$), this refers to a single activity or a single work package (here to the activity with the number 1.4.2). Without a subscript index, the values always refer to the entire project.
These newly introduced terms are discussed in detail below and illustrated using three examples.

Example

Earned value of a building construction project: An office building is being constructed as part of an investment project. The budget at completion (you remember, this is the total cost of the project according to the cost plan) is BAC = 2,000,000 US$. During the construction of the foundation, some things go wrong. On the date of completion of the foundation, the actual cost of construction is AC = 250,000 US$. Up to this point, the work was originally only planned to cost PV 200,000 US$. However, the value of the completed foundation has not increased as a result of the 50,000 US$ rework carried out and still amounts to EV = 200,000 US$. In other words, applying the definition of EV: The planned value (200,000 US$) of the work actually performed (foundation) is the earned value EV = 200,000 US$ even if the actual cost AC was higher.

The comparison of planned value and actual cost on a reporting date helps to assess the project status. Compare the two situations in the following examples:

Example

Cost without time reference: The client of a project wants a progress report from the project manager. The project manager knows that 100,000 US$ have already been spent and that the overall project is expected to cost 200,000 US$. Based on this data, it is not possible to predict whether the project will remain on schedule and within budget.

Example

Cost with time reference: The client of a project wants a progress report from the project manager. The project manager knows that 100,000 US$ have already been spent. According to the cost plan, 80,000 US$ should have been spent by today's deadline (reporting

date). The total cost (budget at completion) is 200,000 US$. Since tasks worth PC = 50% of the budget at completion have already been completed, she can draw the following conclusions:
The actual cost and the earned value (50% of the total cost) are both 100,000 US$. This means that, as things stand today, the project will achieve its cost targets. In fact, costs of 80,000 US$ should have been spent by the reporting date. However, the fact that a value of 100,000 US$ has already been created signals to the project manager that the project is ahead of schedule. The exact methodology of the calculation is explained in more detail in the following sections.

The three examples illustrate the possibilities project managers have to analyze the project status if they have determined the right key figures and know how to use them.

Determining the project status

The basis for monitoring and analyzing the progress of a project is to record the current status, or more precisely the **percent complete PC**. There are various methods for this, for example:

- Estimating the percent complete (see fact sheet M.2.1 Estimate progress in Chapter 8)
- Measure the percent complete (see fact sheet M.2.2 Measure proportionalities) using
  - quantities
  - time
  - time in conjunction with estimates to complete (see M.2.3 Determine estimate to complete)
  - secondary figures
- Determine the percent complete using
  - weighted milestones (see M.2.4 Determine weighted milestones)
  - apply the percent-start/percent-finish method (see M.2.5 Apply percent-start/percent-finish method)

The information in brackets refers to the fact sheets in the toolbox in Chapter 8. The respective methods are explained there, which is why we will not go into more detail here. We assume that we know the percent complete PC by using one of the methods mentioned. Figure 3.49 compares the methods presented for determining the percent complete and the planned costs and lists possible areas of application.

Tip

In practice, you can use different methods to determine the percentage of completion for different activities, work packages and sub-projects. For example, you can analyze short tasks using the simple, imprecise but absolutely objective percent-start/percent-finish method. You can determine the percent complete of longer tasks either by measuring or using the weighted milestones.

	Expert estimate	Quantity-related proportionality	Time-related proportionality	Secondary proportionality	Weighted milestones	0-100 method	50-50 method
Objectivity	low	very high	low-medium	depends on secondary value	high	very high	very high
Accuracy	low	medium	low-medium	depends on secondary value	high	low, tends to underestimate	low
Complexity of application	low	medium	medium	depends on secondary value	medium-high	low	low
Suitability for	short, non-critical activities	activities with measurable quantities	non-critical activities	non-critical activities	all activities	short activities	short activities

Figure 3.49: Comparison of methods for determining the percentage complete.

With the methods mentioned above, we are able to determine the project status. This is important for the following analysis, but is not yet sufficient. In addition to the methods for the quantitative collection of the key figures explained, we must also record the qualitative progress of the project deliverables.
This refers to the quality check of the results produced, for example the quality of developed software. In the toolbox in Chapter 8 you will find some methods for this, such as the verification of the project deliverables (fact sheet M.2.7) or its validation (fact sheet M.2.8). The suitability of the methods depends on the deliverables. For this reason, it should already be determined in the definition and planning phase how the results are to be qualitatively evaluated later.

Analyze the project status

The determined values for the percent complete, the planned value and the actual costs can be visualized in a suitable form so that

- you as a project manager have a good **overview**
- the determined status can be analyzed and **control measures derived from it**
- possible bottlenecks and the **optimization potential can be identified** and
- the **project progress** can be comprehensively **communicated** to the project team and other important stakeholders.

The right form of visualization depends on the project you are working on. In a small project with a team that sits close together, other methods make sense than in a large international project with several hundred team members. Methods for visualization and analysis include

- the **earned value analysis,**
- the **analysis of the work breakdown structure**
- the **control of the schedule on the Gantt chart** and
- the **milestone and cost trend analysis**.

All methods can be found in the toolbox in Chapter 8. Here we will briefly focus on the **earned value analysis** as the most powerful, but also the most complex method. The earned value analysis is an integrative method. It serves not only to determine the progress of the project but also to analyze it. It includes the work performed and its value, the planned costs and the actual costs.

The prerequisite for using the method is a time-resolved cost plan with a total cost line, as you have already become familiar with. The total cost line on a certain reporting date is the *planned value PV* of the project. In the earned value analysis, we compare this with the *actual cost AC* and the value of the work actually completed, the so-called *earned value EV*.

The PV, AC and EV values are monitored regularly by the project manager. They can be easily displayed in a diagram. You can see an example of such a diagram with additional explanations in Figure 3.50.

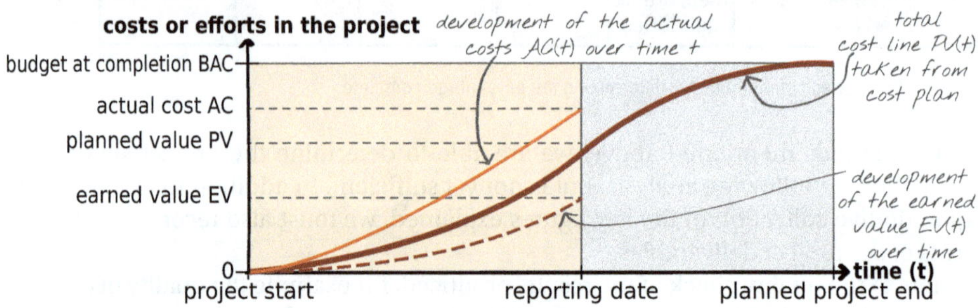

Figure 3.50: Typical illustration of the key figures planned value PV, actual costs AC and earned value EV for the earned value analysis on a chosen reporting date.

Tip

Instead of costs, other effort-related figures can also be compared. Examples are planned working hours PV, actual working hours AC and working hours EV, which correspond to the amount of deliverables actually completed. However, PV, AC and EV must have the same unit (costs in US$, working hours, etc.) so that the analysis presented below can be conducted.

This means that the earned value analysis can also be used in non-monetary projects.

If the project runs as planned, the PV, AC and EV curves are identical up to the reporting date. You can obtain the data for plotting the curves from the following sources:

- You obtain the planned value PV from the total cost line of the cost plan.
- You receive the actual cost AC from the accounting of your project. You can either track these yourself or receive the values from your project controller.
- The earned value EV corresponds to the planned costs of the actual work. Accordingly, these can be calculated from the product of budget at completion BAC, i.e. the planned total costs, and percent complete PC, which refers to the actual work already completed: EV = PC *x* BAC.

The last point is probably not immediately obvious, which is why we will discuss it in more detail. The earned value EV corresponds to the planned costs of an actual work completed.

The actual work completed can be represented as a percentage of the total work and thus by the percent complete PC. We have referred to the total costs as budget at completion BAC. This allows us to calculate the product:

$$EV = PC \times BAC$$

Conversely, if the earned value EV and budget at completion BAC are known, the percent complete PC can be calculated:

$$PC = \frac{EV}{BAC}$$

Figure 3.51 illustrates the calculation of the earned value EV and the percent complete PC by means of an example.

ID	Name	BAC_i	PV_i	PC_i	EV_i
1.1.1	Start project	30,000	10,000	100 %	30,000
1.1.2	Control project	50,000	30,000	50 %	25,000
1.1.3	Close project	10,000	0	0 %	0
1.2.1	Analyse feasibility	100,000	80,000	100 %	100,000
1.2.2	Collect requirements	80,000	80,000	100 %	80,000
1.2.3	Prepare design	80,000	80,000	100 %	80,000
1.3.1	Develop hardware	200,000	180,000	80 %	160,000
1.3.2	Develop software	250,000	220,000	50 %	125,000
1.4.1	Verify system	100,000	0	0 %	0
1.4.2	Validate system	100,000	0	0 %	0
	Sum	1,000,000	680,000		600,000
	actual percent complete PC of the project			60 %	
	planned percent complete PC at reporting date			68 %	

EV_i results from line-by-line multiplication of BAC_i and PC_i

results from $\frac{\Sigma EV_i}{\Sigma BAC_i}$

results from $\frac{\Sigma PV_i}{\Sigma BAC_i}$

Figure 3.51: Calculation of the earned value EV and the percent complete PC of a project: For each activity, the budget at completion BAC_i, the planned value PV_i and the percent complete PC_i are determined. PC_i can be derived using one of the methods mentioned before, for example the weighted milestones method. From this, the earned value $EV_i = PC_i \times BAC_i$ of each activity can be calculated. The actual percent complete of the project is then calculated by dividing the sum of the individual earned values and the budget at completion. In addition, the planned percent complete can be calculated as indicated in the figure.

The example shows that it is possible to analyze the project status by calculating the actual and planned percent complete: In the example, the project should have been completed to 68% by the reporting date, but it is actually only 60% complete.
However, the earned value analysis can do even more: by calculating the **cost variance CV** or the **cost performance index CPI**, it is possible to analyze the status of the project costs. The **schedule variance SV** or the **schedule performance index SPI** can be used to analyze the schedule of the project.
The data and its interpretation are explained in more detail in Figure 3.52.

Cost analysis			
Cost variance $CV = EV - AC$			
Cost performance index $CPI = \frac{EV}{AC}$			
The cost variance and cost performance index compare the earned value EV with the actual costs AC.			
The following statements can be derived:			
The project is on budget as planned.	EV = AC	CV = 0	CPI = 1
The project costs more than planned.	EV < AC	CV < 0	CPI < 1
The project costs less than planned.	EV > AC	CV > 0	CPI > 1

Schedule analysis			
Schedule variance $SV = EV - PV$			
Schedule performance index $SPI = \frac{EV}{PV}$			
The schedule variance and schedule performance index compare the earned value EV with the planned value PV.			
The following statements can be derived:			
The project is on schedule as planned.	EV = PV	SV = 0	SPI = 1
The project is slower than planned.	EV < PV	SV < 0	SPI < 1
The project is faster than planned.	EV > PV	SV > 0	SPI > 1

Figure 3.52: Explanation of the cost and schedule analysis. It is usually sufficient to calculate either the variances (absolute values) or the performance indices (relative values). By calculating these, statements can be made as to whether the project is more expensive or cheaper, slower or faster or whether it is progressing exactly as planned.

Example

Let's analyze the cost and schedule performance of the project discussed in Figure 3.51. Remember: The EV of this project was calculated with EV = 600,000 US$. Assume, that we have learned from the project controller that AC = 500,000 US$ have already been spent on the project:
The cost performance index is:

$$CPI = \frac{EV}{AC} = \frac{600.000 \text{ US\$}}{500.000 \text{ US\$}} = 1,2 > 1$$

The project is therefore significantly cheaper than originally planned.
The schedule development index is:

$$SPI = \frac{EV}{PV} = \frac{600.000 \text{ US\$}}{680.000 \text{ US\$}} = 0,88 < 1$$

The project is therefore slower than planned. The reason why the project is cheaper and slower must be found out by analyzing the root causes. The earned value analysis only provides the figures, not their final interpretation.

Forecast

In a further step, the data can now be used to create forecasts. You should be aware that this look into the future is created using data from the past and therefore only allows limited predictions to be made about the further course of the project.

*The **estimate at completion EAC** is the forecast of the project progress at the planned completion date.*

Indices can be assigned to differentiate between expected total costs and expected completion date. EAC_c then stands for the expected total costs and EAC_t for the expected completion date.

The **linear forecast** assumes that past developments will continue proportionally in the future. For example, if personnel costs have increased in the past due to wage increases, these will also affect the project in the future. The **project costs** are forecast using the cost performance index:

$$EAC_c = \frac{BAC}{CPI} = BAC \times \frac{AC}{EV}$$

The **project duration** is forecast in the same way using the schedule performance index:

$$EAC_t = \frac{\text{planned project duration}}{SPI} = \text{planned project duration} \times \frac{PV}{EV}$$

With the **additive forecast**, it is assumed that the reason for a deviation from the plan in the past was of unique nature and will not occur again. For example, if there have been additional costs due to the failure of an unreliable supplier and no similar problems are expected in the future, the additive forecast is used.

The forecast for project costs is made using the cost variance:

$$EAC_c = BAC - CV = BAC + AC - EV$$

Warning

Whenever specific key performance indicators are collected, there is always a risk of blindly relying on them. You should therefore familiarize yourself with the limitations of the earned value analysis and take these into account when intervening in the project on the basis of the acquired data:

- We have implicitly assumed that work performed and costs are proportional – if this is not the case, the earned value analysis provides misleading values.
- When calculating the key performance indicators, we rely on collected data, which in turn is subject to uncertainties. In determining the percent complete, for example, we have found that this can only be estimated with a limited degree of accuracy in some cases. The accuracy of the actual cost AC must also be examined. The earned value analysis can only be as accurate as the data we use to calculate it.

Managing progress

The project plan has been approved by the client or project sponsor. The client expects you to carry out the project as planned. A good plan is a good basis for project execution. However, the novelty and unique nature of projects and the large number of interested or affected persons (stakeholders) can lead to deviations from the plan.

During the controlling phase of projects, the project manager needs to identify deviations from the project plan and analyze their root causes. If the analysis shows that the project objectives are at risk, countermeasures to achieve the objectives must be defined, introduced

and monitored. Analyzing the causes of deviations can also lead to questioning the current objectives and readjusting the project and its objectives. This may be the case, for example, if resource availability or costs have changed to such an extent that an economically justifiable project result can no longer be achieved.
The measures for achieving the objectives, the objectives and corresponding changes to the plan become an integral part of the existing project plans. The earlier such measures are initiated, the more time they have to take effect and the smaller the measures can therefore be.
Deviations during project execution are quite common due to the unique nature of projects. Reasons for deviations are, for example

- changes in the general conditions, e.g. due to changes in legislation, rising raw material prices, budget cuts, etc.
- the insight that an initially planned solution cannot be achieved as intended or
- changed customer needs or market requirements.

A change always means that rework is required - even if only the plans or requirements documents need to be updated. This creates new work.
As a project manager, you want to conduct the project in such a way that the client is satisfied at the end. This means that you cannot generally oppose changes. However, you are the person who points out the consequences of changes. These consequences can be higher costs or a longer project duration, for example. Considering the consequences is part of the decision as to whether a change should be made or not.
The project manager can use **scenario planning** to examine the effect of potential changes or control measures and make a good decision based on this.

> ***Scenarios** are models of the future; they are intended to help illustrate the potential future course of a project. They go further than the pure forecast values of the earned value analysis. Soft factors and additional estimates can also be incorporated into the development of a scenario.*

Typically, at least three scenarios are developed: a *pessimistic*, an *optimistic* and a *probable* or realistic scenario. The optimistic scenario is sometimes called the best-case scenario, while the pessimistic scenario is called the worst-case or base-case scenario. However, these terms can be misleading, as the real extremes (best or worst case) are not usually examined.
Scenario planning is explained in the toolbox in Chapter 8. There you will also find descriptions of the control measures:

- change in resources
- reduction of effort
- increase in productivity
- improvement of process quality and collaboration
- change of the project scope.

These help to reduce deviations from the plan and achieve the most important objectives.

Tip

If you can already foresee major changes at the start of the project, you should consider an agile or at least iterative process model for project management. You learned about these in Chapter 2. We go into this in more detail in Chapter 4. If this is not possible, you should pay particular attention to risk and change management, which we will discuss later in this chapter under the ongoing project management tasks.

Reporting

Reporting informs key stakeholders about the status of the project. Reporting can be informal or formal. Examples of *informal* reporting are spontaneous discussions about the status of the project and the posting of information and project successes on the bulletin board. Other project marketing measures such as electronic newsletters can also be understood as informal reporting.

The *formal* reporting typically uses standardized communication channels, for example with the help of a progress report. Important recipients of this report are the client of the project and the portfolio manager or the project management office (PMO). The client is usually particularly interested in the qualitative development of the project deliverables and the achievement of the project objectives. Depending on who is in charge, either the portfolio manager or the PMO, will be interested in any existing changes, resource demands and project risks.

Tip

Agree early on with the relevant stakeholders, especially the client, how often and to what extent the project status should be reported. Too frequent reporting causes unnecessary effort due to the preparation of data with little progress between the reporting dates. If reporting is too seldom, deviations from the plan can lead to the client being poorly informed, the need for control being recognized too late, and the project objectives not being achieved. All of this can lead to serious conflicts.

Often, regular status meetings are scheduled at which the project manager reports to the client, the portfolio manager or the PMO using standardized reporting forms (progress reports). For example, it can be agreed that the individual project managers report once a month.

In addition, there may be extraordinary status meetings, for example at important milestones, at transitions from one project phase to the next, when a risk occurs or when conflicts or escalations arise.

> ***Progress reports** are report forms that summarize the project status on a specific reporting date.*

Other terms for the progress report are *traffic light report, one-pager, status report* or *project report*. Many managers prefer a condensed version with highly aggregated information.

The (often complex) project status is to be summarized in the form of a traffic light: Green for "everything as planned", yellow for "deviations from the plan but with measures to achieve the plan" and red for "deviations from the plan without the possibility of achieving the objectives".
This form of communication is difficult, especially if the choice of color is not clearly defined. If you set the project traffic light to green, you initially have peace of mind from the customer. However, if problems arise later, you will be asked why you did not report them earlier. If you choose a red color code, you will quickly be seen as an alarmist who does not have his project under control.
A good progress report therefore provides more meaningful information than the presentation of a status traffic light alone, see Figure 3.53.

Project name: *TechExpo2025*	Project ID: *16-0733*	Reporting date: Jun. 22nd 2025	Traffic light:
Project manager: *Meike Schmitt*	Core team: *Design:* *Bernd Heider*		Deviations without measures to achieve the objectives
Brief description: *Construction of a prototype of a new injection molding system for the SG2025 trade fair*	*Requirem. engin.:* *Kim Richards* *System test:* *Claudia Gross* *Technical author:* *Ralf Weber*		Deviations with measures to achieve the objectives (marked) As planned

Top 3 objectives and status:	planned	expected	Reasons for deviations
Cost objective:	*400,000 US$*	*450,000 Euro*	*Change order 15-0732 due to marketing*
Deadlines:	*project completed by Jun 30th 2025*	*Jun 30th 2025*	

Achievements since last report / most recently completed work packages / changes
The specification phase with work packages 1.2.1 to 1.2.8 has been fully completed.
Due to the changed design of the cover (amendment 15-0732), some work had to be repeated.
Time delays can be made up through additional work.

Top 3 risks and countermeasures:

Risk	Impact	PO	AD	Countermeasure	New PO	New AD
Manufacturing tolerance cannot be adhered to due to defective control SW	*Deviation/ delay*	*3-medium*	*4-high*	*Change SW architecture to facilitate later patches*	*3-medium*	*2-low*
…	…	…	…	…	…	…

PO: probability of occurrence AD: amount of damage

Next steps:	Miscellaneous:
Start of the implementation phase with WP 1.3.1 to 1.3.10	*Increased customer inquiries regarding delivery date*

Decisions made at this project status meeting:
The risk management measure to change the software architecture is approved.

Figure 3.53: Example of a progress report. In practice, there are countless variations. The minimum content should be an identification of the project, a comparison of the objectives (deadlines and costs), the last and next steps in the project, the greatest risks and countermeasures.

In addition to information on the project, the most important objectives and their level of achievement are usually documented. In addition, the accomplishments from the previous reporting period are summarized and the greatest risks and their countermeasures are listed. The report is completed by a list of the next steps and documentation of the decisions made at the status meeting.

Project closure

Overview – where are we in the project management process?

You have completed the project deliverables and are about to hand them over to the client. The project documentation must be completed. Experience and project knowledge are edited and archived in such a way that subsequent projects and the line organization benefit from them. Last but not least, the project team and the project infrastructure must be dissolved. The classification in the project life cycle is shown in Figure 3.54.

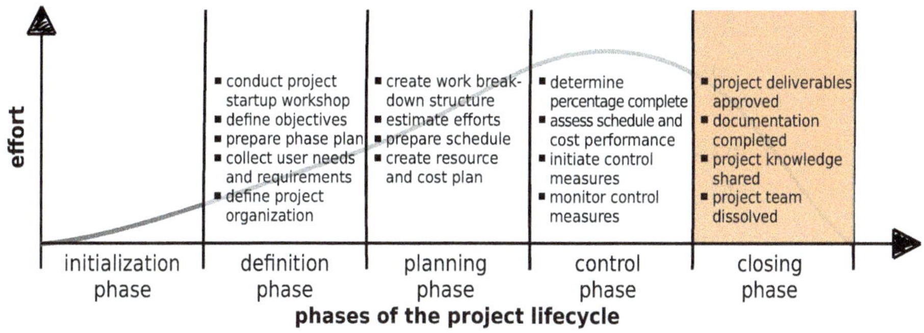

Figure 3.54: In the closing phase, the last deliverables are completed. Finally, the project team is dissolved and the project is completed.

It is not always clearly defined when a project actually ends. Possible interpretations are

- as the project deliverables have been handed over to the client
- as soon as all documents related to the project have been created and released
- as soon as the project team has been dissolved
- as soon as the success of the project has been evaluated.

Example

Suppose a project aims to optimize the production process so that the error rate is reduced by 20%. To achieve this, the production process is recorded and analyzed, sources of error are identified and measures for improvement are derived. The measures are implemented and production is now carried out in accordance with the revised production process. If the optimization measures work as desired, the error rates should decrease significantly. The project manager would like to close the project. However, the client insists on proof that the error rates have actually fallen by 20%. A longer evaluation phase is necessary for this. The project can therefore not yet be completed.

Example

A new, innovative audio system was to be developed as part of a development project. Originally, it was agreed that the project would be completed with the successful transfer to production. As some requirements were not implemented in order to meet the deadline targets, the client requested that the project be continued in order to fulfill these requirements in a new development cycle.

In practice, the examples shown are problematic. Although the project was carefully planned at the beginning and then managed in the control phase so that no major deviations from the plan occurred, it is sometimes difficult for the project participants to complete the project as planned.

There are many reasons for this, for example

- The **client** is **not satisfied** with the result achieved.
- The **client** is **so pleased with the result** that he or she immediately wants to assign new additional tasks to the successful team.
- Those involved want to implement something **additional**, but fear that they will not be approved for **a follow-up project**. They therefore try to extend the existing project.
- There is an interest in the content of the **topic of the project** coupled with the desire to make it **perfect.**

However, as the resources are usually allocated to other projects from the planned end of the project, a project extension affects the start of the other projects.

Tip

Even if project completion is still a long way off when the project starts: agree with the client when and under what conditions the project will be completed. This will make it easier for you to plan and later hand over and approve the project deliverables.

Phases of project closure

The activities in the project closing phase can be divided into four sections, see Figure 3.55.

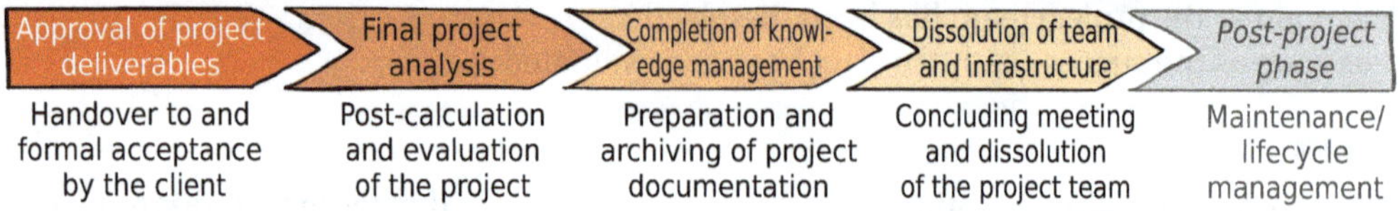

Figure 3.55: The four phases of project completion based on Burghardt [Burghardt, 2018]. The post-project phase is no longer part of the project itself, but takes up the topics of the project and continues them as required.

Handover and acceptance

The client of the project is obliged to accept the project deliverables at the end of the project. In a development project, the project deliverables may, for example, consist of a functional model, individual components, software, documentation, several design drawings or similar. If the project deliverables are not clearly defined at the start of the project, this can lead to problems during acceptance. There may be disagreements between the client and the project team over the question of what is part of the project object and must be handed over accordingly.

The acceptance itself should be well prepared and documented with a **acceptance report.** If similar projects are carried out repeatedly in the company, standardized acceptance protocols in the form of checklists can be helpful.

In the acceptance report it is recorded,

- **what** is handed over (project deliverables)
- **how** it was handed over (free of defects or with defects) and
- **what rework** still needs to be done.

The acceptance report must be signed by the client and project team (usually the project manager in this case) and included in the project documentation.
If the need for rework is identified, you should describe this in a way that is comprehensible to all parties involved, agree on an acceptance criterion and a deadline and also document both.

Final project analysis

The final analysis at the end of the project helps you to evaluate the success of the project. In addition, opportunities for improvement can be identified for future projects. To do this, the project definition, project planning and project controlling must be analyzed. Findings from which future projects can benefit must be shared with them, see Figure 3.56.

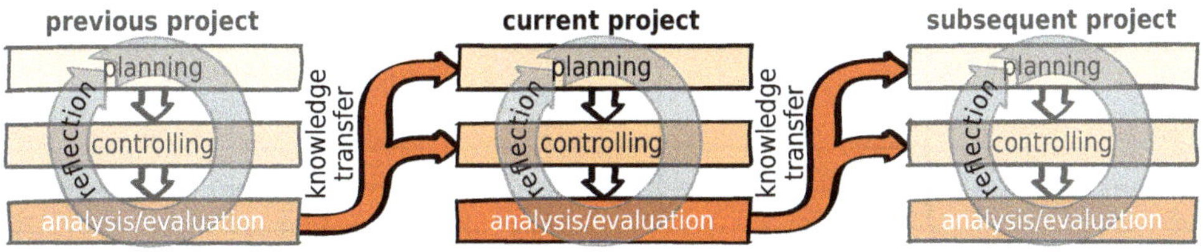

Figure 3.56: Projects can benefit from the experiences of previous projects. These must be analyzed and evaluated. Regular reflections during the project lifetime are helpful in order to gather important findings that can be passed on later. Relevant findings must then be transferred to future projects. This can be done, for example, by revising the process model, creating new templates and checklists or training project staff.

The final analysis includes
- the **post-calculation** of the project
- an **analysis of the project objectives** and
- an **evaluation of the project** by key stakeholders.

In **post-calculation**, the originally estimated and actual costs and efforts are compared with each other. Any deviations are analyzed. A simple example of a post-calculation is shown in Figure 3.57.
The reasons for deviations are sometimes difficult to determine. In the case of larger deviations, more precise analyses can help to determine the root cause, for example by using additional tools such as
- the Ishikawa diagram
- the ABC analysis
- the 5-Why method.

Warning

Do not be satisfied with superficial reasons for deviations. Only when you have identified the *root cause* measures can be developed for future projects to help prevent similar deviations. However, distinguish between real deviation and inaccuracy of the measurement or estimate. We will always have to accept a certain degree of uncertainty in the estimate of projects (remember the unique character).

Post-calculation of the project

work package	planned BAC [US$]	approved change request (CR)	new BAC [US$]	AC [US$]	deviation [US$]	reason
1.3.2 implement database	20,000	CR100053: addition of a further address line in the data model	24,000	26,000	+2,000	late modification of the data model was more complex than estimated
1.3.3 implement printer interface	10,000	-	-	10,000	-	-
1.3.4 design user interface	30,000	CR100054: added requirement that interface can also be used on mobile devices	45,000	40,000	-5,000	purchase of a software library led to simpler implementation
1.3.5 create manual	10,000	-	-	13,000	+3,000	no structured effort estimation
...						
Total						
Assessment summary:						

Figure 3.57: Exemplary excerpt from a simple post-calculation. Deviations can have various causes, such as an incorrect effort estimate or unexpectedly sharp increases in raw material prices during the project. However, they can also result from approved changes. Alternatively or additionally, not only the costs, but also the estimated effort in staff hours can be compared and analyzed.

When defining the **project objectives**, great importance was attached to formulating measurable objectives. The objectives were examined for feasibility and possible conflicts of objectives. Now it is time to evaluate the objectives; the following questions arise:

- Were the objectives actually achieved?
- Did unidentified conflicts of objectives occur?
- Did (sub)objectives have to be postponed in favor of other (sub)objectives?
- What were the causes of any unachieved objectives?

It is also important to identify and analyze the causes of possible deviations from the project objectives.

The success of the project is determined by the appreciation of the most important stakeholders. You should therefore determine the stakeholders' perception of the deliverables and then analyze it. This can be done either as part of an analysis workshop, through informal surveys, structured interviews or questionnaires. The group of participants for the analysis can result from your original stakeholder analysis. You should therefore know who you want to include in the **evaluation**, for example

- the **client** of the project
- important **end customers/users** of the project result

- **suppliers**
- involved **line departments** in the company.

An exemplary questionnaire for determining the assessment of stakeholders can be found in fact sheet C.2.2 of the toolbox in Chapter 8.

Completion of knowledge management

Knowledge management is concerned with making the knowledge and experience of individual employees or entire organizational units available to other employees and organizational units. The challenge with projects is that they are limited in time. At the end, the team dissolves. New teams are formed with different members and new objectives. Knowledge and experience must be saved through this temporal and organizational break.
At the end of a project, all project documents must be **archived**. This includes

- project plans and contracts
- protocols/meeting minutes
- the documentation of the project deliverables, for example design drawings or process descriptions
- the post-calculation
- the evaluation of the project.

Archiving ensures that project events and results remain traceable over the long term - for example, in the event of later problems or further developments of the product.
However, archiving project documentation is not enough to successfully pass on knowledge and experience to participants of future projects. The overall documentation is usually too lengthy to be read by participants of future projects and, above all, to be put to good use. It is therefore important to think about other ways of sharing knowledge and experiences.
The **safeguarding of knowledge** at the end of the project is also referred to as **Lessons Learned**. Methods for sharing knowledge and experiences are

- **wikis** on the intranet of the company
- **checklists**, which are always updated at the end of the project; examples include checklists for potential project risks and stakeholders
- **tables or electronic databases with experiences of effort estimation**, especially of efforts that may also be arising in future projects; examples are efforts for the creation of functional specifications, design drafts or final documentation
- **templates** for the creation of documents, for example for project planning and specifications
- **exchange of experiences** through reports, presentations and meetings
- **updating** the company-specific **process model** according to the experiences of the current project.

Not every method is suitable in every company and project constellation, which is why the transfer of knowledge and experience should be integrated into an overall concept of knowledge management.

Tip

Even the best safeguarding of knowledge is useless if existing knowledge is not recalled at the start of new projects. Get into the habit of using up-to-date templates, checklists and empirical values when estimating the effort involved. Talk to colleagues from previous projects.

Warning

Do not confuse the completion of knowledge management at the end of the project with knowledge management itself. Knowledge management is an ongoing task, which we will discuss later in this chapter. Knowledge management spans the entire project lifecycle and beyond. At the end of the project, we then ensure that knowledge management for this specific project has also been conducted and that all the necessary tasks have been completed. Due to the time limitation of projects, it is important that these tasks are completed by the current team before it is dissolved.

Dissolution of the team and infrastructure

During the steering phase of the project, the project team works intensively on the project deliverables. In the closing phase, more and more work is completed. Employees who no longer have any open work packages are withdrawn from the project and assigned to new projects.

The spirit in the project changes noticeably. Employees who are already working on new projects tell their old team colleagues about new tasks, while only a handful of team members are still working on the current project. For these employees may be uncertain as to where and how they will be assigned once the project is completed. This concern is particularly pronounced or even vital for freelance, non-permanent employees. It can lead to them wanting to keep the project alive for as long as possible.

Warning

Once the project deliverables have been handed over, some members of the project team feel that the project has been completed. Deadline and cost pressures ease immediately. Project managers find it difficult to organize the final work. In most cases, there is also a lack of support from senior management in this phase. After the handover, resources are preferably invested directly in new projects. Overall, there is a risk that the project team will fall apart or at least lose much of its performance.

The consequences of this behavior can be that the project takes longer than planned or that no orderly project completion is achieved at all.

These considerations show that the closing phase is a particularly challenging phase for the project manager from a leadership perspective. They must keep the project team's **commitment at a high level**, while at the same time **releasing** individual **employees** for other projects and **giving** other employees **perspectives** beyond the end of the project together with the resource managers.

Some issues that you should consider as a project manager in this phase are:

- Plan the project completion carefully and coordinate this plan with the employees and their resource managers.

- Before you release resources to other projects: Check that all work has been completed and documented. Knowledge and experience must be secured for future projects.
- Give employees feedback on their contribution to the project. Positive feedback and appreciation in particular are sometimes overlooked. Also pass this feedback on to your employees' disciplinary supervisors.
- Mark the end of the project with a final meeting. This gives structure to the end of the project and ensures a clear finish.

The counterpart to the project start workshop at the beginning of the project is the **project closure meeting** at the end of the project. You should not schedule this too late:
- The project team should still be in place. If the team dissolution is too far advanced, no one will come to the meeting.
- The most important project deliverables should have been completed. You can link the meeting to the handover to the client, for example.
- Open tasks up to project completion should be foreseeable and assigned to the team members.

The purpose of the project closure meeting is as follows:
- analysis and evaluation of the project; every team member should have their say; identify not only what went badly and why, but also what went well and should be transferred to future projects
- safeguarding knowledge and experience
- presentation of the further project plan until the final end of the project
- distribution of remaining tasks
- appreciation of the work performed and emotional conclusion, for example through an appropriate celebration.

Project final reports can document the project in summary form. However, they also require a lot of effort to create and are rarely read.

Post-project phase

As the name suggests, the post-project phase is no longer part of the project itself. The post-project phase usually refers to the utilization or application phase of the project results. Examples are
- the series production and use of a developed product
- the use of optimized company processes
- the use of a new system for a production plant.

Follow-up activities may also arise if the project results need to be revised after the end of the project, for example due to a component being discontinued or a further need to improve company processes.
From the company's point of view, despite a completed project, there is still work to be done in one form or another. Depending on the scope of the project and the company, this work can be organized in different ways, for example
- completely in the line, for example as part of product management or production or
- as a lifecycle or maintenance project.

A less favorable alternative is the permanent continuation of the current project in order to carry out the activities of the post-project phase. Such projects have little support within the company, as they are actually completed.
These considerations once again highlight the need for an orderly project closure. Project knowledge and project documentation must be saved in such a way that they can be used in subsequent activities.

Ongoing project management tasks

Overview – where are we in the project management process?

So far in this chapter, you have been introduced to the individual project management phases in chronological order. In addition to tasks that can be clearly assigned to individual phases, there is also a whole range of continuous and overlapping tasks that we will now look at in more detail, including aspects of

- stakeholder and risk management
- project marketing
- quality management
- document and configuration management as well es and change control
- contract and claim management and
- knowledge management.

Stakeholder and risk management

Projects are unique endeavors with limited resources. We can therefore transfer our experience to new projects only to a limited extent. In addition, a project is carried out in a specific environment. *This environment consists of people and groups of people* whose individual goals do not always coincide with the goals of the project.
Based on these considerations, we can state that there are influencing factors that have a positive or negative impact on project success:

- stakeholders and
- opportunities and risks.

Let's take a closer look at the terms:

> ***Stakeholders** are persons or groups of persons who are interested in the project, who are involved in the project, who can influence the project or who are or who feel affected by the project.*

> ***Risks** and **opportunities** are potential, unplanned events that are characterized by a probability of occurrence and an impact on the project objectives. Risks counteract the project objectives, i.e. the project suffers damage. Opportunities support the objectives.*

Stakeholders can be the cause of risks and opportunities and can themselves influence them both positively and negatively. However, risks and opportunities can also be of a purely factual, economic or technical nature.
We will only talk about stakeholders and risks below. Just as we try to *prevent* the occurrence of a risk, we can also try to *promote* the occurrence of an opportunity. The methods for this are basically the same as for risk management
The aim of stakeholder and risk management is to

- **identify** all relevant stakeholders and risks
- **assess** stakeholders and risks in terms of their significance and impact on the project
- plan and implement **measures** and to monitor and control their **effectiveness** so that stakeholders and risks do not jeopardize the success of the project.

The processes of stakeholder and risk management are very similar, see Figure 3.58.
First, the stakeholders and risks relevant to the project must be identified. You should proceed as follows:

- *1st step creativity phase*: Stakeholders and risks are identified through the use of creativity techniques.
- *2nd step systematic review phase*: Go through the project documentation to date, including the plans already prepared, and examine the extent to which stakeholders are involved there and risks may arise.
- *3rd step systematic control phase*: Use checklists and questionnaires from past projects to check whether stakeholders and risks have influenced the success of the project that you have not yet identified.

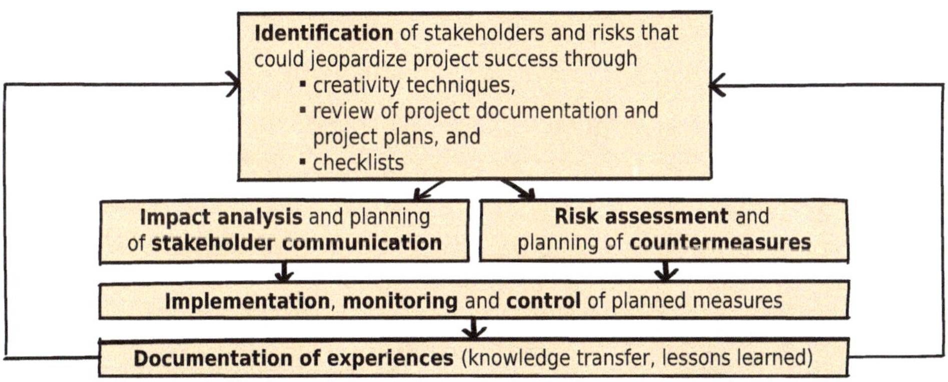

Figure 3.58: Stakeholder and risk management process.

Tip

Checklists are valuable tools for learning from past projects. They help us not to forget things in future projects. However, always use checklists *after* creativity techniques and the systematic review of the project documentation. Otherwise, there is a risk that the team members' commitment to identification will diminish. Team members may rely purely on the checklists. Project-specific stakeholders and risks that are not included in the checklists are forgotten.

Overlooked stakeholders and risks can jeopardize the success of the project. It is therefore important to first identify as many potential stakeholders and risks as possible. Involve your project team in the identification process. The team knows the respective areas of the project in greater detail than you alone and can therefore make valuable contributions.

STEEP analysis is a mixture of brainstorming and mind mapping. STEEP stands for **S**ociological, **T**echnological, **E**conomical, **E**nvironmental and **P**olitical. After an initial brainstorming session, identified stakeholders and risks can be assigned to one of this five areas of STEEP. If no stakeholders and risks have been identified for individual areas, a second iteration of creativity can be used to search for additional stakeholders and risks for these areas. The categorization helps to search the entire project environment for stakeholders and risks. Figure 3.59 shows an example of a STEEP analysis.

Once potential project stakeholders have been identified, their significance for the success of the project must be investigated. This is done in the **impact analysis**. Communication measures are then planned and implemented for the stakeholders that could jeopardize the success of the project.

As part of the impact analysis, you examine which stakeholders can support the project and which are critical or even opposed to the project. To do this, you determine for each stakeholder

- the degree of impact
- the type of impact
- the influence or power of the stakeholder on the project
- the potential for conflict and
- the stakeholder's presumed expectation of the project.

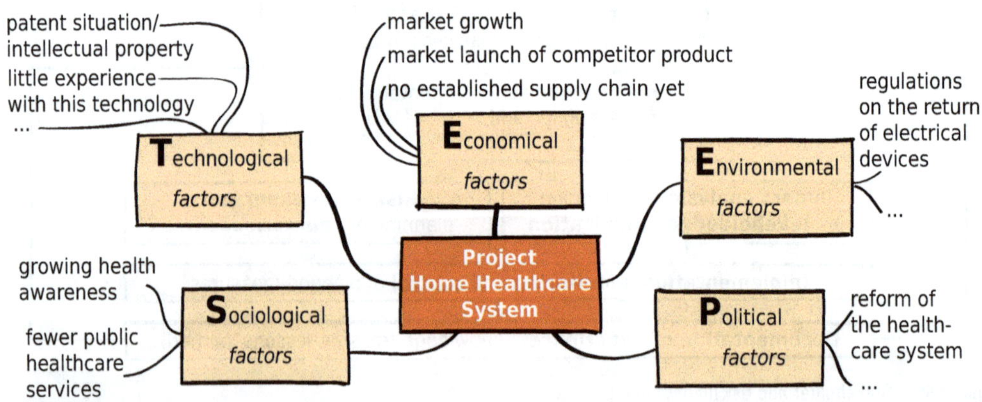

Figure 3.59: STEEP analysis of an exemplary project. The five areas of STEEP help to consider the entire project environment as far as possible and to identify potential stakeholders and risks.

The data can be summarized in a table. An example is shown in Figure 3.60. The impact analysis should always take into account the perspective of the respective stakeholder.

The stakeholders can then be presented in a stakeholder portfolio and prioritized in this diagram, see Figure 3.61.

The portfolio view helps to categorize stakeholders according to their potential risk to the project. Critical stakeholders are those,

Name of the stakeholder	Degree of impact	Type of impact	Influence/ power	Potential for conflict	Assumed expectation of the stakeholder
marketing manager	high	negative	medium	high	believes that the customer does not like the product design
component supplier A	medium	negative	low	high	concern about not receiving an order for series production

Figure 3.60: Example of an impact analysis in tabular form based on Patzak and Rattay [Patzak and Rattay, 2017].

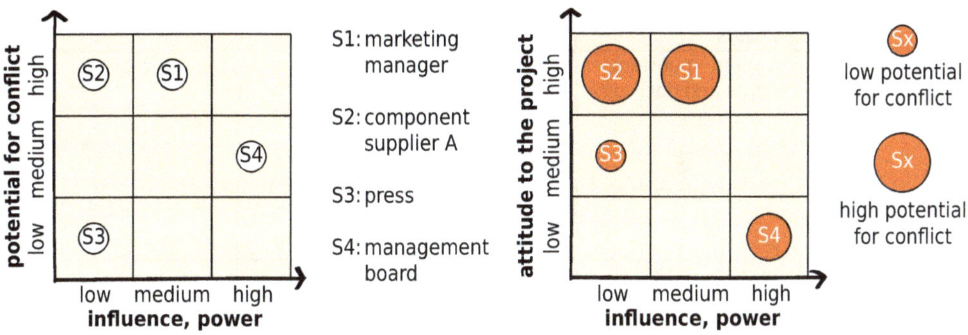

Figure 3.61: Two examples of a portfolio representation for stakeholders. On the left, the stakeholders are arranged along the axes of influence and conflict potential. The further the stakeholders are arranged at the top right, the more they can jeopardize the success of the project. On the right, a portfolio was drawn with the axes of attitude towards the project and influence. As a third dimension, the potential for conflict was marked by circles of different sizes.

- that have a negative attitude towards the project
- who have a lot of influence on the project or comparatively a lot of power in the company and
- who are prone to conflict.

You must now focus on these stakeholders. You plan measures for them that will ensure that the anticipated negative effects on the success of the project are avoided or that the desired positive support from the stakeholders is provided. The measures must be implemented and checked for effectiveness.

Warning

You have become familiar with a very detailed variant of the impact analysis. Depending on the stakeholder structure and the complexity of the project, simpler analyses may also be sufficient. In the simplest case, you simply determine which stakeholders are opposed to the success of the project (opponents) and which promote the success of the project (promoters).
As with all project management methods, the effort involved in the method must be in line with the result to be achieved.

When planning measures, you can choose between three basic communication strategies:

- participative strategy, i.e. participation and involvement of stakeholders
- discursive strategy, i.e. balancing and negotiating different perspectives and opinions
- repressive strategy, i.e. use of power to enforce the interests of the project.

Figure 3.62 briefly summarizes the three strategies. You can derive specific measures from these strategies: Check…

- whether short-term immediate measures are necessary
- whether precautionary and long-term measures should be considered and
- what effects the selected measures will have on the long term.

	Participative communication strategy	Discursive communication strategy	Repressive communication strategy
Features	involvement or participation of stakeholders	striving for a balance between the various interests	use of power
Typical procedure	informing about important developments and decisions participation in project work participation in decisions	conducting negotiations use of conflict management and negotiation methods	selective disclosure of information disinformation creating fait accompli only pretending participation pressure and threats
Opportunities	valuable input from stakeholders	sustainable balance of interests	no longer a threat to project
Threats	delay through participation	balance with compromise/ disadvantage for the project	permanent disruption of communication with stakeholders

Figure 3.62: Stakeholder communication strategies.

Warning

Repressive measures can permanently disturb cooperation. With internal stakeholders, important suppliers or the public, such measures can lead to lasting damage to communication and cooperation. You should therefore always check whether the planned measures are appropriate.

The planned measures for stakeholder communication can be documented in a **communication matrix**. The minimum contents of such a matrix are typically

- the stakeholders
- the measures and
- the persons responsible for implementing the measures.

In addition, the content and scope of the measures and their frequency can be planned in the communication matrix. If the measures result in larger scopes and costs, for example for image or marketing campaigns, these must be taken into account in the work breakdown structure, schedule, resource and cost plan. Figure 3.63 shows an excerpt from a communication matrix as an example.
The starting point for the following consideration of risk management is the list of identified risks. We will assess these risks and then focus on the risks that most jeopardize the success of the project. You need to plan measures to mitigate these risks. These risk management measures will either reduce the likelihood of them occurring or the damage they can cause to the project.

Stakeholder	Measure	Content	Scope	Frequency	Responsible person
marketing manager	involvement in project and product definition	acceptance of customer requirements represented by the marketing manager	approx. 1-2 hours	weekly	project manager
component supplier A	announcement of supplier selection	component supplier A will not supply components for the company's products	1 page	once	project manager

Figure 3.63: Excerpt from a communication matrix. A participative communication strategy was selected and planned for the marketing manager and a potentially repressive one for the component supplier. If a long-term cooperation with the supplier is desired, the negative decision must be supplemented by a long-term perspective for future procurement contracts.

For larger and more complex projects, risk management begins with a risk management plan as part of the project plan. This defines

- the **probability of occurrence** of risks
- the **damage** that a risk can cause
- which **risks** are still **acceptable** and which are **unacceptable**
- which **methods** are used for risk identification, assessment and planning of measures
- which **persons** are involved in risk management and who is responsible for it
- how the **risks are documented** and reported and
- how the **experience** gained in risk management will be **passed on** to future projects.

The definition of the probability of occurrence, the damage and the risk acceptance is important in order to be able to evaluate risks consistently over the duration of the project. The other components of the plan are usually only necessary for larger and more complex projects.

Example

Example risk assessment: The probability of occurrence of the risk of a delayed delivery of a component is assessed as high at the start of the project. After a few weeks, the project team looks at the risks again and rates the probability of occurrence as medium, even though nothing has changed in the general conditions. The client of the project is pleased with the overall reduction in project risk and asks the project manager for the reason for the reduction in risk.

In the example, a decrease in risk is reported, although nothing has changed in the general conditions. Due to a lack of definition of the term "high probability of occurrence", the risk has simply been reassessed and the impression has been created that the risk has been reduced.

Definition of the probability of occurrence

You can either specify the probability of occurrence of the individual risks *in percent* or in the form of *probability categories*:

0 - no risk

1 - low probability of occurrence

2 - medium probability of occurrence

3 - high probability of occurrence

For each category, you should briefly define how it is defined in a way that the project team can understand. You can specify percentages or give examples. For example, you can define that a low probability of occurrence corresponds to a maximum of one in 1,000 possible cases.

Tip

Risks with a probability of occurrence of 50% and more should no longer be dealt with as part of risk management. The probability of occurrence is so high that action is absolutely necessary. Some project managers no longer refer to risks with a probability of occurrence of 20 to 30% as a risk, but rather as a problem.

Definition of damage

The damage that a risk can cause to your project is usually of a financial or time-related nature. You can therefore try to specify the damage as a monetary amount or as a project delay. As both are not always easy to determine, you can define the damage in damage categories.

Warning

Most risks can only be roughly estimated in terms of their probability of occurrence and potential damage. If you specify these values in percentages, monetary or time units, you could suggest that you can predict both quite accurately. The same applies if you determine a large number of probability and damage categories. Check how accurately you can really estimate both and then choose an appropriate scale.

Risk matrix

In the risk matrix, you determine which risks are acceptable for you and for which risks you must plan countermeasures. You can see an example of a risk matrix and the definition of probability and damage categories in Figure 3.64.

Definition of probability and damage categories

category	probability of occurrence	damage (costs or delay)
5 - very high	> 30 %	> 100,000 US$ > 20 days
4 - high	20 %-30 %	up to 40,000 US$ up to 20 days
3 - medium	10 %-20 %	up to 30,000 US$ up to 10 days
2 - low	1 %-10 %	up to 20,000 US$ up to 5 days
1 - very low	0 %-1 %	up to 10,000 US$ up to 1 day

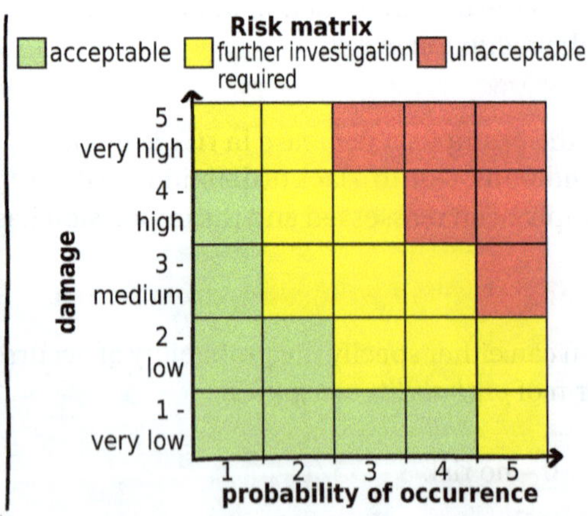

Figure 3.64: Left: Example of the definition of probability and damage categories. Damage usually manifests itself in higher costs or delays. Right: Example of a risk matrix that defines which combinations of probability of occurrence and damage are acceptable or unacceptable or need to be investigated in more detail.

The damage categories defined in the diagram depend on the respective project, its environment and the size of the company. In a small organizational project, 20,000 US$ additional costs can mean a very high damage. In an international construction project with a budget of several million US$, the same amount is likely to be assessed as minor damage.

Risk index

The risk index is usually calculated in addition to or instead of the risk matrix:

> *The* ***risk index*** *is the product of the probability of occurrence and damage category. It can be calculated for each individual risk. The sum of the risk indices of all individual risks is the overall risk index of the project.*

The assessment of the probability of occurrence and the damage must be reviewed regularly and updated if necessary. Depending on many factors, both influencing factors can change. This also changes the risk index.

Example

At the start of an organizational project to plan and host an open-air concert in early summer, the risk of bad weather is considered low. However, one week before the event will take place, the weather turns unstable with a strong tendency to thunderstorms. The probability of occurrence is revised significantly upwards and planned countermeasures are implemented.

The overall risk index is also suitable for reporting purposes. If it is calculated regularly and plotted along a time axis in a diagram, the development of risks can be monitored quite easily.

Methodology

Think about how you want to identify and assess risks and how you want to plan subsequent measures, for example with the help of creativity techniques and the STEEP analysis.

People

The project manager is often in charge of risk management and involves the project team in identifying, evaluating and planning measures. Depending on the scope of the project, you should plan the time and effort required for risk management in order to ensure that employees have enough time to complete their tasks. Changing perspectives helps to identify as many risks as possible. You should therefore integrate various stakeholders into risk management wherever possible.

Warning

In some projects, risks are trivialized instead of managed. A typical mistake is for the project manager to prepare the risk assessment alone, which is then only formally confirmed by the project team. Careful risk management requires open discussion, critical questioning and the realization that not everything can be predicted and planned. Only if you take this seriously your risk management will be effective.

Documentation and reporting

Do not overwhelm the client with a confusingly long list of risks. Instead, pick out the most threatening risks for the project, including their assessment, and document these together with planned countermeasures in the project progress report. This will show that you are practicing focused risk management and know how to reduce risks accordingly.
Two typical errors in project reports are

- The project team creates the impression that there are no risks. It wants to make it clear that everything is going according to plan and that the tasks are under control. In fact, it documents the opposite.
- The project team documents risks in the project report. However, either too many risks are reported, so that the focus is lost on the essentials, or no countermeasures are reported. This gives the impression that the team has no plan to control the risks.

Risk assessment and risk mitigation

The minimum contents of the risk management plan are the definition of the probability of occurrence and the damage as well as the definition of the risk matrix. Once this has been defined, the identified risks can be consistently assessed.

> **Tip**
>
> To really reduce a risk, you should identify its root cause. Then you can plan an appropriate measure that specifically counteracts this cause. Otherwise, you run the risk of only mitigating the symptoms of the cause.

When assessing a risk, you determine its probability of occurrence and the potential damage caused by the risk. You can either do this qualitatively with a categorization or quantitatively by directly specifying percentages and costs or time delay. An example is shown in Figure 3.65.

Qualitative risk assessment

risk	root cause	P	D	RI
Risk number 1: 1 week delay due to staff shortage	parallel project with the same bottleneck staff	4	3	12

Quantitative risk assessment

risk	root cause	P	D	ExV
Risk number 1: 1 week delay due to staff shortage	parallel project with the same bottleneck staff	25 %	25,000 US$	6,250 US$

P: probability of occurrence D: damage RI: risk index RI = P x D ExV: expected value ExV = P x D

Figure 3.65: Example of a qualitative and quantitative risk assessment with calculated risk index or statistical expected value.

To compare the risks, you can calculate the **risk index** for the **qualitative assessment**, see Figure 3.65 on the left, or the **expected value** for the **quantitative assessment**, see Figure 3.65 on the right. Both are calculated by multiplying the probability of occurrence by the damage. The higher the risk index or the expected value, the greater the danger for the project. Threshold values can be defined for the risk index or the expected value, above which risk control measures are recommended or mandatory.

Warning

The calculation of the expected value can lead to a misleading interpretation of the risk: In the example of the Figure 3.65, the expected value EW is 6,250 US$. However, this value will never be charged to the project. Either the risk occurs, in which case additional costs of 25,000 US$ are expected in the project, or the risk does not occur and no costs are incurred. In this respect, the expected value is a purely calculatory figure for comparing the combination of probability of occurrence and level of damage.
The sum of the expected values of all risks, on the other hand, provides an overview of the potential damage – assuming that not all risks will actually occur.

The identified and assessed risks can be entered in the defined risk matrix, see Figure 3.66. Similar to the risk index, the risk matrix helps to determine the risks for which measures are recommended (yellow area) or mandatory (red area). The risk matrix also provides a clear overview of all individual risks and helps to identify risk clusters. The calculation of the risk index for the risks in the risk matrix is illustrated on the right-hand side of the illustration and the total risk index of the project is shown as the sum of the risk indices of the individual risks.

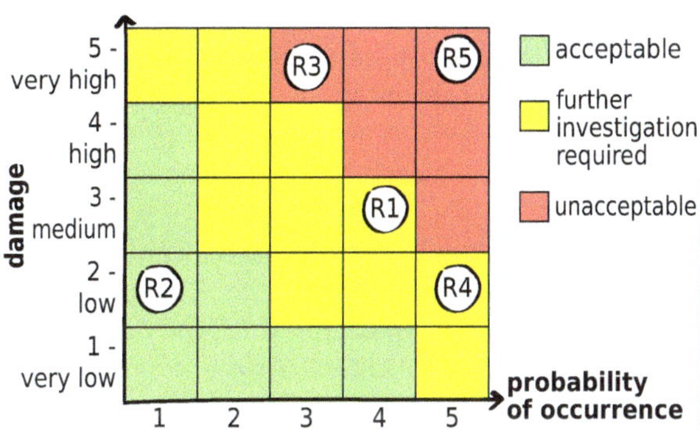

risk	probability of occurrence	damage	risk index
R1	4	3	12
R2	1	2	2
R3	3	5	15
R4	5	2	10
R5	5	5	25
total risk index: 64			

Figure 3.66: Left: Risks R1 to R5 in the risk matrix as defined in the risk management plan. Risk R2 has been assessed as acceptable. Countermeasures must be planned for risks R3 and R5. Risks R1 and R4 should be examined more closely, countermeasures are advisable. Right: Calculation of the risk indices of the individual risks and the total risk index.

In order not to jeopardize the success of the project, countermeasures should be defined for risks that have been assessed as unacceptable:

- Very high **risks** (high probability of occurrence and high damage) should be **avoided** wherever possible, for example by using a less risky technology.
- In some cases, **risks** can be transferred, for example by passing them on to suppliers or by signing up for insurance.
- **Risks** can be **mitigated** by reducing the probability of occurrence or the damage.
- Smaller **risks** can be **accepted** if they cannot cause significant damage.

Accepted and unidentified risks together constitute the **residual risks** of the project. In order to check the success of the measures, the risks are reassessed taking the measures into account and the new risk index or the new expected value is determined. A continuation of the example from Figure 3.65 can be seen in Figure 3.67.

Qualitative risk assessment						**cost of**	**new**		
risk	**root cause**	**P**	**D**	**RI**	**countermeasure**	**measure**	**P**	**D**	**RI**
Risk number 1: 1 week delay due to staff shortage	parallel project with the same bottleneck staff	4	3	12	coordination of the project plans for the 2 projects	4 h of work equals 600 US$	2	3	6

Quantitative risk assessment						**cost of**	**new**		
risk	**root cause**	**P**	**D**	**ExV**	**countermeasure**	**measure**	**P**	**D**	**ExV**
Risk number 1: 1 week delay due to staff shortage	parallel project with the same bottleneck staff	25 %	25,000 US$	6,250 US$	coordination of the project plans for the 2 projects	4 h of work equals 600 US$	5 %	25,000 US$	1,250 US$

P: probability of occurrence D: damage RI: risk index RI = P x D ExV: expected value ExV = P x D

Figure 3.67: Planning measures and reassessing a risk. In practice, you carry out either the qualitative or the quantitative assessment.

Example

- For a trade fair booth, an elaborate and movable construction is planned for an impressive product presentation. The risk of an error during implementation is considered to be very high. In order not to jeopardize the product presentation, a standard construction is used instead and the **risk** is thus **avoided.**
- When moving, the **risk** of furniture being damaged is **transferred** to the transport company's insurance.
- To reduce the **risk** of a resource bottleneck, two employees are always trained for all critical tasks in the project. This **reduces** the **probability** of a bottleneck occurring.
- There is a threat of rain showers at a company's summer party. To **limit the damage** to the party in the event of rain and thus reduce the risk, pavilions are rented and set up as a precaution.
- A video is to be made for an interactive support website for a product. An employee reports that the completion of the video may be delayed by a few days, as the video server has to be updated and unforeseen technical difficulties may arise. However, the basic functionality of the website is still available without the video. This **risk** is **accepted** by the project team.

Contingency and management reserve

Planning risk management measures is not enough: the measures must be monitored and checked for effectiveness. If the desired effect is not achieved, new measures must be planned.

One way of monitoring the development of risks and reporting on them as part of progress monitoring is to calculate the risk index. If the risks are regularly reassessed and the risk indices of the individual risks are added up and plotted along a time axis, the total should decrease over the duration of the project. This enables you, as the project manager, to check how large the remaining risks are and whether the measures are having the desired effect overall.

Risk management measures usually cost money or generate efforts in the project. The total amount of costs, effort or time required for this is usually called **contingency reserve**. It corresponds to a cost, effort or time-related buffer that is planned and monitored in the project to manage **known risks**.

As there are also **unknown risks** in every project, for which no specific risk management measures are planned, a second cost, effort or time-related buffer can be planned for this: the **management reserve**.

Successful project-oriented companies deal transparently with such buffers, plan them to an appropriate extent and monitor their consumption.

Warning

Many clients or managers don't want to hear about planned buffers. If a project manager plans for them, they are immediately cut back or canceled altogether. Later, you are surprised that foreseen or unforeseen risks occur and their consequences cost money or time.

It makes more sense to transparently plan, monitor and report on the uncertainties within a project, which always exist due to the degree of novelty. This reduces budget reallocations, crisis discussions and finger-pointing in case a risk occurs.

Figure 3.68 illustrates how the consumption of the contingency and management reserve can be visualized and monitored. The available residual buffer of the contingency and management reserve is recorded along the project's progress axis. At the start of the project, the project progress is still 0% and the entire planned buffer is available. The amount of buffer still required for the two categories is checked at regular intervals. If the remaining buffer is below the black bisector, the respective buffer will be used up faster than the project progresses. The buffer will probably not be sufficient and the project will take longer or require more costs than planned. Above the second bisector, less buffer is required than project progress is made.

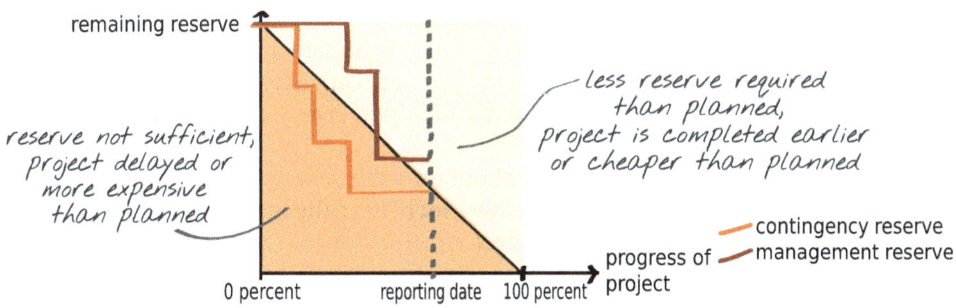

Figure 3.68: Visualization of the contingency and management reserve to monitor and control reserve consumption.

The advantage of this type of monitoring is that buffers that are no longer required can be released. This is particularly interesting for financial resources allocated for this purpose. The money can then be used for other projects and no longer needs to be held back for the particular risk. Conversely, there are still sufficient reserves available for the remaining risks of the project.

Project marketing

An old saying is: "Do good and talk about it." This also applies to projects. Don't just be happy about milestones reached and good interim results, but also report on them. As this involves communication with interest groups, project marketing is actually a sub-area of stakeholder management. However, I will briefly discuss a few specific aspects, here.
As a project manager, you are not only responsible for project definition, planning, controlling and closure, but also for ensuring that the project is perceived by relevant stakeholders in a way that promotes the project. The **objectives of project marketing** are:

- ensuring the success of the project
- increasing customer satisfaction
- increasing the satisfaction and motivation of project employees
- positive influence on the project environment.

Example

The new IviDi2 project to develop a new in-vitro diagnostic analysis device was launched with great enthusiasm. The head of development emphasized the strategic importance of the project at the start and provided the project with the planned resources.
The subsequent monthly project reviews initially went well. However, problems then increasingly emerged due to difficulties with a supplier. At the same time, another business unit requested the start of another project to develop a new patient monitor. It is said that customers are rushing to the competition because their product has a more modern user interface.
A few months later, the development manager begins to gradually take resources away from the IviDi2 project and allocate them to the patient monitor development project.

Example

As part of a project to build a paint shop for an external client, minor technical problems arise after an initially good start. The project manager regularly reports these to the client in a technically neutral manner. Although the problems are manageable and the project is still on time and on budget, the client increasingly perceives the project as poorly managed.
The client complains to the management about the project manager and demands more professional project execution. The client does not believe the management's assurances that the project is on schedule and on budget.

Below you will get to know some **typical methods** or media of project marketing. The table in Figure 3.69 provides an overview with the main areas of application.
Lectures/talks and publications at conferences and trade fairs are suitable for presenting (interim) results of the project to a broad professional audience of experts.
Especially in research projects, **science communication** is becoming increasingly important. The COVID-19 pandemic has shown how important it can be to present and communicate the latest findings to relevant stakeholders in an understandable way. Today, there are

Method	Effort	Frequency		Predominant suitability	
		once	regularly	within the company	external to the company
lecture/talk	+++	X		x	X
science communication	+++	X	X		X
information event	++	X	X	X	x
milestone meeting	++		X	X	X
newsletter	++		X	X	X
personal conversation	+++	X	X	X	x
press release	+	X	X		X
project identity	++		X	X	X
bulletin board	++		X	X	
phone hotline/chat group	+++		X	x	X
wiki/project website	+++		X	X	x
social media/networks	++		X	x	X

+	little effort	x	partially suitable
++	medium effort	X	suitable
+++	high effort		

Figure 3.69: Overview of the presented methods of project marketing and some characteristics.

experts in science communication who can present even complex issues clearly to the target group.

Information events inform internal or external stakeholders about the project. This often happens right at the start of the project. Information events can be organized as pure presentations. However, they can also involve participants in the form of workshops or surveys.

Achieved **milestones** and interim results can be celebrated. The setting ranges from simple announcements to small receptions and large celebrations.

Newsletters are a relatively simple way of making information available to many stakeholders simultaneously and regularly. Newsletters can be sent electronically, for example by email, or in paper form as a flyer. Due to the more cost-effective and simpler way of sending them, the electronic version is particularly common today. The ease of use of newsletters has led to a veritable overload of newsletters, which in turn means that we do not read all of them.

Personal conversations with key stakeholders help to explain the project, gather feedback and involve stakeholders in general. Don't leave these to chance, but think about who you want to talk to, about what and when. You should also find a suitable discussion format. Depending on the stakeholder, a formal meeting with a planned date or a seemingly chance encounter may be more effective. Of course, you will have to help a little to ensure that the coincidence actually happens.

Press releases reach a broad public. However, the dissemination of information is rather unspecific. Not all recipients of information are stakeholders in the project. Conversely, there is no guarantee that all relevant stakeholders will receive the message.

You can try to present your project like a brand. To do this, create a **project identity** by presenting a uniform image to the outside world. This includes an appealing and memorable project logo and a catchy project name. However, both must also be used consistently. Try to establish your project as a successful brand that makes an important contribution to the company, is well managed and is a pleasure to work on.

Even in the age of email and the internet/intranet, conventional **bulletin boards** still have an important function: they provide information. Set up in the right places in the company, preferably in high-traffic areas such as near coffee machines, they reach many employees. However, to ensure they are noticed, they must also be updated regularly.
For external projects, setting up a **phone hotline** or a **chat group** in one of the well-known **messenger services** can help to present the project positively. One example could be a hotline for concerned and affected residents living near a major construction project. However, the effort involved is considerable. With chat groups, you also need to make sure that they are not overloaded with annoying or inappropriate messages.

Warning

It is hard to imagine many areas of our lives without the use of messenger services. Whether school class groups, club activities or simply among friends – hardly any group today can do without its own chat group. Caution is advised in a professional context: Chats do allow quick accessibility. However, because they focus on short texts, they are a limited communication medium. Misunderstandings or conflicts often arise due to incorrect or inappropriate communication in the heat of the moment. In order to make the most of the advantages, it is advisable to define rules of conduct and insist that they are adhered to within the team.

A **project website** on the internet or intranet can help to communicate important information about the project and promote its positive perception. The integration of a **wiki** can also be an important building block in the context of knowledge management. Important information and experiences are stored in the wiki and can be easily accessed by others.
Today, there are many so-called **social media or networks**. The various providers differ in the way they pass on information. Examples include networks that focus on creating small websites, publishing short messages, sharing photos or distributing videos. For the professional context, there are dedicated platforms that report on companies, projects or individual employees. Well-made content can achieve a very high coverage. However, this also increases the risk that the smallest mistakes in communication will be picked up by critics and turn the actual intention of the message to be communicated into its opposite. Therefore, proceed carefully when using social media and networks and consult experts to ensure that your messages reach the desired target group and are received according to your intentions.
Like all project management activities, you should plan project marketing. Random marketing campaigns generate short-term attention at best, but not the desired long-term success. However, they can also have the opposite effect and annoy stakeholders. The **project marketing plan** should therefore define the following aspects:

- the concrete goals of project marketing
- the measures derived from them and
- the resulting work packages.

To avoid countless parallel plans, these work packages should be included in the work breakdown structure. This means that they are also included in the schedule, the resource plan and the cost plan, as well as in the monitoring of the project.

Quality management

Quality management in the context of project management is more than the definition of objectives for performance and quality. Some aspects with special reference to project management will now be presented.
Let's start with two definitions derived from DIN EN ISO 9000 standard:

> ***Quality*** *is the degree to which an object fulfills certain requirements.*

> ***Quality management*** *comprises the definition of quality policy, the derivation of quality objectives, their planning and control as well as the continuous improvement of quality.*

Quality management affects projects in two dimensions:

- quality management in relation to *project management*, its processes, roles and tasks and
- quality management related to the *project deliverables*, with the aim of identifying and fulfilling all requirements for these.

The quality of project management has a considerable influence on the quality of the project deliverables.

Goal

Quality management in projects aims to

- ensure that no requirements for the project are missed
- implement all requirements (remember the definition of quality!)
- minimize risks that may arise during implementation
- involve stakeholders in such a way that the success of the project is not jeopardized and
- hand over the project deliverables to the client in the desired quality at the end of the project.

The last point is the primary goal of quality management in projects: **The project deliverables are to be handed over in the required quality to the client who is satisfied with the project deliverables and the course of the project.**

PDCA cycle

Quality management is a continuous process of improvement. Deming and Shewhart have described this in the so-called **PDCA cycle**. PDCA stands for **P**lan, **D**o, **C**heck, **A**ct, see Figure 3.70 (left).

Quality management system

The call for continuous improvement stems from the realization that complex systems such as company processes are difficult to set up in an optimal state from scratch. It is more realistic

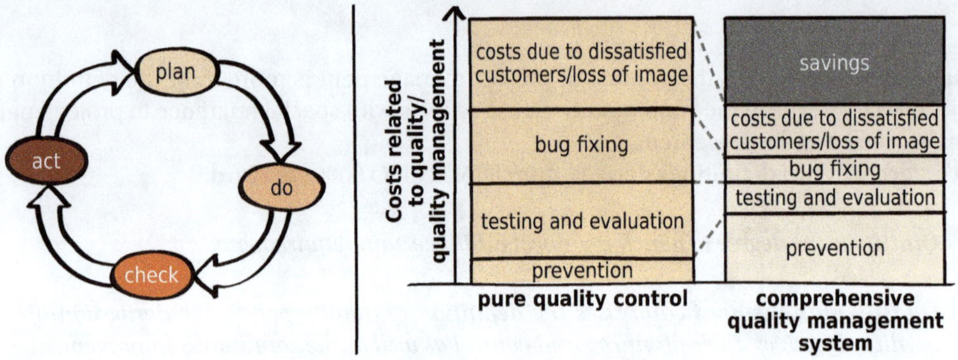

Figure 3.70: Left: The PDCA cycle according to Deming and Shewhart. Quality-relevant aspects are planned (plan), implemented (do), reviewed (check) and controlled (act) through corrective and preventive actions. On this basis, the continuous improvement plan is revised and the cycle begins again. Right: Cost savings related to quality and quality management through a comprehensive quality management system compared to pure quality control.

to scrutinize and continuously improve processes based on a certain status quo. This requires a quality management system that encompasses all of the company's organizational units. In addition, all employees must have a common understanding of the importance of quality. The costs incurred by the improvement are offset by savings through lower bug fixing costs, see Figure 3.70 (right).
I will briefly outline the four sub-tasks of quality management, quality planning, quality control, quality assurance and quality improvement *with reference to project management.*

Quality planning

During quality planning, you define

- what quality requirements you or the client have for the project
- how you measure and evaluate these requirements
- how (with which methods) and by whom (resources) the subsequent steps of quality control, assurance and improvement will be conducted.

The planned tasks must be incorporated into the respective work packages of the work breakdown structure.

Example

In the planning phase of the project, the conditions under which a supplier's software is integrated into the overall system are defined. A criterion is defined:

- A maximum of five errors rated as non-critical may be open.

Quality control

In quality control, the achievement of the targets defined in quality planning is regularly reviewed. In the event of deviations, measures are introduced to meet the targets.

Example

To ensure that the supplier's software from the previous example achieves the desired quality at the scheduled deadline, the open errors are continuously monitored and recorded, see Figure 3.71.

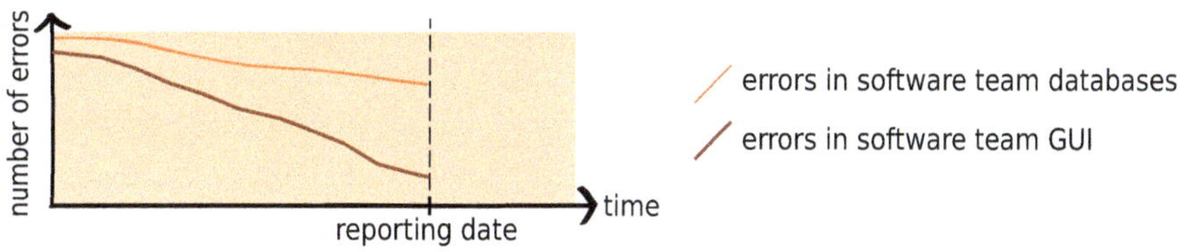

Figure 3.71: Exemplary progression of the number of open errors of two software teams. In the example, the number of open errors in the graphical user interface (GUI) team decreases significantly faster than the number of open errors in the databases team.

The diagram shows that the part of the software created by the GUI software team has significantly fewer open errors than the part of the software created by the database team. This causes the project manager to critically analyze the error history. He realizes that the reason for the sharp decline in GUI errors is not due to a higher quality of the software, but to a decrease in testing intensity. The GUI was hardly tested, but open errors were consistently processed and resolved.
The mere consideration of the number of open errors is therefore misleading if it is not checked at the same time whether tests are actually carried out to detect errors.

Quality assurance

Quality assurance comprises all planned and systematic activities that are necessary to achieve quality requirements. A typical activity is the incorporation of new findings into the quality management system. This prevents current quality problems from recurring in the future.
The documentation in the quality management system builds trust in the company's quality policy.

Example

The observations described in the previous example cause the project manager and the quality manager to define how software quality at suppliers should be evaluated in the future. Therefore, the template for new supplier contracts is changed. For new contracts, the number of permissible open, non-critical errors must be defined and it must be specified that the software must be fully tested at least once.

Quality improvement

Quality improvement aims to increase the quality level. Examples include defining a lower number of defects or achieving a defined defect level more quickly.

Example

As the project manager from the previous examples was not satisfied with the supplier's software quality for his project, he considers how this can be improved. For this purpose, he defines coding standards that the supplier must comply with in future. He also requests comprehensive code reviews for the next milestones according to the four-eyes principle, in which the program code is checked by at least one other expert.

Document management

Many documents are usually generated in projects, such as the initial project brief, requirements documents, plans, reports, change requests and orders, contracts, design and construction documents, program codes, marketing brochures, etc. The list is far from exhaustive. In addition, documents go through various stages (draft stage, released document, revised document, etc.), which further increases the number of documents.

Warning

At the start of a project, most employees are full of enthusiasm and want to start working on the project as quickly as possible. Documents such as plans and specifications are stored unstructured in a file folder. Typical file names are then Schedule_Project-XY_final.xlsx or Schedule_Project-XY_New.xlsx (which one is the current document?). After a few weeks at the latest, when the number of documents has continued to grow, they lose track and can no longer find important documents. They may continue to work with the wrong, out-of-date documents.

Three important terms in document management are:

> *A **document** contains information that is summarized in the form of text, images or video and audio recordings in a self-contained form. Documents are usually stored in paper form or electronically as a file or website.*

> *The term **documentation** refers to the complete life cycle of documents, from their creation and revision to their archiving and destruction.*

> *The **project file** refers to the set of all project-related documents.*

The aim of **document management** is to make documents available with reasonable effort. To achieve this, documents are labeled, registered and archived.
Today, database-supported software is usually used to store and retrieve documents. Good software for document management

- allows documents to be stored and accessed securely from any location (e.g. access via web interfaces)
- offers a rights management system that individually rules who can create, view, change and delete documents
- includes the possibility to edit documents simultaneously

- assigns version numbers to documents and makes it possible to view older documents or earlier versions of documents and
- creates backups of the saved documents to prevent data loss.

Smaller projects can manage without such software. Instead, documents can be stored in paper form in the project manager's office or electronically on a shared cloud storage space or a network hard disk. Make sure you have adequate data backups.

As a project manager, you should create a **project file** at the start of the project. This defines how documents are created and archived in the project. Today, it is common practice to provide documents electronically. A shared drive on the intranet or a secure cloud enables all team members to access documents and store their own documents. Without rules, however, document storage quickly becomes confusing.

You should therefore define the structure and organization of the project file. This includes

- an overview of all required documents
- a scheme for naming files and documents
- an overview of the folder structure of the project file
- a procedure for creating and approving new documents.

Warning

Storing files in a shared electronic drive is the simplest, but also the most dangerous way of storing documents. Without a rights management system, any employee can overwrite, change, delete and move files. There is now freely available or inexpensive software for document management, the use of which increases the security of the project file.

We will now look at a few tools for document management.

A **document overview list** defines, among other things, which documents are to be created in the project, who has to create them, whether there are templates for creating them and where they are archived. Figure 3.72 shows an example.

ID	name of document	author	due date	status	version	approved by	template	archiving
1.0	project brief	program mgr.	Mar 31st 2025	approved	1.2	management board	1.0_proj_brief_tmpl_V2.5.docx	/phase0_Init/0_pm/
...	...	...	...	...	...	...	...	...
1.6	req. specs.	req. engineer	Jun 15th 2026	draft	0.7	program manager	1.2_req_specs_tmpls_V3.0.docx	/phase1_definition_0_pm/
...	...	...	...	...	...	...	...	...

Figure 3.72: Exemplary excerpt from a document overview list.

The document overview list can also be used for project controlling by indicating the expected creation date, the status of the documents and the version number. As a project manager, you therefore have an overview of all documents and their status.

Documents should be labeled appropriately so that you can retrieve them later and distinguish current versions from outdated versions. Documents are usually labeled with the following information:

- project name and project number (if available)
- document type (e.g. protocol, report, plan)
- author and approving person (these should not be identical)

- creation and release date, date of validity if applicable
- status of the document (e.g. draft, approved, obsolete)
- version number
- number of total pages so that the completeness of the document can be checked
- unique document number
- archiving location
- if there are several versions: a change history of the most important changes from version to version

Released documents should be stored in a file format that cannot be changed accidentally, for example in Postscript (.ps) or Portable Document Format (.pdf). For additional protection, such documents can be signed with a verifiable digital certificate.

In smaller projects with limited complexity, it is usually sufficient to appoint an author for each document. The document is then usually approved by the project manager and stored in the project file.

The author and the approving authority should always be different people. This corresponds to the four-eyes principle. Involving an additional person can reduce errors in documents. The approval should be noted in the document.

More complex documents can also be subject to a **review process**:

- While the document is being created, it has the status "draft".
- Once the author has finished creating the document, they change the status to "in review" and send the document to peers for review.
- Changes are discussed and incorporated if necessary.
- The document is then submitted to the approving authority, which approves the document and changes the status to "approved".

Such reviews can be helpful for plans, requirements and design documents, for example, and help to improve the quality of the documents.

Another document management tool is the **document planning matrix**. It helps to determine the need for documents. Based on the defined project deliverables, typical document types are checked for relevance and entered into the document planning matrix. Figure 3.73 shows an example of such a matrix. The documents to be created can either be marked with

type of document	object / deliverable to be documented				
	casing	board/controller	operating software	...	overall system
requirements specification	not required	not required	1.1_func_spec_template_V2.0.docx	...	1.2_req_spec_template_V3.0.docx
technical design	1.4.1_component_design_template_V1.0.docx	1.4.1_component_design_template_V1.0.docx	1.4.2_software_architecture_template_V1.0.docx	...	1.4_system_design_template_V2.0.docx
...	...	...	...	...	...
user manual	not required	not required	5.2_sw_manual_template_V3.0.docx	...	5.1_user_manual_template_V4.0.docx

Figure 3.73: Example of a document planning matrix. The document types are entered in the rows of the left-hand column and delivery objects of the project are entered in the other columns. If a document type is required for a specific delivery type, the applicable template is listed in the corresponding field of the matrix.

a cross in the matrix or specified as template documents, as in the example. These make it easier for the author to create the document.

Configuration management

Document management helps us to bring order to the multitude of documents in the project. Configuration management has a similar objective, but is broader in scope and not limited to documents. As the term may be less common, I'll start with a few definitions:

> *A **configuration** describes the functional and physical characteristics of a product, a service or an object in general.*

> *A **configuration unit** fulfills one or more of the characteristics defined in the configuration.*

> ***Configuration management** ensures the transparency, traceability and consistency of all functional and physical characteristics of an object throughout its entire lifetime.*

There is a dedicated standard for configuration management in the form of EN ISO 10007.

Example

A manufacturer of sports equipment sells a slackline set consisting of a 10 m long and 45 mm wide slackline webbing, two slack ratchets, two ratchet paddings, a transportation bag and assembly and operating instructions. The configuration of the product consists of the functional and physical description of the product, i.e.

- the exact designation of the product components including serial numbers
- the characteristics of the product, for example the load capacity of the slackline webbing and the slack ratchets
- associated manufacturing instructions, construction drawings and design details, but which are not delivered to the end customer
- packaging information and
- certificates and documents proving the safety of the product.

The **objectives of configuration management** already result from the definition of the term. In order to achieve the desired transparency, traceability and conformity of requirements and features, the following process is common:

- **configuration management planning** – definition of the procedures, methods and responsibilities to be applied
- **configuration identification** – definition of a reference configuration and its description
- **configuration control** – controlling and monitoring changes to the configuration
- **configuration status accounting** – ensuring traceability by documenting the reference configuration and all changes released as part of the monitoring process

- **configuration auditing** – checking whether a product complies with its requirements and the documented configuration.

Figure 3.74 shows the process in context.
You can see interfaces to document management and change management from these process steps. In addition, agreed configurations form the basis for subsequent claim management within contract management.

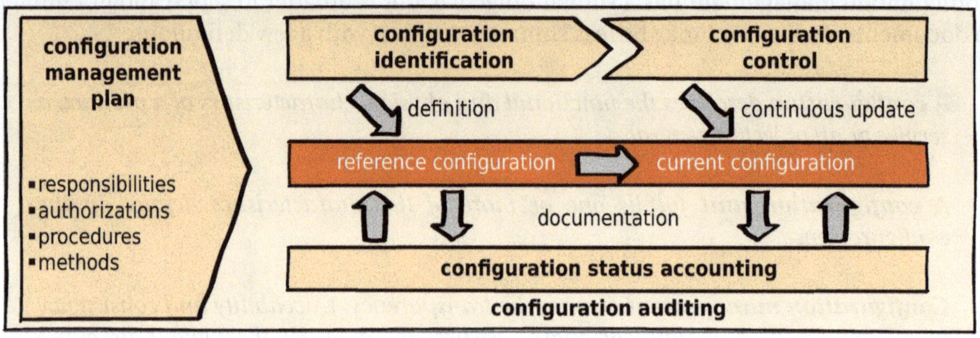

Figure 3.74: Overview of the individual aspects of configuration management.

The **configuration management plan** is part of the project planning. As a minimum, you should specify in it

- who is responsible for configuration management and what authority is assigned to this person
- how configuration identification, control and monitoring, accounting and auditing are carried out
- which components of the project deliverables are generally subject to configuration management
- how configurations are clearly labeled and versioned.

The last point involves defining a labeling system for the components of the configuration to be identified and how future changes in the labeling system will be taken into account.

Example

The manufacturer of the slackline set from the previous example defines the following labeling system for its products as part of the configuration management plan: NUMBER_X_VERSION.

- NUMBER is replaced by a unique identification number. This is structured hierarchically/decadally to indicate which configuration units belong together.
- X is replaced by P (product), C (product component), I (production instruction) etc.
- VERSION is replaced by a version number, for example 1.0. In the case of subsequent changes, only the number after the separating dot is incremented if compatibility with the associated other configuration units is maintained. For major changes without maintaining compatibility, the version number before the dot is incremented.

During **configuration identification**, a reference configuration is defined. This includes all product components, services, plans and documents associated with the configuration.
The reference configuration is the basis for the rest of the product lifecycle. Changes always refer to this basis and must be traceable back to this reference configuration with the help of configuration control and status accounting.

Example

The manufacturer of the slackline set presented in the previous examples specifies the following reference configuration:

1000_P_1.0 Slackline set

1000_Z_1.0 Slackline set - proof of compliance with relevant standards

1000_V_1.0 Slackline set - product description and packaging information

1100_C_1.0 Slackline webbing with a length of 10 m and a width of 45 mm

1100_I_1.0 Slackline webbing - manufacturing instructions

1200_C_1.0 Slack ratchets

1200_I_1.0 Slack ratchets - manufacturing instructions

1300_C_1.0 Transport bag

1300_S_1.0 Transport bag - product description of the supplier

1400_C_1.0 Assembly and operating instructions

...

Configuration control is closely linked to change management. If a component changes, this must be indicated in the configuration. Otherwise, it is not clear exactly which product is affected in the event of complaints or the supply of spare parts, for example.
The **configuration status accounting** ensures that the reference configuration and subsequent changes are documented in a traceable manner. You should document the original reference configuration in a document in the project file. Later changes to the configuration, stating the date of the change, the reason for the change and the effect of the change, then lead to an update of the document.
The **configuration audit** is used to check whether a product corresponds to the documented configuration. A distinction is made between

- *functional configuration audits* to check the functional characteristics and
- *physical configuration audits* to check the compliance of the physical characteristics with the documented requirements.

As with all audits, you should ensure that the auditors do not audit their own work. The configuration audit should therefore not be carried out by the person who is responsible for configuration management or who created the configuration.

Example

The slackline set that you already know from the previous examples is to be revised. The manufacturer wants to respond to customer requests for a longer webbing. In future, the set will be supplied with a 15 m long slackline webbing. The slack ratchets do not need to be replaced for this. The transport bag is also large enough to accommodate the longer strap. However, the assembly and operating instructions must be adapted and the proof of compliance with relevant standards must be reissued. The configuration is changed as follows:

1000_P_1.1 Slackline set

1000_Z_1.1 Slackline set - proof of compliance with relevant standards

1000_V_1.1 Slackline set - product description and packaging information

1100_C_1.1 Slackline webbing with a length of 15 m and a width of 45 mm

1100_I_1.1 Slackline webbing - manufacturing instructions

1200_C_1.0 Slack ratchets

1200_I_1.0 Slack ratchets - manufacturing instructions

1300_C_1.0 Transport bag

1300_S_1.0 Transport bag - product description of the supplier

1400_C_1.1 Assembly and operating instructions

...

Change control

Even in well-planned projects with carefully defined and analyzed objectives, it is highly likely that changes will occur during the course of the project. The causes of changes in the project include

- changes to the boundary conditions, for example new laws, standards and directives
- new technologies or discontinuation of components that should become part of the product
- changed customer requirements, for example inspired by new competitor products or changed markets
- problems with the implementation of the planned concept.

Changes can be *necessary* and make the success of the project possible in the first place. They can be *meaningful* and make the product or service to be developed better and more competitive. However, changes can also cause serious difficulties for projects if they are introduced in an uncontrolled manner and implemented without knowledge of the effects of the changes.

Change control ensures the transparent and controlled integration of changes into the project. For this purpose

- the need for changes is identified
- the changes are described
- the effects of the changes are determined
- the change requests are evaluated and
- the changes are approved, implemented and verified if necessary.

Tip

Establish an orderly change management process even in small projects. Otherwise there is a risk that creative employees will simply implement changes in the form of product improvements without your knowledge. The consequences of the changes for the project scope, the project costs and the schedule may not be sufficiently investigated and the success of the project may be jeopardized.

A typical change control process is shown in Figure 3.75.

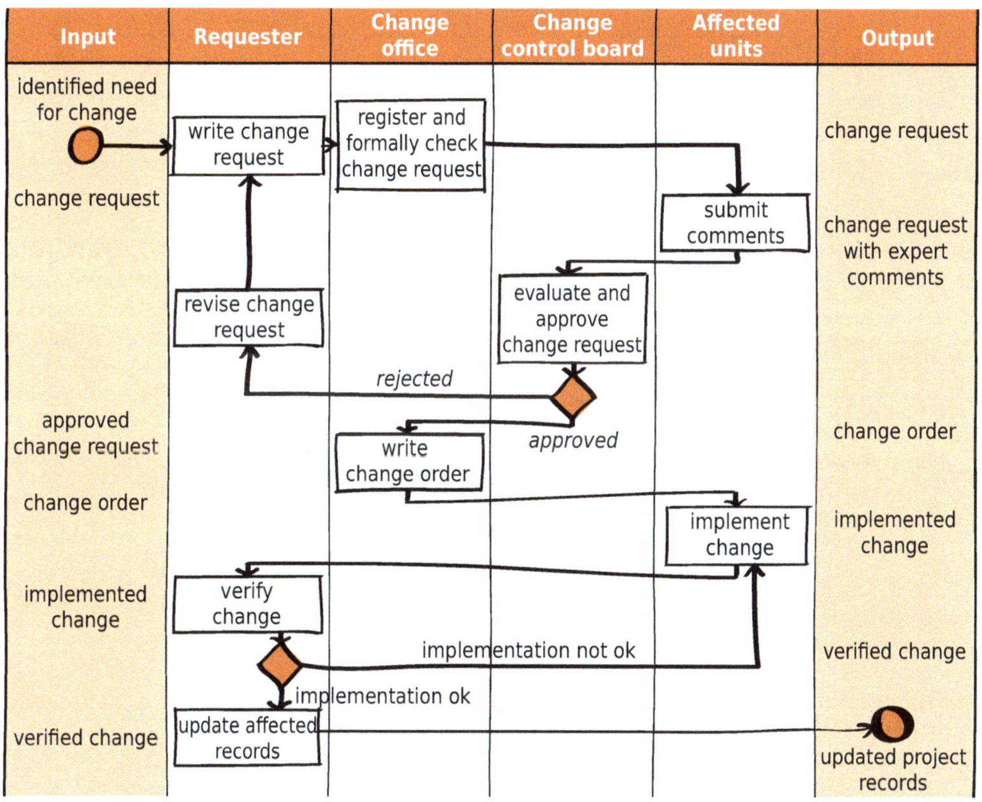

Figure 3.75: Overview of change control process.

If a need for change has been identified, the person making the request will submit a change request. A formal **change request** is shown in Figure 3.76. The change office checks whether it has been filled out completely and formally registers the request.

<table>
<tr><th colspan="3">Change Request</th></tr>
<tr><td rowspan="4">to be completed by the requester</td><td>Requester:</td><td>ID of request:</td></tr>
<tr><td colspan="2">Configuration unit to be changes:</td></tr>
<tr><td>Description and scope of change:</td><td>Reason for change:</td></tr>
<tr><td colspan="2">Effects (on deliverables, costs, schedule, effort):</td></tr>
<tr><td rowspan="2">to be completed by affected units</td><td colspan="2">Affected units:</td></tr>
<tr><td colspan="2">Comments:</td></tr>
<tr><td rowspan="2">to be completed by change control board</td><td>Need for the change:
low ○ ○ ○ ○ ○ high</td><td>Urgency of the change:
low ○ ○ ○ ○ ○ high</td></tr>
<tr><td colspan="2">Decision of change control board:</td></tr>
<tr><td rowspan="2">to be completed change office</td><td colspan="2">Result of verification:</td></tr>
<tr><td colspan="2">Change order approved:</td></tr>
</table>

Figure 3.76: Example of a form for requesting a change.

In order to assess the requested change and its effects, affected units are asked for their comments.

Tip

Record the effects carefully! This includes recording the significance of the change for the project deliverables. Typical questions to be clarified are

- Which components or parts of the project are affected?
- Does the change affect existing products or services, for example identical parts that are used in several products?
- Which interfaces between components and products are affected?
- Will the product be easier, more difficult, cheaper or more expensive to manufacture as a result?

In addition to effects on the project deliverables, effects on the project itself must be examined. What does the change mean for adhering to the project budget and schedule?

The **change control** board then decides whether the change should be implemented immediately. If the change is important but not urgent, it can be proposed to other projects in the future.

An approved change request leads to a **change order**, in which all affected departments are informed about the change and receive work packages on how to implement the change. The implementation of the change is then confirmed. If the confirmation is successful, all documents in the project file are updated and the change is completed.

How formally you organize the change control process in your project depends on the complexity of the project and the size of the project team. In the following, you will learn about some aspects that have proven successful in organizing change control. Adapt these to the needs of your project.

Tip

Very few projects fail because change control is too formalized. Conversely, however, many projects fail because they have implemented changes in an uncontrolled manner and have not considered all the effects of the change.
So if you are unsure, it is better to start with a slightly more formal process and then adapt it before you jeopardize the project's success through a lack of change management.

In simple projects, you can register and manage changes on a pinboard, a Kanban board or in an electronic file, such as a table in a spreadsheet software program. Every change request that deviates from the work breakdown structure and the reference configuration should be recorded and its effects evaluated.
In more complex projects, the use of special software for change management is recommended. This allows electronic change requests to be configured. Using integrated workflow management or a ticket system, the change request is forwarded from the requester via the change office to the change control board for processing. Each instance adds information that is fully recorded by the software.
Depending on the scope of the project, **change requests** can be made informally, pinned to a pinboard as notes or entered in an electronic table. Forms help the applicant to collect and enter all the important information so that a decision can be made quickly and competently on the request.
In many projects, the project manager is also the **change office**, which is responsible for organizing change control and formally checking and finalizing requests. In project-oriented companies, there may also be separate change offices, or change control may be taken over by the project office.
The **change control board** decides on the acceptance and rejection of an change request. As the effects of a change can be quite extensive, it is advisable to staff the change control board with appropriate experts. In technical development projects, the change control board is often chaired by the project manager or a person appointed by the project manager. Representatives from development, production, customer service, marketing, sales and quality management can supplement the board and assess the potential impact of the change on their respective areas. Additional experts can be invited as the situation demands. Try to keep the group of people as small as possible and as large as necessary. If you forget important stakeholders, the change may not be accepted. This leads to time-consuming follow-up discussions. In addition, missing stakeholders can lead to negative effects on the project being overlooked. Conversely, every unnecessary participant makes the change conference longer and more expensive.
The meetings of the change control board should be carefully prepared. Requests should be checked for completeness beforehand.

Depending on the scope and frequency of change requests, meetings of the change control board can be scheduled on an event-driven basis or at regular intervals. Before project planning is completed and a reference configuration is defined, there will be few or no change requests. Experience shows that more change requests are submitted as the project progresses. The reason for the increase is usually due to problems in the implementation of the project that lead to changes in the solution.

Contract management

As soon as organizations outside the company are involved in the project, for example suppliers and development partners, agreements are made in the form of contracts. However, cooperation between departments within the company can also be governed by project agreements, service level agreements or other contracts. Project managers are usually not trained legal experts. However, you should have some basic legal knowledge and knowledge of contract management in order to be able to work together with the legal department or the procurement department.

> ***Contracts*** *govern the relationship between two or more parties. They define what is to be provided by whom to whom and which rules apply. Contracts are the basis of legal transactions.*

> ***Contract management*** *is one of the tasks of project management. It serves to prepare, conclude and process contracts.*

Contract management supports you in achieving the project goals by

- concluding the right contracts in a form that serves the project
- checking compliance with the relevant parts of the contract during project management and
- carefully documenting contract deviations and demanding corresponding claims from the contractual partners.

The last point is already part of claim management, which is closely linked to contract management and the basics of which you will also learn about in this chapter.

> *A* ***claim*** *is a contractual demand that can be made on the other party of the contract. This requires that the claim is already defined in the contract, in an amendment agreement, or an approved change order.*

> *The task of* ***claim management*** *is to monitor deviations and changes to contracts. These are evaluated with respect to their consequences for the project and, if necessary, enforced against the contractual partner in the form of additional claims.*

Contracts are signed between the client and the contractor. Contracts can therefore work in two directions in a project:

1. Projects are engaged by a client.
2. Projects in turn commission suppliers and are themselves clients to them.

Contracts can be concluded *informally*, in the form of a *written document*, *orally* or *electronically*.
However, to make it easier to enforce contracts later, it is best to put them in writing. If a contract is agreed verbally, the contents should then be documented and signed by the contracting parties.
The following **minimum contents** should be documented in a contract:

- the exact name of the **parties to the contract** with address and authorized representatives
- a **preamble** explaining the initial situation is optional, but can be helpful for further understanding
- a precise description of the **object of the contract** and the associated **terms of payment**; terms that are not generally understood should be defined in the contract to avoid misunderstandings
- **prices** and **terms of delivery**, including delivery times and place of delivery (who delivers to whom and where)
- **conditions of acceptance**
- **legal consequences** in the event of violations of the terms of contract.

Tip

The last point is sometimes neglected in the misguided belief that an amicable agreement can be reached in the event of a dispute. In fact, it is usually easier to settle and document the consequences of violations of the contract by mutual agreement at the start of the contract. This reduces the risk of later conflicts.

Contracts can be concluded as an **individual contract** or as a long-term **framework contract**. Framework contracts usually contain general agreements on payment and delivery terms as well as consequences in the event of violations of the terms of the contract. Individual agreements are concluded in individual contracts that refer to the framework contract with regard to the aforementioned general provisions.
There are different types of contracts based on national laws. These can differ from country to country, which is why we can only go into general aspects here.

Example

A company regularly purchases development deliverables from another company. There is a framework agreement that regulates

- the payment terms and conditions
- the type and scope of the documentation of the development services and
- the legal consequences of violations of the contract.

If the company wishes to have specific development deliverables agreed, a contract for work and services is concluded for this purpose, which refers to the framework agreement. The development deliverables and delivery dates are agreed in the contract for work and services. The framework agreement and the contract for work and services together form the binding legal framework for the collaboration.

Contract management usually goes through the **phases** of negotiation, contract signing, execution and conclusion.

In the **negotiation phase**, the contract is drafted and prepared for signing. Sometimes the negotiation phase is accompanied by a documented **declaration of intent** (also: letter of intent). Please note that such letters of intent usually have little or no legal force.

In principle, there is **freedom of contract** within legal limits. The legislator gives the contracting parties the freedom to decide whether and, if so, what should be contractually regulated. A contract usually consists of a **offer**, which defines the service or work to be provided by one contracting party, and a **offer acceptance** by the other contracting party. In order for an offer to come into effect in the sense of a contract, the party making the offer must demonstrate a **will for legal binding**. Formulations such as "non-binding offer" or the mere sending of a price list do typically not indicate any intention to be legally bound.

Contracts are usually not written by project managers, but by the company's legal department or commissioned lawyers. Nevertheless, as a project manager, you should check the contracts as part of the **contract analysis**, because:

- Deadlines agreed in the contract must fit in with your project schedule.
- The content of the contract and corresponding services must match the deliverables of the project and the project work packages.
- Costs and possible penalties have an impact on the project costs and must be taken into account accordingly.
- The contract may contain risks for the project, for example in the form of exit clauses or limited liability assumptions.

The contract is concluded in the **contract signing phase**. A contract is concluded through the *unconditional acceptance* of an offer.

Example

A company asks a supplier of electronic components for a quotation for an order quantity of 10,000 units. The supplier submits a binding offer at a price of 75,000 US$. This price is too high for the company. It returns the signed offer, but changes the price to 70,000 US$. As the supplier's offer was not accepted unconditionally, no contract was concluded. Instead, the company has submitted a counter-offer at a price of 70,000 US$. If the supplier accepts this counter-offer, the contract is concluded.

Caution: Silence by the supplier cannot typically be interpreted as acceptance of the counter-offer.

Offers can be limited in time. The offeror's intention to be legally bound expires after this period.

In the **execution phase**, you must check the fulfillment of the contract. Deviations must be recorded and complaints made. One option for complete documentation is a *contract diary*. In this diary, you record all activities and contract-related correspondence. This includes, for example, meeting notes to document verbal communication, all correspondence (letters, e3mails, chat logs), photos of good or damaged deliverables, documentation of service disruptions, acceptance reports, changes related to the contract and subsequent claims.

> ***Default in performance** arise from violations of obligations by contracting parties. Defaults can be, for example, incomplete deliveries, delays or poor performance.*

If there is a default in performance, you must grant the other party a reasonable period of time for **supplementary performance**. If the other party does not comply with this, you can:
- conduct a **fulfillment by yourself** and charge the other party for this
- **demand a reduction** of the costs or
- **terminate the contract**.

Agreed **contractual penalties** and justified **claims for damages** can also be raised.
Pay attention to **statute of limitations** when asserting claims. Many national laws specify corresponding time limits. A period of 3 years is common, for example. This begins at the end of the year in which the claim occurred and the defendant party to the contract became aware of it. Other periods may apply in special cases and contracts.
The contractor has the **right of acceptance** of the deliverables of the contract by the client. The acceptance should be recorded and open points and defects documented. A deadline for correcting these should be set. If necessary, you should set an additional acceptance date to check that the defects have been eliminated.
Acceptance can take place in stages, for example after delivery, after a trial run or before the expiry of warranty periods.
The usual rule for acceptance is that
- payments become due
- the risk is transferred from the contractor to the client, i.e. the risk associated with accidental deterioration of the service is transferred to the client
- the burden of proof for defects is transferred from the contractor to the client and
- warranty periods for defects begin.

As national laws sometimes specify different regulations, you should always familiarize yourself with the country-specific laws.

Knowledge management

You have already learned that projects are temporary endeavors. It is therefore particularly important that the knowledge acquired in projects is not also temporary, but is passed on to important stakeholders. In order to better understand and differentiate between the tasks associated with knowledge management, we will first define a few terms:

> *An **information** is an intangible good that has a value for the recipient (of the information).*

Information is transmitted from a sender to a recipient via a channel (speech, writing, etc.). The recipient can use the information to derive actions and make decisions.

> ***Data** are characters (numbers, letters, etc.) that describe a certain issue and are administered on a data carrier (paper, hard disk, etc.).*

Projects typically generate a lot of data. Examples include requirements, plans and protocols. The data can represent information, i.e. represent a value for the reader of the data.

***Knowledge** is based on information and data and enables a person to solve problems.*

We often use our own experiences for our knowledge, for example information about past successes and failures. In order to manage knowledge, we therefore need to collect, process and utilize (valuable) information and data. Knowledge management is a very powerful management tool, as it enables organizations to learn from experience, avoid mistakes and continuously improve. Despite this theoretically plausible insight, knowledge management is often neglected in projects. Reasons for this include that:

- it partly lacks the infrastructure to collect, process and utilize knowledge without great effort and at the right time and
- knowledge management predominantly has a long-term effect, i.e. beyond the end of the project. The project participants therefore often do not feel any immediate benefit for their own project.

In some projects, lessons learned are therefore written down at the end of the project in an alibi-like manner, which then disappear into a drawer. Their potential for future improvements therefore remains unused. For knowledge management to be effective, however, it must be firmly anchored in the project lifecycle and continuously practiced.
The typical components of comprehensive knowledge management are illustrated in Figure 3.77.

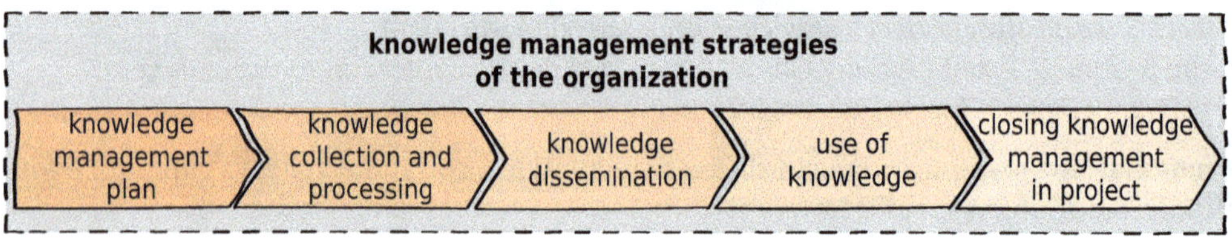

Figure 3.77: Knowledge management in projects.

Knowledge management in projects should be based on the company's **knowledge management strategy**. This strategy can include objectives and general conditions as well as cross-project contexts. Establishing knowledge management throughout the company also ensures the necessary support from the company management. Objectives can, for example, relate to specific areas of knowledge, exchange formats or the consideration of specific target groups.

Example

In its knowledge management strategy, a company defines that knowledge management should in particular promote the exchange of information across the boundaries of specialist departments. It also requires that certain data, such as the accuracy of effort estimates and the handling of risks, be collected and taken into account when starting new projects. Project teams must set out in their knowledge management plans how they intend to achieve these goals in their projects.

The **knowledge management plan** defines at project level how the strategy is implemented in the project. To do so, responsibilities can be named, tools to be used, such as specific software and document templates, can be specified and processes to be followed can be defined. Work packages can also be derived from the knowledge management plan and incorporated into the work breakdown structure.

Warning

The knowledge management plan is not necessarily one of the standard plans in projects. Nevertheless, it can provide valuable inspiration for effective knowledge management. However, the best plan is useless if compliance with it is not monitored and controlled.

Example

In the knowledge management plan, a project team specifies that it will store all minutes in a wiki software in the company cloud. The minutes can be searched using a full-text search. Easy to complete templates are provided for creating new minutes, so that the effort involved is minimal.
The knowledge management plan also specifies that a project reflection always takes place on the last Friday of the month. A duration of approx. 1 hour is recommended. In this reflection, the cooperation within the project team is to be analyzed and opportunities for improvement derived. The results are also documented in the above-mentioned wiki software and the initiated measures are controlled via it.
The plan also explains how important data is collected and also documented in the wiki software. For example, standardized work packages that regularly occur in projects are characterized using defined parameters and the required effort is stored as reference for future projects.

With **knowledge collection and processing**, you record the relevant information and data. Think about a format and a process for this that makes it possible to retrieve and use it later in a meaningful way.
The **knowledge dissemination** ensures that the recorded knowledge is distributed. This can be done actively or passively. In active knowledge dissemination, relevant stakeholders are informed directly, for example through a newsletter or a presentation. In passive dissemination, knowledge is made available for retrieval by stakeholders. They can then access it as required.
Collected and disseminated knowledge is pointless if it is not used. Only **knowledge utilization** avoids recurring errors and the continuous improvement of working methods in the company. Therefore, integrate the use of existing knowledge firmly into your routines. One of the best ways to do this is to integrate knowledge into document templates and software that are to be used anyway. For example, if certain tasks are repeatedly forgotten by project participants, these tasks can be integrated into the template for creating a work breakdown structure.

Tip

In practice, knowledge management often fails because it is cut back in favor of other activities due to a lack of priority in the hectic daily project routine. Therefore, set yourself realistic goals that you can also implement on the long-term. Create a user-friendly infrastructure in the form of good templates, processes and supporting tools so that the effort required to carry out knowledge management remains low. Finally, you should always remember that knowledge management is only useful if knowledge is not only archived, but also retrieved.

At a glance

- Plan-based, traditional project management refers to a plan-oriented project management in which deviations from the plan are minimized during the control phase.
- A process model combines methods and elements, processes and phases for a standardized project process.
- Well-known process models for plan-based project management include the waterfall model, the V-model, concurrent engineering and the spiral model.
- All plan-based process models have in common that they use plans to map the future course of the project up to the end of the project.
- Most plan-based process models structure the project into an initialization phase, a definition phase, a planning phase, a control phase and a closing phase.
- There are suitable project management methods for each phase, which were presented in this chapter.
- In addition, there are methods and ongoing tasks that cover all phases, such as stakeholder and risk management, project marketing, quality, document, configuration management, change control, contract management and knowledge management.

Practical examples and case studies

Below you will find some examples or case studies that illustrate the practical application of plan-based process models. Some case studies have key questions, the answers to which can be found at the end of the book. You should answer the key questions in the middle of the case study before reading on.

Case study 3.1: Exercises for project initialization and definition

Exercise 3.1.1

Can you name important obstacles when a project begins without a defined start? What are the immediate problems for the project manager?

Exercise 3.1.2

Think about a case in which a combination of top-down and bottom-up preparation of a hierarchy of objectives can be useful.

Exercise 3.1.3

Assign each of the following two sentences to either the user needs or requirements specification and give reasons for your assignment:

- The system must be able to print customer addresses on labels.
- The system has a USB 3.0 port to which printers can be connected.

Exercise 3.1.4

Evaluate the wording of the following terms and correct them if necessary:

- Milestone: Project plan
- Task: Hire architect
- Phase: Project closing phase

Exercise 3.1.5

Briefly explain the relationship between the phase model and the phase plan.

Exercise 3.1.6

Give an example of a project type that typically uses an functional project organization.

Exercise 3.1.7

Why should you create a RACI matrix and why should you do this at an early stage?

Case study 3.2: Exercises for project planning

Exercise 3.2.1

You are planning a two-day company outing as a project. It should include a joint arrival and departure, a group activity and a dinner together. The following tasks have been identified for the project:
"Select restaurant", "Book music band for dinner", "Select rafting organizer", "Conduct rafting", "Conduct safety briefing", "Carry out dinner", "Book bus company", "Arrival", "Departure", "Buy drinks and snacks for arrival and departure".
Create a complete object-oriented work breakdown structure.

Exercise 3.2.2

Calculate the following network diagram and mark the critical path.

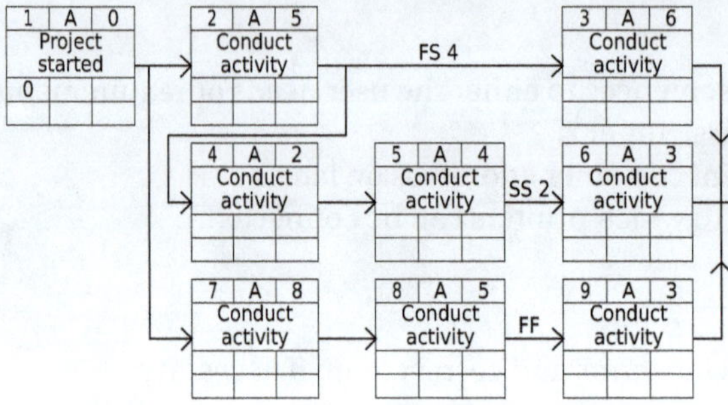

Exercise 3.2.3

Convert the network diagram from exercise 3.2.2 into a Gantt chart (earliest positions only).

Exercise 3.2.4

Calculate the following network diagram and mark the critical path. The colored markings are only required in the following exercise 3.2.5 and can be ignored for the calculation of the network diagram.

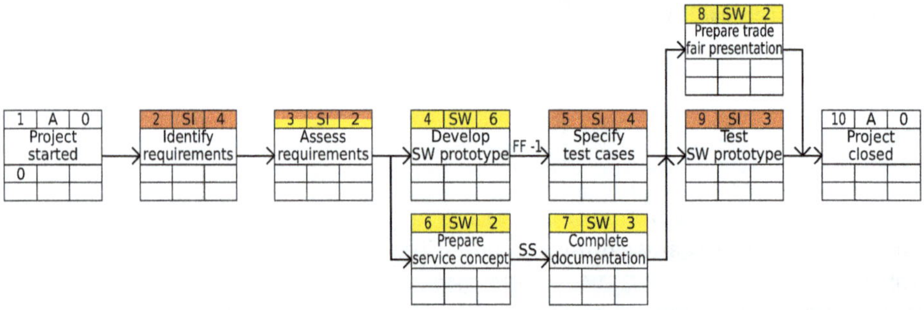

Exercise 3.2.5

Assume that the duration of the activities in the network diagram of exercise 3.2.4 is given in working days. Also assume that one software developer works in each of the activities marked in yellow and one systems engineer works in each of the activities marked in orange. In processes with both color markings, one software developer and one systems engineer are working. One working day corresponds to eight hours of work and 100 US$ are charged per hour per employee.
Based on this information and using the data from the network diagram in exercise 3.2.4, create a time-resolved resource and cost plan.

Exercise 3.2.6

Give a practical application example for linking two activities for each of the relationships start to start, finish to finish and start to finish.

Exercise 3.2.7

The development department of a company has a total budget of 10,000,000 US$. This is divided into three projects and so-called overheads (for general infrastructure, line functions, etc.) as follows:
Project A: 2,000,000 US$
Project B: 4,000,000 US$
Project C: 1,000,000 US$
Overheads: 3,000,000 US$
Name at least three alternatives to include overhead costs in the project cost plan and calculate the respective share.

Case study 3.3: Exercises for project control

Exercise 3.3.1

Consider the following work packages of a project:

- WP1: Creation of a functional specification based on a specification sheet comprising 200 requirements
- WP2: Creation of a complex prototype with software, hardware and documentation components
- WP3: Creation of a complex prototype with software, hardware and documentation components, divided into 20 separate activities

Which method would you use to determine the percentage complete based on the information provided?

Exercise 3.3.2

Assume a project consists of three work packages whose respective percentage complete PC_i and budget at completion BAC_i are given as follows:

- WP1: $PC_1 = 50\%$ $BAC_1 = 2{,}000$ US$
- WP2: $PC_2 = 20\%$ $BAC_2 = 5{,}000$ US$
- WP3: $PC_3 = 80\%$ $BAC_3 = 1{,}000$ US$

What is the overall percent complete PC of the project?

Exercise 3.3.3

Look at the following diagram of a milestone trend analysis, see also fact sheet M.3.3 in the toolbox in Chapter 8, and explain what risk to the project you can see.

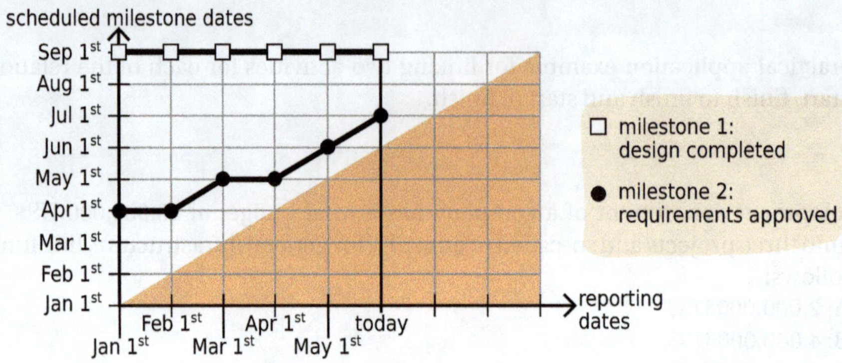

Exercise 3.3.4

Look at the following diagram of a cost trend analysis, see also the fact sheet M.3.15 in the toolbox in Chapter 8. Briefly explain and interpret the development of cost items 1, 2 and 3.

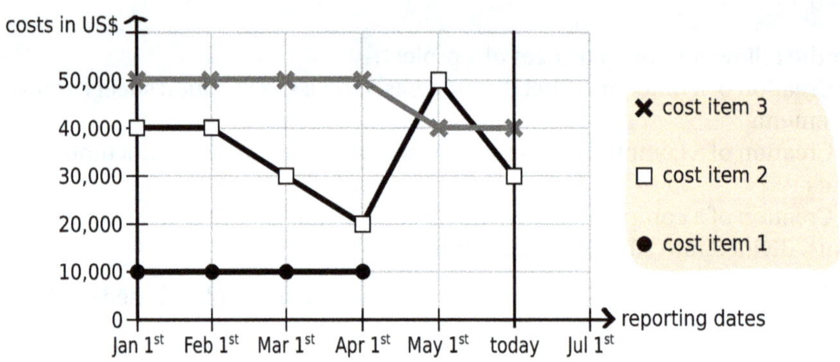

Exercise 3.3.5

The total costs of a project are budgeted at 100,000 US$. The project controller reports to the project manager costs of 80,000 US$ already spent. According to the plan, however, 40,000 US$ in costs are still pending and will be charged as the project progresses. Concerned, the project manager determines the percentage of completion of the project and calculates a value of 50%.
Calculate the cost and schedule development index and interpret the values obtained.

Exercise 3.3.6

A development project is significantly behind schedule shortly before the planned end of the project. The client wants to reduce this delay by providing the project manager with additional staff for his project team. Why can this measure actually increase the project delay in this situation?

Exercise 3.3.7

Take a look at the following illustration. The three milestones MS 1, MS 2 and MS 3 are listed along the time axis. The vertical axis shows the probability of the milestone occurring on the date entered.

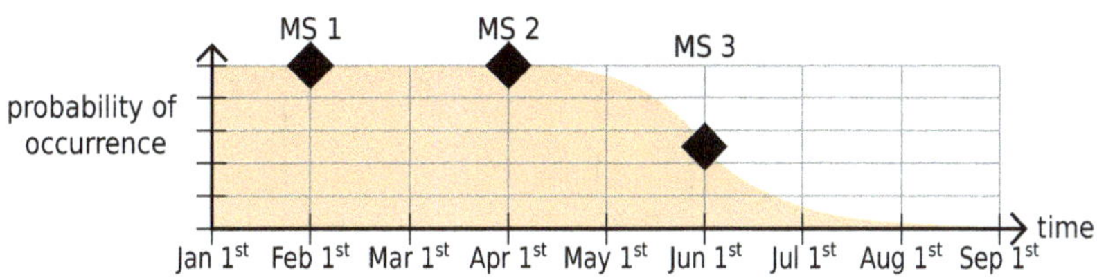

Milestones MS 1 and MS 2 have already been passed, which is why the probability is given as 100% in each case. For milestone MS 3, it can be seen that the milestone will occur on June 1st with an estimated probability of 50%.
This type of milestone reporting is not very common, but offers advantages in certain situations. Can you think of such a situation? How can this type of presentation be perceived by the report recipients (clients, portfolio managers, etc.)?

Case study 3.4: Exercises for project closure

Exercise 3.4.1

You have identified important project risks as part of project risk management. How can you pass on this experience to future projects?

Exercise 3.4.2

An employee suggests the following question for the evaluation questionnaire for the project's customer. How do you rate this question? Would you include it in the questionnaire?

4. How satisfied were you with the supplier management and contract administration?						
very satisfied	++	+	o	-	--	not at all satisfied

Case study 3.5: Exercises for ongoing project management tasks

Exercise 3.5.1

What advantage does the combination of brainstorming and STEEP analysis offer compared to simple brainstorming when identifying stakeholders and risks?

Exercise 3.5.2

Consider the following risk and its countermeasure and assessment. Would you implement the measure or not? Give reasons for your decision!

risk	root cause	P	D	ExV	countermeasure	cost of measure	new P	new D	new ExV
Reworking/ changes due to errors discovered in the test phase	New technologies and complex design	15 %	20,000 US$	3,000 US$	Build a prototype first	Personnel and material costs 5,000 US$	5 %	20,000 US$	1,000 US$

P: probability of occurrence D: damage ExV: expected value ExV = P x D

Exercise 3.5.3

The expected value of all risks in a project is 100,000 US$. The measures for risk management cost 25,000 US$. The measures can reduce the expected value of the reassessed risks to 40,000 US$. What cost-related contingency reserve and what cost-related management reserve are you budgeting for?

Exercise 3.5.4

How can the introduction of a comprehensive quality management system result in savings?

Exercise 3.5.5

Why should the archiving location and the total number of pages be indicated on documents?

Exercise 3.5.6

How does documentation and configuration management relate to each other?

Exercise 3.5.7

Think of one example of a change that is rated as absolutely necessary but not urgent and one that is rated as not necessary but very urgent.

Exercise 3.5.8

When a medium-sized plant engineering company retrospectively analyzes its most recent projects, it discovers that many projects have failed to meet their profitability targets. The investigation into the causes reveals that the reason for this is due to many changes that the company implemented at the customer's request during the project period. These were introduced at the project management level and implemented by the responsible engineers in the project teams without further consultation. The engineers justify this with the desire to offer the client a good service and to work in a customer-oriented manner. In fact, the customer rated the cooperation as very constructive and courteous.

Based on this information, how would you rate change control? What opportunities for improvement do you see?

Exercise 3.5.9

A supplier submits an offer to a company for a specific development service (contract for work and services). The supplier limits the offer to three months after the company receives the offer. After four months, the company accepts the offer. Does this constitute a contract? If yes, why? If not, how could a contract be concluded?

4 Agile project management

In this Chapter ...

- you will learn about agile project management,
- you will learn how this differs from plan-based process models and
- you will acquire the skills to apply Scrum, Kanban and other agile process models in your own projects.

Agile manifesto and agile values

A quick look into the past

Anyone who has already been involved with agile project management knows that there is no such thing as (one) agile project management. There are agile values and principles as well as a philosophy of agile action. And then there are many different so-called agile process models, methods and tools that can be used in project management. But can process models, methods and tools really be agile? Actually, it should probably be called agility-promoting project management. This is because certain process models, methods and tools can promote the application of agile values and principles. In the end, however, it is always the people involved who decide how agile the work actually is. Since "agility-promoting project management" is a rather cumbersome term, I also use the short form "agile project management" - but now you know what I mean.

The probably most well-known representatives of agile project management are **Scrum** and **Kanban**. However, there are also other process models, such as DevOps, Extreme Programming, Crystal, Feature-Driven Development and Test-Driven Development.

What we understand by agile project management today has existed for much longer than the term itself. Many ideas that we have known since the middle of the 20th century under the buzzword "lean" are important cornerstones of agile project management. This is why we also discuss **lean project management** in this chapter and look for similarities and differences to agile project management.

The publication *The New New Product Development Game*, which appeared in the Harvard Business Review in January 1986 [Takeuchi and Nonaka, 1986], is regarded as an important milestone in the development of today's understanding of agility. Interestingly, the terms agile or agility do not appear anywhere in the text, but the term *Scrum*, which Takeuchi and Nonaka derived from rugby, does. In rugby, scrum stands for a crowd. With the term, the authors wanted to emphasize the importance of *interactive cooperation and self-organization of a team* during product development.

Scrum as an agile process model, as we know it today, was largely shaped by Ken Schwaber and Jeff Sutherland and presented at the Object-Oriented Programming, Systems, Languages & Applications (OOPSLA) conference in 1995 [Schwaber, 1997]. This was followed in 2001

by *Agile Software Development with Scrum*, the first book on Scrum published by Schwaber and Beedle [Schwaber and Beedle, 2002]. In the same year, the *Manifesto for Agile Software Development* was published [Beck et al., 2001], to which Schwaber, Sutherland and Beedle also contributed.
Today, various agile process models are in use. In a study conducted by the German Association for Project Management, the participating companies rated Scrum as by far the most important process model for their own application [Komus and Kuberg, 2015]. It was followed by Kanban, Extreme Programming, feature-driven development, lean project management and other process models.
In this chapter, you will get to know the basic values of agile project management. Afterwards, the three representatives mentioned below will be presented:

- Scrum
- Kanban and
- lean project management.

In addition, you will get to know the scaled agile frameworks which enables you to use Scrum also in large projects. We will also look at other agile process models in a profile-style compilation. This will give you a good basis for using the agile process model that best suits your project.

Foundations of agile project management and agile manifesto

We learned about plan-based process models in the previous chapter. They attempt to predict the entire project lifecycle, map it to a plan and then follow it sequentially (one after the other) or, if necessary, incrementally (in partial steps). Agile thinking assumes that the project lifecycle cannot be predicted with sufficient accuracy at the start of the project and that mapping it in plans is therefore not a promising approach. In agile project management, we assume that the project result suffers if the project process is removed from any flexibility to meet customer requirements through rigid plans. Wysocki summarizes this in his lifecycle models as shown in Figure 4.1 [Wysocki, 2019].
In agile process models, planning is always iterative (in small steps). The entire project lifecycle is only planned very roughly, if at all. There is only ever a detailed plan for the near future. In an extremely agile variant (bottom line in Figure 4.1), even the project objectives are iteratively adjusted or detailed. Agile process models are therefore not considered plan-based, even though we will see later that plans also exist in agile project management. However, these usually do not extend very far into the future. Projects have a great deal of flexibility in their implementation to optimize the project deliverables in line with customer requirements.
A further distinction between plan-based and agile project management can be made by interpreting the iron triangle of project management, which is explained in Figure 4.2.
Agile project management has its roots in agile software development. The **values** documented in the Manifesto for Agile Software Development are still valid today and have long since overcome the limitation to software development projects. The authors and signatories value

- **individuals and interactions** over processes and tools
- **working software** over comprehensive documentation

- **customer collaboration** over contract negotiation and
- **responding to change** over following a plan.

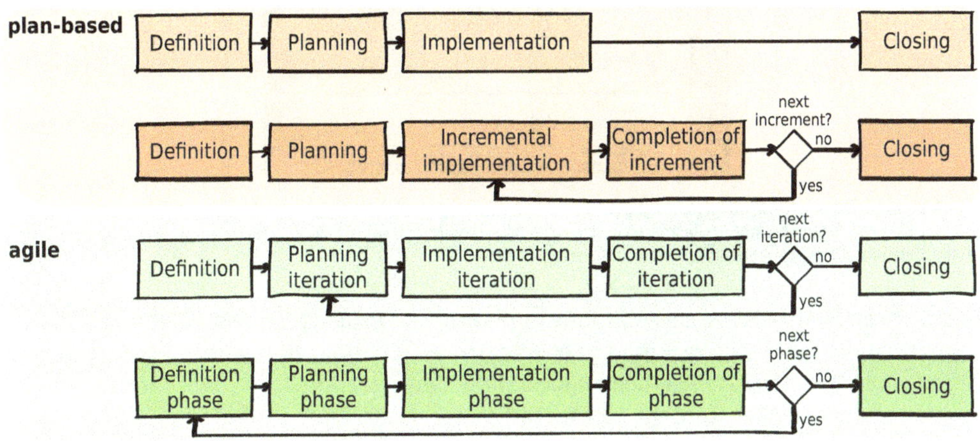

Figure 4.1: Plan-based (top) and agile (bottom) project lifecycle models according to Wysocki [Wysocki, 2019]. Plan-based projects define objectives and map the future in plans, which are then pursued in one piece (process at the top) or in steps (increments, second process from the top) until the planned project deliverables are completed. Agile projects plan iteratively, i.e. in small steps (third process from the top). The planned step is then completed and the next step is planned until the entire project deliverables have been completed. Alternatively, the project can be broken down into phases (lowest process) and a separate objective can be defined and separate plans created for each phase. The project deliverables are then prepared phase by phase.

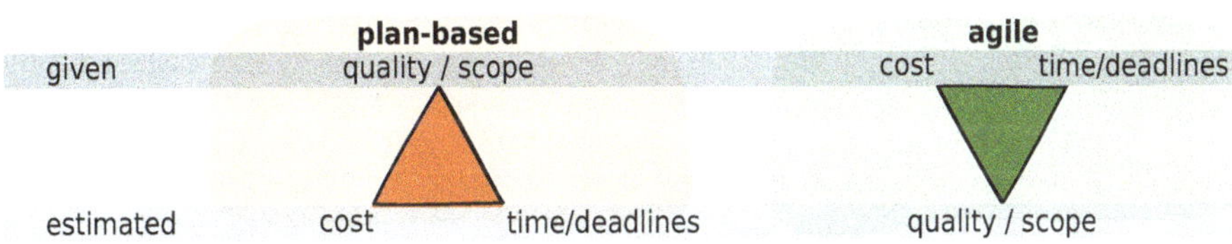

Figure 4.2: Iron triangle in plan-based and agile project management. In plan-based project management, the quality- and scope-related objectives are typically given by the client. Based on these, the costs and deadlines are estimated. In agile project management, the scope is negotiable and is defined and planned iteratively. For this, there are (usually) a fixed number of iterations and clearly defined resources (and therefore costs).

You can confidently generalize the term "functioning software" used back then by functioning product or result in the sense of the customer.
The agile values form the core of agile action. Figure 4.3 illustrates the connection between agile values, principles, methods and process models, which we discuss in detail below.

Warning

Applying agile methods or process models without taking agile values and principles into account does not lead to agility! We will discuss this aspect several times!

Twelve principles were developed and published in the agile manifesto to concretize the agile values and to help implement them in everyday project management. These can be seen together with examples from the often plan-based corporate practice in Figure 4.4.

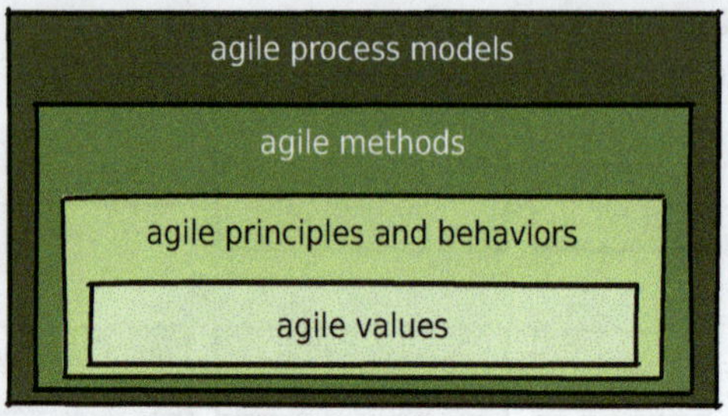

Figure 4.3: Agile values form the core of agile process models. They are concretized through agile principles and practices. Agile methods help us to implement the values and principles in the project. Agile process models combine values, principles and methods into a coherent project management system.

The practical examples are observations of plan-based projects and serve to illustrate how these principles can be disregarded in everyday life. Of course, there are also many positive examples of projects that have been successfully completed using a plan-based approach, resulting in a satisfied customer.

Tip

Even though this is primarily about agile project management, the presentation of agile values and principles will be followed by a look at agility that goes beyond the individual project.
Agility is a term with positive connotations: among other things, it stands for dynamism, speed, flexibility and customer orientation. It is therefore not surprising that many companies describe their management style or even their entire company as agile. Agility is in fashion.
At the operational level, there is another observation: many projects are labeled as agile simply because they work according to agile processes. In fact, a project that works according to plan-based processes, but whose project manager follows the values of the agile manifesto, can be much more agile than a project that follows agile process models (e.g. Scrum), but where the employees are stuck in traditional management structures.
True agility in the sense of the agile manifesto requires the consideration of two further aspects:

- The values and principles of the agile manifesto should be deeply integrated into the company's actions.
- Each company must find its own pragmatic way to implement these values and principles and bring them in line with economic and organizational boundary conditions.

The last point is particularly important: this book places great emphasis on the practical implementation of successful project management. In addition to the necessary theoretical principles, you will therefore always find a number of practical examples to give you ideas for your own application.

Agile working can have a profound impact on an organization and, for example, directly influence the way in which value is created and the nature of the business model. One example of this is the **(No) Money for Nothing, Changes for Free** principle, which many (but by no means all) agile companies have adopted and which I will briefly explain below:
The principle "(no) money for nothing" means that the client of the project may terminate the project after each iteration (agile projects are usually implemented iteratively in two or more iterations). This also applies if further iterations have actually been agreed. Payment is then only made for what has been created up to the (early) termination. If necessary, a certain amount can be agreed as compensation for the planning of future iterations. For the client,

Principles of agile project management	Frequent practice in plan-based projects
■ Our highest priority is to satisfy the customer through early and continuous delivery of valuable software (or in general: deliverables). ■ Deliver working software (or in general other deliverables)frequently, from a couple of weeks to a couple of months, with a preference to the shorter timescale.	■ Plan-based projects are plan-driven: The clients tell us at the start of the project what they want. The specified project deliverables are then implemented according to the prepared plans.
■ Welcome changing requirements, even late in development. Agile processes harness change for the customer's competitive advantage.	■ The plan-based approach leads to a negative attitude towards change. Changes to the project scope also require changes to existing plans.
■ Business people and developers must work together daily throughout the project. ■ Build projects around motivated individuals. Give them the environment and support they need, and trust them to get the job done.	■ Development teams are reorganized during the project lifecycle or work on several projects simultaneously (matrix organization). Due to changing assignments, the sense of own responsibility for the project deliverables can suffer. Typical countermeasure: stronger controlling.
■ The most efficient and effective method of conveying information to and within a development team is face-to-face conversation.	■ Plan-based process models focus on workflows. In everyday business life, e-mails sometimes replace personal conversations.
■ Working software (in general: deliverables without errors) is the primary measure of progress.	■ Project progress is often measured in terms of costs or resource utilization.
■ Agile processes promote sustainable development. The sponsors, developers, and users should be able to maintain a constant pace indefinitely.	■ In practice, many project managers are given great responsibility and little authority. This leads to project delays and escalations.
■ Simplicity - the art of maximizing the amount of work not done - is essential. ■ Continuous attention to technical excellence and good design enhances agility.	■ Complexity is supposed to be made manageable by plans which is not always successful. Late changes or compromises in quality can be the result.
■ The best architectures, requirements, and designs emerge from self-organizing teams. ■ At regular intervals, the team reflects on how to become more effective, then tunes and adjusts its behavior accordingly.	■ Lessons learned and project analysis at the end of the project are typical instruments of knowledge management in plan-based projects. Regular reflections tend to be the exception.

Figure 4.4: Principles of agile project management (left) according to the agile manifesto [Beck et al., 2001] and examples from corporate practice in plan-based projects (right). Agile project management focuses on individuals and interaction. Plan-based project management, on the other hand, focuses on plans and processes. Inexperienced project managers can neglect leadership and communication. This can lead to some of the problems shown in the right-hand column.

this model enables a very flexible way of working with low financial risk. The contractor, on the other hand, must be able to react flexibly to early termination of projects and dynamically transfer project capacities that become available to other projects. However, the ability to do this can represent a major competitive advantage over other companies.

The principle "changes for free" means that the client may make changes to the requirements after each iteration. The changes are free of charge if the requirements to be changed have not yet been implemented and the changed requirements have approximately the same scope as the original requirements. According to the same principle, requirements can also be completely exchanged for other requirements.

These principles can also be used to implement the **agile fixed price**. The agile fixed price is a contract model in which a budget (costs and time) is agreed at the start of the project. Initial considerations and tests of individual requirements can be carried out to ensure that this budget is as reliable as possible. As the budget is then firmly agreed, only the scope of the project can be adjusted as the project progresses. This can be managed using the principle of "(no) money for nothing, changes for free", for example.

These considerations are intended to give an impression that agile working can be more than the mere application of an agile process model. Further suggestions for agile working and specific process models follow in the rest of this chapter.

Scrum

Introduction

The first agile process model we look at is Scrum. I am closely following the Scrum Guide by Schwaber and Sutherland [Schwaber and Sutherland, 2020] and then add some examples of practical implementations. If you already know the basics of Scrum, you can also take a look at the case studies and practical examples below.

Scrum projects are carried out by self-organizing interdisciplinary teams. These teams deliver a product increment at fixed periods of time, so-called *sprints*, which is simply referred to as *increment* in the Scrum Guide. An increment is the result of a sprint. The term "increment" expresses that the overall product is created step by step. At the end of each sprint, an evaluation meeting called a *sprint review* takes place. In this meeting, the increment is evaluated and assessed from the customer's perspective. This is followed by further sprints with new increments until the product desired by the customer is finished.

In **Scrum theory**, the importance of gaining knowledge from experience (empiricism) and avoiding waste by focusing on the essentials (lean thinking) is emphasized [Schwaber and Sutherland, 2020]. The use of experience and increasing the value of the project object can be derived from this. Both can be found in numerous places in Scrum.

Scrum relies on the following **three pillars** to put this theory into practice:

- *Transparency*: The work done in Scrum projects and the artifacts that we will get to know must be visible to relevant stakeholders. Only then will it be possible to review them and ensure their success. This in turn is a prerequisite for avoiding waste.
- *Inspection*: In order to learn from experience and adapt the project object in terms of value maximization, a recurring inspection of the results and artifacts is necessary.

- *Adaptation*: New findings can lead to a need for adaptation. These must be made promptly.

In order to understand Scrum, you will first learn about the following aspects:
- The **Scrum roles** developer, product owner and Scrum master.
- The **Scrum artifacts** product backlog, sprint backlog and increment. Even though the Scrum Guide does not list the release plan and the impediment backlog as separate artifacts, they are widely used in practice, which is why we will also discuss them here.
- The **Scrum events** sprint, sprint planning, daily Scrum, sprint review and sprint retrospective.

For the explanation of Scrum, we are guided by Figure 4.5.

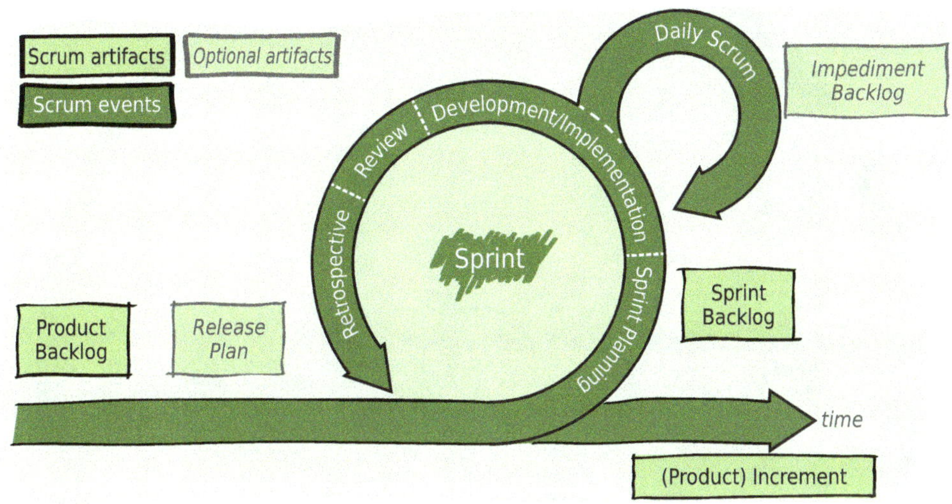

Figure 4.5: Overview of the Scrum events (sprint, sprint planning, daily Scrum, sprint review and sprint retrospective) and artifacts (product backlog, sprint backlog and (product) increment) defined in the Scrum Guide [Schwaber and Sutherland, 2020]. Although the release plan and impediment backlog are not included in the Scrum Guide, they are usually used in practice.

Before we take a closer look at the roles, artifacts and events, let's take a brief look at the **Scrum values** [Schwaber and Sutherland, 2020] defined in the Scrum Guide. These form the basis for joint action:
- *Commitment*: The Scrum team commits to the project objectives and supports each other in achieving them.
- *Focus*: The Scrum team focuses on the work required to achieve the goal.
- *Openness* towards the work to be done and new challenges characterize the joint action.
- *Respect*: The Scrum team respects all people who work together on the project.
- *Courage* to do the right thing and tackle difficult problems are part of daily project work.

We will now look at the effects of Scrum theory and Scrum values on the organization and implementation of Scrum projects using roles, artifacts and events.

Scrum roles

The Scrum team consists of the developers, the product owner and the Scrum master. The term "developer" testifies to the origin of the process model in software development. Generally speaking, the developers are the people who carry out the project in a self-directed manner and complete the tasks. In order to remain consistent with the Scrum Guide, I use the term "developer" here. The responsibilities, authorities and other characteristics can be seen in Figure 4.6.

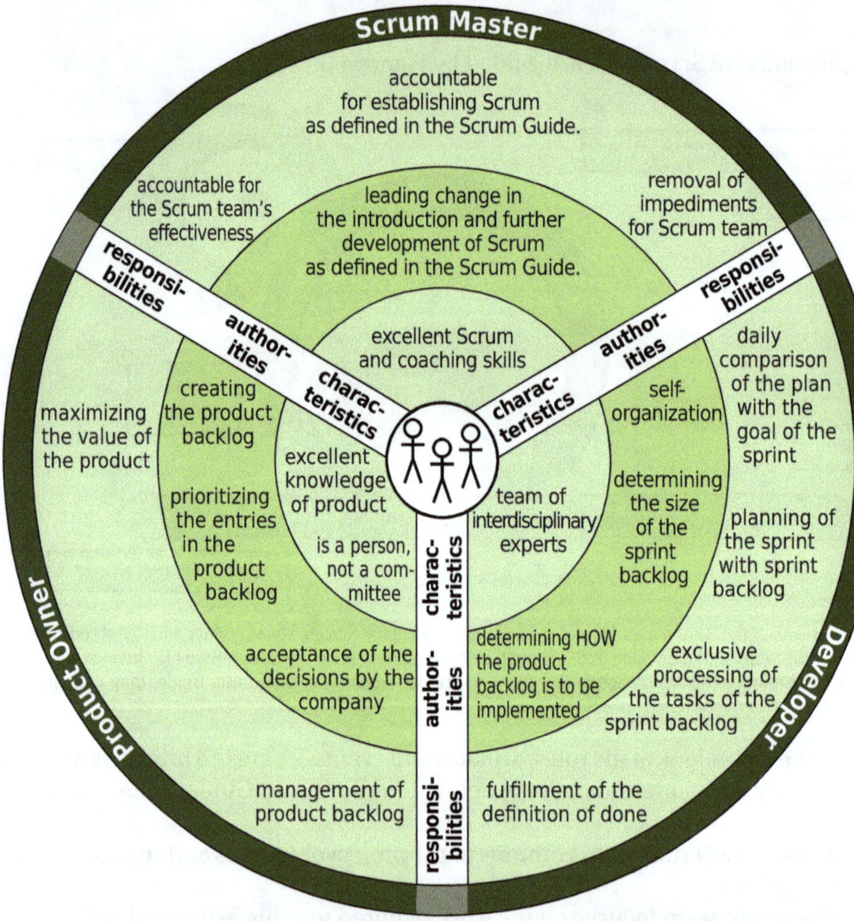

Figure 4.6: Responsibilities, authorities and other characteristics of the three Scrum roles, developer, product owner and Scrum master, who together form the Scrum team.

If you have mainly had experience with plan-based project management, you are probably missing the project manager. *In fact, there is no comparable role in Scrum.*

The entire Scrum team is responsible for the project deliverables, the so-called (product) increment and all product-related tasks. This includes, for example, stakeholder communication, testing and maintenance work.

Developer

The developers consist of interdisciplinary experts. They are responsible for implementing the user needs defined by the product owner in the product backlog. The developers alone determine how this implementation is to take place without any influence from the product owner (or other persons).

The developers organize themselves and work according to the *pull principle.* This means that tasks are not assigned by a project manager or a comparable instance. Instead, developers pick up their tasks from the task list, the so-called taskboard, independently. There is no hierarchy within the developers. The developers should comprise between three and eight people. For smaller teams, the question arises as to whether even a lightweight process model such as Scrum is not too bureaucratic. With larger teams, interaction and self-organization become difficult. In the chapter on scaling Scrum and in the practical examples, you will see an example of how you can deal with this limitation in practice.

Product owner

The product owner is responsible for the commercial success of the product to be developed and tries to maximize this success. He or she represents the relevant customer requirements from a business perspective and summarizes these in a so-called product backlog (i.e. a catalog of requirements). The individual user needs are prioritized in this product backlog and made available to the developers.

The product owner is a single person and not a committee of several people. The product owner represents the customer's interests, but should not be identical to the customer, as he or she is also responsible for the economic success (from the company's point of view) and works closely with the developers. The product owner has clearly defined tasks, such as the initial creation, maintenance and evaluation of the product backlog and the evaluation of the increments. Delegating these tasks to the customer is difficult in practice and critical from an economic point of view, as the company would completely relinquish control of the project scope. The decisions of the product owner must be accepted by the company. The product owner is part of the self-organizing Scrum team.

Apart from the product owncr, nobody is allowed to place requirements regarding the product on the developers. The product backlog mentioned here is a so-called artifact, i.e. an (interim) result in Scrum. I will go into this in more detail later in the chapter on Scrum artifacts.

Scrum master

The Scrum master is an expert in the planning and implementation of projects with Scrum. The Scrum master supports the developers in working effectively (and efficiently) and promotes the project and Scrum to people outside the team. The Scrum masters are both, promoters of agility and coaches. They support the product owner and the developers with all questions relating to the process model.

The Scrum master tries to remove obstacles during development that the Scrum team cannot solve on its own. You can find out more about this when I introduce you to the Impediment Backlog artifact. Across projects, the Scrum master acts as a change agent and drives cultural change towards an agile organization.

In companies with extensive Scrum experience and qualified teams, a Scrum master can coach several projects in the company at the same time. If Scrum has only just been introduced, the role of Scrum master can also completely overload one person. However, the aim should be to make the Scrum master as superfluous as possible and to carry out Scrum projects with qualified product owners and developers.

Other groups involved

Of course, there are other roles in real Scrum projects, but they are not part of the process model in a narrower sense. These roles include, for example, customers, managers, suppliers, teams from other projects, etc. I will discuss some of these roles in the practical examples and exercises.

Summary: Scrum roles

- The Scrum team consists of developers, product owner and Scrum master.
- The developers have three to eight members who organize themselves and plan the implementation of the requirements of the product backlog independently via the sprint backlog.
- The product owner is responsible for the economic success and strategic contribution of the product to be developed and summarizes the necessary user needs and requirements in the product backlog.
- The Scrum master is the Scrum expert, coach and escalation authority for problems that the Scrum team cannot solve on its own, but otherwise holds back as much as possible.

Scrum artifacts

Artifacts in the context of process models are intermediate or final results. In Scrum, these results are used for transparency, inspection and adaptation during the project lifecycle.

Product backlog

The first artifact presented here is the product backlog. In Scrum, this is where the requirements are listed:

> *The **product backlog** is the only source from which the Scrum team derives its work. It contains the overall product objectives and other entries in the form of prioritized requirements to achieve the product objectives. The product owner is responsible for the creation and continuous maintenance of the product backlog.*

Unlike the requirements specification in plan-based project management, the product backlog does not claim to be complete. It is a living collection that is updated and detailed during project implementation. Part of the product backlog are the overall product objectives, which provide the Scrum team with a guideline during the course of the project. The product objectives are substantiated by requirements, which are the main component of the product backlog. The continuous maintenance of the product backlog is also known as **product backlog management** or **product backlog refinement**, see Figure 4.7.

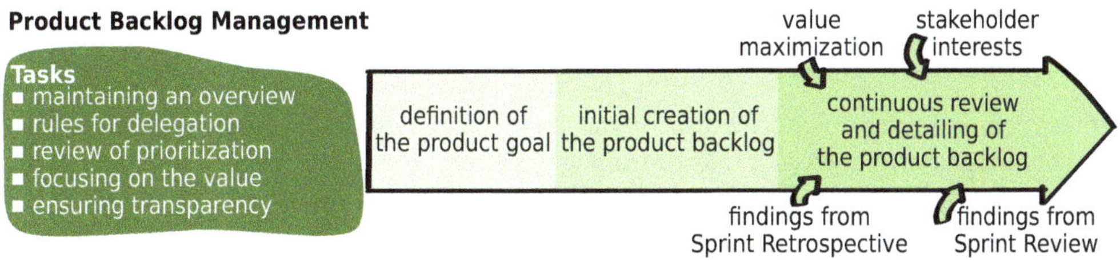

Figure 4.7: Overview of product backlog management: After the initial creation, the product backlog is continuously reviewed, revised and detailed. New findings on the value optimization of the product, from the sprint review and the sprint retrospective as well as from other stakeholder feedback are incorporated here.

One of the tasks of product backlog management is to maintain an overview of the product backlog. As an evolving collection of requirements, the product backlog can quickly become confusing. Organizing it according to specific requirement areas and priorities promotes clarity. The product owner is responsible for the product backlog. However, the individual tasks can be delegated. The prioritization of the entries in the product backlog can change over the course of the project and should therefore be reviewed regularly. The individual entries should also be analyzed again and again to determine whether they actually increase the value of the product. As the sole source of work, the transparency of the product backlog is particularly important. Relevant stakeholders need access to the information in the product backlog.

In Scrum, the requirements contained in the product backlog are usually formulated as user stories and epics.

> *A **user story** is a requirement that is written from the perspective of a specific role, often that of the user.*

A typical formulation is: "As a user, I want to be able to print documents with just one mouse click."

> *An **epic** is a large, still vague user story. The requirement can only be roughly outlined and its size (complexity) cannot be reliably estimated.*

Epics can serve as a placeholder or memo for ideas or rough requirements. Before they can be implemented and estimated in advance, however, they must be concretized further. An example of an epic is "Enable cancellation". This could later be developed into a user story. The user story derived from the epic would then read, for example "As a visitor to the event, I would like to be able to cancel my participation."

User stories can be formulated using the schema illustrated in Figure 4.8.

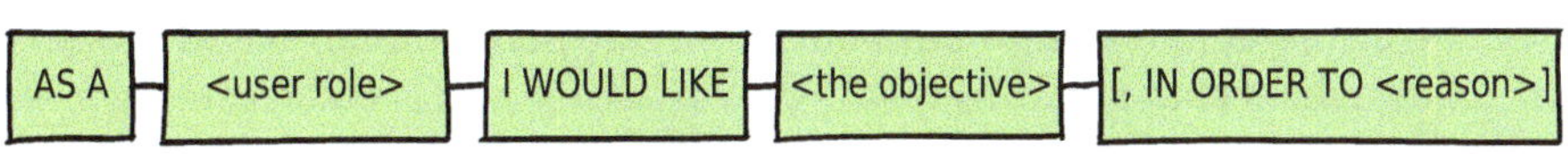

Figure 4.8: Scheme for the formulation of user stories based on Cohn [Cohn, 2004].

Example

Example of a formulated user story:
As a user, I would like to be able to change user data retrospectively in order to correct errors in the user data.

Similar recommendations apply to user stories as for normal requirements of plan-based project management (SMART): User stories …

- … should be **independent** of each other.
- … are **negotiable**.
- … have a **value** for the customer.
- … are **estimatable** with regard to their size.
- … should be **small**.
- … should be **testable**.

The initial letters of the terms marked in bold result in the acronym INVEST and can serve as a mnemonic [Cohn, 2004].
User stories can be collected on metaplan cards (cards made of paper or cardboard on a pinboard or whiteboard) and gradually provided with additional information to make them more concrete, see Figure 4.9. Alternatively, there is now also software that can manage electronic cards. Options for electronic and online-supported implementation are available, for example, with

- Jira and Confluence from Atlassian
- virtual (online) whiteboards, for example Conceptboard, Mural or Miro
- Office software, such as Microsoft 365, LibreOffice, OnlyOffice, Google Docs, Apple iWork.

In addition, there are countless other providers for the organization of Scrum-typical tasks, such as Trello, Asana, Zenkit and Meistertask.

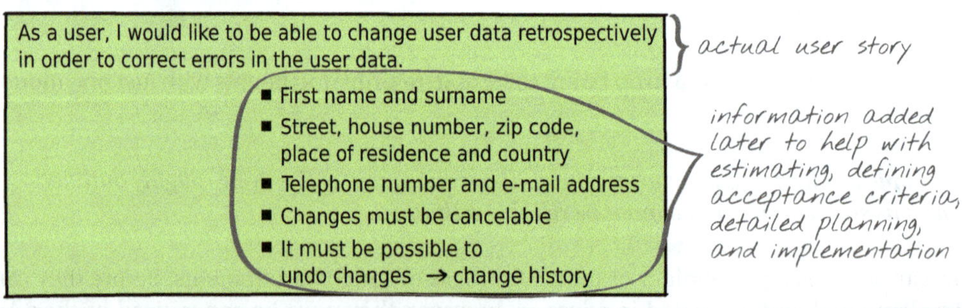

Figure 4.9: Example of a user story with explanatory additions.

Especially at the beginning of a project, some requirements are still unclear. You should avoid this kind of situation in plan-based projects. If requirements are unknown or incomplete, the plans based on them are prone to uncertainty. Assumptions are then often made that later turn out to be incorrect and lead to deviations from the plan. In plan-based projects, very good risk management coupled with equally good change control is crucial in such cases. In practice, both are sometimes neglected, which then leads to problems with plan fulfillment.

In agile projects, uncertainties are accepted and clearly documented. Instead of a fully formulated user story, an epic is included in the product backlog. In other words: the product backlog consists of both user stories and epics!
The user story from Figure 4.9 would simply be called *user management* as an epic. Epics help to ensure that nothing is forgotten and can be understood as placeholders in the product backlog that need to be worked out in detail later.
Scrum deliberately refrains from estimating the (implementation) effort of a user story in staff days. Instead, the *size* of the user stories is estimated. This is a measure of the complexity of a user story. The unit of this size is called story points. If a user story has twice as many story points as another user story, it is twice as complex. You already know that the product backlog is the responsibility of the product owner. However, the size estimation is carried out exclusively by the developers. We will go into this in more detail later.

> ***Story points*** *are a measure of the size of a user story in Scrum. The size reflects the complexity of the requirements documented in the user story and is independent of the effort required. The latter also depends on the skill of the person working on it; this is not taken into account when estimating the story points.*

Warning

The size of a user story depends on its complexity, for example how the function to be implemented interacts with different components of a system. The size is independent of whether the user story is implemented by an experienced or inexperienced developer. This type of estimation decouples the complexity of a user story from the ability of the developer.
It would not even be possible to take into account the skills of the developers, as with Scrum it is not known in advance who will complete a certain task later. This is different in plan-based projects with concrete resource plans.

The prerequisite for estimating the size of user stories is that there is a common understanding of the complete implementation of a user story. Usually, implementation also includes all testing and documentation work. The definition of what exactly is part of the implementation is called **Definition of Done**.

> *The* ***definition of done*** *is the formal description of the finished implementation of a product backlog entry. When the definition of done is reached, this implementation supplements the previous increment.*

A user story is only fully implemented when all the work associated with the user story has been completed and integrated into a functioning increment. Consequently, all of this work must also be taken into account when estimating the size. Depending on the user story, different aspects must be taken into account when implementing it, see Figure 4.10.
In Scrum, the size of the user stories is often estimated using planning poker, which is explained in the toolbox in Chapter 8 under *P.2.7 Playing planning poker*. The story points are then assigned with numbers that are roughly derived from the Fibonacci sequence with 0, 1, 2, 3, 5, 8, 13, 20, 40, 100.

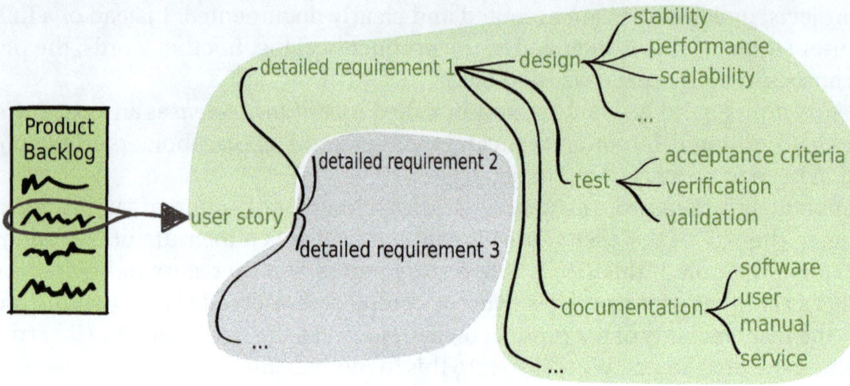

Figure 4.10: Considerations regarding the definition of done: Detailed requirements, designs, tests and documentation may be necessary for the implementation of a user story. Once implementation is complete, no further work on the user story should be necessary.

The **prioritization** is based on the business value and customer importance of the respective user story. There are various options for the prioritization itself, such as

- the pure **ranking** of the user stories from the most important to the least important user story
- the **assignment of points**: the smaller (or larger, depending on the convention) the number of points, the more important the user story
- the **classification**, for example according to the *MuSCoW principle* in *Mu*st have, *S*hould have, *Co*uld have and *W*on't have.

Tip

Make sure to prioritize high-risk user stories appropriately. For high-risk user stories, it can be worthwhile to reduce the risk quickly by implementing the user stories at an early stage and having them tested and evaluated by the customer as part of the sprint review. However, whether a risk reduction can be achieved in this way must always be checked on a case-by-case basis.

The higher the priority of user stories, the more detailed they should be. An epic is too imprecisely described to be implemented in the next sprint, even if it has a high priority.
In practice, it is often difficult to prioritize requirements that are important for the company itself, like the infrastructure, the system architecture, maintenance or production of the product. You can see an example of how this can be handled in the practical examples and case studies below.
There are many different options for the **organization and presentation of a product backlog**: The collection of all user stories and epics on metaplan cards on a real or virtual pinboard is certainly typical. There, the cards can be provided with a size estimate and hung up in prioritized order. Everyone in the team can immediately see the number of requirements still to be processed. One of the advantages of virtual or online pinboards is that they can be archived more easily or older versions can be restored. Online pinboards can also be viewed and edited across different locations. However, a virtual pinboard lacks the attractiveness of a workspace that is very present and easy to edit manually, which is appreciated by many teams.

Instead of using cards on a pinboard, user stories are often saved in lists, for example using a spreadsheet, see Figure 4.11 for an example.

	A	B	C	D	E	F
1	**ID**	**User Story**	**Size [Story Points]**	**Priority**	**Comments**	
2	U5	As a user I would like ...	13	100	-	
3	M3	As a manager I would like ...	20	115	-	
4	P2	As a manufacturer I would ...	8	150	see also Doc. D13	
5	U2	As a user I would like ...	100	200	-	
6	U1	As a user I would like ...	3	205	-	
7	P1	As a manufacturer I would ...	40	210	-	
8	M2	As a manager I would like ...	8	213	-	
9	U4	As a user I would like ...	5	250	align with product T20	
10	M1	As a manager I would like ...	1	260	-	
11	S1	As a service technicians I ...	40	270	-	
12	U3	As a user I would like ...	20	280	approval pending	
13						

Figure 4.11: Administration of a product backlog in a spreadsheet.

Tip

Even if the use of user stories and epics in the product backlog is common, requirements can also be formulated in other ways. The Scrum Guide itself remains quite vague here [Schwaber and Sutherland, 2020]. As a product owner, you should find a pragmatic way to define and maintain the requirements transparently and comprehensibly in the product backlog.

Summary: Product backlog

- The product backlog contains the product goal and the requirements for achieving the product goal.
- The requirements are often formulated as user stories (elaborated requirements) or epics (placeholders for requirements to be elaborated later).
- All requirements are estimated by the developers in terms of their size and prioritized by the product owner.
- As part of product backlog management, the product backlog is continuously updated and concretized.
- Good user stories ...
  - are not too big (no epic!)
  - are estimated in terms of their size
  - have a fixed definition of done
  - have a justification for their existence or their contribution to increasing the value of the product and can therefore be prioritized.

Sprint backlog

The user stories of the product backlog explain the requirements of the customer or user. These must now be used to describe specific tasks that must be carried out by the developers to fulfill the requirements. This is done at the beginning of each sprint as part of sprint planning. The requirements to be implemented in the following sprint are selected from the product backlog (entries in the product backlog, usually user stories) and **tasks** are formulated for implementation. When dividing a user story into tasks, the size of the user story, specified in story points, is distributed to the individual tasks. Each task thus also receives a size estimate in story points. All tasks of a sprint are collected in the sprint backlog.

The ***sprint backlog*** *contains the sprint goal of the current sprint, the entries (user stories) selected from the product backlog for implementation and the tasks derived from them to achieve the sprint goal.*

While the product backlog states *what is required overall*, the sprint backlog specifies *what is to be implemented in the current sprint* and *how this implementation will take place.*
When selecting the user stories to be implemented, the developers must know their **velocity**.

The ***velocity*** *is the number of story points that the team can complete in one sprint.*

Newly formed teams do not initially know this value and have to make an assumption here. If the developers have already worked together, they know how many story points were completed in the previous sprint and can use this value for the current sprint. If improvements were agreed during the previous sprint retrospective that suggest an increase in velocity, this can be taken into account when determining the velocity for the new sprint.
The velocity is therefore an important key figure that the team can use to measure its own performance.

Example

A Scrum team has a velocity of 60 story points. As part of the planning for the current sprint, user stories with 60 story points are selected and converted into tasks. These are then worked on.
The team was successful and worked efficiently. In the end, an additional user story with a scope of 10 story points could even be completed. This means that a total of 70 story points were implemented in this sprint. If the team is of the opinion that it will be able to implement this quantity in future, it will specify 70 story points as the new velocity.

Warning

Inexperienced teams tend to overestimate their velocity and take on too much. At the end of the sprint, less is achieved than planned, which can frustrate both the team itself and the customer. In the practical examples and case studies, you will see how an inexperienced team can proceed in order to achieve an initial success and then continuously improve further.

The sprint backlog can be handled as a list like the product backlog. However, it is more useful to turn it into a real working tool and convert it into a taskboard and supplementary burn-down chart.

> *The **taskboard** is used to visualize the sprint backlog in Scrum. The user stories of the current sprint are broken down into tasks and usually ordered according to their processing status in columns "to do", "in progress" and "done".*

The taskboard can be created as a pinboard or electronically (in the simplest case with a spreadsheet). A physical board (pinboard, whiteboard, metaplan board, etc.) definitely has advantages in small teams that work together in the same room:

- The developers conduct the daily Scrums (short daily meetings) directly at the board; the board becomes a working medium and planning tool.
- Creating transparency: everyone can see the project status.
- The taskboard can foster the team's identity and provide motivation; everyone in the team can see their contribution in the context of the team's overall work and monitor progress.

The board should be supplemented by the sprint goal. If this is documented at the top of the board, it is even easier to recognize the relevance of the individual tasks for achieving the sprint goal. An example of a taskboard can be seen in Figure 4.12.

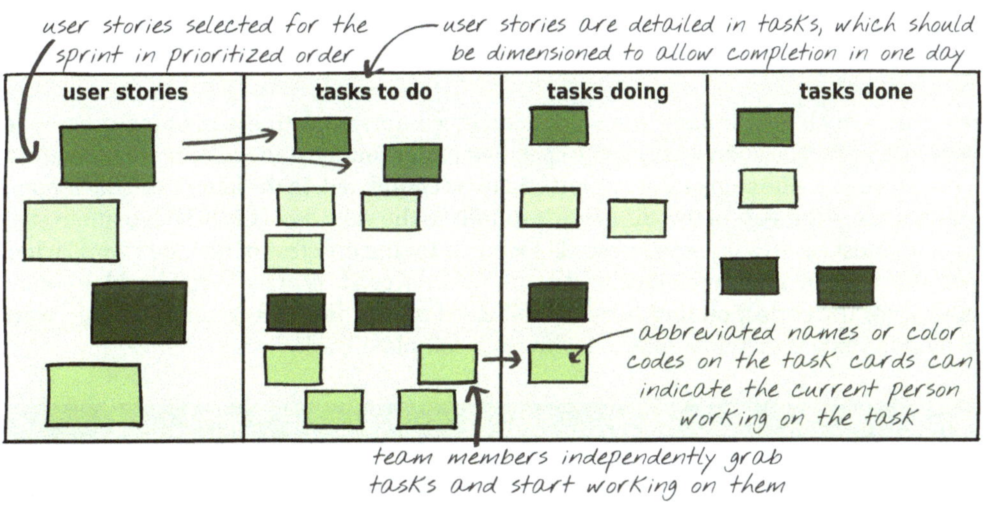

Figure 4.12: Taskboard with user stories and work status.

Tip

As an alternative to the taskboard, you can also work with a Kanban board, which has further advantages. You will get to know the Kanban board later in the explanations of Kanban.

Tip

Try to size the individual tasks so that they can be completed in approximately one day. This will make it easier for you to recognize later if there are problems when processing a task. These are then manifested by the task remaining in the column "in progress" and hanging there longer than usual.

The sprint backlog can also be additionally turned into a **burndown chart**. In the burndown chart, the outstanding work is plotted against a timeline. The timeline extends from the start of the sprint to its end. An example of a burndown chart can be seen in Figure 4.13.

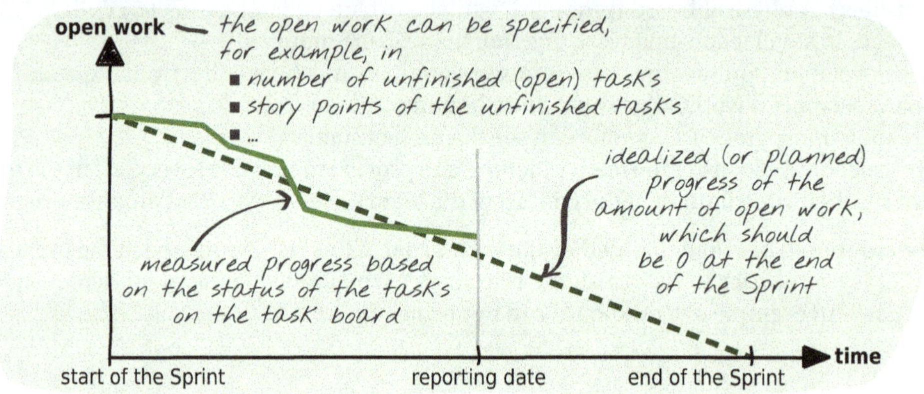

Figure 4.13: Example of a burndown chart with idealized and actual measured progression of open work up to the reporting date.

The burndown chart compares the work actually completed with the planned progress of completed work. The open work can be specified either simply as the number of tasks still to be processed or as the sum of story points still to be completed. In the latter case, the amount of open work at the start of the sprint corresponds to the velocity of the Scrum team in story points. In most cases, a linear decrease is assumed for the progress of the open work, which is entered in the diagram as an idealized graph. Whenever a task of the sprint backlog is completed, it (or the corresponding story points) is subtracted from the amount of open work. The plotted value is therefore reduced with each completed task.

Warning

Especially at the beginning of a sprint, the team may lag behind the idealized, linear progress until full working capacity is established and the first tasks are completed. This can lead to dissatisfaction on the part of the product owner, who might feel that the sprint goal is at risk. Some teams postpone the start of the linear decrease of open work further to the right in order to avoid coming under pressure to justify themselves right from the start.

Summary: Sprint backlog

- The sprint backlog contains the sprint goal, the user stories of the product backlog selected for implementation in the current sprint (*what* is desired) and the tasks derived from them (*how* the requirements are fulfilled).

- The sprint backlog is created by the developers during sprint planning.
- The creation and management of the sprint backlog are the responsibility of the developers and support them in their daily work.
- Typical visualization aids of the sprint backlog are the taskboard and the burndown chart.

(Product-)Increment

At the end of each sprint, there is a new (product) increment or, at some point, the final product.

> *An* ***increment*** *is a usable and tested (intermediate) result of a product. Increments build on each other in such a way that they represent a step towards achieving the product goal.*

An increment is therefore the (partial or intermediate) result of a sprint. This can be, for example

- a new software
- a new hardware
- a new system or end product
- a concept for a process improvement or an event
- an investment object etc.

It is important that every single increment is functional, usable or at least presentable. The product owner or customer must be able to evaluate the new increment. Only then will you be able to obtain feedback, refine epics and user stories and thus reduce risks. The knowledge gained is then used to plan the next sprint. It can be deduced from these considerations that new results (in the sense of completed tasks) can only become part of the increment once the definition of done has been fulfilled.
The demand that an increment should be usable or presentable has an impact on the selection of user stories for the current sprint. The user stories must always be selected in such a way that together they result in something presentable.

Example

Assume that some tasks are still open towards the end of a sprint, but that there is only little time left. Then the tasks that contribute to a functioning increment should be prioritized.

- The final product is created incrementally with Scrum.
- A new increment is available after each sprint.
- The team must ensure that each increment can be tried out and evaluated by the product owner and/or the customer.
- The experiences of the product owner and/or customer with the increment are incorporated into the planning of the next sprint.

Release plan

The release plan is a rough overview plan that shows which increment is to be created in which sprint. It therefore takes into account

- the priority of the entries in the product backlog (user stories)
- their estimated size
- the velocity of the team and
- the strategic and economic constraints of the product.

Even if the release plan is not an artifact mentioned in the Scrum Guide [Schwaber and Sutherland, 2020], it is important in practice. The product owner needs an idea of how many sprints it will take to create coherent increments. It would be naive and not justifiable from an economic point of view to simply start developing without knowing when a product is likely to be ready for the market. The example of a release plan can be seen in Figure 4.14.

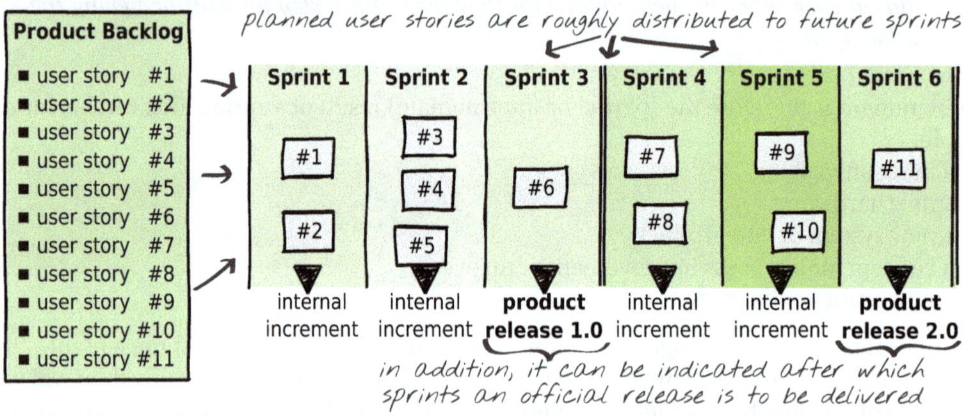

Figure 4.14: Release plan. After sprint 3 and sprint 6, official product releases are planned that will go into distribution. The other increments are used for internal evaluation and further development.

Warning

The release plan is usually created by the product owner. It includes not only the current sprint, but also future sprints for which the product owner does not yet have a firm commitment from the developers regarding the respective increments (and the underlying sprint backlogs). The release plan is therefore only a rough plan.

Impediment backlog

The Impediment Backlog also does not appear in the Scrum Guide as an "official" artifact, but it has great practical significance:

> *In the **impediment backlog**, the Scrum master notes all impediments that the team encounters while working. It is the Scrum master's responsibility to take care of these obstacles.*

The impediment backlog is usually filled with impediments that were identified, e.g. during the daily Scrum. The removal of these obstacles can also be noted in the impediment backlog and reported to the team in the next daily Scrum.
Some Scrum masters divide the impediment backlog into a strategic and an operational section. In the *strategic* or long-term section, the Scrum master notes perceived obstacles that are of a general nature, such as a lack of support from the company's internal clients during the sprint reviews.
In the *operational* part, short-term obstacles affecting the current Scrum project are noted. The impediment backlog helps to ensure that no impediments are forgotten and can also support the long-term safeguarding of experience.

Scrum events

The Scrum events describe the individual steps of Scrum [Schwaber and Sutherland, 2020]. Figure 4.15 provides a classification including the roles and artifacts already presented. The individual events and relationships are explained below.

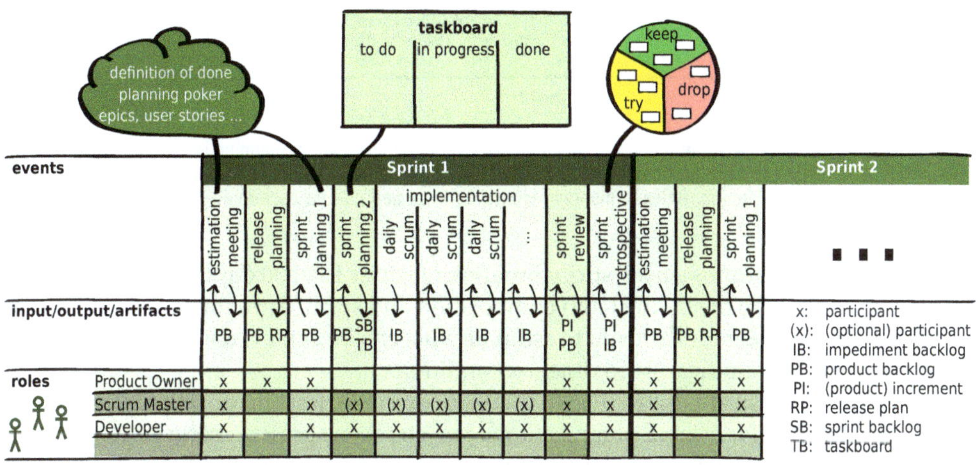

Figure 4.15: Overview of the Scrum roles, their responsibilities for individual Scrum Events and the chronological order of the events within a sprint. Scrum Artifacts are intermediate or final results and at the same time input or output variables of the Events.

Sprint

Let's first clarify the term "sprint":

> *A* ***sprint*** *is a fixed period of time in which the requirements for achieving the sprint goal are implemented and an increment is created. In practice, typical sprints usually last between 14 and 30 days.*

Like all Scrum events, the sprint is subject to the timebox principle:

*A **timebox** is a fixed period of time in which something is to be completed. The work ends at the end of the specified time, even if the task is not completed.*

A sprint begins with its planning. This is followed by the implementation of the planned user stories. At the end of the sprint, the achieved increment (result of the sprint) is checked in the sprint review and internal processes are optimized in the sprint retrospective. This is immediately followed by the next sprint, which starts again with planning. There is therefore no time span, events or actions between the sprints. Figure 4.16 provides an overview of the individual steps within a sprint.

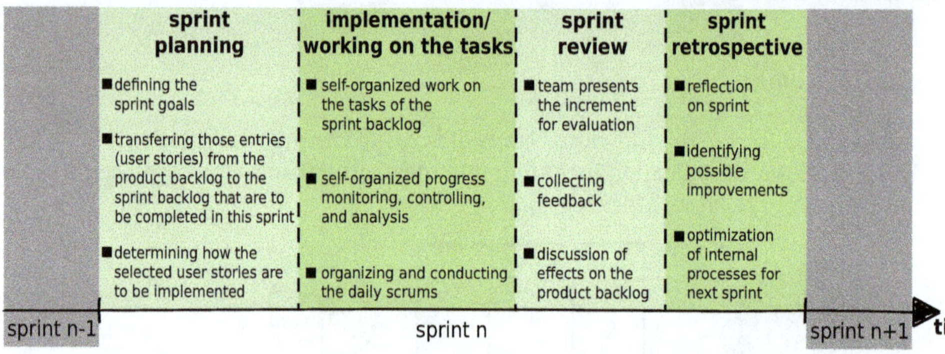

Figure 4.16: Processes within a sprint n that are repeated in all sprints. Part of sprint planning is also the (update of) size estimation of the entries in the product backlog. Optionally, this estimation can be carried out in a separate estimation meeting prior to the sprint planning. Release planning is also optional and can be performed in parallel by the product owner.

The literature often assumes a fixed sprint duration of either 14 or 30 days. Scrum teams work together on a long-term basis. The team composition and sprint duration are therefore constant. This makes it easier for the team to estimate what can be achieved in a sprint compared to other projects without these fixed timebox conditions.

In practice, also other sprint durations are used. For example, a sprint duration of seven days may be more suitable for a small organizational project, as progress is made at shorter intervals and customer feedback can be evaluated early on. In some large projects, on the other hand, progress in 30 days may be so slow that such short sprints lead to unnecessarily high administration overhead. The sprint duration can be extended to 60 days instead. In any case, the duration should be kept constant throughout the duration of the project. Ideally, it should be identical and constant across comparable project types.

Tip

Find out the right sprint duration for your area. The 30 days recommendation is certainly a good starting point. However, depending on the progress made in your projects, consider whether you need more frequent feedback in order to work efficiently and effectively and eliminate risks.

Conversely, you should not choose sprints that are too short. This increases the administrative overhead of sprint planning, sprint review and sprint retrospective and shortens the time available for the actual work on the tasks.

Warning

The Scrum Guide [Schwaber and Sutherland, 2020] recommends a maximum sprint duration of 30 days. In fact, a longer duration should be chosen with caution. With long sprints, the planning horizon is longer (plan-based), feedback is less frequent and the ability to change is reduced. In other words, adherence to agile values can become more difficult.

A *sprint goal* is specified for each sprint and corresponding user stories are selected from the product backlog. Each sprint can therefore be considered a (sub)project with its own magic triangle of costs, time and work. In contrast to plan-based projects, however, the time remains constant. It corresponds exactly to the sprint duration and is non-negotiable. The team size and therefore the (personnel) costs are also constant in agile projects. The quality of the requirements implementation is also non-negotiable, but the number of entries to be implemented from the product backlog is. If unforeseen difficulties arise during implementation, the quantity is reduced in consultation with the product owner.
No changes may be made during the sprint that could jeopardize the achievement of the sprint goal. New findings resulting from the ongoing implementation can be used to revise the product backlog or to further specify its entries. The scope of the sprint can also be renegotiated with the product owner as long as this does not jeopardize the sprint goal. If the sprint goal cannot be achieved, only the product owner has the authority to cancel a sprint.

Summary: Sprint

- The duration of the sprints is constant.
- A sprint goal is defined for each sprint.
- A sprint begins with sprint planning and ends with sprint review and sprint retrospective.
- For control in the event of unforeseen difficulties ...
  - No compromises are made in terms of quality.
  - The sprint duration is not changed.
  - The number of entries to be implemented from the product backlog can be changed.

Sprint planning

A prerequisite for the success of a sprint is the definition of what needs to be done in this sprint. That is why every sprint begins with sprint planning.

*The **sprint planning** is intended to answer two questions:*

- *What is the goal of the sprint and which entries in the product backlog (user stories) should be implemented to achieve the goal?*
- *Which tasks have to be completed to implement the selected entries, i.e. how is the implementation to be carried out?*

Like the sprint, sprint planning is also subject to the timebox principle: The Scrum Guide [Schwaber and Sutherland, 2020] recommends eight hours of sprint planning for a sprint of 30 days. In practice, the two questions mentioned above are answered in two consecutive meetings called sprint planning 1 and sprint planning 2, see Fig. 4.17.

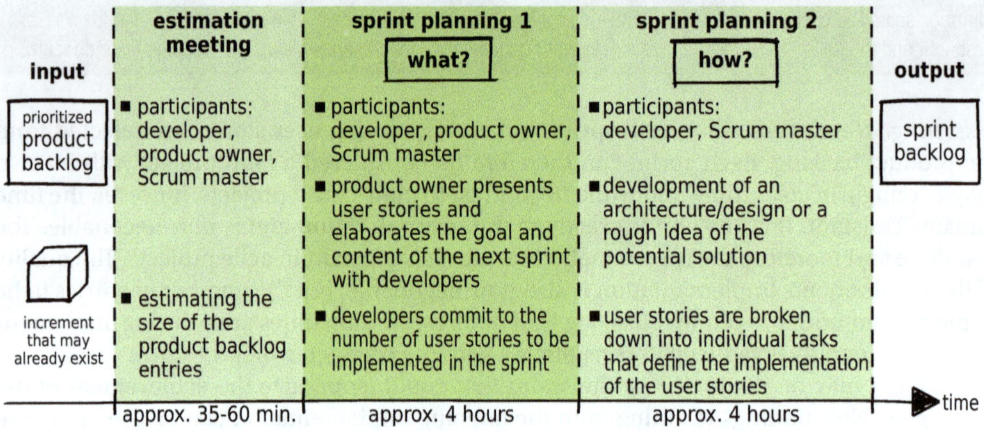

Figure 4.17: Sprint planning overview: The entries (usually user stories) of the prioritized product backlog must be estimated by the developers in terms of their size (complexity). This can be done in a separate estimation meeting or at the start of sprint planning 1. Based on the estimated and prioritized product backlog and any existing increments from the previous sprint, the goal of the sprint and its content are defined. The developers then derive the tasks to be completed to achieve the sprint goal, which then form the sprint backlog.

The prerequisite for planning the sprint is an existing product backlog, i.e. a catalog of requirements. In contrast to the requirements catalog of plan-based projects, the product backlog is not necessarily complete. Sprint planning is always based on the current version of the product backlog.

The product backlog consists of the entries collected and prioritized by the product owner, i.e. the requirements, which are usually formulated as user stories. The size of the user stories is estimated by the developers. It should preferably take place in a separate **estimation meeting**, but can also be carried out at the start of sprint planning 1.

> *In the **estimation meeting**, the developers familiarize themselves with the current status of the product backlog. They estimate the size of the entries in the product backlog. The product owner is available to answer questions regarding the entries.*

For each sprint, the product owner defines a goal that represents the focus of the sprint. Like all goals, the sprint goal should also be formulated SMART, where SMART stands for specific, measurable, achievable, realistic and time-bound.

Developers know from the past how many story points they can implement in a sprint. In Scrum, this quantity of story points is called *velocity*. With this and the knowledge of the priority and the estimated size of the entries, the developers can now determine the user stories to be implemented in the next sprint, see Figure 4.18. If no empirical values are available, an assumption must be made for the velocity. Calibrating the size estimate is helpful for this: a user story is selected from the user stories that the developers assume can be completed by one developer in one day. All other user stories are estimated relative to this user story and the number of story points that the team can complete in one sprint is derived from this.

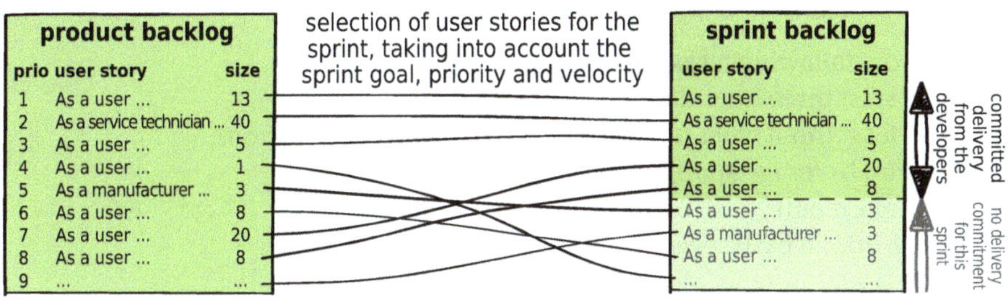

Figure 4.18: The user stories to be completed by the team in the current sprint are selected from the product backlog. These user stories are transferred to the sprint backlog. This is further detailed later in sprint planning 2 and the user stories are broken down into individual tasks.

When selecting the entries in the product backlog for the sprint, the developers take into account

- the goal of the sprint
- the priority of the user stories
- the size of the user stories in story points and the scope of story points that can be implemented in a sprint (= velocity) and
- the context of the user stories so that a coherent, evaluable increment is created at the end of the sprint.

It is also advisable to address particularly *high-risk entries (user stories) in the product backlog in an early sprint* in order to reduce the corresponding risk as early as possible. Risks can, for example, be the technical risks of a user story or arise from a customer request that is still unclear. To assess the risks, their probability of occurrence and the severity of the damage that may occur can be estimated. In addition to a priority, some teams also assign a risk index to each user story. This assesses the risk of the user story relative to the risks of the other stories. Before the developers finally confirm the scope of delivery for the planned sprint to the product owner, they should have a brief discussion on their own. The acceptance of Scrum depends heavily on how reliably it is delivered. In the long term, it does not help those involved if more is promised prematurely or under pressure from the product owner than can be delivered later.

Once the developers have agreed on the scope of delivery and communicated this to the product owner, sprint planning 1 comes to an end. This is followed by **sprint planning 2.** This usually takes place on the same day after a short break. Only the developers and, if applicable, the Scrum master are present at sprint planning 2. It has proven to be a good idea for the product owner and, if necessary, the customer to be available in case of questions.

In sprint planning 2, the developers break down the selected user stories into tasks to be completed. This also includes accompanying activities, such as the creation of necessary documentation or testing, as illustrated in Figure 4.10. The tasks are now stored in the actual sprint backlog or transferred directly to a taskboard.

Example

From the user story
- As a visitor to the festival, I would like to be able to choose from various drinks to quench my thirst.

for example, the following tasks could be derived:
- Specify at least three non-alcoholic beverages to choose from.
- Specify at least three types of alcoholic beverages to choose from.
- Commission beverage suppliers.
- Create drinks menu.
- Program drinks in cash register system.

Summary: Sprint planning

- The following applies to Sprint planning 1:
  - Participants: Product owner, developer, Scrum master
  - Definition of the sprint goal
  - Presentation of the entries in the product backlog by the product owner
  - Selection of the entries that promote the sprint goal and should become part of the increment to be implemented
  - Determination of the definition of done
  - Firm commitment from the developers about the scope of the sprint.
- The following applies to Sprint planning 2:
  - Participants: Developer, optionally Scrum master
  - Derivation of tasks that lead to the implementation of the selected entries of the product backlog
  - Preparation, e.g. creation of the taskboard and/or burndown charts, as a working basis for the sprint

Daily Scrum

During the sprint, there is a short meeting, the so-called daily Scrum.

> *The **daily Scrum** is the short daily meeting of the developers. It is always held at the same time as a regular meeting. The daily Scrum lasts a maximum of 15 minutes. There is a brief discussion about what has been achieved since the last meeting, what is to be achieved until the next meeting and what obstacles there are on the way or who can help to solve the obstacles.*

The daily Scrum is the central instrument for
- status monitoring and analysis
- identification of obstacles
- synchronization of tasks in the team
- initiation of control measures and
- planning the next 24 hours (until the next daily Scrum).

The daily Scrum should take place *daily at the same time* to structure the work and provide orientation for the team. The meeting should be short and effective so that as few additional developer meetings as possible are necessary. The Scrum Guide [Schwaber and Sutherland, 2020] specifies a maximum duration of 15 minutes, i.e., as with all Scrum events, the timebox principle applies here. The daily Scrum usually takes place directly at the taskboard, which you have already become familiar with when working on the sprint backlog.
The daily Scrum is organized and held by the developers. In addition to them, the Scrum master and product owner can also participate as developers if they are actively working on sprint backlog items themselves.
If the developers identify problems that they cannot solve on their own, these are passed on to the Scrum master. The Scrum master takes up the problems, documents them in the so-called *impediment backlog* and takes care of suitable solutions. In inexperienced teams, the Scrum master can initially support the daily Scrum as a moderator.
A typical agenda for a daily Scrum can be seen in Figure 4.19. Each developer answers the four questions listed for their area of work.

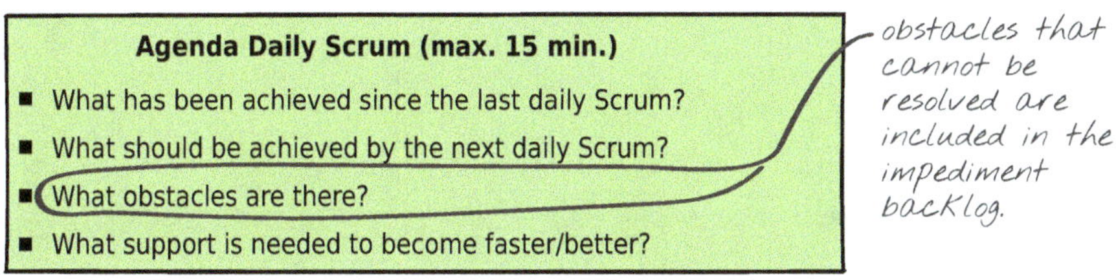

Figure 4.19: Typical agenda of a daily Scrum.

Tip

The daily Scrum is the central control instrument in Scrum. It should therefore be given a high priority. The timely and complete attendance of the developers is just as important as effective organization. The absence of a developer can block the progress of the entire team, as collaboration issues may not be resolved.

An interesting method for enriching the daily Scrum is the goal survey [Kniberg, 2012], see Figure 4.20. At the beginning of the daily Scrum, the developers write down on a flipchart how realistic they think it is to achieve the sprint goal. The subjective survey promotes two beneficial aspects:

- By concentrating on the goal, it is regularly brought to mind. This prevents losing sight of the big picture.
- If there are doubts about the achievement of the goal, these can serve as an entry point for analyzing the causes and solving the problem at an early stage.

The daily Scrum is not the only meeting of the developers. It is certainly possible to arrange further meetings, for example to solve problems, for longer-lasting design discussions or for joint concept creation. For the daily Scrum, stick to the scheduled timebox of 15 minutes and the agenda suggestion from Figure 4.19. Then arrange separate meetings for all additional coordination.

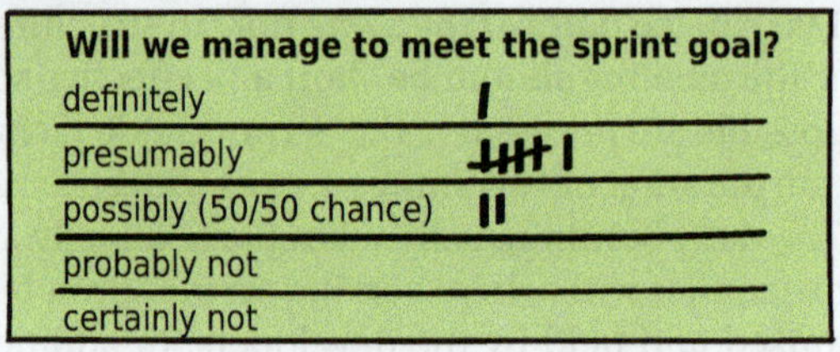

Figure 4.20: Assessment of the achievement of the sprint goal: All developers document their own estimation by drawing a mark in the corresponding field on the flipchart.

Summary: Daily Scrum

- The daily Scrum is a short daily meeting at the same time every day.
- Participants: Developers; Scrum master and product owner can participate if they are actively working on sprint backlog items.
- Good practice: Conducting the daily Scrum at the taskboard.
- Each developer reports on what has been achieved since yesterday, what is to be achieved today, what obstacles there are and what support is needed.
- The Scrum master provides support for problems that the developers cannot solve themselves and maintains the impediment backlog for this purpose.

Sprint review

At the end of the sprint, a new increment (intermediate or final product result) is created from the results of the completed tasks. The increment fulfills the sprint goal and represents the implementation of the sprint backlog. The increment is checked in the sprint review and made available for use.

> *In the **sprint review**, the increment is presented and evaluated by the Scrum team. Participants are the Scrum team and other stakeholders who may be invited by the Product Owner, for example customers. The sprint review is a working meeting that is subject to the timebox principle.*

Warning

The *definition of done* has already been discussed. For an increment to be created, the necessary tasks must be completed. This can include tests, documentation and integration tasks. In a pure software project, for example, an executable program must be compiled. In a mixed hardware and software project, the software must also be integrated into the developed hardware. In an organizational project, all organizational elements must be available in a form that the customer can understand.

A sprint review can follow the process shown in Figure 4.21.

At the end of the sprint review, the Scrum team has new insights into the acceptance of the current increment, can assess risks for further development and plan countermeasures.

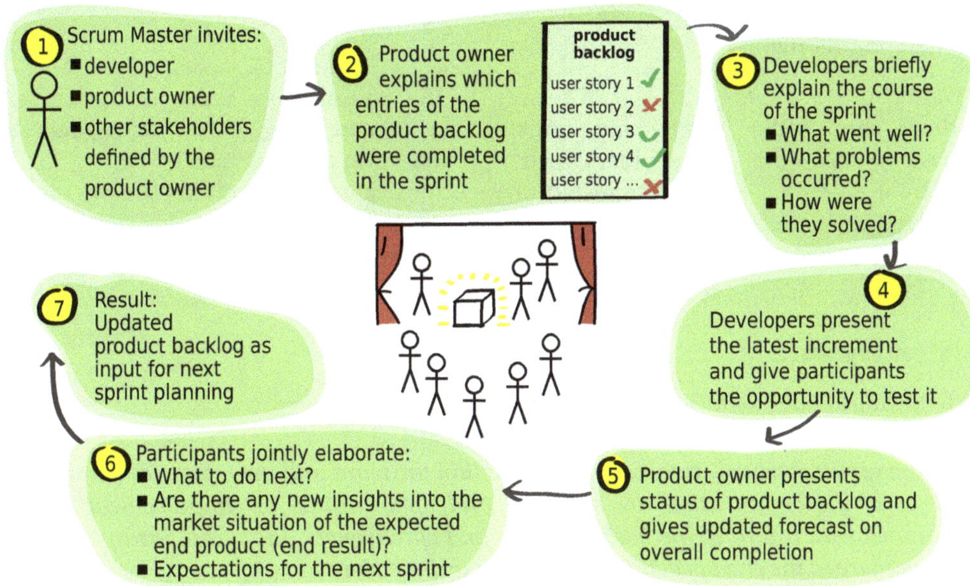

Figure 4.21: Sprint review process. The current increment is presented and assessed by the participants of the sprint review. Consequences for further development are discussed and the product backlog is revised accordingly. The next sprint planning is then carried out on this basis.

The product backlog is updated, i.e. new entries are added if necessary, existing entries are revised and prioritization is reviewed.
If the product owner is satisfied with the increment and it fulfills all the entries in the product backlog that are considered important, the Scrum project is completed (after the subsequent sprint retrospective). No further sprint is required.

Warning

As part of the sprint review, check whether the project deliverables have been fully processed in line with the project objectives or the product objectives. In some projects, product owners or other stakeholders are always demanding new things. Further sprints are added and the end of the project is delayed. Scrum aims to deliver good products and satisfy customers. This should not be misunderstood as an invitation to keep continuing the project indefinitely.

The duration of the sprint review should be appropriate to the complexity of the increment and the project environment. A superficial presentation without the opportunity to discuss weaknesses and opportunities for improvement and to work out consequences for the next sprints does not help. Conversely, if the sprint review is too long, it will be perceived as lengthy and superfluous by those involved. The Scrum Guide recommends a duration of 4 hours for a sprint duration of 30 days [Schwaber and Sutherland, 2020].

Summary: Sprint review

- Participants are the entire Scrum team (developers, product owners and Scrum masters) and other stakeholders invited by the product owner (interested or affected persons, such as customers).
- The Scrum team presents the current increment and gives the participants the opportunity to evaluate it.
- The evaluation results are used to update the product backlog so that it can be considered in the next sprint planning.

Sprint retrospective

The following applies to the sprint retrospective:

> *In the **sprint retrospective**, cooperation within the team and the interfaces to other areas of the organization are analyzed and opportunities for improvement are planned with regard to increasing quality and effectiveness. The sprint retrospective takes place at the end of each sprint after the sprint review.*

The retrospective also serves to benefit the customer, as the team becomes faster, the results achieve a higher quality and overall motivation increases. The sprint retrospective also contributes to the *lessons learned* and secures and analyzes the experience gained.

The Scrum master usually invites the team to the sprint retrospective. Its duration is also subject to the timebox principle and should be chosen so that there is sufficient time for a well-founded analysis and the development of improvement measures. For a sprint duration of 30 days, the Scrum Guide recommends a retrospective lasting approx. 3 hours [Schwaber and Sutherland, 2020].

As a starting point for the retrospective, positive and negative aspects of the sprint can be collected in a team brainstorming session together with ideas for future improvements and categorized according to the keep-drop-try scheme shown in Figure 4.22.

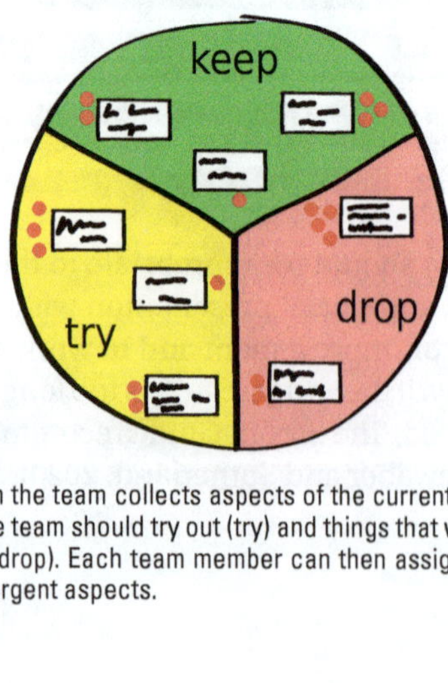

Figure 4.22: Keep-drop-try: Everyone in the team collects aspects of the current sprint on metaplan cards that they liked and should be kept (keep), ideas that the team should try out (try) and things that were not good or did not work and should therefore be dropped in future sprints (drop). Each team member can then assign priorities, for example using red sticky dots to highlight the most important or urgent aspects.

The collected and categorized contributions are then evaluated together as a team and different points of view are discussed. Aspects that one team member liked may be disliked by another team member. The evaluation can be completed with a prioritization. A common procedure here is that each team member can allocate a certain number of points to the collected contributions. This can take the form of red sticky dots on a metaplan board, for example. The contributions that were rated most important or most urgent, i.e. for which the most points were awarded, are now considered further. In the next step, concrete measures for change are derived from these contributions and passed on to the responsible persons for implementation.

Warning

A prerequisite for a successful retrospective is an atmosphere of trust. If individual team members have the impression that the content of the retrospective will be used against them at some point or that their contributions will not be taken seriously, the retrospective will lose its effect. Affected team members will only address superficial and euphemistic issues that do not contribute to a profound improvement.

As part of the retrospective, key figures can also be recorded and compared with previous sprints. Some key figures are also suitable for exchanging experiences with other projects and teams in order to derive success factors. Typical key figures are:

- the number of unresolved error messages
- the time taken per troubleshooting
- the throughput.

Krebs also suggests evaluating unusual indicators such as work ethic using a point scale, for example [Krebs, 2008]. This takes into account the particular importance of self-organization and commitment in agile projects.

Tip

For a purposeful and efficient retrospective, all the necessary data about the sprint must be available. You can use the impediment backlog for this purpose. You can also keep a chronological sprint diary. This helps you to review the past sprint along its timeline and to better remember problems, solutions and best practices. A retrospective should therefore always be well prepared and the necessary facts should be bundled together.

Summary: Sprint retrospective

- Participants are the entire Scrum team (developers, product owner and Scrum master).
- Plan a sufficient duration so that profound improvements can be developed.
- Create an atmosphere of trust.
- Prevent disruptions from people outside the team.
- Collect what was good in the sprint, what was bad and what opportunities for improvement you see.
- Derive concrete measures for improvement and prioritize them.

Scrum summarized

- The Scrum team consists of the developers, product owner and Scrum master.
- Scrum artifacts are the product backlog, the sprint backlog, the increment, the release plan and the impediment backlog.
- The product backlog is created by the product owner and contains the product goal and the requirements for the project scope, which are usually formulated in the form of prioritized user stories.
- The sprint backlog is used to plan the current sprint. The entries in the product backlog are broken down into individual tasks to be worked on and placed on a taskboard.
- The increment is the result of a sprint. At the end of all sprints of a project, the increment corresponds to the completed project deliverables.
- The release plan is an overview plan that roughly shows how many sprints are planned and in which sprint which entries of the product backlog are to be implemented.
- In the impediment backlog, the Scrum master collects obstacles that the developers cannot remove themselves. This is the task of the Scrum Master.
- The Scrum events follow the timebox principle and consist of the sprint, sprint planning, daily Scrum, sprint review and sprint retrospective.
- Sprints pursue a defined goal and are constant in length.
- In sprint planning, the goal of the sprint is defined and the entries in the product backlog that are to be implemented in the current sprint are selected. To do this, the entries are broken down into individual tasks.
- Daily Scrums are short daily meetings at the same time every day. They take place directly at the taskboard and are used for internal team coordination and to identify and solve problems.
- At the sprint review, the Scrum team presents the current increment and gives the participants the opportunity to test it. The results are used to update the product backlog.
- During the sprint retrospective, the Scrum team collects information on what was good and what was bad in the sprint. Specific measures for improvement are derived and their implementation is planned.

Scaled Scrum

Introduction

The Scrum Guide recommends the use of Scrum only up to a team size of 10 people [Schwaber and Sutherland, 2020]. This is perfectly adequate for smaller projects. However, there are also many projects that require much larger teams in order to be able to work on the project deliverables. In practice, you can deviate from the above-mentioned limit of 10 people and form teams with more people. Above a certain size, however, it becomes difficult to follow the agile values and principles:

- Individuals are at risk of getting lost in the size of the team. Structures and processes are needed in order to be able to interact properly. If the structures and processes are poorly designed, project management can become bureaucratic.
- In order for the project result to work, special coordination mechanisms are required in larger teams. Otherwise, there is a risk of errors at the interfaces of the partial results that together form the overall product to be delivered. These coordination mechanisms can be documentation-intensive and therefore also increase bureaucracy.
- Cooperation with the customer can become more difficult as direct contact becomes more difficult in a larger team.
- A large team may not be able to react to changes as quickly as a small team due to the need for coordination.

Depending on the qualifications of the Scrum team, the project environment and the project deliverables, larger teams can also work well and agilely. The use of Scrum in such teams is also referred to as **scaled Scrum** and refers to the scaling of the team. Several sets of rules have been established for concrete implementation. The three best-known frameworks are Nexus, SAFe and LeSS, which you will learn about in this chapter.

Warning

Experts are quite critical of the scaling of Scrum. Opponents of scaling argue that simple and direct collaboration in the sense of the agile manifesto suffers as a result of scaling. Supporters, on the other hand, argue that from a certain amount of project deliverables, the teams must automatically become larger. Even if a scaled Scrum process model is then no longer as simple and dynamic as in a small team, implementation can still be more agile in many cases than if a plan-based process model is chosen.

Nexus

Nexus is probably the most lightweight set of rules for scaled Scrum. It was developed by Ken Schwaber and published free of charge via Scrum.org [Schwaber, 2021]. Nexus attempts to preserve the basic ideas of Scrum with as few additional structures and processes as possible while still enabling good collaboration in larger teams. The basic features of Nexus are explained using Figure 4.23.

Scaling at Nexus is achieved by having up to nine individual Scrum teams in parallel, each with a maximum of 10 people. These are coordinated by modified and additional roles as well as additional artifacts and events. In simple terms, there are simply several Scrum teams working together in parallel as part of a larger overall project.

Nexus roles

A Nexus project has a single **product owner** who is responsible for the cross-team joint product backlog. So even with Nexus, there is only *a single source of work*. All Scrum teams work with this one product backlog. The product owner is a member of the Nexus integration team, which you will get to know in a moment.

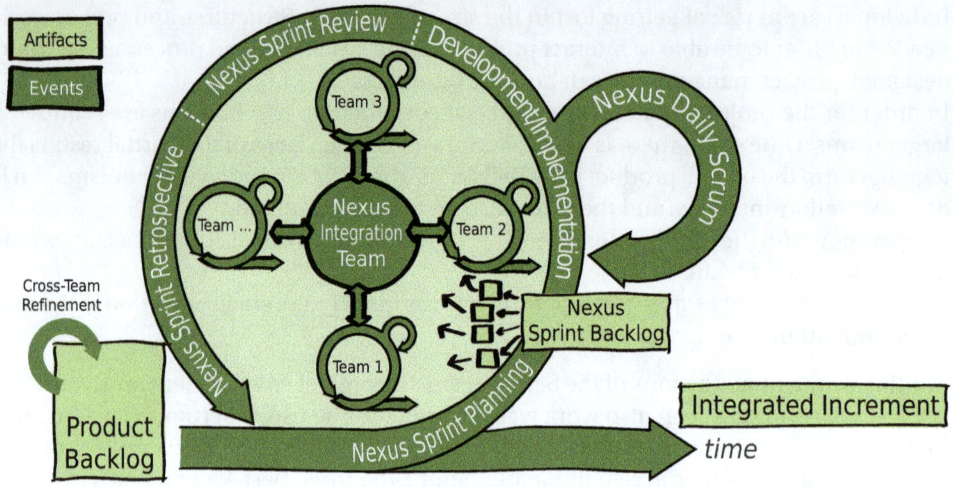

Figure 4.23: Scaled Scrum according to Nexus in the overview based on [Schwaber, 2021]. Nexus connects up to nine individual Scrum teams, which are labeled team 1 etc. in the figure, through additional or changed roles as well as additional artifacts and events.

Another member of the Nexus Integration Team is a **Scrum master**. This person is responsible for ensuring that Nexus is understood and applied. The Scrum master can also be the Scrum master of one or more of the maximum 9 Scrum teams.

The **developers** each work in one of the maximum 9 Scrum teams.

The **Nexus integration team** consists of the product owner, the aforementioned Scrum master and one representative from each of the Scrum teams. The Nexus integration team is responsible for ensuring that a common, integrated, cross-team increment is created at least once per sprint. This is called an *integrated increment* in the Nexus Scrum Guide. The work in the Nexus integration team is prioritized over the work in the individual Scrum teams. This ensures that higher-level issues are dealt with before work is carried out at (individual) team level.

Nexus artifacts

In Nexus, there is a single **product backlog**. As with Scrum, this consists of a product goal and requirements in the form of user stories and epics. As the product backlog contains entries for each of the up to nine Scrum teams, the product owner has to find ways to identify and minimize cross-team dependencies.

The **Nexus sprint backlog** is the higher-level backlog for the current sprint. It consists of the Nexus sprint goal, the selected entries of the product backlog for this sprint and the tasks derived from them. It can be supplemented with notes on cross-team dependencies.

Each of the maximum of nine Scrum teams works with its own **sprint backlog**. These team-specific sprint backlogs must be aligned with the higher-level Nexus sprint backlog.

The **integrated increment** is the combined sum of all partial results of the individual Scrum teams. This is evaluated in a joint Nexus sprint review. There is a corresponding *definition of done* for the integrated increment.

Nexus events

The **sprint** in Nexus corresponds to the sprint in Scrum and the same principles apply with regard to the timebox principle. At the end of the sprint there is the integrated increment.
The **cross-team refinement** is a continuous activity. The aim is to assess which entries in the product backlog can be processed by which of the individual Scrum teams in the future and to identify and minimize dependencies between the entries. The group of participants is not defined in detail and may change over the course of the project.
The current sprint is planned during the **Nexus sprint planning**. Participants are the product owner and representatives of the individual Scrum teams. The following results are available at the end of the event:

- the Nexus sprint goal based on the product goal and aligned sprint goals for each Scrum team
- a cross-team, shared Nexus sprint backlog that makes cross-team dependencies transparent
- individual sprint backlogs for each Scrum team.

The **daily Scrums** of the individual Scrum teams are supplemented by a **Nexus daily Scrum**, in which cross-team aspects are discussed and possible problems in the integration of the individual partial results of the Scrum teams are identified. Each Scrum team delegates a suitable representative to the Nexus daily Scrum.
The **Nexus sprint review** evaluates the integrated increment at the end of the sprint. It is carried out jointly across teams. In Nexus, there are no team-internal sprint reviews in the individual Scrum teams.
The final reflection is divided into two parts: In the **Nexus sprint retrospective**, cross-team issues and improvements are reflected and discussed. The aim is to continuously optimize the higher-level work. Each Scrum team also conducts its own **sprint retrospective** for further team-internal improvement.

SAFe

SAFe stands for **S**caled **A**gile **F**ram**e**work and was created by Dean Leffingwell [Leffingwell, 2011]. Similar to Nexus, scaling is achieved by coordinating several agile teams working in parallel. However, the teams are not restricted to Scrum. Rather, other agile process models, such as Kanban or Extreme Programming, can also be used. There are several scaling levels for SAFe:

- **Essential** is the basic configuration and already contains many of the SAFe rules.
- **Large solution** is for particularly large teams and contains additional roles and structures.
- **Portfolio** offers additional elements with strategic alignment of an entire portfolio of projects and products.
- **Full** includes all configurations for company-wide management with scaled teams.

Compared to Nexus, SAFe allows even larger projects with more than 100 teams. This is made possible by significantly more extensive rules with more roles, events and artifacts. A full description of these would go beyond the scope of this chapter. The following sections therefore only outline the basic features for an initial introduction to SAFe. A comprehensive source of further information is the website https://www.scaledagileframework.com [Leffingwell, 2011].

Core competencies and basic structure of SAFe

SAFe defines seven core competencies for agile working with SAFe:

- **Lean-agile leadership:** Leading those involved while adhering to lean and agile values and principles.
- **Team and technical agility:** The individual teams work according to agile values and principles and strive for technically high-quality customer-oriented solutions.
- **Agile product delivery:** Products are delivered continuously and in a customer-oriented manner. Delivery takes place as so-called *program increments* (PI), which have a fixed duration (timebox principle) of eight to twelve weeks. The program increments are divided into *iterations*, typically lasting two weeks.
- **Enterprise solution delivery:** To enable a large number of teams to jointly develop complex solutions, company-wide coordinated processes must be established to promote agile working methods.
- **Lean portfolio management:** Traditional project management is replaced by portfolio management that follows lean and agile principles. Those involved think in terms of value streams and take strategic, investment-related, financial and application-oriented aspects into account.
- **Organizational agility:** Companies must align their entire organization with lean and agile values and principles in order to be able to react quickly and flexibly to change.
- **Continuous learning culture:** Learning and development are an integral part of everyday work.

I explain the **application of SAFe** using the agile release train shown in Figure 4.24.

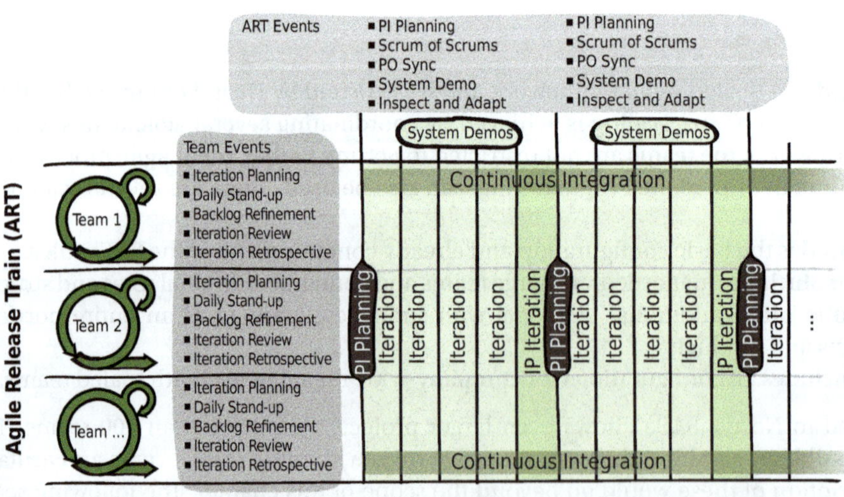

Figure 4.24: Scaled Scrum according to SAFe. The central element is the Agile Release Train (ART).

> *The **agile release train** (ART) consists of several agile teams. These iteratively develop value stream-oriented solutions with other stakeholders, deliver them and operate them if necessary.*

The agile release train links the work in the individual teams. These teams do not necessarily work according to Scrum, but can also use other agile process models. This is why SAFe does not use many Scrum-specific terms. Instead, many Scrum artifacts and events are used in an abstract way. The teams structure their work in synchronized iterations. These iterations are comparable to the sprints in Scrum with a fixed duration of two to four weeks. An iteration is performed by *each team separately* and follows the following sequence:

- **iteration planning** comparable to sprint planning
- **daily stand-up** comparable to the daily Scrum
- **backlog refinement** comparable to Scrum
- **iteration review** comparable to the sprint review
- **iteration retrospective** comparable to the sprint retrospective.

Cross-team coordination of the overall project takes place at the level of the agile release train (ART). In the ART, five iterations are usually combined into a so-called **program increment**. Depending on the duration of the iterations, a program increment takes around eight to twelve weeks.

The last iteration of a program increment is called **innovation and planning iteration** (IP iteration) and differs from the normal iterations in terms of its purpose and process. The IP iteration fulfills the following tasks:

- providing a time buffer for tasks that were planned but could not be completed in the previous iterations
- time for the development and implementation of innovations that are otherwise given too little time in the course of the project
- time for continuous learning and overarching knowledge transfer
- planning of the subsequent program increment
- implementation of the “Inspect and Adapt” event, which is explained in more detail below.

The **agile release train** contains the following events:

- **Program increment planning:** The program increment is planned based on the requirements of the requirements catalog referred to as *program backlog*. The planning results include the goal of the program increment and the commitment and rough planning of the content to be implemented by the individual teams.
- **Scrum of Scrums:** The Scrum of Scrums is a weekly cross-team coordination meeting to make transparent dependencies between the teams’ tasks and obstacles to collaboration. Participants are usually the release train engineer, whose role can simply be described as the cross-team Scrum master, and the Scrum masters of the individual teams. The meeting follows the timebox principle with a fixed duration of 30 to 60 minutes.
- **PO Sync:** The PO Sync meeting also takes place weekly with a fixed timebox of 30 to 60 minutes. Participants are the product owners of the individual teams and the product manager. The latter can simply be referred to as the cross-team product owner. In the PO Sync, questions regarding the content of the requirements are discussed, the program

backlog is refined and updated and the achievement of the program increment targets is reviewed.
- **System demo:** In the system demo(nstration), the individual results of the individual teams are integrated as far as possible after each sprint and evaluated as an overall result. The system demo takes place after each iteration.
- **Inspect and adapt:** At the end of the last iteration, all (partial) results are integrated into the program increment (PI) and evaluated as part of the PI system demo. This evaluation is part of the inspect and adapt meeting, in which process and quality-related issues are also discussed, similar to a cross-team retrospective.

Roles

The individual teams consist of the familiar roles of product owner, developer and Scrum master. If the teams do not work according to Scrum, but for example according to Kanban or Extreme Programming, comparable roles are defined. There are three SAFe-specific roles for the overarching coordination of the agile release train (ART).
The **product manager** is responsible for the overall product and its success. They act in a customer-oriented manner and know suitable methods, such as design thinking, to gather creative new ideas.
The **system architect** is responsible for the system architecture. They define the context of the solution, important interfaces and coordinate technical aspects of cross-team development.
The **release train engineer** is responsible for the so-called *agile release train* (ART), which is used to organize continuous development. They support the individual teams in making their contribution to the ART and act as a coach. In this way, the release train engineer takes on a role similar to that of the Scrum master, but at a cross-team level.

Summary

The components and processes of SAFe presented here are only an excerpt. SAFe describes itself as a framework. Within the defined framework, flexible specifications can be made that increase with the size of the project.
In addition to the description of roles, events and artifacts, SAFe attaches great importance to competencies, values and principles that the people working within the framework should always take into account. Due to the complexity of the framework, the introduction of SAFe in a company is considered a challenge. As is so often the case, a step-by-step approach can be useful here in order to adopt the many behaviors and elements of SAFe, consolidate them in day-to-day work and then deepen them further.

LeSS

LeSS stands for **Large** **S**cale **S**crum and was designed by Craig Larman and Bas Vodde [Larman and Vodde, 2013]. It is a framework that allows the processing of large projects with several Scrum teams. The name *LeSS* already indicates that the framework tries to get by with as few additional roles, events and artifacts as possible. LeSS contains
- principles
- rules

- guides and
- experiments.

The LeSS principles

LeSS defines ten principles. These are not a guide to using LeSS, but embody the basic attitude that people involved should have when using LeSS. The principles are [Larman and Vodde, 2013]:

- **Large-scale Scrum is Scrum:** Even if LeSS can be applied to several teams, the framework is understood as Scrum.
- **More with LeSS:** The aim is to achieve lean processes, less bureaucracy, an agile mindset and adherence to lean principles in order to create more added value.
- **Systems thinking:** The product is understood as a system and developed in accordance with the principles of systems engineering and systems management. The system dynamics are assessed, modeled if necessary and taken into account during implementation.
- **Lean thinking:** Acting in accordance with the lean principles is an integral part of LeSS. This includes avoiding waste, focusing on the value stream and continuous improvement. You can find out more about lean in a separate section on lean project management.
- **Empirical process control:** Like Scrum, LeSS relies on the shortest possible iterations in which valuable increments are generated. Process control is achieved through transparency and self-organization of the team.
- **Transparency:** Transparency is an important prerequisite for control in self-organized teams.
- **Continuous improvement towards perfection:** Teams working according to LeSS create economically successful products that delight customers.
- **Customer-centric:** LeSS wants to ensure that the relationship to the customer and their needs is maintained despite scaling.
- **Whole product focus:** The team does not think in terms of individual parts of a product, but keeps the big picture in mind. The customer expects a complete product, possibly with accompanying services. Even if individual people only ever make limited contributions to this overall product, the focus remains on the latter.
- **Queuing theory:** Queues should be avoided through good workflows. Otherwise, there is a risk of harmful multitasking, expensive storage of unfinished results and an overload of the people involved. Some of these aspects can also be found in critical chain project management, which I will introduce to you later in the chapter on agile project management.

The LeSS rules

LeSS is suitable for projects with two to eight teams, each of which should not exceed the maximum size of 10 people recommended in the Scrum Guide [Schwaber and Sutherland, 2020]. If even more teams are required, there is the LeSS Huge extension, which I will briefly outline later. But first, let's take a look at the basic LeSS rules using Figure 4.25.

With LeSS, the individual teams work as self-organized teams according to Scrum. Each Scrum master can support one to three teams in applying the LeSS (and Scrum) rules. One of the Scrum masters is responsible for the cross-team perspective.

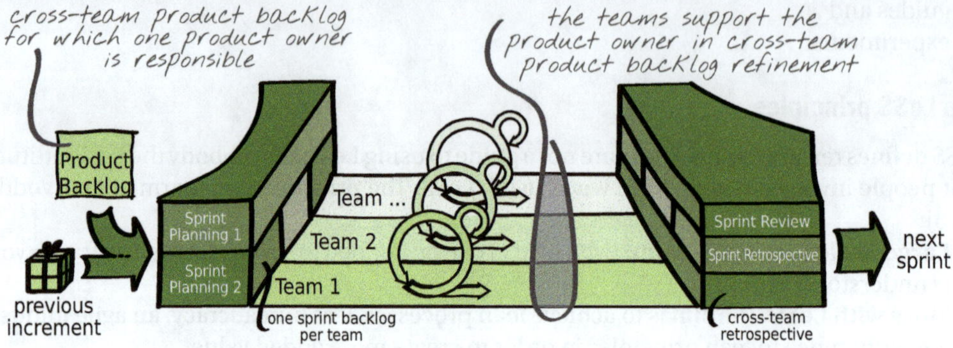

Figure 4.25: Scaled Scrum based on LeSS according to [Larman and Vodde, 2013].

The requirements for the project originate from a single, cross-team product backlog for which a single product owner is responsible. The teams support the product owner in the continuous, cross-team product backlog refinement.

The sprints are synchronized across teams, i.e. they start and end at the same time for all teams. Sprint planning 1 is carried out across teams. Each team appoints suitable representatives for this purpose. Together with the product owner, they define the sprint. The detailed planning of the sprint takes place in sprint planning 2, which is carried out separately by each team and results in team-specific sprint backlogs.

At the end of the sprint, there is a cross-team review. The sprint retrospective is divided into a joint, cross-team part and a team-specific part with a corresponding focus on either cross-team or team-specific aspects.

LeSS guides and experiments

When Larman and Vodde designed LeSS, they conducted many experiments on agile working. They encourage people to carry out their own experiments in the application of LeSS and thus adapt the way of working to the respective context and environment. The experiences are documented in guidelines. These do not have the character of fixed rules, but are intended to provide suggestions for application in practice.

Examples of guides include recommendations for the introduction of LeSS or backlog management. The experiments provide impetus to try things out, such as certain methods for conducting cross-team sprint reviews. If you want to find out more about guidelines and experiments, you can find several hundred examples in the books by Larman and Vodde [Larman and Vodde, 2008], [Larman and Vodde, 2010], [Larman and Vodde, 2016].

LeSS Huge

If more than eight teams are required, LeSS offers the *Huge* extension. The key feature of LeSS Huge is that requirements are divided into areas. Each area has its own area-specific product owner, the so-called area product owner. The area product owners are coordinated by a single superordinate product owner. A requirements area should be worked on by four to eight teams.

Tip

Don't take the limit of eight teams too seriously. Depending on the specific constellation and environment, more than eight teams can still work reasonably without the LeSS Huge extensions. Conversely, in other situations the Huge extensions can be useful even with fewer than eight teams.

Summary

The previous sections on Nexus, SAFe and LeSS have given you an insight into some of the most common ways of scaling Scrum. In addition to these, there are others, such as Scrum@Scale. The basic idea is similar in all frameworks: There are several small teams that are synchronized in a coordinating way. By having several teams working together, more people can be involved and thus larger tasks can be solved. The various frameworks differ from each other in detail. LeSS and Nexus in particular are considered minimalist and very close to the agile Scrum ideal, while SAFe is associated with significantly more extensive rules. Due to the complexity of scaled (large-scale) projects, the introduction of these frameworks should be well planned and supported. Extensive previous experience in agile working is an important success factor.

Kanban

Introduction

The term "Kanban" comes from the Japanese and can be translated simply as "card". Kanban originated in the middle of the 20th century when Toyota developed a method to increase the flexibility and efficiency of its production. It was only much later that David J. Anderson transferred the ideas of Kanban to the management of projects. Anderson was inspired by principles from lean, theory of constraint and, of course, the Kanban cards.

The theoretical foundations of Kanban are described by **four core principles** and **six practices**. This shows that Kanban follows an evolutionary change management approach and aims to continuously improve the way we work.

In the following sections, you will learn about the four core principles and six practices. Then, as with Scrum, we will explore these in greater depth using practical examples and case studies. We will repeatedly make comparisons with Scrum. First, however, we will take a look at the roles and organizational aspects of Kanban.

Roles and organizational aspects

Unlike Scrum, Kanban does not prescribe processes and structures in detail. There are no specific roles or meetings, such as the daily Scrum. There are also no fixed development times in the sense of a sprint.

However, this does not mean that such process and structural elements cannot be integrated into Kanban. Many teams that work with Kanban meet daily at the Kanban board to discuss the current situation of the work packages and to discuss similar questions as in the daily Scrum.

The pull principle on which Kanban is based promotes self-organization. It means that employees pull the tasks themselves instead of being assigned to them by the project manager. Unlike Scrum, many projects that work with Kanban are coordinated by a project manager. However, if they are used in Kanban, project managers have a different role than in plan-based projects. This person often takes on a role that is a mixture of traditional project manager, Scrum master and product owner. They ask the critical questions at the Kanban board, identify bottlenecks and eliminate them for the team. They are the interface to management outside the project and escalation instance for the team. They give the team similar freedom in the assignment and processing of work packages as is the case with teams working according to Scrum.

Core principles

The four core principles of Kanban are:
- Start with what you are currently doing.
- Strive for incremental, evolutionary change.
- Respect current processes, roles, responsibilities and titles.
- Promote leadership and accountability at all levels of the organization.

The first three principles are closely linked. They make it clear that when introducing and applying Kanban, you can always start where you are at the moment. On this basis, the process and organizational structure is then improved in line with the basic principles and by applying the practices. The first and third principles call for an examination of the status quo. Building on this, the second and fourth principles encourage us to change things for the sake of improvement.

Start with what you are currently doing

Kanban starts where you are with your project. This makes the introduction of Kanban easy, but requires that you really know where you stand. The questions you should be able to answer for this are:
- Who is involved in the project and how intensively? Who has what responsibilities and authorities?
- How does the project work? Which process models and processes are used? Are there specific processes required by the company?
- Which tools are used?
- How long does it typically take to process work packages in the project?
- What are the strengths and weaknesses of the current project management?
- What motivates us to think about changes now?

Tip

In this book, I also use the term "work package" in the context of Kanban to describe self-contained tasks. Alternatively, the terms "user story" and "task", which are commonly used in Scrum, could be used. Here, however, Kanban is deliberately differentiated from Scrum and described as an independent process model. In practice, Kanban can be used independently or combined with Scrum or plan-based process models to create hybrid project management process models.

Strive for incremental, evolutionary change

If the current status is known, improvements can be developed. The throughput time of the work in the project is a very important criterion for measuring the success of Kanban. Measures must be taken that allow you to regularly record the lead time in order to make process improvements visible. As we will need the term "lead time" more often, I will introduce it and the related term throughput here:

> *The* ***lead time*** *or duration is the pure time required to complete a work package or activity, taking into account the resources available.*

> *The* ***throughput*** *is the quantity of work packages that can be completely processed within a predefined time.*

Kanban relies on changes being implemented step by step. It focuses on evolution rather than revolution. The aim is not to lose important stakeholders and to take everyone along on the path of change. In addition, major changes carry the risk that their consequences are not fully understood and the expected improvement does not happen at all.
Critics of evolutionary change fear that the small steps towards change will not enable any major innovative changes. Instead of striving for a global optimum, local optimization is used. Successful managers are aware of this weakness of evolutionary change and regularly reflect on the status of their company or project. In addition to evolutionary improvements, they check whether it is even possible to gradually approach the optimum or whether major, revolutionary changes are necessary.

Warning

With the principle of incremental evolutionary change, Kanban follows the Plan-Do-Check-Act principle, i.e. the continuous planning of improvements, their implementation, success monitoring and control. However, some improvements require radical steps and radical changes. Make sure that you accompany these with good change management. An improvement is and remains an improvement only if it is implemented and applied in the long term. To achieve this, all relevant stakeholders must be involved.

Respect current processes, roles, responsibilities and titles

One consequence of the two core principles outlined above is that current processes, roles, responsibilities and titles must be respected. You can easily imagine what happens when changes are poorly initiated and communicated. Affected employees will switch to an attitude of refusal and block the planned changes wherever they can. There are always employees who are satisfied with the status quo and will not miss an opportunity to undermine your change.

Nevertheless, it may be useful or necessary to change processes, roles, responsibilities and titles in order to improve the organization and its procedures. However, follow the second core principle and take an incremental and evolutionary approach to these changes. Justify the change with solid facts and involve all the people affected by the change.

Changes to the organizational structure and processes usually fail not because they are too complex, but because the people affected are not involved in the change process and consequently fight the change to the utmost.

Promote leadership and accountability at all levels of the organization

Living agility means, among other things, transferring responsibility from individual managers to teams and giving them the freedom to organize themselves. Kanban emphasizes this aspect in the fourth principle. In contrast to Scrum, Kanban leaves the specific form to you. I will therefore return to the agile management concept in general terms. This should form the basis for implementation in your projects.

Two provocative statements:

- The nightmare of many agile advocates are teams that only work to instructions, have forgotten how to think for themselves and (have to) ask their superiors about every little detail.
- The nightmare of many traditionally minded managers are teams that question guidelines, demand responsibility and authority and want to organize themselves and their work.

Both statements are generalizations, but are intended to illustrate the fundamentally different mindset of agile management. If you introduce Kanban as an agile process model, you will have to strengthen leadership and responsibility. There are team members who are happy if they don't have to take responsibility. Other team members would like to take on responsibility but are prevented from doing so by their managers. Still other team members want to and are allowed to take on responsibility. However, they may lack the qualifications to fulfill this responsibility.

The transformation to an agile company must be well prepared. Employees must be qualified and established management structures must be broken up with the appropriate sensitivity. Agility does not necessarily lead to disempowered managers and teams without leaders. However, agile management will change the way we work together. It will go hand in hand with more freedom, duties and responsibility for the teams.

Warning

Do not underestimate persistent forces when changing established management structures. Managers who are unwilling to relinquish responsibility and authority can very quickly thwart your plans to introduce agile management, seek allies and cause the introduction to fail.

Tip

Consider the benefits to managers of promoting team accountability and empowerment and use these in communication. Managers can benefit from the successes of their teams, become more strategically involved and have more time for leadership tasks if they can rely on their employees.

The four basic principles presented here should help you to understand the guiding principles of Kanban. They are still too vague for everyday project work. There are six practices that we will now look at.

Practices

The six practices of Kanban are:
- Visualize the work (and its flow).
- Limit the amount of work in progress.
- Measure and manage flow.
- Make process rules explicit (clear and known).
- Develop feedback loops.
- Perform collaborative improvements.

Visualize the work (and its flow)

The first practice is sometimes also called *visualize the workflow*. Unlike other progress boards such as the taskboard in Scrum, the aim is not to show the status of a work package, but its flow in the project. Figure 4.26 illustrates the typical structure of a Kanban board using an example.
Starting from a pool of tasks to be processed, important processing stations are displayed as separate columns. Each work package passes through these until it is completed at the end. Typically, work packages are grouped together if they can be processed smoothly by the same employee.
The major advantage of this type of visualization over the purely status-oriented visualization with "in progress" and "completed" is that it makes the **workflow** through the processing stages visible and *possible obstacles can be identified and localized* more quickly. If a work package is stuck in a stage for a long period of time or if many work packages accumulate in a certain stage, there might be a bottleneck. The Kanban board helps to identify such bottlenecks, resolve them and steer them towards a uniformly optimal throughput time.

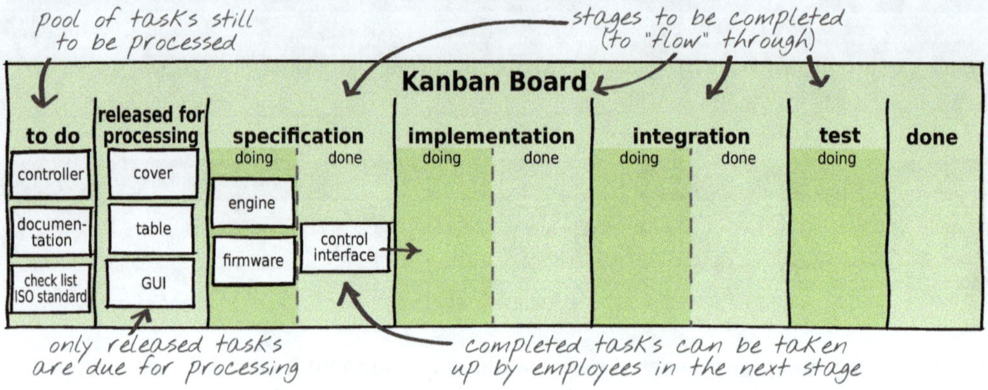

Figure 4.26: Structure of a Kanban board with processing stages to visualize the tasks passing through them.

The columns *to do* and *released for processing* shown in Figure 4.26 are not always used in practice. However, there are two reasons for using them:

- The *to do* shows which work packages are still waiting for the team. This makes it easier to classify work that has already been completed and work that is still open.
- The intermediate instance *released for processing* prevents work packages from being started in an uncontrolled manner. Only high-priority work packages that are due for processing may be processed. This also supports the second practice, which you will learn about below.

Tip

The attempt to create an analogy to Scrum leads to the following comparison: The "to do" and "released for processing" columns can be understood as simplified forms of a backlog. The to do roughly corresponds to the product backlog in Scrum and the released column roughly corresponds to the corresponding sprint backlog.

Warning

Similar to Scrum, you should also agree on what you mean by completed in Kanban, i.e. define a *definition of done*. Otherwise, you run the risk of not being able to correctly assess the project status, as the status "completed" can be interpreted differently by different employees.

Kanban also often involves self-organizing teams, even if this is not required in this clarity. Similar to Scrum, Kanban relies on the pull principle. This means that team members pull upcoming tasks to themselves and then work on them.

The Kanban board can be implemented as a physical board on a pinboard or a whiteboard. Many teams prefer this type, as the project status is always available and the Kanban board is both a status report and a meeting place for discussions. However, electronic web-based boards are ideal for distributed teams.

Tip

There are now many providers of boards that can be used online. If you are interested and want to choose one, you should make sure that you can design the board according to your needs and equip it with the rules that are important to you. Some providers only offer column-based boards where only the column heading can be changed. At the other extreme are flexible whiteboards that can be designed in any way you like, but do not have any automatic electronic rules. In between, there are offers where at least some of the following Kanban practices can be defined as rules.

Limit the amount of work in progress

Many project managers are familiar with the situation where many tasks are started but only a few are completed. The reason for this can be intensive multitasking, i.e. working on several tasks at the same time. Goldratt [Goldratt, 1997] presents the hypothesis that multitasking can be exaggerated. Employees get lost or need mental time to switch from one task to the next, etc.

Kanban emphasizes the value-adding focus on the right work at the right time and therefore requires the amount of work started to be limited to a reasonable level. The limitation should

- enable focused, value-adding work by all team members on the one hand, but also
- ensure that no one in the team, and this applies in particular to critical bottleneck resources, is without work because previous work was not completed on time.

Kanban therefore recognizes that a certain amount of "work inventory" must be carried out in order to keep employees reasonably busy. However, Kanban wants to ensure that work is not started in a completely uncontrolled manner. Do not start any work if you do not need the result at the moment. Such work prevents you from doing what is important for your project *now*. It is important for the project manager and the team to continuously approach the optimum work limit. The limit can change during the project.

Tip

If you are not sure how to set a reasonable limit, you should start with a low limit and increase it later. Conversely, if you start with a limit that is too low, you run the risk of creating additional bottlenecks from the outset and reducing throughput.

Limitation can be implemented using personalized markers on the Kanban board, for example. Each employee marks the work packages they are working on the Kanban board with their name abbreviation or a colored sticky dot. Each team member may assign a maximum of a predetermined number of these markers at a time. New work packages may only be started once another work package has been completed and the sticky dot has been removed. The example is illustrated in Figure 4.27.

Alternatively or additionally, it can be specified that only a certain number of tasks may be pending in a certain processing stage. For example, it can be agreed that in the column "Implementation" in Figure 4.27 a maximum of five tasks may be in progress at the same time. However, this type of limitation does not necessarily limit the workload of individual employees. Instead, this limit ensures that the work is distributed evenly across several

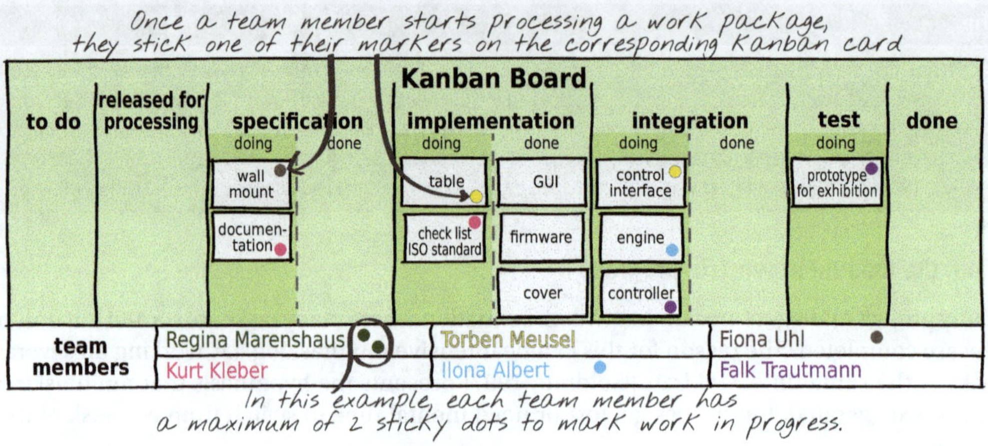

Figure 4.27: Exemplary implementation of work limitation using a limited number of colored sticky dots to identify team members.

processing stages. This can promote the flow of tasks through the project. However, it is important to check which type of limitation is appropriate for each project.
If the effort required for the individual work packages differs greatly, limiting the number of work packages started may be too imprecise. Alternatively, a limitation based on the effort estimate underlying the work packages may be a good alternative.
The integrated visualization of

- work packages
- processing stages to be passed through and
- responsible team members

promotes the identification of bottlenecks. If work packages accumulate at a processing stage or with a specific team member, the reasons can be investigated in a targeted manner. Other team members with comparable skills may be able to provide temporary or permanent support in order to resolve a bottleneck. Alternatively, the bottleneck worker can be relieved of non-value-adding activities in order to be able to process the value-adding work packages more quickly and thus resolve the bottleneck.

Example

Suppose you are in a restaurant and have ordered a burger. The restaurant's kitchen can prepare 30 burgers per hour. To deduce a waiting time of 2 minutes (60 minutes divided by 30 burgers per hour) from this would be naive, as you are probably not the only customer. If a total of 30 guests are waiting for their burger, this results in a throughput time of one hour (30 burgers divided by 30 burgers per hour). The same applies to your work packages: The more you process at the same time with the same number of team members, the longer it will take to process each individual work package. Now, of course, you can say that it is not the processing time of each individual work package that matters, but the processing time of the overall project. This is countered by the fact that the parallel processing of many work packages leads to so-called harmful multitasking (you will learn more about this in the chapter on lean project management), which in turn leads to mental setup times and slows down the project. In addition, the completion of a work package also means that the associated risks are eliminated. The faster you complete high-risk work packages, the fewer residual risks your project will have.

Measure and manage flow

The central metric (measure) in Kanban is the lead time of a work package until it is completed. This must be optimized. Another important metric is the waiting time of a work package with status *done* of a processing stage until it is picked up by the next stage and processed further.
The idea behind these considerations is that the faster a work package can be completed, the earlier risks to the overall project can be eliminated. It is difficult for a project manager to identify true obstacles and bottlenecks if work packages regularly have long waiting times due to dependencies on other work packages.

Warning

When dimensioning the work packages, special attention should be paid to two aspects for a meaningful measurement of the lead time:

- The work packages should be subdivided to such an extent that they can be processed by individual team members.
- The work packages should be subdivided to such an extent that there are as few dependencies on other work packages as possible.

Depending on the task, it may therefore make sense to break work packages down to the functional level to be implemented. In other words, there are no work packages on the Kanban board, but rather individual product functions to be specified, implemented and tested.
It is important that the lead time of the work packages (or functions etc.) is measured regularly. Inexperienced project managers and teams only start measuring when delays have already occurred. However, it is then difficult to make a comparison with other lead times. Therefore, at the start of the project, think about what metrics are important for you and your project, for example:

- the *lead time* of a work package or function: the faster a work package can be processed across all processing stations, the faster the project can usually be completed
- the *throughput*, i.e. the number of work packages per time: the more work packages can be processed per time, the faster the project can usually be completed
- the *waiting time* of work packages before the work is taken over by the next processing stage: the shorter the waiting time, the fewer work packages are in expensive "inventory", i.e. the more work packages are worked on productively
- the *cumulative flow*, i.e. the number of work packages at a certain processing stage.

Figure 4.28 shows an overview of the aforementioned metrics with an exemplary progression. Do not select any metrics that you do not use for control purposes. Each metric with the associated measurements increases the administrative measurement effort. Depending on the project, additional or different metrics may be useful, for example *number of work packages per team member*.

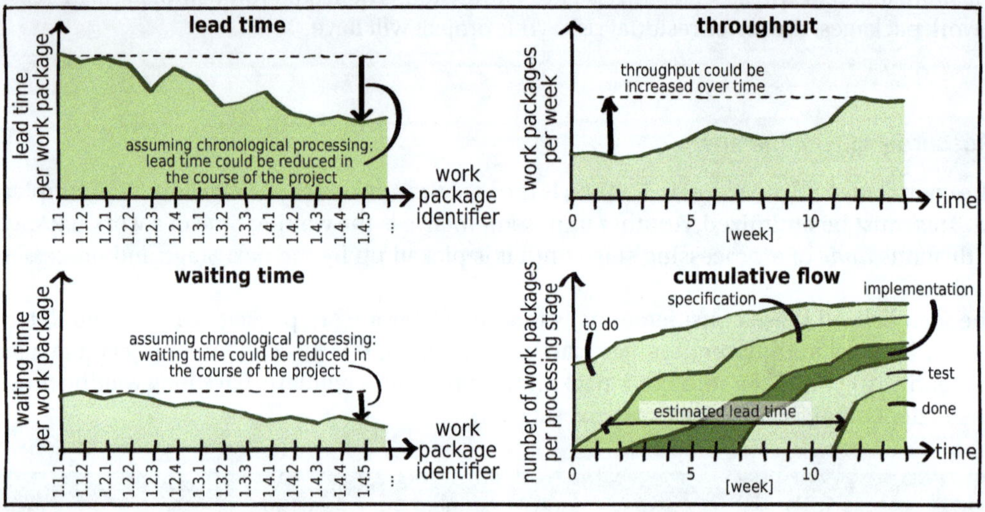

Figure 4.28: Exemplary representation of various Kanban metrics: The lead time is usually recorded per work package. The throughput indicates the number of completed work packages per time. The waiting time indicates how long a work package has to wait on average between processing in two processing stages, and the cumulative flow shows the cumulative total of work packages in a processing stage.

Kanban aims to optimize the workflow through the project. As a result, you should always consider how you can improve the lead time or throughput and reduce the waiting time of work packages. The cumulative flow helps you to identify bottlenecks. The cumulative flow chart can also be used to estimate the lead time. For example, Figure 4.28 shows a cumulative flow diagram at the bottom right. In this diagram, work packages that went into specification in the first week are completed in the eleventh week. This results in a lead time of 10 weeks. We look at an example of bottleneck analysis using the cumulative flow chart in the practical examples.

Make process rules explicit (clear and known)

This practice is actually a matter of course - but is not always implemented in practice: Everyone in the team must know the rules of cooperation. These include, for example:

- start, duration and location of meetings
- rules on when cards may be moved on the Kanban board
- rules for limiting the amount of work started
- communication rules
- priorities
- definition of Done
- stakeholders of the project.

The rules also form a basis from which improvements can be made to project management. For example, the rules for limiting work can be adapted if there is a concern that the current rules are hindering the flow of work rather than promoting it. A fixed set of rules forms the starting point for changes. The effects of rule changes can be measured and their success evaluated.

Make sure that everyone in the team knows the rules. Otherwise you will stir up resentment in the team if individuals do not follow the rules. In the long term, motivation and team cohesion will suffer.

You should therefore take rule violations seriously. If an employee is always late for a meeting, this disturbs the other employees and reduces added value because the others have to wait or things that have already been discussed have to be repeated.

Warning

Get to the bottom of rule violations and be careful with rash accusations. Employees who regularly arrive late may be late because they have another meeting with their line manager beforehand, the end of which they cannot influence.

In any case, look for the causes of rule violations, eliminate them and then implement the rules or revise them transparently for everyone so that they can be adhered to again. You can write down the most important rules directly on the Kanban board, for example.

Develop feedback loops

Feedback loops give the project manager or team the opportunity to measure progress, identify obstacles and solve problems. Typical information that should be reported (and processed) is:

- processing status of the work packages and their throughput times
- waiting times of work packages and/or team members
- quality status of the work packages
- obstacles to further processing
- ideas for optimizing your own processes
- etc.

For small teams, a **short daily meeting in front of the Kanban board** is a suitable instrument for discussing the aforementioned points. Removing obstacles may also require feedback from and coordination with people outside the project. Regular meetings, often called jour fixes, with line and program managers can be used for this purpose. The project manager or a representative of the project team should therefore maintain a good network with decision-makers outside the project.

Tip

Encourage open and timely feedback in your team through a good feedback culture. If employees are labeled as naysayers as soon as you openly address obstacles and potential problems, this leads to a culture of hiding and concealment. The problems will still surface – just much later. Then the solution may be much more expensive.

Perform collaborative improvements

The results of the feedback must be used to make improvements. Anderson recommends using established models and theories such as bottleneck theory (also known as the theory of constraints) or lean approaches to avoid waste [Anderson, 2010].

Warning

Project managers should constantly try to optimize processes within the team. The starting point for Kanban is the current way of working, current roles and responsibilities (first core principle of Kanban). It is therefore necessary to check whether communication within the team is working, whether the hierarchies are compatible with agile principles and whether more or fewer meetings make sense. It is also important to question whether the right people are taking part in the meetings. Obstacles and bottlenecks must be identified and removed at an early stage. Even when the project is going according to plan, there is still a need for action: the next possible hurdles must be identified and monitored with foresight. Potential for improvement must be identified and knowledge – including positive experiences and success factors – must be saved for future projects.
However, project managers who are purely involved in technical matters and work on technical work packages themselves often have no time (or believe they have no time) for improvements. The consequences are borne by the entire team or the company.

The emphasis on *collaborative improvement* is a direct consequence of the fourth core principle of Kanban: *Promote leadership and accountability at all levels of the organization.* Only employees who take responsibility for themselves, their own work, the overall project and the company will actively initiate and shape improvements. Such behavior must be demanded and encouraged. It puts high demands on employees who need leadership qualities beyond their professional qualifications and who are willing to contribute these without looking exclusively at the next step of their career.
This is one of the reasons why the introduction of Kanban is more than just switching project management from bar charts to Kanban boards. You can see an example of the introduction of Kanban below in the practical examples and case studies.

Comparison of Kanban and Scrum

Kanban, like Scrum, promotes agility. Scrum is known for its incremental approach in sprints. For many, the first thing that comes to mind with Kanban is the characteristic optimization of lead time, flow and the limitation of work in progress. The table in Figure 4.29 shows further aspects of Kanban and Scrum in comparison.
In practice, aspects of Kanban and Scrum are also used together under the name ScrumBan. This is a hybrid approach - in the sense of this book.
In particular, the practices

- **visualize the work (and its flow)** and
- **limit the amount of work in progress**

are also frequently used in projects working according to Scrum. As Kanban itself does not specify roles and communication methods like the daily Scrum, many teams working according to Kanban are guided by structures that they are familiar with from either Scrum or

Comparison of Kanban and Scrum		
	Kanban	**Scrum**
agile approach	yes	yes
limitation of work in progress	yes, continuously on the Kanban board	yes, but iteratively per sprint
team organization	self-organizing	self-organizing
time schedule	not defined, flow-oriented	discrete time in sprints (timebox principle)
important optimization criterion	lead time	velocity
roles	not defined	team, Scrum master, product owner
effort/size estimation	not defined	prescribed
changes	allowed at any time	only allowed in subsequent sprint; current sprint is not changed
visualization	flow-oriented Kanban board	status-oriented task board
persistence	Kanban board remains in place during the project lifecycle	task board is deleted after each sprint

Figure 4.29: Comparison of Kanban and Scrum in extracts based on [Kniberg and Skarin, 2010].

plan-based projects. In the practical examples and case studies at the end of the chapter, you will get to know a few examples of the design of roles in Kanban projects.

Kanban summarized

- Kanban is based on four core principles:
  - Start with what you are currently doing.
  - Strive for incremental, evolutionary change.
  - Respect current processes, roles, responsibilities and titles.
  - Promote leadership and accountability at all levels of the organization.
- For concrete application in projects, Kanban proposes six practices:
  - Visualize the work (and its flow).
  - Limit the amount of work in progress.
  - Measure and manage flow.
  - Make process rules explicit (clear and known).
  - Develop feedback loops.
  - Perform collaborative improvements.

Lean project management and theory of constraints

Lean management combines principles and methods for efficient, waste-free design of the value chain. Many of the ideas originate from the production system of the Japanese car manufacturer Toyota, which began its triumphant advance in the middle of the 20th century. Today, **lean management** is also used in non-production areas of a company, such as purchasing or project management.
So what is a system from production management doing in a book on project management? And why is it included in a chapter on agility?

Well, the Scrum and Kanban process models presented so far not only implement the values of the agile manifesto, but also some lean principles. They are considered lean and lightweight, attempt to focus on value-adding team activities and avoid unnecessary plans and cumbersome project structures. However, despite many similarities, lean project management should not be equated with agile project management.
Based on the widespread agile process models Scrum and Kanban, we are now generalizing and learning about lean principles and methods that help us to carry out projects lean and efficiently.
Understanding the theory of constraints, which is based on the ideas of Eliyahu Goldratt [Goldratt and Cox, 1984], will help us with implementation. Even if this should not be equated with lean management, the ideas contained therein will help us with concrete implementation and avoiding waste.

Lean principles

The literature on lean management contains various detailed principles that embody the lean philosophy. We will focus on the following principles and look at their implementation in the project:
- focus on the customer
- identify the value stream
- implement the flow principle
- use the pull principle
- strive for perfection.

Avoiding waste is a goal often associated with lean. In projects, waste manifests itself, for example, through
- poorly defined requirements
- erroneous or unnecessarily complex solutions
- frequent task switching (harmful multitasking)
- unnecessary interfaces in the project and between the project and other stakeholders
- unnecessary waiting times
- project bureaucracy
- unnecessary revisions of the project object.

Before I explain the lean principles mentioned at the beginning in detail and look at specific implementation options in projects, I will briefly discuss Goldratt's theory of constraints and critical chain project management. Both will help us below to develop concrete examples of lean project management implementation.

Theory of constraints

The core statement of the theory of constraints is quite simple:
The throughput of a system is determined by its bottleneck.
An example in a specific project situation illustrates this: If all functions developed in a project have to be tested and only one person is available for this, this limits the throughput

of the project. In this case, it is of comparatively little use to specify or develop the functions more quickly as long as the bottleneck is somewhere else.
For a project manager, specific instructions for action can be derived from this observation: The first step should always be to identify the bottleneck in the project, optimize its load, take further measures if necessary to relieve the bottleneck from unnecessary tasks, and expand the capacity of the bottleneck if required, see Figure 4.30.

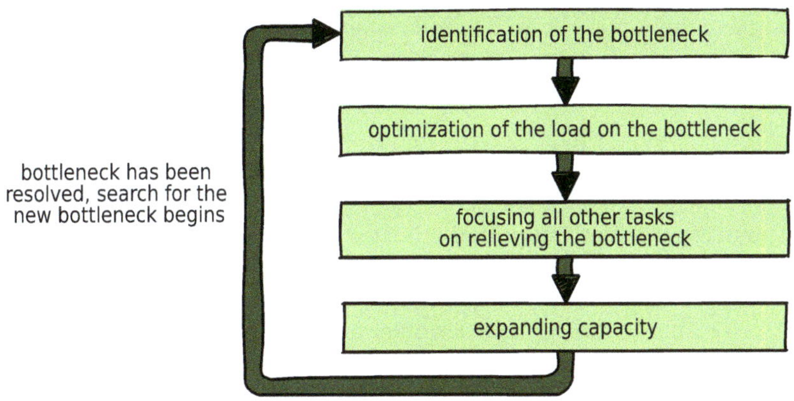

Figure 4.30: Process of optimizing the bottleneck to increase throughput in projects.

Identification of the bottleneck

Once a project reaches a certain level of complexity, it is difficult for the project manager to keep an eye on all work packages and team members at the same time. It is therefore important to focus and intervene where there is a bottleneck that is limiting throughput:

- Take signals such as waiting times from team members who are waiting for deliveries from other team members seriously.
- Check whether there are team members complaining about overload.
- Compare planned efforts and throughput times with actual efforts and throughput times. To visualize the flow and throughput times, you can use a Kanban board or traditional bar charts and action item lists, for example.

Optimization of the load of the bottleneck

Once you have identified the bottleneck, you can optimize its workload. It is important to observe the lean management principles here: The point is not to give the bottleneck as much work as possible, but to use the labor of the bottleneck worker to create as much value as possible:

- Activities that do not contribute to the success of the project should be omitted.
- Tasks that can be performed by other team members who are not themselves a bottleneck should be assigned to them.

Warning

Make a clear distinction between optimizing workload and overloading or overworking. The latter leads to frustration, reduces motivation and will exacerbate the bottleneck in the long term, as commitment will decrease.

Focusing all other tasks on relieving the bottleneck

The team member you have identified as a bottleneck limits the speed at which you can successfully complete the project. It therefore does not make sense to invest a lot of energy in optimizing all other team members – with two important exceptions:

- Team members who can relieve the bottleneck by taking over tasks should be appointed for this purpose.
- Team members who supply the bottleneck worker with results must ensure that the bottleneck does not have to wait for these results.

The second point is particularly important: the bottleneck limits the processing speed of the project. It would be negligent if the bottleneck itself cannot work and the situation is further exacerbated simply because other team members, who are not themselves a bottleneck, cannot provide their results to the bottleneck worker on time.

Expanding capacity

Regularly check whether there are short-, medium- and long-term measures to expand the capacity of the bottleneck. Examples of this are:

- relieving the bottleneck worker by having other team members take over tasks
- improving processes or the tools used so that the bottleneck worker can work more efficiently
- the qualification of team members so that they can take on tasks that previously only the bottleneck worker could perform
- the employment of additional team members or the procurement of additional resources who can perform the tasks at the bottleneck.

Once you have gone through these steps, you have probably resolved the bottleneck in your project and a different resource, i.e. team member or machine, is limiting the progress of the project. A new bottleneck has arisen. You should therefore regularly check where there are new bottlenecks. Proceed with these as outlined in Figure 4.30.

Critical chain project management

The consistent application of the theory of constraints in practice already helps to avoid waste and increase throughput in the project. It is therefore an important pillar of lean project management. Goldratt has further substantiated the application of this theory in the concept of the **critical chain** [Goldratt, 1997]. The critical chain is defined as follows:

> *The **critical chain** is the sequence of tasks that represent the actual bottleneck, taking into account the resources actually available in the project.*

I am supplementing Goldratt's recommendations with a few additional aspects.

Considerations for estimating expenses

In order to stay within the planned deadline and cost targets at the end of the project, some project managers are tempted to include a buffer or reserve in their effort estimates. The same applies to team members who prefer to play it safe and give higher estimates for the effort than is actually necessary. But the reverse is also true: People overlook important aspects or are generally too optimistic about the future processing of a work package and underestimate the effort to be determined. Both underestimation and overestimation are bad in terms of lean management, as they create waste, see Figure 4.31.

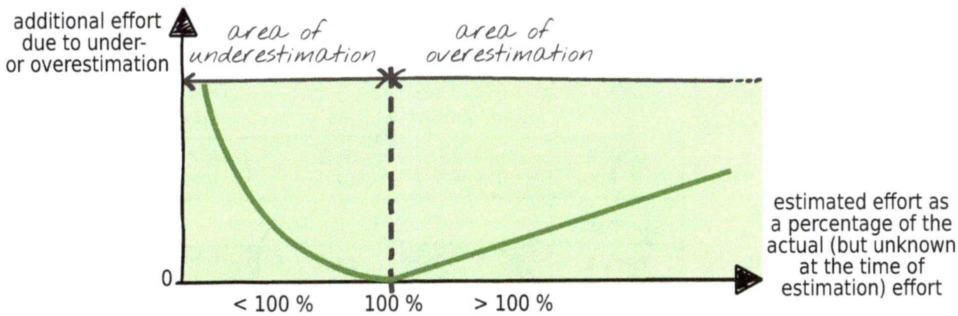

Figure 4.31: If the actual effort required to perform a task is underestimated (< 100%) or overestimated (> 100%), additional effort is required, which increases the total effort of the project [McConnell, 2006].

How often do you find that team members hand in their tasks earlier than agreed? Frequently? Congratulations – in practice, however, this tends to be the exception. In fact, most team members have many tasks to complete at the same time. If they notice that the completion of one work package is progressing faster than necessary, they switch to other work packages that seem more urgent. C. Northcote Parkinson described this relationship as follows in the **Parkinson's law**, which was later named after him:

> *Work expands so as to fill the time available for its completion.*

Parkinson's law is closely linked to the **student syndrome**, which states that students tend to always leave annoying things like exam preparation to the last minute. To the students' credit, however, it must be added here that this type of behavior is also common among non-students and can also be observed in projects.

The more time you give team members to work on tasks, the more time they will need to do so. This approximate linear relationship is shown in Figure 4.31 in the area of overestimation (> 100%). Parkinson's law and the student syndrome are always particularly problematic when risks suddenly occur: There would actually have been enough time to deal with them. However, as time was wasted in the sense of Parkinson's law, this time is now missing as a buffer for the risk that has occurred.

Example

Assume you have a project with four tasks to be processed sequentially. The actual, but a priori unknown effort of all four tasks is 10 days each, see Figure 4.32 above. The estimated effort of the four tasks is 7 days, 12 days, 8 days and 13 days, see Figure 4.32 below. During processing, team members cannot be faster than the actual effort required. Task 1 was incorrectly estimated at 7 days, but requires 10 days, see Figure 4.32 in the middle. Task 2 was incorrectly estimated to take 12 days, although it could have been processed in 10 days. According to Parkinson's law, however, the team member will take the 12 days once estimated, so that the actual effort will be 12 days. The same considerations as for activities 1 and 2 apply analogously to activities 3 and 4. Due to the effects described, the project in this example takes 5 days longer than actually necessary. This does not take into account additional delays caused by any risks that may occur.

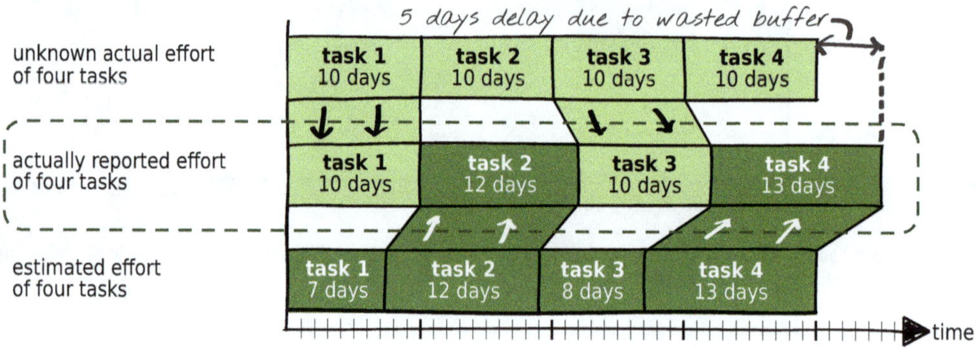

Figure 4.32: Consequences of Parkinson's law and the student syndrome: In practice, employees take as much time to complete a task as was originally estimated. If the task was underestimated, they work on it longer than planned. These two effects lead to an prolongation of the project.

The previous example shows that projects can be delayed simply due to incorrect estimates. Although the estimated total effort and the actual effort of the work packages are each 40 days, the project takes 5 days longer than planned. After these considerations, let's look back at Figure 4.31. We now know the consequences of Parkinson's law and the student syndrome. But what if we underestimate the **effort** (< 100%)? The figure predicts a more than proportional increase in additional effort in the project. Anyone who has ever managed a real project can probably guess the reason for this: If there is a threat of delay compared to the plan based on estimated values, hectic activities set in:

- plans need to be updated and
- emergency meetings are scheduled with the aim of speeding up the project. Initially, however, they tie up resources that cause costs without contributing to the progress of the project.

Conclusion: Take the topic of effort estimation seriously and try to estimate the effort as well as possible. In Chapter 8 you will find suitable methods that you can use for this. The three-point estimate is simple, quick and good, while the Delphi method, for example, is more accurate but more complex. An explanation of these and other estimation methods can be found in the toolbox in Chapter 8:

- P.2.1 Perform expert estimation
- P.2.2 Estimate by analogy
- P.2.3 Perform three-point estimation (PERT)
- P.2.4 Perform parameter-based estimation
- P.2.5 Apply function point method
- P.2.6 Estimate user story
- P.2.7 Play planning poker

Avoiding harmful multitasking

Imagine two tasks of 10 staff days each. If two qualified team members start working on each one, it will take 10 working days each. If you only have one person available who is qualified for both tasks, you have to decide whether you want to process the tasks sequentially, i.e. one after the other, or in parallel.

It often sounds tempting to start several tasks in parallel, as this gives the feeling that something is happening, i.e. the project is progressing. In fact, working on multiple tasks in parallel creates waste, see Figure 4.33.

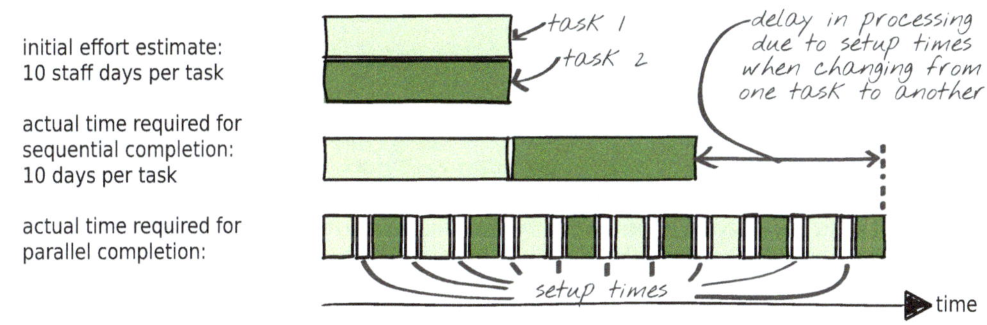

Figure 4.33: Harmful multitasking leads to delays and overload.

You're probably familiar with this situation: there are several tasks on your desk, for example the documents to prepare for two different exams or two work packages for your manager, who wanted the results already "yesterday" as usual. We can't really work on the tasks in parallel. We start with one and switch to the next task after a certain amount of time. At some point, we switch back to the first task. We then have to find out where we need to continue working. We need a setup time or switching time to move from one task to the next. This was probably not taken into account in the original workload estimate. However, since the time of the test or the handover to the manager is fixed, we have to work more in order to finish on time despite the additional setup time.

In the long term, this extra work can lead to an overload. In any case, it is waste in the sense of lean management and should be avoided whenever possible.

Tip

As a project manager who acts according to the principles of lean management, you try to reduce the parallel processing of tasks by one employee as much as possible in order to prevent waste.

Working with buffers

We have already talked about Parkinson's law and the student syndrome. We know that overestimating effort leads to waste. But we also know that we ourselves like to overestimate the effort involved in tasks in order to protect ourselves against the unforeseen. We call this protection **safety buffer**. Depending on the project risks, these are important and correct. However, they must be placed correctly so that they do not fizzle out due to Parkinson's law.

Tip

A major challenge for project managers is to motivate employees and reduce the number of individual safety buffers in favor of important project buffers. This requires trust on the part of the project manager and responsibility on the part of the team members. If team members have to expect negative consequences if deadlines are not met, they will inevitably schedule personal safety buffers and, in accordance with the student syndrome and Parkinson's law, deliver later than they actually could. Besides, no one wants to undermine their own estimate. Otherwise, there is a risk that the project manager will no longer trust it in future projects.

As a project manager, you need *buffer sovereignty* – in consultation with the team, of course. If you take away all safety buffers from team members and punish them at the end of the project for not meeting deadlines, you will lose this buffer sovereignty and team members will always overestimate their efforts in future in order to protect themselves personally against this punishment. You will then lack meaningful buffers for dealing with risks that arise. Of course, you could build in additional buffers, but this will unnecessarily delay your project and make it more expensive.

Let's assume that you manage to obtain net estimates from the team members and also an estimate of how much safety buffer is required for each process. You could then plan these safety buffers at the critical points in the project. This can be achieved with the three-point estimate, which is also known as Program Evaluation and Review Technique (PERT). See Figure 4.34 for an example.

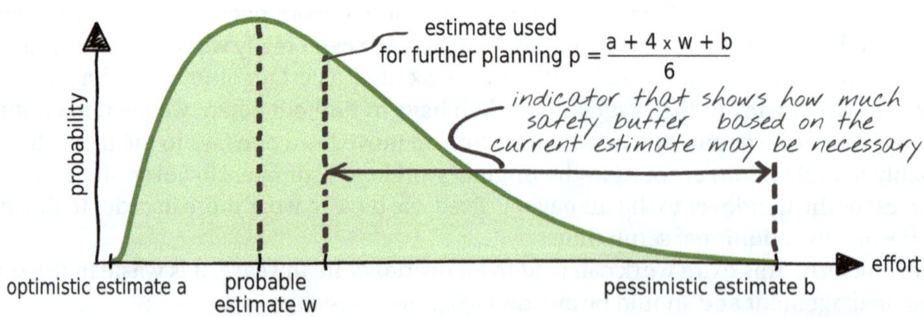

Figure 4.34: Optimistic, probable, pessimistic and estimate used for further planning based on the beta distribution.

Experts not only estimate one value for the effort, but instead an optimistic, a probable and a pessimistic value. A value p used for planning can then be calculated on the basis of the beta distribution. The formula for calculating p is shown in Figure 4.34. The range between the pessimistic and optimistic value or the value used for planning is an indicator of how much uncertainty and therefore risk there is in the estimated task and how much safety buffer should be planned for. Whether you use the difference between the pessimistic value b and the probable value w or a different value depends on the risk awareness and the ability to manage these risks. However, it would be negligent to ignore a large difference in the estimated values and just hope that the estimated value p used for planning can somehow be realized.

Example

A project manager receives the values a = 10 staff days, b = 40 staff days and w = 25 staff days from his software developer as estimates for the implementation of a certain functionality. As the pessimistic value b differs significantly from the optimistic value a, the project manager questions the estimate. In a discussion with the software developer, the project manager identifies an unclear requirement as the cause for the large difference between a and b. After the requirement has been specified together with the customer, the estimated values are a = 10 staff days, b = 18 staff days and w = 13 staff days. The project manager adopts the planned value $p = (4 x 13 + 10 + 18)/6 = 13$ staff days according to the formula shown in Figure 4.34 and plans an additional 5 staff days buffer, which corresponds to the difference between the pessimistic value b and the planned value p.

Tip

Always question uncertain estimates. Try to reduce large differences between the pessimistic and optimistic estimates by analyzing the causes and taking appropriate measures. Be aware that the difference between the estimates will have an impact on your project schedule or your project costs. Depending on how accurately you want to predict the duration and costs, you must take action if the estimated values differ. You should plan concrete measures to reduce the uncertainties at the latest if there are deviations of 50%. Conversely, deviations between the pessimistic and optimistic values of 10% to 20% are already very good. You can incorporate the respective residual uncertainty as a risk in your risk management and include it as a safety buffer in your project planning, e.g. as contingency or management reserve.

In his book *Critical Chain* [Goldratt, 1997], Goldratt gives two very simple rules on how the aforementioned safety buffers should be planned into the project:

1. Identify the critical chain in your schedule, i.e. the tasks on which the team member identified as a bottleneck is working. Remove all safety buffers in these tasks, reduce them by half and then schedule them at the end of the project.
2. Determine all task chains that deliver to the critical chain. These task chains are also called the feeding chain. Remove all safety buffers in these tasks, cut them in half and schedule them at the point in the plan where the feeding chains merge into the critical chain.

The two rules are explained using an example in Figure 4.35.

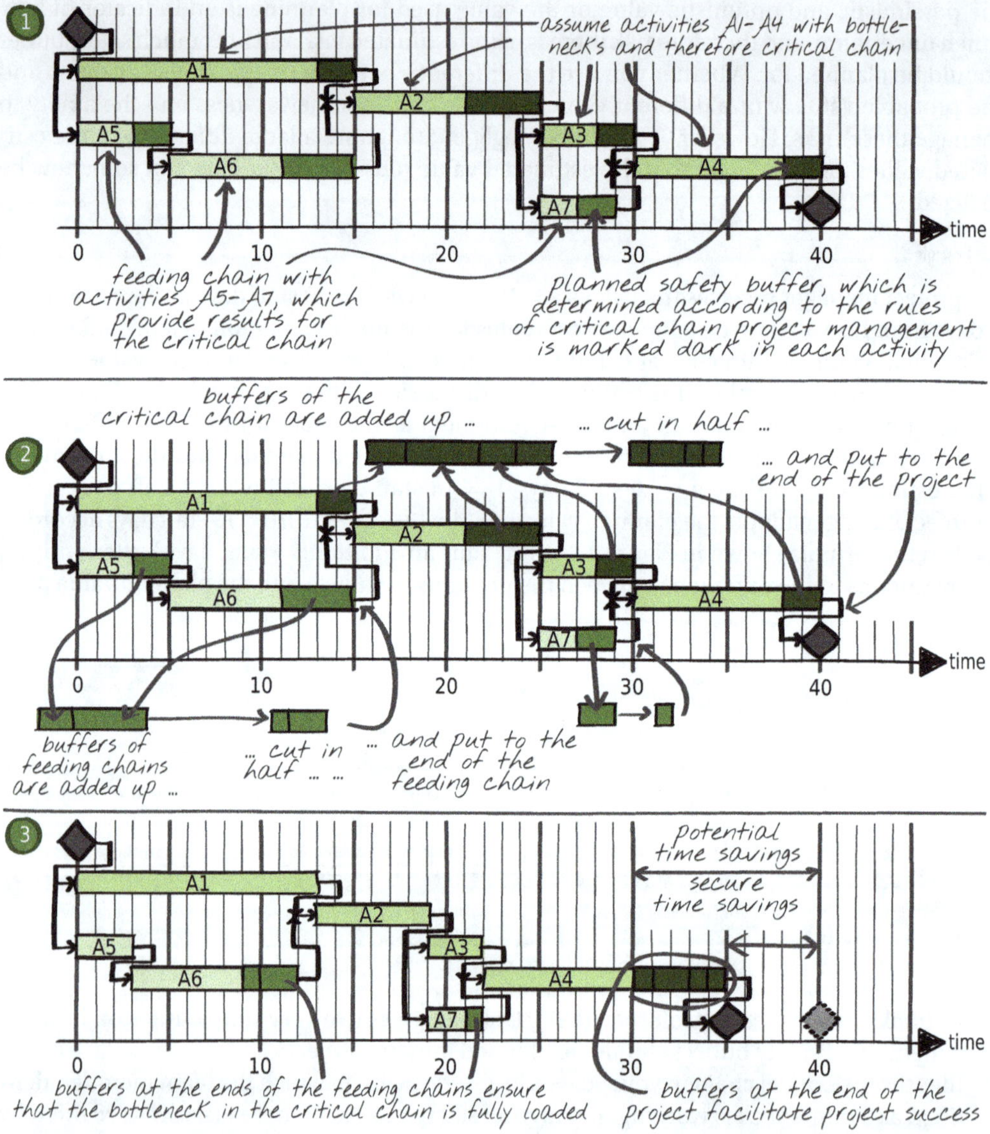

Figure 4.35: 1) Gantt chart of a project. Each activity contains a safety buffer marked in dark. The employee representing the bottleneck works on activities A1 to A4. 2) The buffers of the critical and feeding chains are each added up separately, halved and moved to the end of the respective chains. 3) Adjusted Gantt chart: The project duration is reduced from 40 to 35 time units. If the planned safety buffer is not required at the end of the project, it can even be reduced to just 30 time units.

Tip

Please note: By deliberately placing safety buffers at the critical points of the project, you can not only shorten the project duration, but also increase the probability that you will actually be able to meet the scheduled deadline. This may sound contradictory at first, as a more ambitious plan usually goes hand in hand with a greater risk of delays. However, this does not apply if you manage to avoid the effects of Parkinson's law and protect the project against any risks that arise by cleverly placing buffers.

In practice, of course, you can deviate from the rules:

- If you do not want to secure the end of the project, but an intermediate date such as a trade fair visit or a delivery to a customer, do not place the buffer of the critical chain at the end of the project, but before this intermediate deadline.
- The rule of halving the accumulated buffer is based on the observation that buffers are associated with uncertainties and therefore risks. The simplified assumption for practice is: only every second risk will occur, so only half of the buffer is needed. Whether the reduction by 50% is suitable for your projects depends on your need for safety or the specific risks associated with the buffers. Of course, you can decide individually for each process whether you want to take over the buffer completely or reduce it proportionately.

Overall, it can be observed that projects using critical chain project management are 20% to 30% faster or cheaper than conventionally planned projects. So far, we have focused on a single project. In practice, however, it is quite common that dependencies between projects arise as a result of employees working on multiple projects, even if they have nothing to do with each other in terms of content. These dependencies are particularly dangerous if employees who are identified as a bottleneck work on several projects at the same time and may also be the bottleneck in several projects. In such cases, Goldratt recommends **staggering of projects** by the program or portfolio management. The aim is to minimize dependencies and harmful multitasking of bottleneck resources.

Warning

On paper, the parallel processing of several projects may seem attractive because they deliver results quickly if they are completed on schedule. However, if the projects are interdependent due to shared bottleneck resources, it is unlikely that they will be completed on schedule. Without appropriate measures, *all* projects will be delayed. When releasing projects, good program and portfolio managers therefore ensure that dependencies due to bottleneck resources are largely avoided. To this end, projects can be staggered so that projects with the same bottleneck resources do not run in parallel. Alternatively, projects can be decoupled by changing the allocation of resources.

Applying lean project management

We have developed some basic principles that we now want to apply in order to put lean project management into practice. The following sections provide examples of how this can be implemented. Unlike Scrum, for example, there is no fixed set of rules that tells you exactly how to proceed in order to apply the lean principles.

Focus on the customer

Start with the obvious and first identify and characterize the customer of your project. This sounds trivial - but we often start a project and think we understand the customers without having dealt with them and their interests.
The next step is to identify the customer's requirements systematically and as completely as possible. Scrum chooses the product backlog for this and promotes an iterative approach. If other process models are chosen, this possibility of structured updating of requirements may be missing. Many projects fail not because the customer's requirements are too high, but because requirements were not recorded completely or comprehensibly.

Tip

When determining requirements, do not forget any important stakeholder groups, but go through the product life cycle consistently. Think about which stakeholders come into contact with the product and when, and what requirements they have of it. Again and again, products that are actually considered finished have to be revised because they are difficult to produce or cannot be maintained cheaply. Stakeholders from the production and service departments were obviously not sufficiently involved in the requirements analysis. Late reworking of products is wasteful and therefore not in the spirit of lean management.

A lack of customer understanding can lead to both
- *missing requirements* and to
- *unnecessary requirements*, which the customer does not value at all.

The latter case is also mockingly referred to as the development of golden taps. This refers to functionality that makes the product more expensive without any added value for the customer.

Identify the value stream

Do you know the value stream of your project? In project management, the value stream refers to all value-adding and non-value-adding processes and tasks that are necessary for the implementation of the project deliverables. As soon as you know these stream of tasks, you can start to separate value-adding from non-value-adding tasks and reduce the latter. It is important that you eliminate waste as far as possible. This occurs in projects in the following form, for example:
- bureaucratic processes
- too many meetings
- the storage of work packages that have been started but not yet completed
- the use of inefficient tools
- the use of unqualified employees
- unspecific objectives for tasks
- unnecessarily high documentation effort
- a lack of knowledge management, which leads to similar errors being made again and again.

You don't have to set the entire project management machinery in motion for every small and medium-sized project. Conversely, it is negligent to forget important steps such as determining the requirements, risks and stakeholders.
Good communication does not mean that all employees have to sit in a long meeting every day. Make sure that the right people talk to each other and create communication channels or media through which important information can be distributed efficiently. A project information board at the coffee machine can sometimes be more efficient than a daily status meeting in which no problems are solved.

Tip

Regularly check whether your employees have the necessary skills and tools to complete their current tasks. A lack of knowledge or inefficient tools are demotivating and lead to a waste of important resources.

Get into the habit of questioning the work that is done in the project and the processes for completing this work. In many companies, inefficient working methods have crept in, which are adopted and accepted without reflection from project to project.
When avoiding waste, it is important that you eliminate its *cause* and do not keep fighting symptoms. Tools for identifying these causes include the 5-Why method or the cause-effect diagram (Ishikawa diagram). Both methods are presented in the section *Practical tools for daily project work* in Chapter 3.
You have probably noticed that many of the aspects of the lean principle fit perfectly with agile values and principles. Less project bureaucracy, avoidance of waste, efficient meetings and appropriate documentation are also characteristic of agile projects based on Scrum and Kanban.

Implement the flow principle

Take a look at your project schedule and focus on the critical chain. Then ask yourself what obstacles and risks exist that could prevent the work on the critical chain from being processed as quickly and efficiently as possible. You need to focus on these in particular in order to increase the work flow and throughput of the project.
The ideas of critical chain project management discussed above, taking into account the bottleneck theory, can be used to optimize throughput:

- Avoid harmful multitasking.
- Reduce personal safety buffers, which are distributed across the entire project and instead place sufficient buffers at the critical points of the project. This refers to important milestones or the handover of results from non-critical resources to critical bottleneck resources.
- Decouple projects as much as possible. This applies in particular to interlinked projects with employees who represent a bottleneck in several projects at the same time. Otherwise, your entire project portfolio will come to a standstill at some point.
- Observe whether bottlenecks occur again and again with the same employees and take countermeasures by:
  - providing employees with better tools to work more efficiently

- training employees on how to work more efficiently
- relieving employees from non-critical tasks,
- qualifying other employees so that they can also take on the work of the bottleneck resource in the future
- hiring additional employees to relieve the bottleneck resource.

The flow principle probably reminds you of Kanban. This is not surprising, as Kanban and Lean have their origins in production and are used together there. At this point, you will once again recognize the similarities between lean and agile project management.

Use the pull principle

The pull principle in production and logistics means that work is pulled from the end of the process.

Example

A company operating according to the pull principle aligns its product creation with customer demand. The customer requests a product via the sales department, the sales department requests the creation of the product from the production department, the production department requests the materials required for production, and so on. The customers are therefore the trigger and pull on the product they want to purchase. In contrast, with the push principle, product creation is initiated by the company and the product is then pushed onto the market.

The pull principle can also be applied to projects: in traditional plan-based projects, the project manager assigns tasks to team members, which corresponds to the push principle. In agile projects and especially in lean project management, employees pull tasks to be completed. For example, if there are several tasks on the taskboard, employees who have finished one task independently pick up the next task and start working on it. They select the work in such a way that the project can be completed as efficiently as possible. The pull principle means that you need employees whom you can trust and who take responsibility, combat waste and optimize the value stream with a high level of commitment. You need employees who do not do things by the book, but who critically question the status quo, recognize improvements and initiate their implementation.

Team members will only develop these qualities in a correspondingly strong corporate culture. If a different philosophy is exemplified by senior management, the project manager will find it difficult to motivate team members to act independently.

Among other things, the following are supportive:

- the delegation of authorities and responsibilities
- the continuing qualification of employees
- the appreciation of employees and their results
- good communication
- opportunities for participation to contribute improvements.

Strive for perfection

Lean management approaches are based on the strive for perfection. This includes:

- continuous further development of employees
- the delivery of high-quality products and
- the continuous improvement of processes.

For employees, **continuous development** means learning from mistakes and avoiding similar mistakes in the future. In other words, knowledge management plays a central role and must be promoted accordingly.
The real challenge here lies in practicable, sustainable implementations. Comprehensive project experience reports to secure knowledge are bureaucratic and are neglected after a brief initial euphoria.
Collaborative approaches such as

- a wiki or
- professional social media

on the company intranet promise the participation of everyone and little waste due to complicated processes. However, it is not enough to create media to secure knowledge if these are not consistently used to retrieve knowledge.
It is not enough to anchor the promotion of continuous learning in the project, as a lot of knowledge is lost at the interface from one project to the next. Consequently, structures must be created that allow employees to retain knowledge for the company across projects. This includes, for example, knowledge about

- efforts and costs of certain recurring work packages
- typical project risks and successful countermeasures
- success factors in stakeholder communication
- strengths and weaknesses in cooperation with certain suppliers.

Let's look at the example of effort and costs: the creation of a requirements specification or the final documentation are typical work packages that have to be completed in many projects. When estimating the effort required for these packages, you can benefit from empirical values from comparable previous projects. However, a project manager must be able to call up this experience from previous projects at the start of the project. It is wasteful and therefore not in the spirit of lean management if she has to laboriously search through old project plans for the figures she is looking for. However, if the company has standardized work breakdown structures in which the final documentation work package always has the work package number 1.5.3, for example, and if the estimated and actually required efforts and costs are maintained by the project managers in a shared database or spreadsheet, the search is quick and easy.

Tip

Check which knowledge helps project managers and team members with project planning and implementation. Be aware that pure knowledge storage without knowledge retrieval is a waste and always look for practicable solutions for knowledge storage and retrieval that can also be used in the long term in the hectic day-to-day running of a company.

In projects that follow lean principles, the focus is on the customer. They expect a **product of high quality**. If you implement the lean principles correctly, the product will satisfy the customer and achieve the agreed quality.
Waste occurs, among other things, when

- the product has additional functionality that the customer does not value
- the product has to be reworked because the product does not achieve the agreed quality
- customers are lost because they are dissatisfied with the product.

One of the first principles that prospective project managers learned in the past (and in some cases still learn today) is the iron triangle of project management consisting of costs, deadlines and quality. If something is changed in one of these three corners, this has an impact on the other corners of the triangle. If time is running out at the end of the project and the product delivery date is at risk, either the costs can be increased in order to meet the deadline with more resources, or quality can be compromised.
Applying lean management means focusing on the customer and delivering products that meet their quality standards. This should be the guideline for your decision on how to deal with the limitations of the magic triangle.
The principle of **continuous improvement** is established in almost all quality management standards. Improvement is often divided into the four steps **p**lan, **d**o, **c**heck and **a**ct, see Figure 4.36. This is also referred to as the PDCA or Deming cycle. The American William Edwards Deming was one of the founders of the Toyota production system in the 1950s.

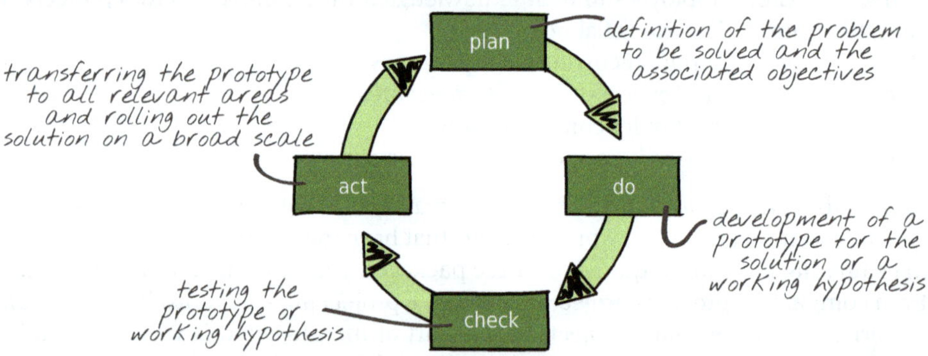

Figure 4.36: PDCA cycle: A plan is developed to solve a problem, initial prototypes or hypotheses are created, checked and implemented if successful. The cycle then starts all over again with the identification of further improvements.

Lean project management summarized

- Important principles of lean project management are:
  - focus on the customer
  - identify the value stream
  - implement the flow principle
  - use the pull principle
  - strive for perfection.
- Critical chain project management, which is derived from the theory of constraints, helps with implementation in everyday life through
  - the identification of bottlenecks
  - the optimization of the load on the bottleneck and the avoidance of waste
  - the alignment of all other tasks to relieve the bottleneck and, if necessary
  - the expansion of the bottleneck.

More agile process models

Introduction

In the following chapters, I would like to introduce you to some additional process models that promote agile working. As these process models either

- do not cover the entire project management process, such as Design Thinking and Lean Startup,
- are only suitable for specific projects, such as DevOps or
- are not (or no longer) particularly widespread, such as Extreme Programming and Crystal

we will only discuss them relatively briefly. However, knowledge of the basics of these process models should be part of the toolkit of modern project managers. For example, Design Thinking can be used in certain project phases to develop innovative solutions. The DevOps concept helps when considering the entire product life cycle. Aspects of Extreme Programming and Crystal can help to deliver good results in pure software projects. And Lean Startup promotes the implementation of innovations in practice.

Design thinking

Design thinking is a process model for problem solving, idea generation and evaluation. Even though design thinking does not constitute a complete project management process model, it can be a valuable asset for projects when it comes to developing and implementing innovative ideas in an agile manner.

Design Thinking is based on the assumption that an interdisciplinary team can come up with good ideas in a creative environment. Design Thinking is usually moderated or accompanied by a Design Thinking coach, often also referred to as a **facilitator**. The process of a design thinking project consists of a **macro cycle**, the steps of which follow recurring **micro cycles**, see Figure 4.37.

The macro cycle begins with the search for divergent alternative solutions. The team is asked to collect inspiring ideas and build up a large solution space of different alternatives, see point (1) in Figure 4.37. Encouraging divergence is intended to increase creativity.

The features critical to the success of the product are then identified (2) and comparisons are made with products and solutions on the market (3). Based on the previous considerations, a dark horse prototype is now designed (4). This should be surprising and creative. It is helpful if limiting restrictions or technical framework conditions are not taken into account. The funky prototype (5) should go one step further in the direction of innovation. It has no limits in terms of feasibility or cost-effectiveness. Even if the funky prototype can probably never be realized in this form, it can be a source of inspiration for new solutions or at least provide creative partial solutions for the final result to be derived later. It is important that the funky prototype solves the identified problem creatively and innovatively in the interests of the future users.

After each of the steps (1) to (5) outlined so far, but at the latest after step (5), the so-called vision prototype (6) is created in the so-called groan zone. It should be so concrete that it can be evaluated by potential users. Depending on the evaluation results, a step is taken back to (1) to (5) or towards convergence via steps (A) to (D). The vision prototype should be able to

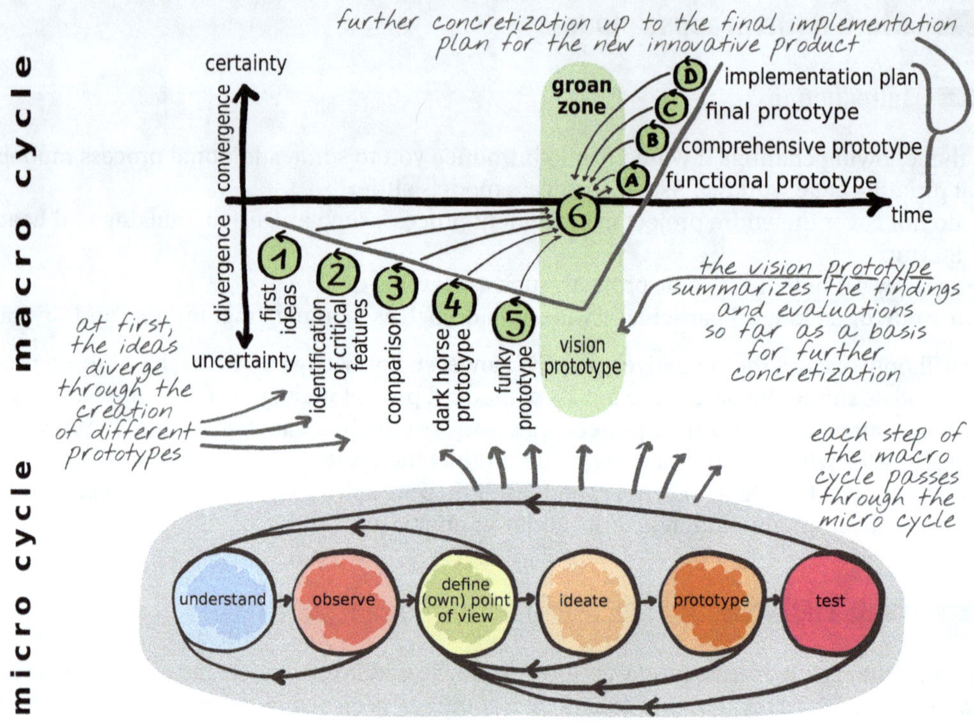

Figure 4.37: Overview of the process of a design thinking project: In the macro cycle, innovative ideas are first conceived and tested as prototypes 1-6. This is then used to develop an implementation plan for a new product using realistic prototypes A–D. The individual steps of the macro cycle pass through the so-called micro cycle.

generate attention and inspire potential users. However, as a transition from divergence to convergence, it must also take feasibility into account.

If the evaluation of the vision prototype was successful, it is developed into a functional prototype (A). The comprehensive prototype (B) and the final prototype (C) are used for further development up to the implementation plan (D). If the tests of steps (A) to (D) are not successful, for example because potential users are not satisfied with certain components of the prototypes or the implementation plan shows that certain aspects cannot be reasonably realized contrary to the previous consideration, it is possible to switch back to steps (1) to (6). Each of the briefly outlined steps (1) to (6) and (A) to (D) follows the micro cycle.

The micro cycle, see the lower part of Figure 4.37, begins with the development of an understanding of the problem, possible causes of the problem and the identification of affected groups of people (e.g. future customers of a new product). In the next step, we should form our own picture and observe. We try to understand the problem and supplement the problem description with our own impressions. This enables us to define our own point of view for further work. The knowledge gained so far is collected, analyzed and weighted. The next step is to find ideas that solve the identified problem. As with any brainstorming session, it is also desirable in this phase to first collect many different ideas before these are later categorized and prioritized. We develop promising ideas into prototypes. Prototypes can be simple handwritten drawings, photorealistic representations, videos and graphic or physical 3D models, for example. The aim is to make the ideas as realistic as possible. This makes the final step of the micro cycle, testing, easier. In Design Thinking, we want frequent and meaningful feedback on our ideas.

Tip

Design Thinking can be easily integrated into other process models, for example as a particularly agile first creative phase or to determine the requirements for the project object. If you would like to find out more about Design Thinking and helpful tools for its application, you will find a beautifully illustrated and informative description in the book *The Design Thinking Playbook* [Lewrick et al., 2018].

Lean Startup

Lean Startup is a process model that aims to test an idea on the market as quickly as possible with the least possible use of resources and then establish it. It is therefore not a typical process model for project management in general. In the context of agile product development, however, Lean Startup has become quite popular and is therefore part of the toolkit for agile working.

The Lean Startup process model is outlined in Figure 4.38. A product idea forms the hypothesis whose success needs to be tested on the market. An initial prototype, the so-called minimum viable product (MVP), is created for this purpose.

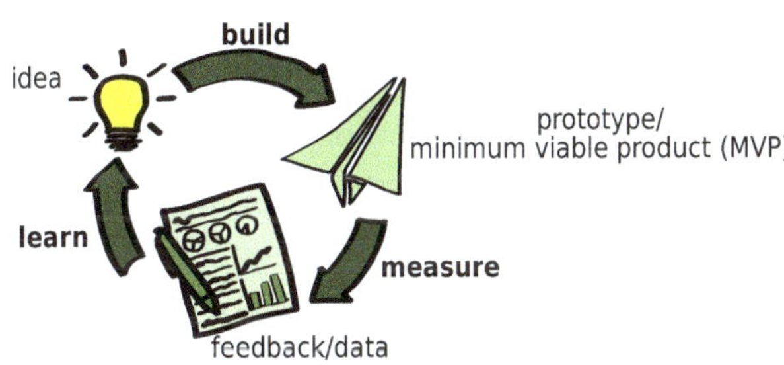

Figure 4.38: Overview of the Lean Startup process model.

> *A **minimum viable product** is a product that offers the minimum number of features to be usable and valuable for the customer.*

Warning

The minimum viable product must be developed with careful consideration. It should be as simple as possible so that its acceptance on the market can be tested as quickly as possible. At the same time, it must be attractive. If it is not accepted due to bugs or unusable functions, the idea will be wasted and customers will be annoyed. It will then be difficult to place the same idea on the market again.

Example

The car manufacturer Tesla is a well-known example of the use of Lean Startup. Its first product was a roadster, which was designed to show how attractive electric vehicles can be. It was produced in small numbers on the basis of an existing vehicle (Lotus Elise).

The acceptance of the minimum viable product on the market is carefully measured. This includes, for example, product functionality, distribution channels, price, product maintenance issues and accompanying services. The collected data is analyzed and the necessary conclusions are drawn. This allows the product to be developed further.
Lean Startup is accompanied by principles designed to promote the entrepreneurial spirit and focus on the customer. It includes:

- Entrepreneurs are everywhere: Promoting people with an entrepreneurial spirit even in traditional organizations.
- Entrepreneurship is management: Aligning management with the needs of product innovation and entrepreneurship.
- Validated learning process: The creation and careful review of the minimum viable product is seen as a value-adding process. The feedback of the knowledge gained is essential for long-term success.
- Accounting of the innovation (innovation accounting): The product lifecycle is monitored and controlled. Particularly in the initial phase, when it is often not yet possible to generate profits, it is important to establish non-monetary controlling.
- Build, measure, learn: Continuation of the iterative product development process, as illustrated in Figure 4.38.

If you would like to find out more about Lean Startup, you will find a comprehensive description in the book by Eric Ries, who played a key role in shaping Lean Startup [Ries, 2017].

DevOps

DevOps is a process model for the **Dev**elopment and **Op**eration**s** of software. Accordingly, the name is made up of the abbreviations Dev for development and Ops for operations. The basic idea behind DevOps is to reduce the gap that occurs in practice between agile software development with regular product increments and software operation, which is geared more towards stability.
I will introduce you to the individual steps of DevOps using the software life cycle shown in Figure 4.39. The development of the software is planned, implemented as program code (code), build to form the overall system (build) and then tested.
The operation of the software begins with the release. This is followed by deployment to the users and actual operation, which is monitored and analyzed. The findings from this are used to plan the next release.
The methods used during the lifecycle include continuous integration and continuous delivery, i.e. the continuous creation of new software code, its integration into the overall system and automated testing. The aim is to achieve an overall system that is always stable and continuously improving. This requires effective version and configuration management as well as

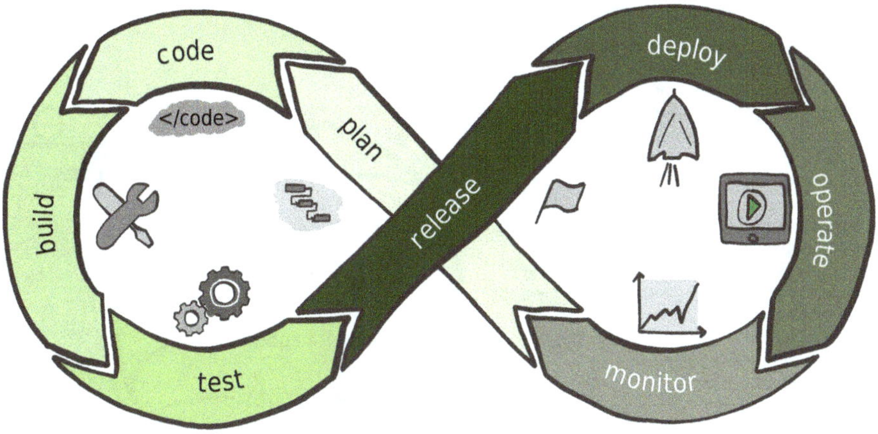

Figure 4.39: DevOps connects the development of software (light green with black lettering) with its operation (dark green with white lettering).

an infrastructure that enables regular updates to the overall system during productive operation. In order to be able to respond flexibly to the results of monitoring and quickly integrate aspects that are important to the user, the actual development and programming is carried out agilely in incremental steps.

Tip

The DevOps process model is now standard in many software projects, as it promotes the short-cycle development, delivery and improvement of software. Many products containing software, such as smartphones, computers and now even cars, regularly receive software updates that either eliminate errors or security gaps or provide new functions. DevOps can be combined well with other process models such as Scrum and Kanban.

Extreme Programming

Extreme Programming (often referred to as XP) is a process model that originated in software development. It is based on the work of Kent Beck, Ward Cunningham and Ron Jeffries and was developed in the 1990s. The core values of Extreme Programming are courage, respect, communication, simplicity and feedback. Similar to Kanban, Extreme Programming also consists of principles and practices. However, the formulation is much more specific. The principles include a focus on quality, humanity, self-equality, efficiency, reflection, small steps, mutual benefit, diversity and the pursuit of improvement. The main practices according to [Beck and Cynthia, 2004] are briefly summarized in Figure 4.40 (left).

Extreme Programming uses many planning and feedback loops, which take place at different intervals and are shown in Figure 4.40 (right). A release plan is created at intervals of one or more months, which is detailed in an iteration plan. Various tests and meetings at shorter and shorter intervals promote the quality of the results and the exchange of information.

Extreme Programming is suitable for pure software projects in an agile environment. It has some similarities with Scrum and Kanban, which is why similar recommendations apply as

Important practices of Extreme Programming

- **Spatial proximity of workplaces:** Workplaces are open (open space) and close to each other.
- **Team:** Projects are staffed in such a way that all disciplines relevant to the project are available.
- **Informative workplace:** Important aspects of the work should be visualized at the workplace.
- **Energetic work**: The team only works for as long as its energy reserves allow it to work productively.
- **Pair programming:** Two programmers work on the code simultaneously at one workplace.
- **Stories**: Functionality is described as stories.
- **Weekly cycle**: Each week begins with a review and the definition of the stories that are still open.
- **Quarterly cycle**: Rough planning of the "big picture" takes place on a quarterly basis.
- **Slack/relaxed work**: There is room for the unforeseen in every cycle.
- **Ten-minute build**: The entire system can be built and automatically tested in around ten minutes.
- **Permanent, continuous integration**: New modules are integrated into the system as quickly as possible.
- **Test-driven development**: Individual modules and their interaction are permanently tested.
- **Incremental design**: The system design allows incremental work and subsequent changes.

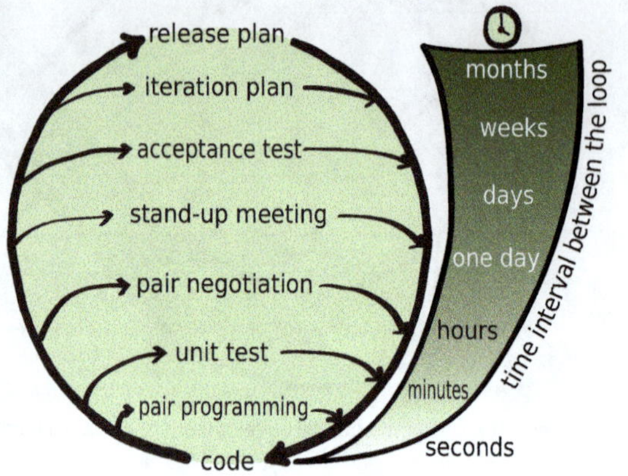

Figure 4.40: Left: Main practices of Extreme Programming. Right: Feedback loops and their timing.

for these process models. In contrast to Scrum and Kanban, however, Extreme Programming offers concrete assistance for the development of software and thus supports the application in such projects. If you would like to learn more about Extreme Programming, you can find a comprehensive introduction in the book *Extreme Programming Explained* [Beck and Cynthia, 2004]).

Crystal

Crystal represents an entire family of process models for software development, which was coined by Alistair Cockburn. The individual versions of the process models are marked with colors (clear, orange, etc.) and are based on the number of people involved and the level of risk. The simplest form is called Crystal Clear. The idea behind the color-coded levels is that larger and riskier projects require more extensive project management than small projects with fewer risks. Crystal is also based on principles. These include:

- informal exchange of knowledge through physical proximity and corresponding freedom for communication
- personal safety through open communication without disadvantages for people who express criticism
- continuous improvement through regular collection of suggestions, their evaluation and implementation
- incremental work with frequent deliveries at regular intervals of a few months or weeks
- regular integration of the program code into the overall system
- automated testing during the entire development process

- focused work without harmful multitasking and disruptions
- cooperation with users to clarify detailed questions and evaluate the results
- configuration management or version management.

Like Extreme Programming, Crystal is also primarily suitable for software projects. Compared to Extreme Programming, it is less strongly regulated and therefore tends to allow greater flexibility in application. If you would like to find out more about Crystal, you should read the description by Cockburn [Cockburn, 2004].

At a glance

- Scrum is one of the most widespread agile process models. Scrum defines:
  - roles (developer, product owner and Scrum master)
  - artifacts (product backlog, sprint backlog, product increment, release plan and impediment backlog) and
  - events (sprint, sprint planning, daily Scrum, sprint review and sprint retrospective).
- With Nexus, SAFe and LeSS, there are several frameworks for applying Scrum in large teams.
- Kanban is based on four basic principles:
  - Start with what you are currently doing.
  - Strive for incremental, evolutionary change.
  - Respect current processes, roles, responsibilities and titles.
  - Promote leadership and accountability at all levels of the organization.
- Kanban proposes six practices for its application:
  - Visualize the work (and its flow).
  - Limit the amount of work in progress.
  - Measure and manage the flow.
  - Make process rules explicit (clear and known).
  - Develop feedback loops.
  - Perform collaborative improvements.
- Important principles of lean projcct management are:
  - focus on the customer
  - identify the value stream
  - implement the flow principle
  - use the pull principle
  - strive for perfection.
- Critical Chain Project Management, which is derived from the theory of constraints, helps with day-to-day implementation by
  - the identification of bottlenecks
  - the optimization of the workload of the bottleneck and the avoidance of waste
  - the alignment of all other tasks to relieve the bottleneck and, if necessary,
  - the expansion of the bottleneck.

Depending on the environment, it may also be worth taking a closer look at other process models such as Design Thinking, DevOps or Lean Startup. These can help to master certain situations in terms of agile working.

Practical examples and case studies

Below you will find some examples and case studies that can help with the practical application of agile working. Some case studies have guiding questions, the answers to which can be found at the end of the book.

Case study 4.1: Sprint backlog

As part of sprint planning, developers and product owners agree on the sprint backlog. As usual, the developers commit to requirements to the extent of the velocity known from previous sprints. In the burndown chart, the open work at the start of the sprint therefore corresponds to the velocity, see Figure 4.41.

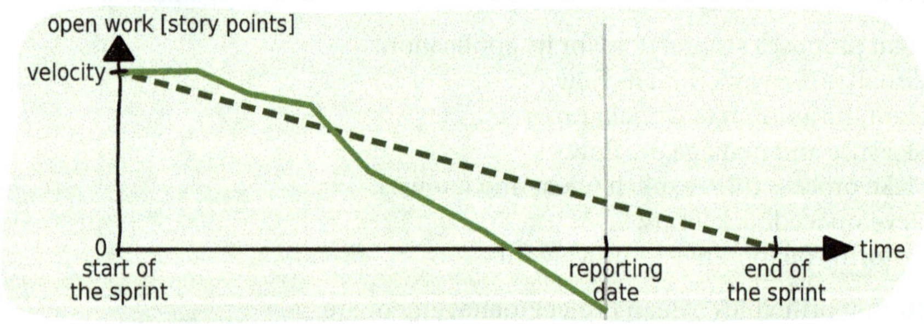

Figure 4.41: Burndown chart with planned progress of open work (dashed line) and actual progress (solid line) up to the reporting date.

The idealized, planned progress of the open work up to the end of the sprint is shown by the dashed line in the figure, as is the actual, measured progress up to the current reporting date.

Guiding question

Describe the actual development up to today's reporting date and interpret it. In doing so, also address the actual amount of open work on the reporting date itself.

Case study 4.2: Synchronization of several teams

The Film Experience Inc. develops software for movie productions and TV studios. The text and sound software department consists of 40 developers. The department is further subdivided into the following groups

- system design and architecture
- graphics and user interfaces
- images and movies
- sounds
- databases and components.

The company has five products on the market. One or two projects are regularly run for each of these products, either to further develop the product or to customize it. Recently, the decision was made to switch development to Scrum.

The Scrum Master is Carl Wessels, who previously managed traditionally organized projects in the company, but has trained as a Scrum master in recent months. One of the challenges he sees when switching to Scrum is the size of the team. He immediately rejects the idea of turning the entire department into a single Scrum team, as

- the team is too large to hold daily Scrum meetings together on taskboards and the communication would suffer in general and
- there are several projects, which means that all 40 developers in the department would have to regularly discuss the project status on several taskboards, which does not seem efficient to him.

Instead, Claus Wessels is examining two alternative structures:

- Group-specific teams: Each group within the department is organized as a Scrum team.
- Project-specific teams: A Scrum team is formed across all groups for each project.

However, he sees risks with both alternatives.

Guiding question 1

What are the risks of the two alternatives?

Despite the risks, Carl Wessels opts for the second alternative and creates project-specific teams. However, he makes a few modifications to reduce the risks:

- There must be at least one member of each group in each team. This ensures that each project-specific team can work on all tasks of a project with its own employees. As the system design and architecture group only comprises three people, people must work on more than one project there.
- In order to keep group-internal synergies, there are group-specific daily Scrums in addition to the project-specific daily Scrums. Carl Wessels places these at the end of the project-specific daily Scrums. However, it turns out that daily group-specific coordination is not necessary. The group-specific daily Scrum meetings, which are henceforth called synergy Scrums, will in future only take place on Wednesdays after the project-specific daily Scrums, see Figure 4.42.

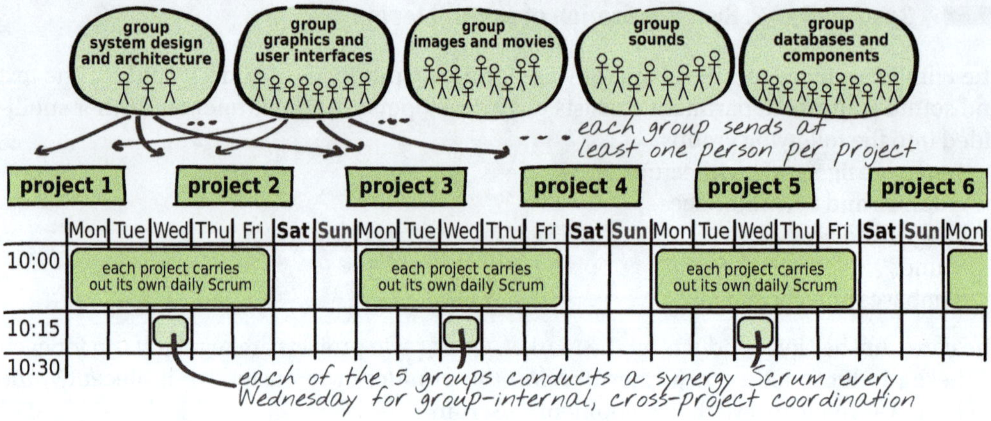

Figure 4.42: Project-specific daily Scrums and group-internal synergy Scrums, which only take place once a week and enable group-internal exchange between the people involved in the project.

The first projects get started. Carl Wessels is satisfied. The members of the system design and architecture group are not entirely happy as they are not always able to attend all daily Scrums due to their involvement in several projects. At an operational level, however, they find a solution to this by taking turns and attending the respective daily Scrum on a prioritized basis when coordination is required. Carl Wessels is less pleased with the observation that the project content sometimes overlaps – despite the specially introduced synergy Scrums. In these, a lot of time is invested in discussing fundamental issues and relatively little in removing obstacles and exploiting synergies.

Following this observation, Carl Wessels therefore introduced a joint product backlog. This contains the user stories for all projects in a common prioritization. However, the product owners of the individual projects have veto options: They can reject a user story that they do not consider relevant for their product. However, if they consider the user story to be relevant, they must adhere to the common prioritization. This creates cross-project synergies because the individual teams can now implement common functionality, see Figure 4.43. The prerequisite for this is that user stories that are required in several projects are planned in the same sprints and synchronized via the synergy Scrums.

However, it is precisely this synchronization of content that is criticized at the next sprint retrospective. The team from project 2 (see Figure 4.43) expresses its frustration because they had to wait for the teams from other projects to implement user story 3, even though they could start already in the first sprint.

Guiding question 2

How could synchronization be improved while continuing to exploit synergies?

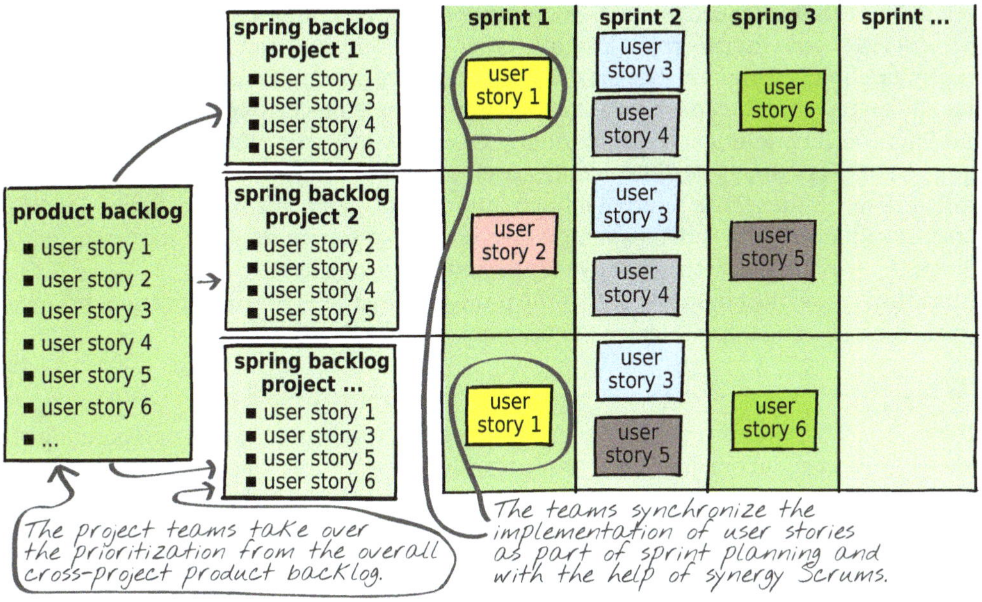

Figure 4.43: Cross-project product backlog. The individual projects plan their sprints, whereby the shared user stories are synchronized to common sprints.

Case study 4.3: Customer and technical design requirements

Scrum focuses on the customer when determining requirements. The user stories that describe the requirements are described from the perspective of a certain role and prioritized by the product owner with regard to economic considerations.

But what happens with technical requirements that affect the long-term profitability and maintainability of a product?

A product consisting of software and hardware is developed as part of a project. The requirements were collected in the product backlog and prioritized by the product owner. During the sprint planning of the first sprint, the developers note that, in their opinion, important requirements for the long-term maintainability of the product and subsequent further development have not been taken into account or have been prioritized too low.

The product owner replies that this was a deliberate decision. In her experience, subsequent further developments often show that scalable and maintainable functions are not used. The company often chooses a completely new development platform, taking new technologies into account, instead of building on existing products. That's why she wants to keep the current development efficient and only implement functions that are actually required by the customer.

This view is certainly justified for some products. It needs to be clarified whether the development platform is not scaled and updated for further developments because it has not been

prepared for this or because this makes no sense from a technological and economic point of view. It is not economic to invest a lot of effort in scalability and maintainability if there are no structures in the company that make it possible to use this functionality later on.
Products that need to be constantly updated, for example because of changing interfaces, legal framework conditions or compatibility requirements with other evolving products, require a correspondingly flexible technical architecture.
Another aspect is the team's internal infrastructure: Self-written auxiliary programs, tools to increase productivity and activities such as cleaning up old source code, etc. require activities that take time in the short term, but usually pay off in the long term.
In the direct economic comparison that the product owner uses for prioritization, however, such architecture and infrastructure requirements usually lose out.

Guiding question

Think about how you can create room for such architecture and technical design requirements. How can you ensure that a product with customer-relevant functionality is developed at the end of the project and that architecture and technical design requirements that are not visible to the customer are implemented at the same time, provided they contribute to the success of the product and the team's ability to work in the long term?

Case study 4.4: Large-scale project with Scrum

Cathy Smith has been assigned to lead a project to develop a new automated transmission for commercial vehicles. She wants to carry out the project as agilely as possible, but has to make adjustments due to the size of the project team. The organizational chart of her project is shown in Figure 4.44. She considers herself to be more of a product owner than a traditional project manager. However, she would like to act as a link between the individual self-organizing teams.

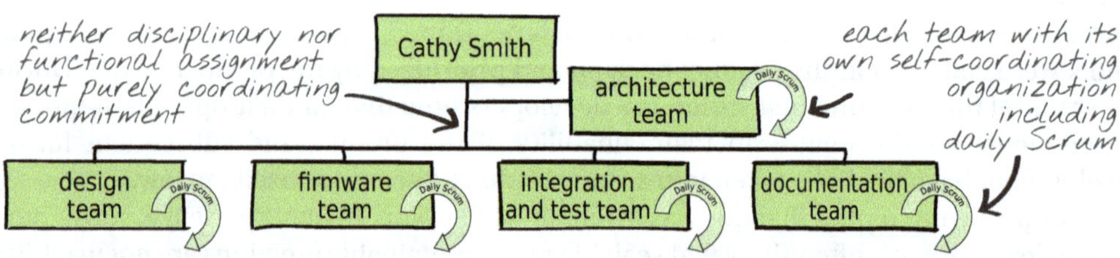

Figure 4.44: Organizational chart of the project to develop a new automated commercial vehicle transmission.

Guiding question 1

Before you continue reading, think about how you would organize the product backlog, sprint backlog and the sprints. Weigh up the pros and cons of different options.

Cathy Smith wants the teams to be able to work as self-organized as possible in line with agile thinking. At the same time, she needs cross-team coordination for the overall project.

If she were to use individual product backlogs for each team, there would be a risk that the teams' development results would not fit together in the end. She therefore chooses a common product backlog and presents the existing user stories on a higher-level project taskboard in prioritized order. For each user story, a primary team is appointed, which is responsible for estimating the size and subsequent coordination with the other teams.
In the sprint planning, the user stories to be worked on by the teams in the next sprint are selected. Tasks for the individual teams are derived from the user stories. The tasks are subdivided in such detail that only one team is involved in the direct implementation of each task. Nevertheless, there are still dependencies between the tasks. It is therefore documented for each task which user story it originates from and which team is responsible for cross-team coordination. The implementation in the project and team taskboard is outlined in Figure 4.45. For example, user story 1 is coordinated by the architecture team. Most of the associated tasks must be implemented by the architecture team. In the example, these are tasks 1.1, 1.2 and 1.3, but other teams must also make contributions. For example, the design team works on task 1.5, but must coordinate its content with the architecture team.

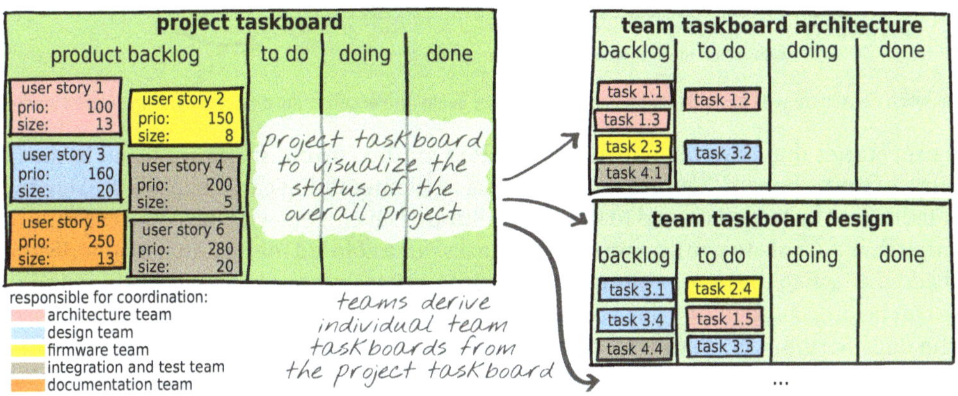

Figure 4.45: Project and team taskboards. The color coding indicates which team is responsible for the cross-team coordination of the user story.

Guiding question 2

How would you organize cross-team coordination?

For coordination within the teams, Cathy Smith relies on the daily Scrums that are common in Scrum. All teams agree to hold them at the same times from 9:45 to 10:00 in the morning. Afterwards, cross-team meetings, so-called **Scrum of Scrums**, take place daily from 10:00 to 10:15. Each team sends one member to these meetings. This is usually a person who is currently working on tasks that need to be synchronized. If several team members are working on tasks that require coordination, several team members can also be sent to the Scrum of Scrums. To ensure that the meeting keeps the character of an efficient and solution-oriented daily Scrum, a maximum of two people per team may be sent.
While the teams meet at their team-specific taskboards for the daily Scrum, the people participating in the Scrum of Scrums work on the overall project taskboard.

Case study 4.5: Daily Scrum

A development team of seven developers is working on the implementation of a new firmware for a measurement system. The team's taskboard is organized as shown in Figure 4.46.

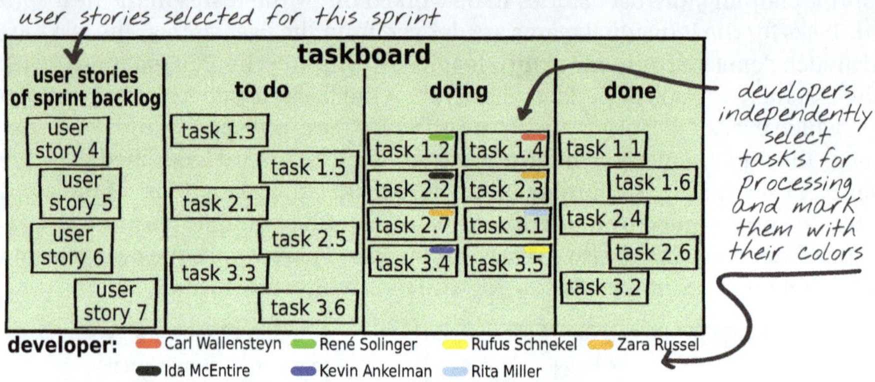

Figure 4.46: Taskboard with assignment of the tasks currently being processed to the developers using colored markers.

The user stories that have been firmly committed for the current sprint are listed in the first column of the taskboard. The user stories whose implementation can begin are broken down into individual tasks and moved to the *to do* column. Developers who have free capacity and feel that they are able to work on one of these tasks put a colored marker on it to identify them and add the task to the *doing* column. In principle, the tasks have been dimensioned so that they can be processed in one day and then placed in the *done* column.

In the daily Scrum, Carl Wallensteyn notices that task 1.4 has already been in *doing* status for two days. When asked about this, he explains that there were some difficulties with the implementation and that he will probably need another two days to complete the task. Two days later, the task has still not been completed and Carl Wallensteyn is not present at the daily Scrum because he has another meeting at the same time. The next day, he explains that the problem is almost solved and that he will complete the task the next day. In fact, he only managed to do this another two days later.

Key questions

- Can the team be satisfied with the way task 1.4 was handled?
- How would you have reacted if you had participated in the daily Scrums?

Case study 4.6: Overload due to incorrect velocity

Viveku Inc. is a contract developer for software and hardware components. Within the software department, it was recently decided to handle all future projects according to Scrum. As the software department comprises around 60 people, several Scrum teams were created, each with a team coordinator. The team coordinators always meet after the daily Scrums and coordinate cross-team aspects. For planning purposes, there is a cross-team product backlog with the requirements of the commissioning company for each order.

Due to the establishment of cross-team team coordinators, additional Scrum masters were not required. However, the teams were trained. Some already have Scrum experience from previous employment at other companies.

Despite this experience, there are considerable difficulties at the end of the first sprint. Not even half of the planned user stories could be implemented. Internal critics of Scrum were immediately confirmed and demanded a return to the V-model, which the company had previously used.

In the sprint retrospective, the introduction of Scrum and an unknown velocity were identified as the main reasons for the unsatisfactory result.

The situation did not look much better at the end of the second sprint either. Again, not all planned user stories could be implemented. In the sprint review, the product owner expresses his anger as he sees the overall success of the project at risk. He calls for more to be done in the next sprint to make up for the delay and get the project back on track for success.

In the teams' retrospective, the lack of experience of some team members with Scrum and a velocity that is still only rudimentarily known are again discussed. In addition, collaboration between the individual teams is more difficult than expected. The teams call on the head of department to urgently install a Scrum master to support the work as a coach when planning the subsequent sprints.

Guiding questions

- What could be other causes of unachieved sprint goals?
- Assume you are appointed as a Scrum master and are to support the planning of the next sprint. Outline how you would help the teams.

Case study 4.7: Agile scaling with Nexus

As a software service provider, SpecLOG Inc. is used to conducting agile projects using Scrum. Accordingly, it goes without saying that all organizational projects also use Scrum. The projects carried out to date usually involve between five and ten people working together in individual Scrum teams. The upcoming tenth anniversary of the company is also to be organized with Scrum. Even during the initial considerations, it becomes clear that significantly more than ten people need to be involved in the project in order to implement the wishes of the management.

These wishes actually exist as a product backlog. The wishes include requirements for marketing, the entertainment program and an accompanying small in-house exhibition. The overarching goals of the company anniversary are:

- celebration of the anniversary as recognition for the company's employees and their families
- addressing a broad public to draw attention to the company and thus arouse interest among potential applicants and customers

- showcasing the range of products and services as part of an accompanying in-house exhibition to attract new customers.

The product backlog comprises three different requirement categories. The managing director therefore proposes three Scrum teams for processing, see Figure 4.47. These teams use the product backlog he has created in a self-directed manner. The managing director sees himself as the product owner, who provides feedback on the status of planning in weekly sprint reviews.

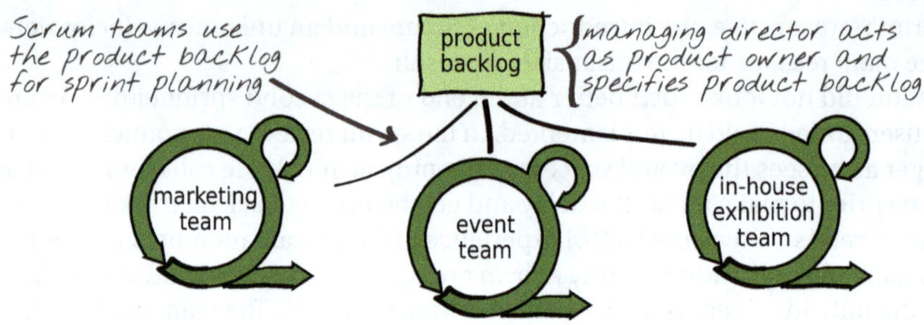

Figure 4.47: Initial organization of the planning of the company anniversary with three independent Scrum teams using a common product backlog.

Guiding question 1

- What problems could arise with the chosen organizational form with three independent Scrum teams?

The first conflict already arises during the first sprint planning: all three Scrum teams want to schedule the sprint planning at the same time at the beginning of the first sprint. As there is only one Product Owner, who cannot be present at three meetings at the same time, the teams finally agree to stagger the sprint planning so that the product owner can attend all meetings. This, in turn, is too long for the managing director, so he decides to attend only two of the three sprint planning meetings.
At the end of the sprint, the sprint reviews provide an uncoordinated picture of the planned event. Some of the available areas were occupied twice by the teams, particularly interesting program items of the general event plan and the in-house exhibition were planned in parallel and employee availability for individual program items was not taken into account.
After a brief consultation, the three Scrum masters of the three teams therefore decide to conduct the following sprint retrospective across teams. It quickly becomes clear that the independent nature of the Scrum teams does not work in the current organizational form. The teams want to see this deficit as an opportunity and carry out a project with Nexus for the first time in the company. If Nexus proves successful, this form of organization could also be used later for the implementation of larger software projects.

Guiding question 2

- How would you implement the current project with Nexus? What changes to the current organizational form are necessary for this?

Case study 4.8: Introduction of Kanban

Carkan Inc. develops tuning solutions for passenger cars. The projects are usually relatively small and are carried out by three to eight people. There is no project management handbook. Nevertheless, project management is relatively standardized due to the company's history. The company's four project managers all use ProjectLibre, an open source software for project planning with bar charts, and carry out the projects according to the waterfall model.

The management is not fundamentally dissatisfied with the project results. Many projects are completed within budget and delays are rare. However, there are also examples where projects were significantly more expensive and took longer than anticipated. The management has observed a work-to-rule mentality among some employees.

The management becomes aware of Kanban through a supplier who delivers software for tuning products. They find the approach fundamentally interesting and are convinced that the projects at Carkan Inc. will also benefit from it. However, the management also knows that employees will have reservations if they change the process model by directive.

A longer change process is therefore planned for the introduction of Kanban. The management is not primarily concerned with switching from bar charts to Kanban boards, but with the transformation to agile thinking and action. The management assigns one of the project managers, Mr. Claas Ryder, who has worked closely with the Kanban-experienced supplier, to develop an agile project management concept for the company.

As a first step, the project manager visits the supplier to learn more about their project management with Kanban. Back at the company, Mr. Ryder convenes several workshops with other project managers, employees and superiors to develop expectations for a successful project management process model. Figure 4.48 shows an excerpt from the discussion results.

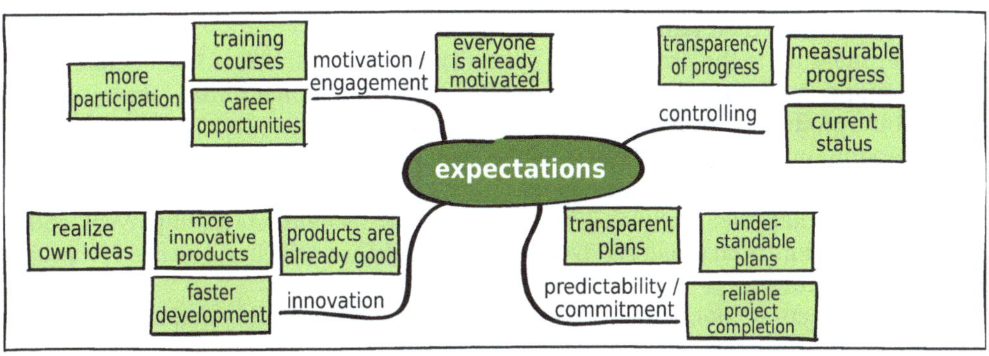

Figure 4.48: Collection of expectations from management, project managers and project team members for improving project management with Kanban.

The contributions indicate that there is dissatisfaction with the current approach. Some expectations do not relate directly to project management, but go beyond this, such as career opportunities that are perceived as limited.

The expectations also show a certain amount of concern among employees about being overwhelmed by new processes. Accordingly, training is desired. For the management, on the other hand, a high level of transparency and an up-to-date project status that can be called up at any time are important. Transparency, in turn, makes some employees uncomfortable. In addition, some employees believe that both the motivation and the products are so good that there is no need for change.

An overall concept should therefore be created for the introduction of Kanban. Unlike Scrum, where roles and processes are clearly defined, the principles and practices of Kanban allow more flexibility in terms of adaptation. The project manager and the management want to take advantage of this: As a first step, Mr. Ryder works with the stakeholders to develop a common set of values. These include values such as

- customer orientation
- commitment
- adherence to deadlines and
- trust.

When defining the values, Mr. Ryder deliberately does not yet want to go into the implementation options with Kanban, but rather create a basis that most employees support. Based on the values, concrete guidelines for Kanban should then be developed. These should be selected in such a way that they are in line with the values in order to make it easier to convince employees that the concept will bring the company forward.

After Mr. Ryder has developed the concept for the new process model together with a core team of important stakeholders, it is discussed in the team and largely approved:

- The existing project managers receive intensive Kanban training. In future, they are to act as Kanban promoters and be given more cross-project tasks in program planning.
- Based on the status quo, the project manager defines goals and participation opportunities for the workforce. In order to take the lean concept into account, the management does not want to create any additional hierarchies. At the same time, it is looking for ways to meet the desire for career opportunities. The project teams are given full responsibility (and authority) for their Kanban-based projects.
- Another measure is the creation of a specialist career model. This is decoupled from the assumption of management responsibility in the traditional sense. Employees who contribute product innovations, take on responsibility in projects and do a good job will receive additional training and can progress from junior developer to senior developer to system architect and principle developer.
- The project management software is replaced by Kanban boards. As all employees work at the same location, each project is given its own whiteboard, which will be used for project planning in future.
- Ongoing projects are not reorganized, but are completed according to the previous procedure. A pilot project is selected, which is carried out using Kanban.
- A Kanban promoter supports the pilot team during implementation and moderates the daily Kanban meetings at the Kanban board in the initial period.

- The pilot project has a duration of 4 months. In order to get the employees of the other, still plan-based projects excited about the new approach, an employee of the pilot project reports every two weeks on their experiences with Kanban so far. These reports serve both to inform and train the other employees and to further improve the processes. The involvement of all employees gives them the (true) impression that they can shape and share responsibility for the processes.

The introduction is generally successful. Individual processes are optimized during the pilot project. For example, the start time of the daily meeting at the Kanban board varies until a time has been found that is acceptable to everyone. The most difficult part of the practical implementation for the teams is limiting the amount of work in progress. Initially, a limit was set that was too low, which meant that individual employees had nothing to do because they had to wait for work packages from other employees. Conversely, there were phases in which there was an overload and too many work packages were started.

Not all employees are able to cope with the increased responsibility of self-organization. In some cases, an improvement has been achieved through further training measures. For other employees, the remaining colleagues have to take on self-organization tasks.

After one year, the management evaluates the projects carried out with Kanban and the attitude of the workforce. Most projects were successfully completed in line with customer requirements and economic targets. Problems with individual projects can usually be traced back to overload situations in which inexperienced teams found it difficult to prioritize tasks correctly. The responsibility for self-organization worked well for most teams. Nevertheless, there are individual employees who only start working on tasks after being asked to do so by other team members.

Based on the findings, the following control measures, among others, will be adopted for future projects:

- The Kanban promoters are continued.
- When appointing new project teams, care is taken to ensure that each project involves employees who have experience with Kanban and have already successfully completed projects.

Case study 4.9: Transferring responsibility to teams

One of the challenges when introducing Kanban in organizations that have previously worked according to traditional, plan-based process models is the transfer of responsibility from individual project and line managers to the team:

- Managers must be willing to hand over responsibility and authority to the teams.
- The team members must be willing and able to take over responsibility and authority.

Warning

Again and again, implementation projects for agile project management fail because managers are unwilling to delegate responsibility and authority. They use every opportunity to more or less subtly create an atmosphere against the new structures and processes. Every mistake is attributed to agility and the previous way of working is portrayed in a glorified and positive light.

As with any change project, it is therefore important to identify the stakeholders affected by the change and their interests and to plan suitable measures. Managers and teams must be convinced of the need for change and recognize the advantages of Kanban. Practical examples from other companies or your own pilot projects can help.

So how can managers and teams be prepared for the transfer of responsibility?

Self-organizing teams do not automatically mean that there are no more managers for disciplinary management tasks. However, their tasks will change: in future, the teams will be entirely responsible for their work. However, the assignment of employees to teams, personnel development and the strategic development of groups and departments will usually remain in the hands of the managers. They must be prepared for the new tasks. They need to be trained to understand Kanban itself and their role in the new organization. Supervisors of employees who do not have the ability to get involved in a Kanban project in a self-organizing way play a key role.

Team members must of course also be trained in Kanban. They must understand that the new authorities of self-organization also come with corresponding responsibility, which they must fulfil. The structures and processes must be familiar, as must the rules for reporting to management. As with any organizational change, positions may be eliminated and new ones added after the change. A balance is not always possible here. It would be foolish to believe that the enthusiasm of individual stakeholders for Kanban (or other agile process models) is so great that they are prepared to give up their own position without a fight.

Case study 4.10: Flow optimization

Xaroy-2 is a project to develop a new plant system with hardware and software components. The project is being carried out using Kanban. A large Kanban board is located in the shared office of the project team. A total of 10 people are working on the project. There is a project manager, Claire Scherr, who takes on coordination and moderation tasks and represents the project towards the company's internal clients. However, she gives the team a great deal of freedom to organize itself.

The project team always meets at 10:00 for a discussion at the Kanban board. The current status of the work is outlined in Figure 4.49.

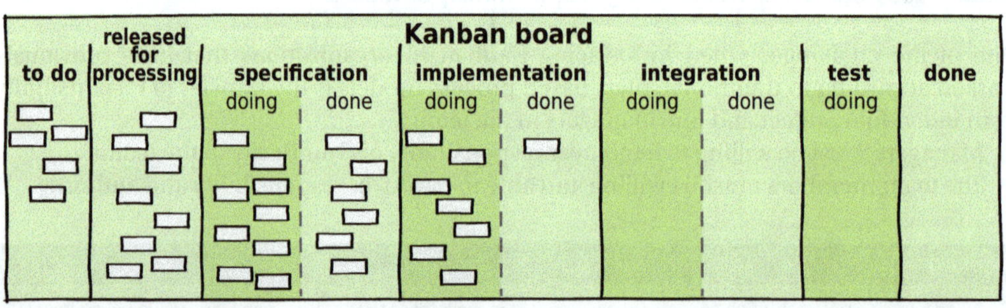

Figure 4.49: Status of the work packages on the Kanban board.

To prevent an overload and unnecessary stockpiling of work in progress, Ms. Scherr has stipulated that each team member may work on a maximum of two work packages at the same time. Everyone adheres to this rule.

Guiding question 1

- What do you notice about the distribution of the work packages?

In the following days, the previous development continues, which Ms. Scherr observes with concern. As an initial control measure, she asks the project team to shift team members from specification to implementation. She hopes that this will allow the work packages to be implemented more quickly. This should reduce the backlog of work packages between the completion of the specification and the implementation. One week after the status outlined in Figure 4.31, the situation shown in Figure 4.50 arises.

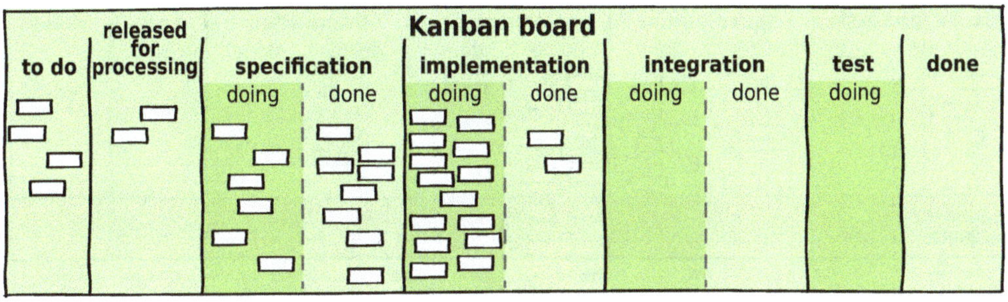

Figure 4.50: Status of the work packages on the Kanban board after another week.

Guiding questions 2

- How do you assess the current situation in the project based on the Kanban board?
- What could be the reason for the slow flow of work?

Together with the team, Ms. Scherr identifies an unexpectedly high number of errors that delay the implementation. Only an error-free implementation allows subsequent integration and testing at overall system level. Happy that at least the cause of the lack of throughput is now known, she and the team consider how they can counteract the problem:

- First, Ms. Scherr and the team introduce a second limit on the amount of work started. Until further notice, each processing stage may contain a maximum of eight work packages.
- Work packages that cannot be completed due to errors are marked with a red bug (sticker).
- Work packages with a red bug must be processed with priority. No new work packages may be added to a processing station as long as work packages with a red marker remain there.

Guiding question 3

- How do you rate the measures taken?

In order to be able to respond better to current needs and prioritization in general, Ms. Scherr adds an emergency lane to the Kanban board. All work packages in which unplanned difficulties arise and which delay the project can be placed on the emergency lane. These must then be processed with the highest priority, see Figure 4.51.
An emergency lane offers the opportunity to carry out individual tasks with priority in an orderly manner. One risk here is that the emergency lane is misused and more and more work packages end up there in order to speed them up. This can lead to work packages only being processed in the emergency lane, which makes a mockery of the emergency lane.

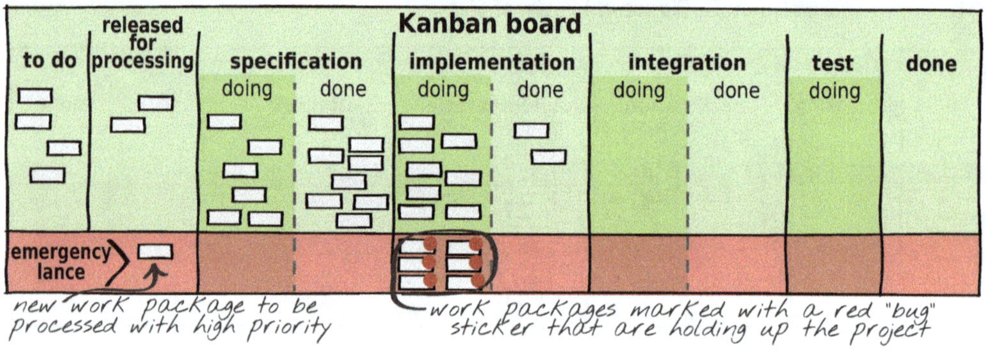

Figure 4.51: Kanban board with additional emergency lane on which high-priority new work packages and work packages that are blocking project progress can be positioned Work packages on the emergency lane must be processed immediately by the team.

Case study 4.11: Cumulative flow

The SpeedUp Navigation project uses Kanban for planning and control. The processing stages are as follows:

- (feature) pool (i.e. work packages to do)
- specification
- coding
- basic test
- online validation.

The cumulative flow diagram after 29 project weeks is shown in Figure 4.52.

Guiding question

- Analyze the project status using the flow chart by making statements about bottlenecks and the development of lead times.

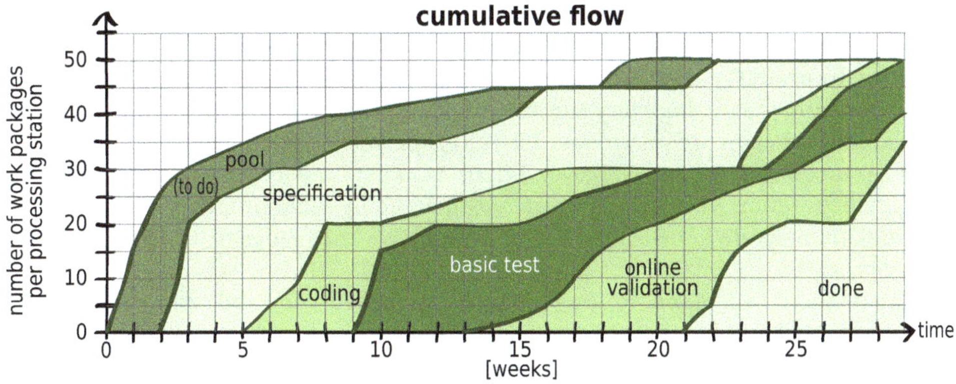

Figure 4.52: Cumulative flow chart of the SpeedUp Navigation project.

Case study 4.12: Optimizing bottlenecks

In a development project that is carried out according to the traditional waterfall model, the *milestone Pa - project plan approved* is due. By this milestone, all of the project's plans must be completed. They are checked by development and program management and approved if suitable. Only then can the next *phase I1 - implementation* of the company's own product realization process begin.

Development projects in this company are organized as a matrix. The project manager is the functional leader of the interdisciplinary team. However, all team members are anchored in their respective groups or departments. Many team members work on more than one development project because they have skills that are needed in several projects, but do not reach the extent that would keep them busy. The organizational chart of the project can be seen in Figure 4.53.

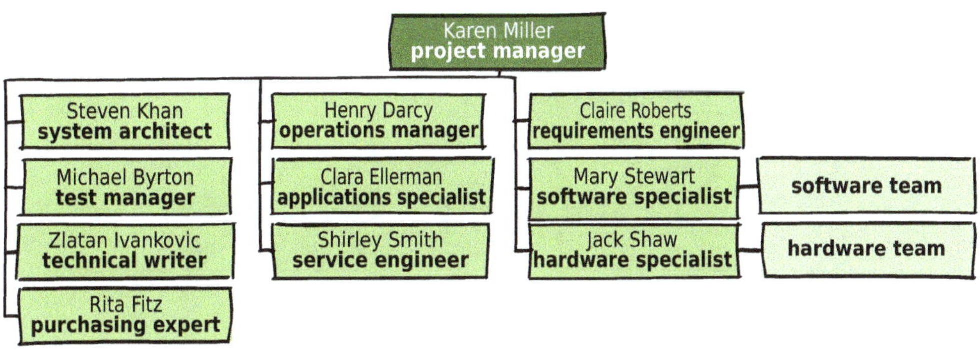

Figure 4.53: Organization chart of the core project team.

The phase transition of the company-specific waterfall model is approved by so-called milestone reviews. In the preparations for the milestone review, in which management checks all the results of the current planning phase, project manager Karen Miller is particularly concerned about the completion of product risk management. By the *milestone Pa*, all product risks must be identified and their extent assessed. Necessary measures to ensure product safety must be planned for the subsequent implementation phase.

Karen Miller has assigned the project's system architect, Steven Khan, to product risk management, as he is the only person with the expertise to assess the potential implications of each risk for the overall system. At the same time, Steven Khan is also working on finalizing the system design and is assisting Mary Stewart and Jack Shaw in planning the implementation of the system architecture requirements. Like all employees, Steven Khan attends the weekly department meetings. Ultimately, Steven Khan's working time for the project is reduced by the recurring questions from the service department about older products. Service employees without architecture knowledge need Steven Khan's knowledge to develop solutions to customer problems.
According to the project planning, the *milestone Pa* is scheduled to take place in two weeks. However, Steven Khan reports that he still needs about three to a maximum of four weeks to complete the product risk management.

Guiding question

- How would you apply lean management principles to achieve milestone Pa on time?

Case study 4.13: Exercise replanning a project according to critical chain project management

Consider the schedule shown in Figure 4.54 in the form of a bar chart. Assume that activities A1 to A6 are processed by the critical bottleneck resource. The dark green bars represent the safety buffers that you quantified using the three-point estimation technique.
Activities A7 to A10 are important supplying activities that do not require bottleneck resources. The safety buffers are marked by a hatched area.

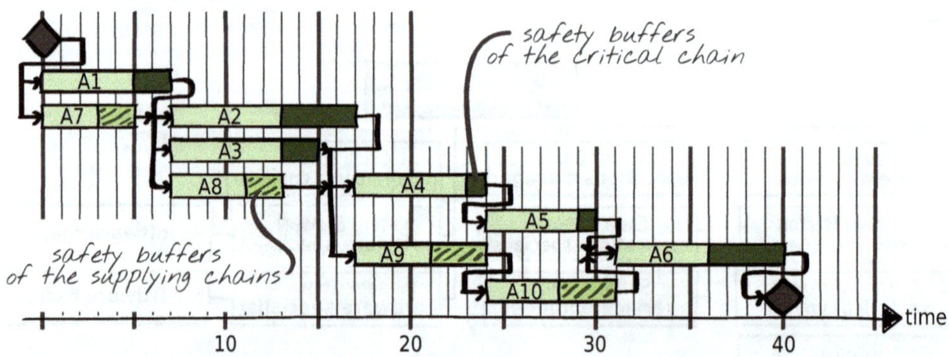

Figure 4.54: Schedule in the form of a bar chart with critical chain A1 to A6 and supplying chains of activities A7 to A10 as well as safety buffer.

Guiding question

- Apply the rules of critical chain project management appropriately and place the buffers accordingly.

5 Hybrid project management

In this Chapter ...

- you will learn what is meant by the term "hybrid project management",
- you will find out what options there are for combining plan-based and agile process models to create hybrid process models and
- you will acquire the ability to recognize for yourself when plan-based and when agile or hybrid processes, methods and behaviors bring advantages for your project.

Introduction

You now know important standards and norms of project management as well as plan-based and agile process models. Whenever two or more different process models are combined, we call them hybrid process models:

> ***Hybrid project management*** *refers to the combined use of methods, roles, processes and phases of different standards or process models.*

In this chapter, we discuss what hybrid project management can look like in concrete terms and in which cases its use makes sense. Standards and process models are important because they give us help and guidance when working on projects. We use them to do the right thing at the right time, for example, to create certain plans or to make arrangements for controlling. Standards and standardized process models also help to define company-specific project management processes and structures and to optimize them. If each project is handled differently, it is difficult to incorporate knowledge about success and failure factors into subsequent projects. Defined standard processes, on the other hand, can be evaluated and improved taking these factors into account.
Examples are:

- Suitable *plans and media* are defined to collect, structure and prioritize goals and requirements and to set deadlines and costs.
- *Risk findings* and their assessment are made available to subsequent projects in the form of checklists.
- *Key figures for controlling* that were not necessary for the success of a project are scrutinized and, if necessary, replaced by other, more meaningful key figures.

- The *structuring of a project* into specific sub-projects to promote communication and collaboration can be adopted in follow-up projects.
- Certain *meetings* such as short daily meetings on certain critical topics (stand-up meetings) can be transferred to other projects if there is a corresponding benefit in one project.

Warning

Conversely, if a standard or an unsuitable process model is chosen that does not fit in with the company and its way of working, the problems multiply. It is not just one project that becomes inefficient and ineffective, but all projects that follow the chosen standard.

It is therefore important not to adopt standards and process models without reflection. In practice, there are often cases where a company has first purchased project management software and then tries to adapt its workflows to the use of the software. More specifically, software is often purchased that supports plan-based project management with bar charts and some controlling functions. Now all project managers are asked to use these plans and functions, even though completely different plans and functions might be optimal.
Jumping to new process models too quickly can also have a negative impact on the project portfolio. Agile project management is considered lean, customer-oriented and modern. However, it is not suitable for every company and every project. Furthermore, an ill-considered switch from plan-based to agile project management can fail if stakeholders are not taken into account. Good change management is therefore important.
Selecting a process model that is suitable for the company is challenging as it depends on many criteria, which we will discuss later. Two simple schemes to support the selection are shown in Figure 5.1. However, these only take into account a fraction of the relevant information or neglect the possibilities of adapting and designing hybrid process models.

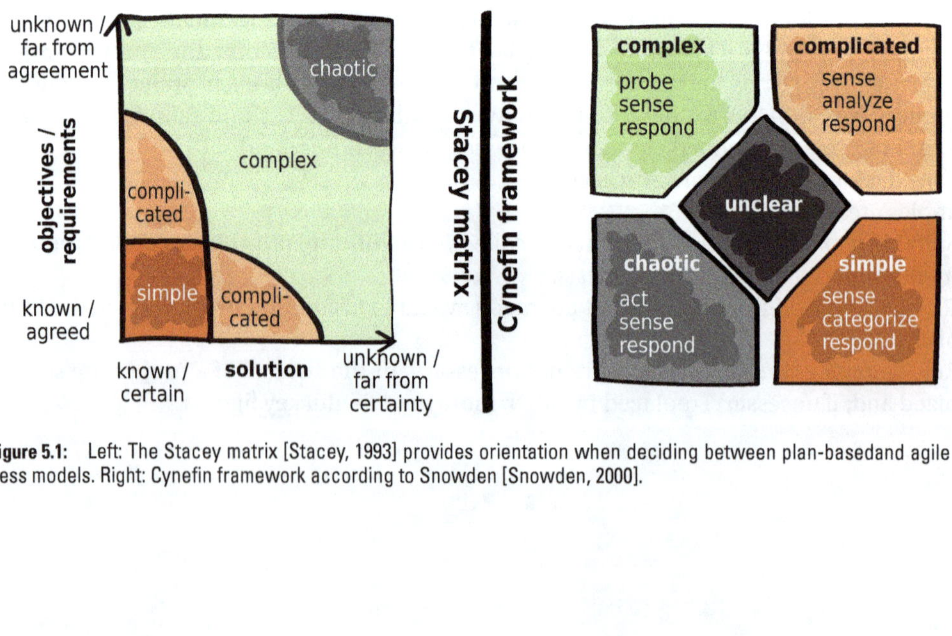

Figure 5.1: Left: The Stacey matrix [Stacey, 1993] provides orientation when deciding between plan-basedand agile process models. Right: Cynefin framework according to Snowden [Snowden, 2000].

The matrix shown in Figure 5.1 on the left is named after Ralph D. Stacey [Stacey, 1993]. In this matrix, the clarity and stability of the objectives and requirements are shown on the vertical axis. The familiarity of the solution is plotted along the horizontal axis. Whether objectives and solutions are known and agreed or unknown and uncertain depends on several factors. In many cases, objectives can be clearly formulated and firmly agreed between the client and contractor through appropriate preparation. Solutions are clear and known if established technologies or processes are chosen. Sometimes, however, it is foreseeable that the client will change the objectives or that so little is known about the objectives that a change is likely. In these cases, the objectives are far from agreement. If new solutions are to be found or new technologies are to be used, these are unknown. Projects can be classified as simple, complicated, complex or chaotic, depending on the combination of clarity of requirements and knowledge of the solution.

Tip

Complicated projects typically have more extensive tasks and more interfaces with the people (groups) and organizations involved (e.g. suppliers) than simple projects. With a little effort, however, the tasks and environment can be planned in advance. With complex projects, on the other hand, the tasks are so extensive or the interfaces so numerous that it is very difficult or even impossible to plan properly in advance.

For simple and many complicated projects, plan-based process models can serve well, as the course of the project can be predicted reasonably well. Agile, iterative process models are often suitable for more difficult, complicated and complex projects. In chaotic situations, the project brief should be clarified first, as chaotic requirements are difficult (or even impossible) to implement even iteratively.

Snowden provides a model that describes the handling of simple, complicated and complex issues with the Cynefin framework [Snowden, 2000], see also Figure 5.1 on the right. This can also be applied to project management.

- Simple projects can be processed by observing (sense), categorizing the project and then responding according to the identified category. In other words, simple projects can be processed on the basis of observation.
- Complicated projects are not so easy to categorize. Instead, Snowden suggests careful analysis, which in most cases will result in project-specific tailoring of project management.
- Complex projects are characterized by the fact that they cannot be planned in detail. Snowden's "trial and error" corresponds to the iterative approach: The project is implemented in iterations. The result is successively assessed and the project is readjusted (responded to) in the next iteration.
- The only thing that helps with chaotic projects is to act, observe (sense) and respond. Chaos is characterized by the fact that two identical actions produce different results, to which you then have to react.

The illustration in Figure 5.2 chooses slightly different axis labels and becomes more specific in the recommendation for the application of a process model. However, this recommendation is only an initial point of reference. Today, there are many examples of successful, agile, complex projects with many participants.

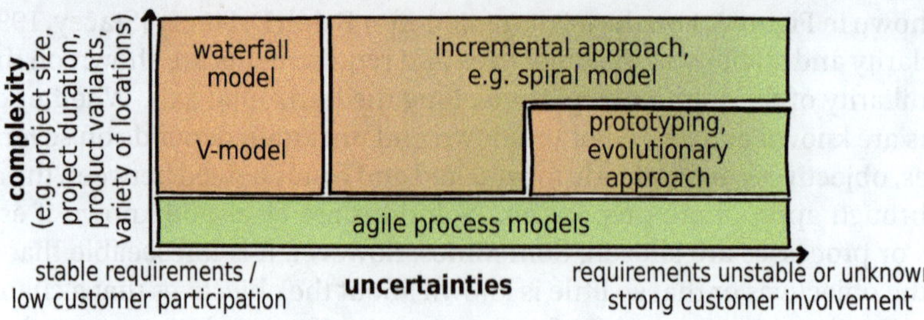

Figure 5.2: Simplified scheme for selecting a plan-based (orange) or agile (green) process model based on the criteria of complexity and uncertainty in accordance with [Ebert, 2019].

The model by Boehm and Turner [Boehm and Turner, 2003] is more meaningful. It takes into account the five criteria:

- personnel (people)
- dynamism (stability of requirements)
- culture of the enterprise
- size (of the project) and
- criticality (of the project deliverables in terms of safety).

In addition to these, other publications use other criteria to choose between plan-based and agile process models. Figure 5.3 shows an overview of 10 common criteria.

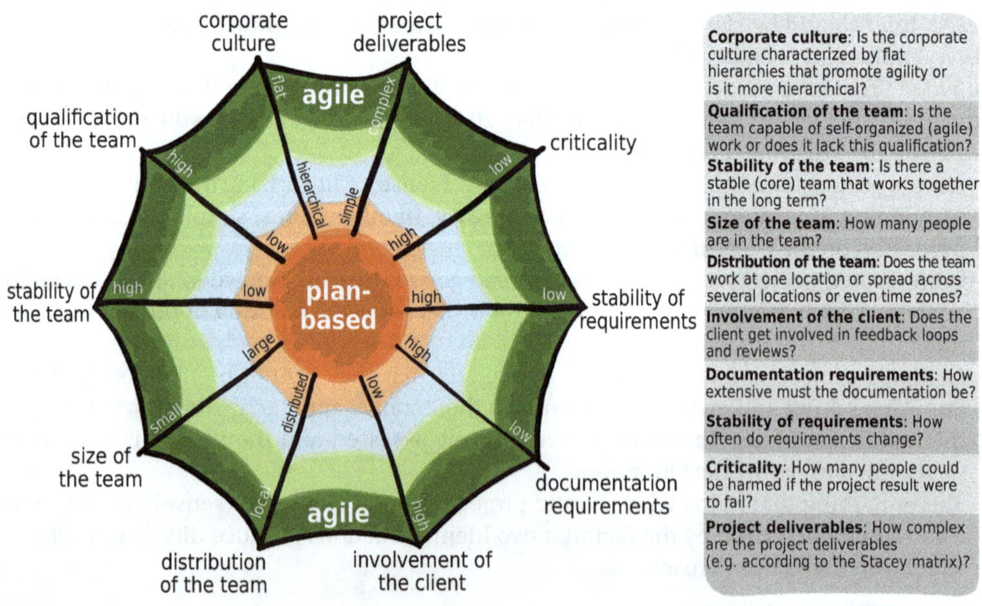

Figure 5.3: Criteria that can influence the decision for the selection of a process model, as a so-called network diagram, also known as spiderweb diagram.

Each criterion can have two different characteristics. Depending on the characteristics, the model then recommends a traditional plan-based or agile approach to the project. Depending on the company and project constellation, there may be other important criteria that influence the decision. It should be noted that not all criteria are equally important or equally influential.

Example

Two examples illustrate that not all of the criteria mentioned in Figure 5.3 are equally important. Some represent a major hurdle for the selection of a specific process model, while others must be taken into account, but can usually be considered well:
If the corporate culture is very hierarchical and characterized by classic controlling, it will be difficult to work according to agile values and use agile process models in line with the agile manifesto.
The documentation requirements, on the other hand, have a rather minor influence: agile process models do not place a particular focus on documentation, but do not prohibit it. If it is important to the customer, agile projects can also be documented accordingly. You just need to pay attention to this when designing the project.

A more detailed example can be found in case study 5.2 at the end of the chapter.
When applying the model, you will encounter several challenges that can only be briefly outlined here:

- There may be other company- or project-specific criteria that need to be identified first.
- The model does not provide a recommendation for a specific process model. Instead, it merely provides a rough guide as to whether a plan-based or an agile process model should be selected.
- In many cases, there will be no clear decision. Rather, it may happen that some criteria speak in favor of a plan-based and some criteria in favor of an agile process model.

Models such as the one in Figure 5.3 can help us to find a suitable process model and adapt it to our needs. This adaptation is also known as tailoring:

> ***Tailoring*** *refers to the individual adaptation of a standardized process model to a specific project.*

In many cases, you will come to the conclusion that a process model suitable for the respective context and your needs has a hybrid character, i.e. it contains components from different process models.

Example

To get a first impression of hybrid project management, imagine a project that is organized according to the plan-based waterfall model. If a sub-project is carried out in an agile manner, for example with Scrum or Kanban, this is referred to as a hybrid overall project.

You will get to know completely different hybrid approaches later on – stay tuned. The starting point for the selection and subsequent adaptation of a standard or process model should be the question of how the company can process the project most efficiently and effectively. The following questions play a role here:

- Do we have large or small, geographically co-located or distributed project teams?
- What qualifications do our project managers and project team members have?
- Do our customers or clients demand a specific process model from us?
- Are there regulatory or legal requirements for a process model (e.g. in the aerospace or medical device industry)?
- Do we have stable or frequently changing requirements?
- What is our long-term strategy?

Further criteria and their comparison with known process models are outlined in Figure 5.4. Depending on the answers to the above questions, you can either select a suitable existing process model, adapt a process model (tailoring) or design your own individual process model from several process models and any existing ideas of your own.

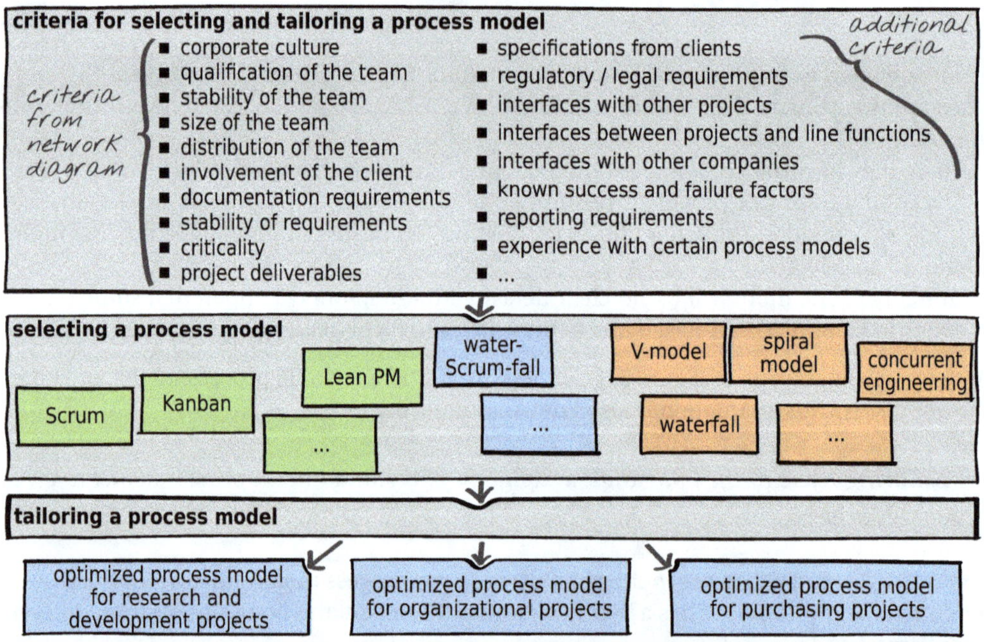

Figure 5.4: Derivation of a company-specific (possibly hybrid) process model (blue) by applying suitable criteria and comparing them with known plan-based (orange) and agile (green) process models. Within a company, there may be different process models for different project types (e.g. research and development projects, organizational projects and purchasing projects).

Warning

Perhaps you are now asking yourself: Do I have to design a separate process model for each project in order to work efficiently and effectively? No, of course not! For the reasons of optimization and knowledge management already mentioned, this would even be disadvantageous. The trick is to use a process model that is suitable for the majority of projects.

Some companies have also established process models that are to be used depending on the type of project. For example, a different process model may apply for development projects in a matrix project organization than for organizational and process improvement projects in a functional project organization.

Example

A company regularly receives orders from a customer who works according to the V-model. Due to the model-integrated emphasis on the traceability of requirements, the customer demands that the company also applies the V-model to supplier projects. However, in order to maintain flexibility, the company allows software sub-projects to use Kanban.

The company also uses Kanban for regular projects to improve production and other company processes.

Other projects, such as purchasing projects, are processed according to a company-specific variant of the waterfall model. Many employees consider the sequential processing of the waterfall to be easy to follow. This is why employees who only work on projects sporadically also get on well with it.

So far in this book, you have become familiar with plan-based project management using the waterfall, spiral and V-models as well as concurrent engineering. You are now also familiar with Scrum and Kanban, two well-known agile process models, and Lean Project Management, another approach that includes agile components.

In many companies, the selection of one of these process models and its subsequent adaptation to the company's individual conditions works extremely well. Sometimes, however, purely plan-based or purely agile process models come up against the limits of existing conditions. This leads to suboptimal project management. It therefore makes sense to examine the combination of different process models according to the motto: choose the best of two or more worlds.

In fact, with careful selection, there is nothing to be said against hybrid process models that combine plan-based and/or agile methods, roles, processes and phases. For example, agile components such as a Kanban board or short daily meetings in the sense of daily Scrums are used in many fundamentally plan-based projects today. Before we discuss further hybrid project examples, however, let's first consider how hybrid process models are created:

- By **sequential combination** of different process models: For example, the planning phase of a construction project can be carried out agilely using Scrum and the subsequent construction phase can be processed according to the plan-based waterfall model.

- Through **parallel combination** of different process models: One example is a development project that is carried out according to the plan-based V-model. One or more subprojects are processed in parallel using Kanban.
- Through **integrated combination** of different process models: In a plan-based project carried out according to the waterfall model, a Kanban board is used for planning and controlling, for example, and the project is managed via short daily stand-up meetings.

Example

In principle, the combination of several plan-based or agile process models also has a hybrid character. For example, consider Figure 5.5, where several plan-based process models are combined.

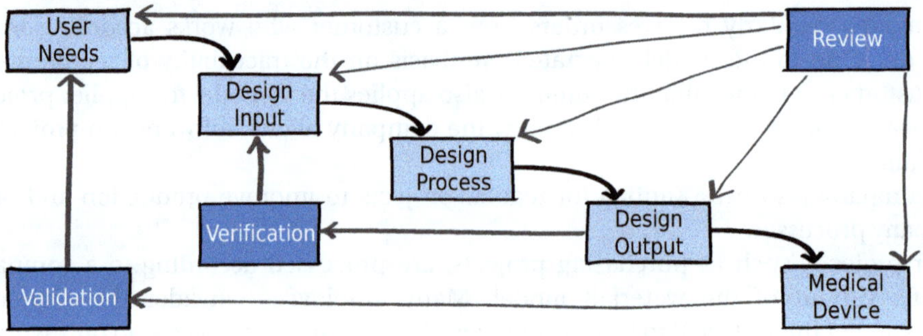

Figure 5.5: Recommendation of the US Food and Drug Administration for a process model for the development of medical devices, taken from [FDA, 1997].

The process model presented there contains elements of both the waterfall model and the V-model. The individual project phases of determining the user needs, deriving the requirements specification (design input), implementation (design process), development of results (design output) and completion of the medical device are run through sequentially according to the waterfall model. The FDA recommends carrying out a review at the end of each phase and uses elements of the quality gate principle for this purpose. In addition, verification should check and prove compliance with the specifications and validation should check and prove fulfillment of the user needs. This verification and validation have been adopted from the V-model. The reason for this combination of waterfall and V-model is the high safety requirements for medical devices: The combination of controlled phase transitions and verification and validation is intended to check several times whether the medical device has been developed safely, effectively and in accordance with regulatory requirements and user needs.

The previous example shows a hybrid process model that is made up of several plan-based process models. In the following, however, we will focus on hybrid process models that consist of plan-based and agile elements.

> **Warning**
>
> Agile project management differs from plan-based project management not only in terms of different processes. In the previous chapter, you learned that agility is characterized by special values. These lead to a different understanding of leadership. The fact that Scrum dispenses with the role of the project manager is just one characteristic of this difference. In practice, these differences lead to major challenges when combining agile and plan-based process models to create hybrid process models. We will look at specific practical examples at the end of this chapter and explore the topic of leadership in an agile and hybrid context in greater depth in Chapter 6.

Development of individual process models for successful project management

Let's go through the process of setting up an individual process model together. You can derive typical phases of the structure quite easily from common problem-solving approaches, for example that of Haberfellner [Haberfellner et al., 2018]. An example is shown in Figure 5.6.

Figure 5.6: Phases of the development of individual process models for successful project management.

Phase 1: Situation analysis

The first step in phase 1 is to identify the need for an individual, e.g. company-specific, optimized and possibly hybrid process model. The need can be identified both by senior management and by employees working on the projects.

> **Warning**
>
> Without the support of senior management, it is virtually impossible to establish a suitable process model for successful project management in the long term. If the need is not identified by senior management themselves, the first step is to win them over to the idea and secure their support. However, gather other supporters around you in addition to senior management. Many change projects fail because, despite initial successes, the opponents of the change ultimately prevail. A comprehensive stakeholder analysis can help you to identify supporters from all affected areas.

During this phase, resist the temptation to develop quick solutions. It is worth describing and analyzing your organization's current situation first. Only then will you have all the information you need to develop a process model that is truly suitable for the long term.
The following key questions can help you with the analysis:

- Why is there a need for change?
- How do we work so far? What roles, interfaces, documentation and reporting obligations are there?

- What has gone well so far, what has gone badly?
- Which stakeholders are there? Who is involved in initiating, defining, planning, controlling and completing the project?
- Who do we work with internally and externally?
- Are there requirements for the way we work, for example from suppliers or regulatory obligations?
- What qualifications do our employees have and what willingness and ability to change do they have?
- Who benefits from the current structures and will be skeptical about change?

Example

The managing director of Resentorp Inc. is dissatisfied with the way projects are handled in the company. These are carried out on a plan-based basis according to the waterfall model. The managing director repeatedly observes delays and cost overruns. Both lead to dissatisfied customers and jeopardize business success. In a large-scale change process, the managing director wants to increase the company's innovative capacity, increase employee commitment and improve the delivery and cost reliability of projects.
The core team of the change project proposes to increase the company's agility and to adapt employee management and the project management process model accordingly. As part of the situation analysis, the current way of working, including its strengths and weaknesses, is first discussed and evaluated.

I'm sure you can think of other questions to capture the situation. Incidentally, the last question is important for the long-term success of the new processes. You will have to deal with skeptics during the subsequent introduction of new processes and structures. It is therefore important to understand their motives and arguments, find compromises and prevent them from falling back into traditional processes and structures.
Discuss this as a team and make sure that you have not forgotten any important aspects. It would be presumptuous to believe that one person alone can think of everything – that's why the entire process of setting up a new process model is a team effort!
Approach the situation analysis in as structured and systematic way. Use suitable tools for this:

- **SWOT analysis:** SWOT stands for strengths, weaknesses, opportunities and threats. For each category, think about characteristics that describe your current situation.
- **Stacey matrix:** With the Stacey matrix, see also Figure 5.1 on the left, you can characterize the general nature of your projects. The more complex the projects are, the sooner you should think about agile process models.
- **Network diagram with criteria:** Create a network diagram according to the template from Figure 5.3 and add criteria that are important in your environment.

An example of a situation analysis can be found in case study 5.2.

Phase 2: Target definition

You now know the situation and can set clear targets for the new process model *and its introduction*. The introduction is emphasized once again, as you are already in the middle of a change project. You won't be able to introduce the new processes and structures later on with a brief announcement from the company management - such attempts are usually doomed to failure due to considerable resistance from the workforce.
Key questions when defining targets include:

- What characterizes successful projects for us?
- What are the resulting targets for project management?
- What are the time and financial constraints?
- Should ongoing projects be converted to the new process model, or will it only be applied to newly launched projects?

Continue to resist the temptation to look for solutions in advance. Narrow targets that unnecessarily restrict the solution domain can prevent the success of the new process model for successful project management, as can very general targets that do not make it clear what is actually to be achieved.

Example

Example: Target definition: As with all project targets, you should also formulate the targets for the change project to establish an individual process model for successful project management in a SMART way, i.e. **s**pecific, **m**easurable, **a**chievable, **r**elevant and **t**ime-bound. Possible targets resulting from the situation analysis are:

- The process model to be created should enable the development projects to be completed after a maximum of one year.
- The development results must have a maturity level of at least 99 points within the project term according to the company's own maturity level model, REPO.
- From July 2026, all newly launched development projects should work according to the process model to be developed.
- By June 2026, all employees in the development, service and production departments should be trained in the new processes and structures.

Example

Example: Limiting the solution domain by defining the target too narrowly: Compare the following two targets:

- The software sub-projects require an interface to the "Software and Systems" department in order to ensure the integration of new software functions into the cross-product and cross-project software platform.
- The software developed in the projects must be able to be integrated directly into the cross-product and cross-project software platform.

Both targets of the latter example serve to ensure that a cross-product software platform can be maintained and further developed. The first formulation also stipulates that this must be

done via an interface between software sub-projects and the "Software and Systems" department, and restricts the development of solutions in terms of processes and structures.
You should prioritize the objectives for the process model to be developed so that you can later decide what a possible compromise solution might look like in the event of potential conflicts of objectives.
At the end of the situation analysis and goal formulation, you should know the criteria for evaluating the process model to be developed. You will later have to use these criteria to select one of the solutions developed.

Example

Let's continue with the example of Resentorp Inc. It has identified a need for change and wants to transform itself into an agile, innovative company.
Based on the situation analysis, common values are first developed. These serve as the basis for the next steps and the design of the future process model. It is important to the core team that the values were developed together with managers and employees. The following main targets are then defined in line with these values:

- The commitment and responsibility of employees in product development should be increased.
- Delivery and cost reliability are to be improved so that the company is known for this in future and can use both as an argument for its own product range in sales.

Phase 3: Search for solutions

You now know the targets and criteria for the new individual process model. You can use these step by step to create a possible solution. The process for finding a solution is illustrated schematically in Figure 5.7.

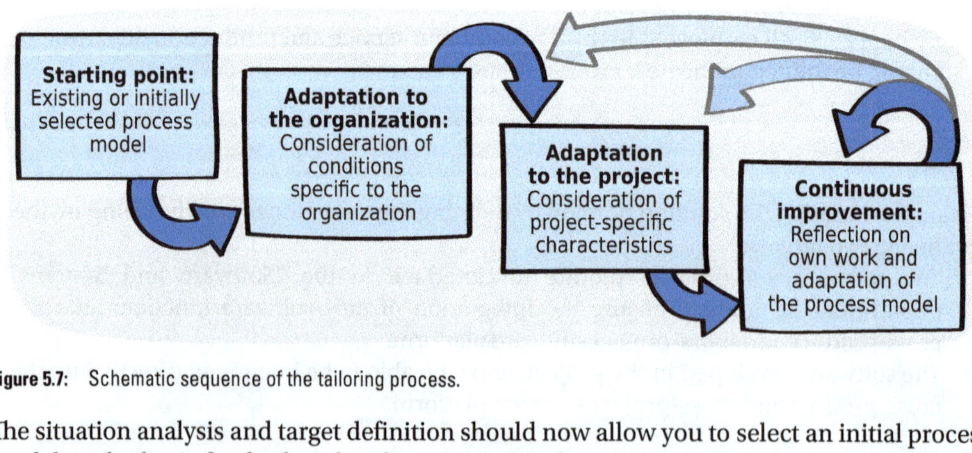

Figure 5.7: Schematic sequence of the tailoring process.

The situation analysis and target definition should now allow you to select an initial process model as the basis for further detailing. For example, you can select the previous process model determined during the situation analysis. If the target formulation already indicates a completely different process model, you can of course also select an alternative process model.

You then take into account the individual conditions of your organization in order to adapt, i.e. tailor, the process model so that it is at least roughly suitable for many projects in your organization.

Example

The situation analysis revealed that you have been working with a plan-based waterfall model to date and that this is causing problems. The main reason you identified was the limited flexibility to respond to the constantly changing requirements of the customers. In the target analysis, you therefore called for a new process model that would allow you to deal with changes much more flexibly. Based on these considerations, you choose Scrum as the starting point for further tailoring. As most of the company's projects consist of at least 30 team members, you select Nexus as the first organization-specific customization. Another customization is the sprint duration, which you set to a relatively long period of 6 weeks due to the characteristics of the project deliverables.

Finally, the process model can be further adapted to project-specific characteristics. This concludes the tailoring process for the time being. However, as tailoring is complex depending on the project environment and affects many stakeholders and workflows, the decisions made should be constantly scrutinized and further improvements identified and implemented. Some process models take this into account automatically when applied correctly. Scrum, for example, provides for such reflections in the regular retrospective. With other process models, you should ensure that you also regularly take time to reflect on and improve your own work. New and significantly changed process models in particular are likely to require improvements.

Example

The Nexus process model selected in the previous example is modified as follows as part of the project-specific adaptation: As the clients have to fulfill certain safety-relevant requirements, they want corresponding verification and validation tests. These tests are to takc place in the penultimate (verification) and final (validation) sprint. For stakeholder communication and progress control, a special Kanban board is used in these two sprints to visualize the completeness of the test. All in all, this results in a hybrid process model based on Scrum, which contains components from the V-model and Kanban.

Try to clearly present the targets and criteria for the new process model and check which plan-based and agile components of known process models fulfill the criteria. You can use the toolbox in Chapter 8 for this purpose, for example. Each method presented there indicates whether it is more suitable for use with plan-based or agile process models.
You can list the identified criteria together with the organization-specific degree of fulfillment in a table and indicate for each criterion whether the degree of fulfillment is more in favor of a plan-based or an agile approach. You can also gather initial ideas on which elements you want to integrate into the new process model or which alternatives you see. You can see an example of such a table in Figure 5.8.
The example in the table shows once again that the development of a process model is not trivial: In this case, volatile requirements speak more in favor of an agile approach, while many other criteria are more in favor of a plan-based approach.

List of criteria:

criterion	fulfillment / measure	degree of agility	proposed processes / methods	alternative
corporate culture	< 4 levels of hierarchy	plan-based agile	reduce 2 levels of hierarchy	
qualification of the team	basic training on agile working	plan-based agile	more training required	
stability of the team	< 2 changes of personnel per project	plan-based agile	further diversification of qualifications	
size of the team	medium	plan-based agile	no need for action	Nexus
distribution of the team	2 sites, USA, France	plan-based agile	virtual stand-up meetings	
involvement of the client	2 reviews per project	plan-based agile	2 sprint reviews with clients, rest without	
documentation requirements	according to order	plan-based agile	preparation of templates	
stability of requirements	no uncontrolled requirements	plan-based agile	free changes in regular review	
criticality	-	plan-based agile	no need for action	
project deliverables	complex	plan-based agile	iterative implementation	
specifications from clients	customer requirements / requirements specification	plan-based agile	change to product backlog	keep existing way of working
regulatory / legal requirements	machinery directive	plan-based agile	no need for action	
...	...	plan-based agile	...	

Figure 5.8: Exemplary overview of criteria that you can use to evaluate the organization-specific process model to be developed. For each criterion, it is indicated how it can be fulfilled in the organization, whether this fulfillment is more in favor of a plan-based or agile approach and which processes and methods you can use to form the individual process model from the information provided.

Warning

You will find many comments and recommendations for plan-based or agile project management in the literature and on the internet. Traditionalists argue that well-managed plan-based projects are very successful in practice. Agile advocates, on the other hand, emphasize the benefits of agility and show examples of how agile project management can be adapted to almost any context.
However, this book is not about which process model is theoretically applicable and fundamentally suitable, but rather about what fits your way of working in a very concrete and pragmatic way under the given conditions in your organization. This is exactly what is needed to achieve optimum efficiency and effectiveness.

You will probably not be able to complete the table in Figure 5.8 right away. You may first need more information about your internal processes and options for plan-based and agile project management, which you will need to collect and evaluate. You may also think of other criteria that you would like to add.
The real challenge comes next: linking selected methods to form a coherent process model and deriving the necessary organizational structures for applying the process model in daily practice.
Organizational frameworks that have configurable processes, methods and roles can help here. Such a framework is shown in Figure 5.9. Organizational frameworks structure systems by breaking them down into individual components and arranging them according to their relationship. The **organizational framework for hybrid project management HyProMM** discussed here consists of three areas:

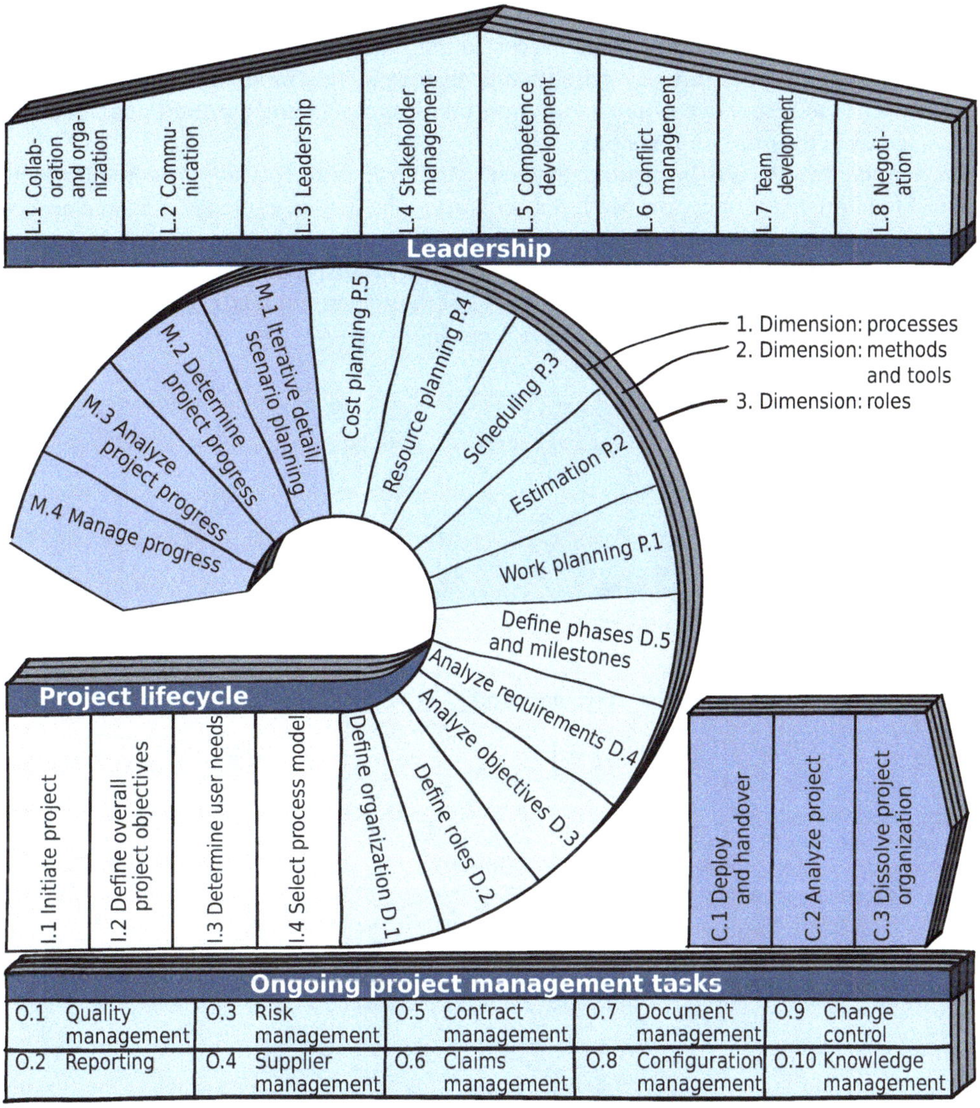

Figure 5.9: Organizational framework for hybrid project management HyProMM, consisting of the three areas of leadership (top), project lifecycle (middle) and ongoing project management tasks (bottom). For each area, there are three views for processes, methods and tools as well as roles.

- leadership,
- project lifecycle and
- ongoing project management tasks.

Each area in turn consists of three dimensions:
- Dimension 1 represents the actual processes.
- Dimension 2 visualizes methods and tools for implementing the processes.
- Dimension 3 contains roles and organizational structures that are necessary for the collaboration and execution of the processes, methods and tools.

Every process, every method, every tool and every role can be plan-based or agile. Figure 5.9 names the respective processes in the process dimension, for example *P.1 Work planning*. However, the figure does not show the modeled process. This modeling is adapted to the conditions of the respective organization and takes into account the methods, tools and roles selected in the other dimensions.

If a new process model is to be created, the areas, dimensions and elements deemed relevant are selected from the framework. Methods, tools and roles are then selected for each process that are to be used in the implementation of the processes.

Take a look at Figure 5.10: This shows the process view of a subsystem of the regulatory framework for an exemplary V-model and Scrum. For both process models, exemplary processes of the regulatory framework are marked.

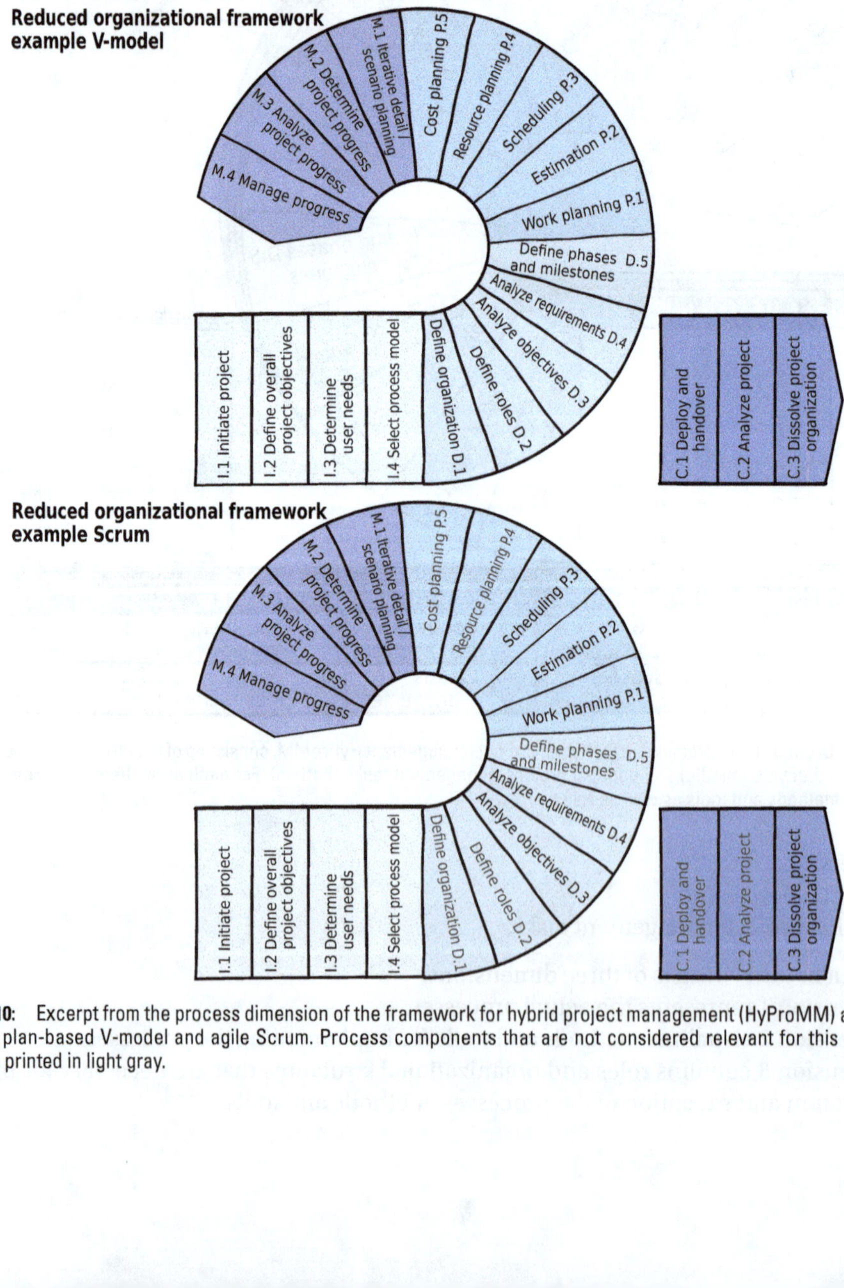

Figure 5.10: Excerpt from the process dimension of the framework for hybrid project management (HyProMM) and reduction to a plan-based V-model and agile Scrum. Process components that are not considered relevant for this exemplary case are printed in light gray.

Methods and tools can now be selected for the identified processes and suitable roles defined. An overview of many of the model's methods can be found in the toolbox in Chapter 8. To illustrate this, Figure 5.11 shows suitable methods and tools for the two selected process models. While the V-model and Scrum still look very similar in the process view of the organizational framework in Figure 5.10, their contents differ significantly in the method and tool view. For example, progress must be analyzed in both process models, which is done in the process section *M.3 Analyze project progress*. However, the methods used for this are different. In the V-model, for example, the *M.3.1 Analyze earned value* can be used. In Scrum, progress analysis is carried out using the *M.3.9 Perform sprint review* and the *M.3.10 Perform sprint retrospective*.

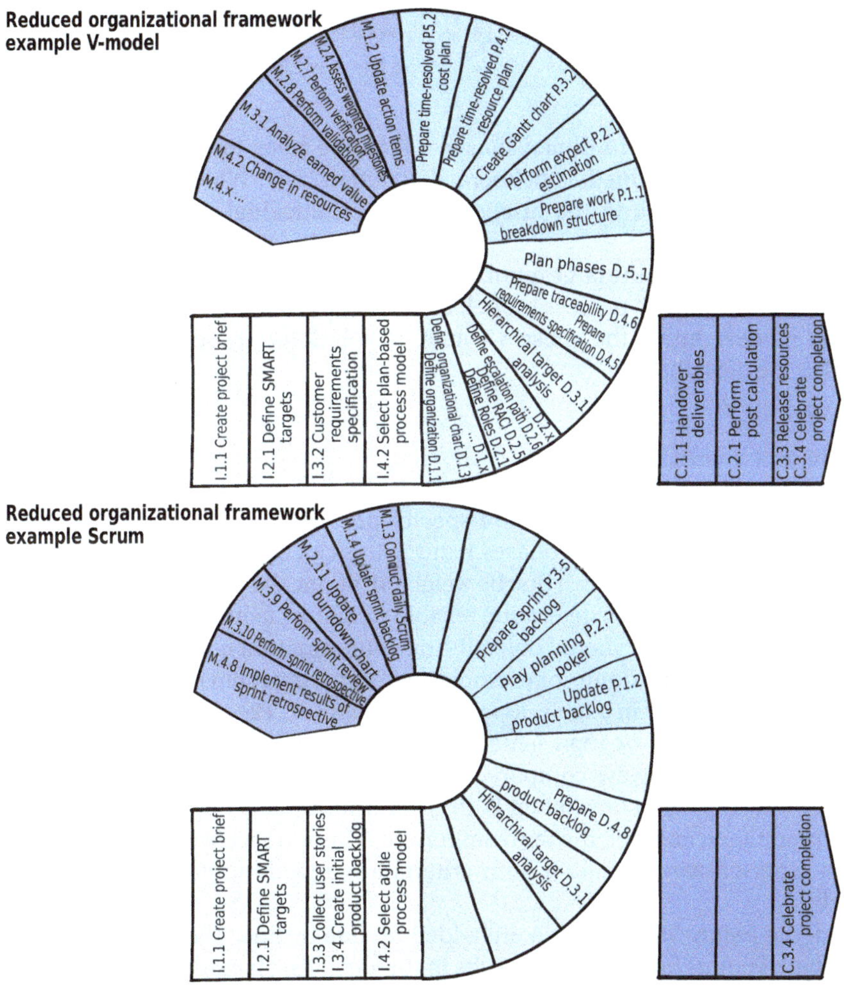

Figure 5.11: Methods and tools for the selected processes of the organizational framework HyProMM of the two exemplary process models V-model and Scrum.

Other characteristic components of the V-model and Scrum can be clearly seen in Figure 5.11:

- The V-model uses many plan-based methods such as work breakdown structures, Gantt charts and earned value analysis. Model-integrated verification and validation is also part of the organizational framework.
- Scrum uses significantly different methods, especially for planning and control. The form of the organizational framework indicates that the definition, planning and control phases can be run through several times. This corresponds to the iterative approach of process models like Scrum.

Tip

The HyProMM framework lists processes, methods, tools and roles. The link between the individual processes and interfaces must then be established. In the example, it must be specified that the processes and methods P and M (planning and monitoring/control) are only run through once in the V-model, whereas they can be run through several times in Scrum. The organizational framework merely indicates this possibility through the circular arrangement.

The presentation of the third dimension, in which the roles for the individual processes, methods and tools are defined, is deliberately omitted here. Which roles are used in each case is explained in detail in the discussion of the individual methods and tools in the toolbox in Chapter 8.

Due to the complexity and the large number of influencing factors and criteria for the creation of an organization-specific process model, it is advisable to conceptualize several competing process models during the phase 3 search for solutions. These are then compared and evaluated in the next phase.

Phase 4: Decision

The competing concepts for organization-specific process models are now evaluated on the basis of the criteria determined in phases 1 and 2. This can be done using a utility value analysis, for example. The criteria must first be weighted, as not all criteria will probably mean the same to you. As it is difficult to determine reliable weighting factors for a large number of criteria from a gut feeling, you should objectify the determination of these weighting factors as far as possible. This can be done, for example, by comparing them in pairs: Write the criteria horizontally and vertically in a table, as outlined in Figure 5.12.

Interview several experts or representatives of all stakeholder groups that are important to you and ask them to always compare two criteria with each other and determine which criterion is more important or whether both criteria are equally important. If one criterion is more important than another, the more important criterion receives two points and the less important criterion zero points. If both criteria are equally important, both receive one point each.

At the end, the points for all criteria are added up line by line. Their share of points in the total number of points for all criteria results in the weighting factor as a percentage.

criterion 4 "size of the team" was rated as more important than criterion 2 "team qualification" (as this can be improved)

criterion	1	2	3	4	5	6	7	8	9	10	11	12	13	line total	weighting in percent
1 corporate culture		2	1	1	2	0	0	1	1	0	0	0	0	8	5
2 team qualification	0		1	0	1	0	0	1	1	0	0	0	0	4	3
3 stability of the team	1	1		1	0	0	0	1	1	0	0	0	0	5	3
4 size of the team	1	2	1		2	1	0	2	2	1	1	1	1	15	10
5 distribution of the team	0	1	2	0		0	0	1	1	1	1	1	1	9	6
6 involvement of the client	2	2	2	1	2		0	0	1	2	1	1	1	15	10
7 documentation requirements	2	2	2	2	2	2		2	2	2	2	2	2	24	15
8 stability of requirements	1	1	1	0	1	2	0		2	0	1	1	0	10	6
9 criticality	1	1	1	0	1	1	0	0		0	1	1	0	7	4
10 project deliverables	2	2	2	1	1	0	0	2	2		1	1	1	15	10
11 specifications from clients	2	2	2	1	1	1	0	1	1	1		0	1	13	8
12 regulatory / legal requirements	2	2	2	1	1	1	0	1	1	1	2		0	14	9
13 interfaces with other projects	2	2	2	1	1	1	0	2	2	1	1	2		17	11
													sum	156	100

scoring 0: less important
1: equally important
2: more important

17 points represent 11 percent of a total of 156 points

Figure 5.12: Pairwise comparison of the criteria to determine the weighting.

taken from pairwise comparison

criterion	weighting in percent	alternative solution 1 (e.g. V-model) score	alternative solution 1 (e.g. V-model) weighted score	alternative solution 2 (e.g. Scrum) score	alternative solution 2 (e.g. Scrum) weighted score
1 corporate culture	5	8	40	5	25
2 team qualification	3	7	21	3	9
3 stability of the team	3	8	24	2	6
4 size of the team	10	3	30	10	100
5 distribution of the team	6	8	48	7	42
6 involvement of the client	10	5	50	7	70
7 documentation requirements	15	5	75	5	75
8 stability of requirements	6	3	18	10	60
9 criticality	4	4	16	8	32
10 project deliverables	10	5	50	10	100
11 specifications from clients	8	7	56	6	48
12 regulatory / legal requirements	9	7	63	6	54
13 interfaces with other projects	11	5	55	8	88
		sum:	546	sum:	709

scoring 0: does not meet criterion
5: does partly meet criterion
10: does fully meet criterion

scoring by experts

Figure 5.13: Comparison of two alternative solutions. Here, the V-model and Scrum are chosen as examples in terms of their suitability for a fictitious company and its criteria. Of course, two hybrid solution alternatives can also be compared instead.

You can now apply these criteria to the solution alternatives found in phase 3 search for solutions, which is illustrated in Figure 5.13. For each criterion, points are awarded for each solution alternative, for example from zero points if the criterion is not fulfilled to ten points if it is completely fulfilled. The points are multiplied by the previously determined weighting factors and the products for each solution alternative are added up. The solution alternative with the higher number of points fulfills the organization-specific criteria better overall.

Warning

We have discussed here a formal and structured way of developing possible process models and comparing and selecting them. A *utility value analysis*, as carried out here, can only ever be a decision template. You must make the decision yourself. This may differ from the recommendation from the utility value analysis, but should be well justified. At the end of the chapter, we will look at a case study in which a organization-specific process model is less structured, but therefore not necessarily less well derived and implemented!

The example presented here simplistically assumes that projects only use one process model (e.g. V-model or Scrum). In reality, there are many projects in practice that combine several process models. We will go into this in detail in the following chapters.

Example

The members of the Resentorp GmbH change project develop several alternative solutions with which the project objectives can be achieved. In the end, the company decides on the following option:

Projects should continue to be carried out sequentially in the future, as this best suits the products to be developed, the working methods and the qualifications of the employees. However, the planning should consistently correspond to the ideas of critical chain project management, including the staggering of projects along the bottleneck resources. The company hopes that this will lead to a significant increase in delivery and cost reliability.

To increase commitment, Kanban boards are used for operational project management, on which all relevant project stakeholders can see the current progress. The flexible use of Kanban boards and the involvement of employees in their design and application are intended to promote a sense of responsibility among employees for the tasks they have taken on.

Phase 5: Introduction and change

You have decided on a new, organization-specific and possibly hybrid process model for successful project management and now want to introduce and apply it. Unfortunately, this phase is often given too little importance. According to the motto "The concept is clear, now everyone just has to work according to it", managers assume that the announcement of new processes and structures will be received with enthusiasm and the change will be implemented. The fact is that change causes some people to worry that they will lose something as a result. There will almost always be people who have to say goodbye to cherished, sometimes comfortable habits as a result of the change and are therefore opposed to the change. The loss may be, for example:

- an informal authority and position in the company
- formal authority through organizational changes or
- expert knowledge that one had regarding the old processes, but which is lost when these processes are changed.

However, it is also possible that other people simply do not grant you the success of being the creator of new processes and structures and are therefore fundamentally opposed to the new.

Warning

Experience shows that many projects that change processes and structures fail during or after the introduction of these processes and structures. Bit by bit or all at once, the wheel is turned back and the old processes and structures are back in place.

A sustainable and therefore successful introduction does not begin and end with the announcement of the new processes and structures. You should bear the following in mind:

- Try to find supporters at an early stage and convince as many stakeholders as possible of the need for change.
- On the one hand, support from senior management is important for the introduction of new processes, but on the other hand, many employees react negatively to innovations imposed from above. As part of change management, you need to address this area of tension.
- Ensure that working with the new processes and structures works and support employees during a sufficiently long transition phase with coaches and mentors.

Change management

At this point, a *brief* introduction to change management is provided. There are numerous models for change management that divide the change process into more or less phases. The best-known models come from Lewin [Lewin, 1947] and Kotter (Kotter 1996) and are summarized in Figure 5.14.

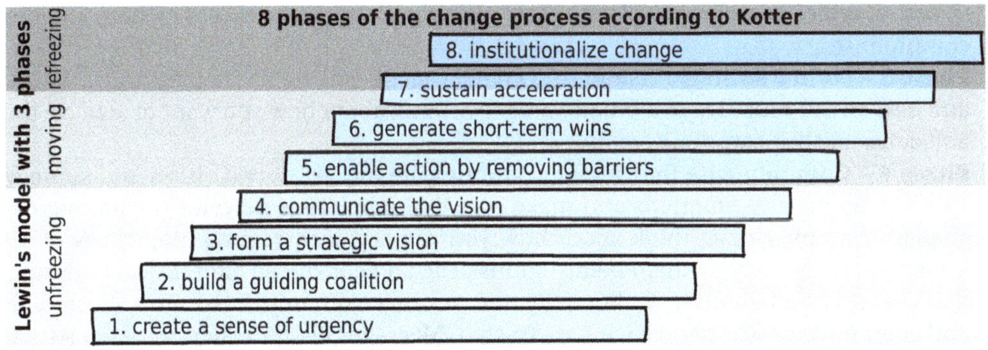

Figure 5.14: Phase models of change management according to Lewin and Kotter.

Lewin divides change into the three phases of unfreezing, moving and refreezing:

- **Unfreezing** is about generating the willingness to change.
- The actual implementation of the change takes place in the **moving** phase.
- The change must then be consolidated and established, which takes place in the **refreezing** phase.

Kotter divides the change process into eight phases, whereby the first four phases largely correspond to Lewin's unfreezing phase. Phases five and six correspond to Lewin's moving phase, while phase eight corresponds to refreezing. Phase seven contains elements of both moving and refreezing, depending on its characteristics. Both models indicate that a change requires preparation and follow-up in order to establish it in the long term. Only when the need for change is recognized will a willingness to implement the change arise.

You've probably heard sayings like "Never change a running system." There are also fears that go hand in hand with change. Change may mean having to give up cherished ways of working. In some companies, one change project follows the next, and each one only serves to reduce the number of employees. Such an environment is then not very suitable for implementing positive changes to shape the future of the company. I want to focus here on changes to introduce new process models for modern project management.

I'll go through Kotter's phases briefly:

- **Phase 1 – Create a sense of urgency:** Think about why there is a need for change and how you can recognize it. Make it clear why the change needs to happen soon, for example because a particularly important project is coming up that cannot be handled successfully with the current way of working or because competitors could launch more innovative products on the market in shorter development cycles.
- **Phase 2 – Build a guiding coalition:** You will not be able to introduce a new process model, including new processes and structures, on your own. You need a team to support you. Management support is just as important as ensuring that the employees who are most affected by the change are properly involved. Make your work easier at the beginning by focusing on employees who are ready for change. In this way, you will gain multipliers and supporters for your project at an early stage. However, also increasingly involve (constructive) critics in order to deal with their arguments and use or disprove them for good compromises.
- **Phase 3 – Form a strategic vision:** As a change agent, you are a visionary and strategist and have to sell your idea to a broad audience. Think about how you want to achieve this and consciously incorporate emotions.
- **Phase 4 – Communicate the vision:** Communicate the developed vision and strategy. Don't spare positive emotions and make sure that employees perceive the announced changes as transparent. Think about how you can regularly remind employees of the vision and do not rely on them being enthusiastic and convinced after the first information event. Pay attention to authenticity: the message communicated must be credible and communicated by people who are trusted. Messages disseminated by management are perceived by some employees as being "imposed from above" and are therefore rejected.
- **Phase 5 – Enable action by removing barriers:** This is where Lewin's Moving Phase begins. There are always obstacles that stand in the way of change. You need to identify these and remove them. This will enable and empower your employees to actually implement the change. The same applies here: the message and action must be credible! A company that claims to sell customer-friendly products but continues to release products with defects for sale will have difficulties implementing change in this area.

- **Phase 6 – Generate short-term wins:** Opponents of change will not stop attempting to undermine the change. Even if you have initiated the right things in the long term, this is not enough to convince all employees. You need short-term successes to underpin your vision and strategy. Otherwise, your concept will not work. So think about how you can design the introduction in such a way that you can demonstrate these successes. For example, applying the new process model to a project in crisis and bringing it to a successful conclusion can be such a success. Plan sufficient resources for training and coaching – because this project must then also be a success.
- **Phase 7 – Sustain acceleration:** Once you have achieved initial success, it is time to roll out the change on a broad front. However, review the experiences from the initial successes and incorporate them into your concept in order to improve it further. Lewin's refreezing phase begins at this stage.
- **Phase 8 – Institutionalize change:** Even after initial successes and the application of the new process model throughout the company, the change is not yet established. The awareness that projects are now running better than before needs to be consolidated. Training courses help employees to overcome uncertainties when applying the new processes. Think about how you can channel and manage the criticism that is sure to arise. But don't close your eyes and ears in the deceptive certainty that the change has already been successfully introduced.

In the practical examples and case studies at the end of the chapter, we will take a closer look at the introduction of a new process model using another example.
In the following chapters, we will now discuss ways of combining plan-based and agile process models in various ways.

Example

The management of Resentorp Inc. decides to introduce a new process model using critical chain project management and Kanban boards based on a sequential approach.
Employees are made aware of the unsatisfactory status quo while the solutions are being developed. The company's successes and failures in recent years are compared with those of competitors. A change project is set up to further develop proposals for change. The proposals are developed with the involvement of the employees. The managing director outlines the vision of an innovative and dynamic company that aligns current products with customer needs and can offer them on the market at good conditions. He emphasizes the importance of the employees for the path to be taken and asks for their support. Respected employees explained to their colleagues how the implementation of this vision promotes the company.
All employees are trained in accordance with the new corporate mission statement; they are shown examples that illustrate how the members of the change project envision the new ways of working and communicating in the company that have been jointly developed. New projects are launched using the new process model and employees are provided with coaches to help them apply and share their experiences.

Summary: Development of a individual process models

- Do not select a new process model without sufficiently analyzing the specific situation of your organization.
- Instead, analyze the situation of your projects and the organization in detail and determine the criteria that are important to you when evaluating a process model.
- Formulate targets that a new plan-based, agile or hybrid process model must fulfill.
- Develop several alternative solutions. Be deliberately creative and include process models, workflows and structures that you do not yet know very well.
- Select the alternative solution that best meets your objectives and criteria.
- Introduce the selected process model sustainably in the company.

Sequential application of different process models

You are already familiar with various plan-based and agile process models. In this and the following sections, we will now look at possible combinations of plan-based and agile process models. There are many projects in which it is worth using both plan-based and agile components.
One way of combining plan-based and agile process models into hybrid process models is the **sequential application.**

Example

The standard procedure for a company's development projects involves an initial concept and feasibility phase. In this phase, the first prototypes are designed and evaluated so that as many risks as possible are removed from the subsequent implementation of the development. The project deliverables usually consist of hardware and software components and are difficult and costly to change.
The company is therefore reluctant to carry out the actual development according to pure agile process models such as Scrum. Instead, development is carried out according to the V-model. In order to eliminate as many risks as possible in the concept and feasibility phase, this is carried out as a Scrum sub-project. Thanks to the strong integration of the customer and other stakeholders, many risks can be identified and the initially volatile (unclear) requirements can be specified to such an extent that the subsequent development is relatively predictable.

The example already shows a trend: upstream agile process models can serve to reduce risk. They are intended to safeguard the subsequent implementation, which takes place according to plan-based process models. This enables the use of agile process models even in industries that tend to be critical of changes as they are difficult or costly to implement. Examples include the construction industry, plant engineering and hardware-oriented research and development projects.

The sequential arrangement also makes it comparatively easy to combine two or more process models. There are clearly defined transition points at which the results of one sub-project are transferred to the other sub-project. An example can be seen in Figure 5.15.

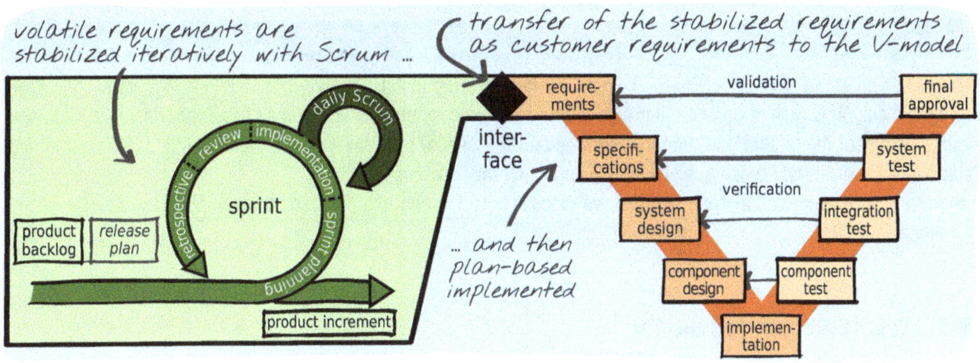

Figure 5.15: Sequential sequence of two sub-projects. First, requirements are concretized as part of a concept and feasibility study according to Scrum and then handed over to a sub-project according to a plan-based V-model.

The representation in Figure 5.15 visualizes the components of the hybrid project using conventional diagrams. The framework for hybrid project management HyProMM presented in this book allows the integrated representation of this hybrid model. The method dimension of this framework is illustrated in Figure 5.16. The advantage of this illustration is that suitable methods, tools and roles can be selected alongside the relevant process areas and combined to create a hybrid process model.

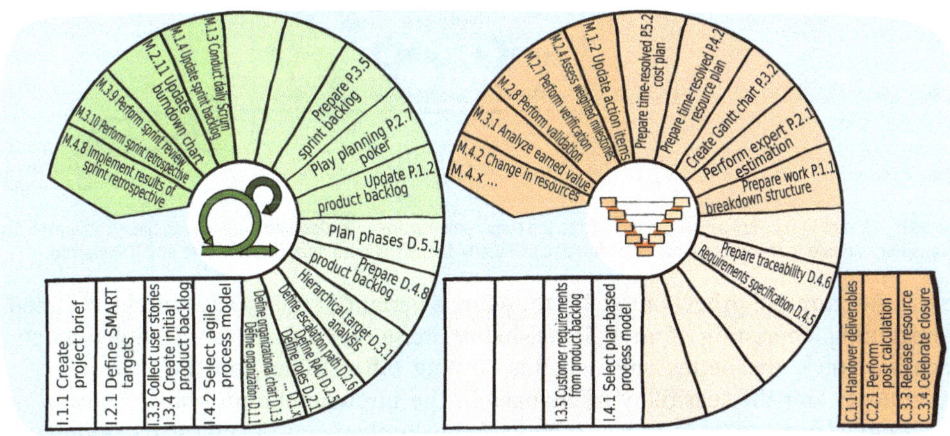

Figure 5.16: Model of the sequential sequence of a Scrum and a V-model project in the method dimension of the organizational framework for hybrid project management HyProMM.

The transition from the agile to the plan-based project phase is modeled in Figure 5.16 by generating the customer requirements from the product backlog in element I.3.5. Here, the requirements stabilized by the agile phase are adopted and implemented in the subsequent plan-based project phase. In the practical examples and case studies, you will get to know the real-life example of the combination of Scrum and the waterfall model in the construction industry.

Warning

We have looked at the process and method dimension here in order to link Scrum with a V-model. In practice, however, the role dimension is also important. Plan-based and agile process models sometimes have very different roles and understandings of leadership. In Chapter 4, you learned that agility requires a new understanding of leadership and collaboration and cannot be achieved by simply applying other processes.
In the practical examples and case studies, we discuss how we can use different roles in hybrid projects and the consequences of this.

Water-Scrum-fall model

A hybrid process model that is quite common is the **water-Scrum-fall model.** The name already indicates that this refers to a combination of the waterfall and Scrum process model. Scrum is embedded in sequentially processed plan-based project phases before and after. Scrum is often used in the middle of the project during the implementation or realization phase, see Figure 5.17.

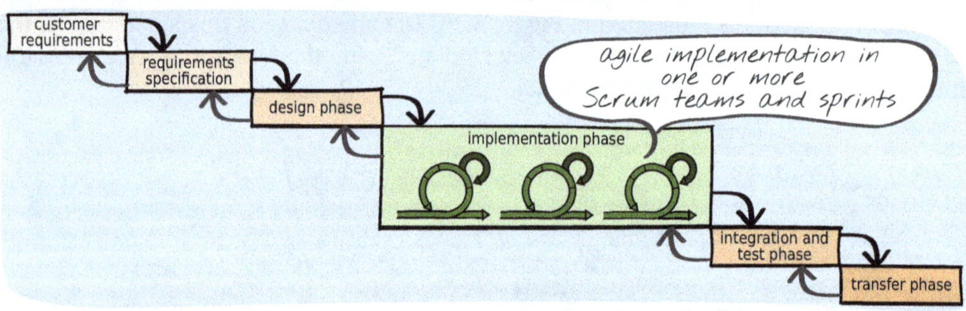

Figure 5.17: Water-Scrum-fall process model: First, the requirements are collected and concretized based on a plan. Then the transition is made to the agile implementation phase. Finally, the results are integrated, tested and transferred.

The water-Scrum-fall process model is therefore a variant of the process model outlined at the beginning, consisting of an agile feasibility phase and a plan-based implementation. Which variant is the better one depends, among other things, on the volatility of the requirements and the feasibility of changes in the implementation phase. Software and organizational processes are easier to change in the implementation phase than foundations cast in concrete or expensive molding and pressing tools.
As with all hybrid and therefore partially agile process models presented here, agility is not created by processes and changed procedures, but must be actively practiced by employees and managers. Interaction with each other, the willingness of the team to take responsibility, the involvement of the customer, the openness to change, etc. must be present.

In practice, it is therefore important to define roles that match the selected process model. For the process model shown in Figure 5.17, for example, the following must be considered: A project manager with a purely plan-based mindset will probably find it difficult to carry out the Scrum phase in the middle in an agile manner. Hybrid process models always require agile people!

One disadvantage of the process model illustrated in Figure 5.17 is that the close involvement of the customer in the agile Scrum phases only takes place at a late stage. The model is therefore based on the assumption that the requirements can be determined well, but that prompt and frequent customer feedback is helpful during implementation.

It is not generally defined which phase or phases in the water-Scrum-fall process model are carried out agilely with Scrum. In principle, this does not have to be the implementation phase. Instead, for example, the first project phase or the design phase can also be carried out agilely with Scrum and in close coordination with the customer.

V-Scrum-model

Another sequential arrangement of plan-based and agile process models is illustrated in Figure 5.18. Here, an agile implementation according to Scrum is embedded in the two branches of the plan-based V-model. This arrangement may seem a little strange at first, as it weakens Scrum's strength of involving the customer at an early stage. We have already made similar considerations with the water-Scrum-fall model. The entire requirements identification and detailed specification is traditionally plan-based and sequential. Only when the design draft has been finalized in detail is it developed in an agile manner and then verified and validated again in a plan-based and sequential manner. Nevertheless, there are constellations in which such an arrangement can be interesting: Projects that work together with other projects or companies that themselves work according to the V-model can define clear external interfaces through such an arrangement.

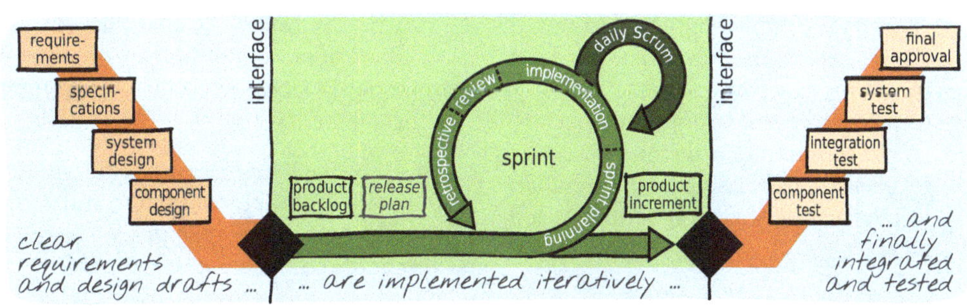

Figure 5.18: Sequential sequence of three sub-projects. First, requirements are determined, specifications drawn up and the design drafted in a plan-based sub-project carried out according to the V-model. This is followed by agile implementation using Scrum. Finally, integration is carried out in another plan-based sub-project based on the V-model and the results are reviewed.

Assume that a company A works strictly according to the V-model and commissions another company B to deliver certain parts of the project object. The project deliverables are defined by company A using customer requirements, and the company expects the contracting company B to provide a corresponding requirements specification and design drafts for approval as well as final verifications and validations. The solution outlined in Figure 5.18 is one of several ways in which collaboration between companies A and B can be organized and company B can at least partially benefit from the advantages of agile development.

Warning

Make sure that the interfaces between plan-based and agile sub-projects are precisely defined and that everyone involved is aware of them. Requirements in the form of user stories differ significantly from typical technical requirements. Make sure that those involved in the project can still understand and correctly translate the requirements. Responsibilities and authorities are also not easy to regulate in detail. For example, what happens to the project manager responsible for the left-hand branch of the V-model in Figure 5.18 after the handover to the agile sub-project, in which there is no project manager according to the strict interpretation?

Companies that use the V-model usually value traceable requirements that can be verified and validated. If agile work is carried out between the two branches of the V, i.e. between specification and design on the one hand and integration, verification and validation on the other, it is precisely this traceability that can get lost. If no precautions are taken, clearly specified requirements are changed in an uncontrolled manner during the agile implementation phase. In the end, seamless and traceable verification and validation will not be possible.
To solve this problem, a "design freeze" can be carried out before handover to the agile implementation phase. This only allows changes if they are documented. Traditional change control management, including the necessary documentation, is therefore incorporated into agile implementation with Scrum. The documented changes make it possible to completely verify and validate any requirements that may have been changed during the implementation phase.

Summary: Sequential application of plan-based and agile process models

- The sequential application of plan-based and agile process models can be used, for example, to:
  - transform volatile requirements into stable requirements using customer-focused agile process models and then implementing these on a plan-based basis or
  - define interfaces to plan-based organized contracting projects and simultaneously benefit from agile iterative implementation.
- One of the challenges is the fundamentally different ways of thinking and acting of plan-based and agile process models, which need to be made compatible.

Parallel application of plan-based and agile process models

Plan-based and agile process models can not only be applied sequentially, i.e. one after the other, but also simultaneously, i.e. in parallel, within a project. This allows different characteristics of sub-projects to be taken into account, for example.

Example

A large development project is carried out in accordance with the waterfall model. The organizational chart is as follows:

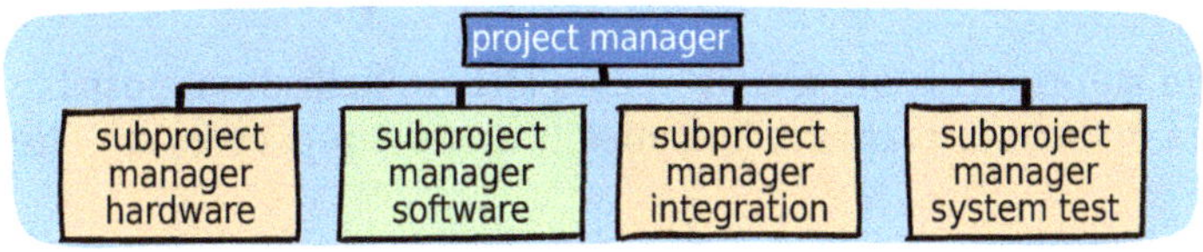

Many parts of the project deliverables are easy to specify and the requirements are largely stable. In addition, not all parts can be changed easily. The software sub-project is carried out in an agile manner according to Scrum. The requirements are much more volatile here. The team is qualified and, with only eight developers, sufficiently small for the application of Scrum.

The situation outlined in the example is typical for the use of hybrid process models with parallel application of agile and plan-based project management. However, there are also other constellations, such as

- agile overall projects with plan-based sub-projects
- agile or plan-based projects in which one supplier works according to the other process model.

In contrast to the sequential application of plan-based and agile process models, plan-based and agile processes, methods and roles are used simultaneously in parallel application. In order to achieve project success, the defined overall project deliverables must be developed at the end. Synchronization points are required throughout the entire project lifecycle so that all parts of the project deliverables fit together in terms of time and content. This poses a particular challenge for project planning and management, as different processes, methods and roles need to be synchronized.

Synchronization of plan-based and agile processes, methods and roles

The synchronization of plan-based and agile processes within a project is necessary to ensure that the desired *project deliverables* are created at the end of the project despite parallel processing.
Take another look at the regulatory framework in Figure 5.9 and the middle section of the project lifecycle. Typically, there is a need for synchronization

- at **project start,** i.e. in the initialization, definition and planning phase
- during **project implementation,** i.e. in the control phase and
- at **project completion,** i.e. in the closing phase.

In the following paragraphs, I look at the synchronization requirements of individual processes in the framework HyProMM and refer to the numbers assigned in the framework, for example *I.1*, *D.2* etc.

Synchronization at the start of the project

Parallelization in sub-projects often only occurs after initialization, possibly only after the project has been defined or planned. This means that the processes are often run through together.

- *I.1 Initiate project*
- *I.2 Define overall project objectives*
- *I.3 Determine user needs*
- *I.4 Select process model*

In the processes *D.1 Define organization* and *D.2 Define roles*, sufficient communication interfaces must be provided between the sub-projects, and clear regulations on responsibilities, authorities and reporting duties must be ensured, for example by defining a so-called RACI matrix (see Chapter 8).

If you combine plan-based process models with agile process models, role conflicts can easily arise. In plan-based projects, there is a more or less powerful project manager role. In agile projects, work is sometimes carried out with or without a project coordinator. Project managers see themselves as technical, sometimes even disciplinary leaders who can use the entire repertoire of participative (collaborative) to authoritarian leadership. A project coordinator tends to be the first among equals, coordinating tasks and supporting the team, but in case of doubt always lets the team decide and take responsibility.

The sharing of responsibility and authority between these fundamentally different role and management models is not easy and should be well prepared so that it works in the hectic day-to-day project work.

Tip

In some hybrid projects with a traditional overall project manager and agile sub-projects according to Scrum, it has proven successful for the overall project manager to take on the role of product owner of the agile sub-projects. This means that the responsibilities and authorities for project management of the sub-project remain within the team. At the same time, however, the project manager is involved in the sub-project and can significantly influence the result by defining and prioritizing the requirements and approving them.

It should be repeated once again at this point: Agility is not characterized by other processes and methods, but by a different mindset, certain values and principles. When setting up a hybrid process model, you want to utilize the advantages of agility in your projects. However, the benefits will only become apparent if you respect the values and principles. Unfortunately, one mistake that can be observed time and again is that companies use agile processes and methods but do not enable their teams to act in an agile manner. So if you want to implement a sub-project in an agile way, you must not put a dominant (sub-)project manager in charge of it.

The process *D.3 Analyze objectives* is comparatively simple, as the definition of objectives is neither specifically traditional plan-based nor agile.
What is more interesting is how the process *D.4 Analyze requirements* is implemented. The requirements must be determined, analyzed and transferred to the respective sub-projects. In plan-based projects, requirements are collected in the requirements and functional specifications, whereas in agile projects they are collected in the product backlog or directly on the Kanban board. While requirements in plan-based projects are often formulated from a system perspective, user stories focus on people and their relationship to the product. In some hybrid projects, requirements are collected in the requirements specification, from which the functional specification and user stories and tasks for agile sub-projects are then derived. This corresponds to *variant 1* in Figure 5.19.

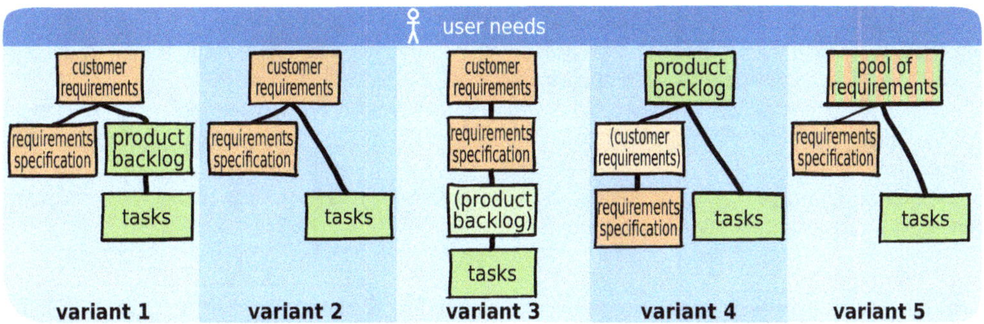

Figure 5.19: Some exemplary variants of dealing with requirements in hybrid projects. Interfaces must be defined between traditionally plan-based (orange) and agile (green) formulated requirements so that nothing is forgotten and a coherent overall set of project deliverables can be created at the end.

It should be noted with this variant that the requirements specification and the tasks should be consistent. Let's assume that a sub-project working according to Scrum creates software. This was initially defined in user stories and concretized in individual tasks. You must ensure that the developed software ultimately works together with the system components specified in the requirements specification.
Alternatively, the intermediate step of a separate product backlog can be omitted so that the tasks are derived directly from the customer requirements. This corresponds to the *variant 2* shown in Figure 5.19. Here, the agile sub-project uses the customer requirements as a replacement for the product backlog and creates the tasks directly from the requirements recorded there.
With *variant 3*, the tasks are either derived directly from the requirements specification, or a product backlog is first created from the requirements specification. In this case, the risk of variants 1 and 2 that specifications in the requirements specification collide with tasks is largely eliminated. The indicated additional creation of a product backlog is actually superfluous and can lead to difficulties, as the product backlog actually contains requirements in the form of user stories, i.e. from the user's perspective, and therefore corresponds more to the customer requirements than the requirements specification.
Instead of starting with a traditional customer requirement specification, the product backlog from the agile world can also be the starting point for further documentation of the requirements. This is illustrated in *variant 4*. Here, the customer requirements should

be considered optional, as the customer requirements and product backlog have a lot in common due to the arguments mentioned in the previous paragraph.
You can also completely detach yourself from the requirements specification or product backlog and generate a common pool of requirements from which the requirements specification and tasks are derived directly. This is shown in *variant 5*. Such a requirements pool can be understood as a hybrid of a requirements specification and a product backlog. Of course, other variants are also possible and can be found in practice.

Tip

The requirements can be controlled and distributed in the project using all the variants presented here. It is important that you define in your projects how to manage the requirements seamlessly and without discrepancies. You must avoid inconsistencies and incompatibilities between the requirements of the plan-based and agile sub-projects.
A particular challenge is that requirements in plan-based projects must be fully determined at the start of the project, whereas in agile projects they can be added to or changed later.
The more documents or databases you create for the compilation of requirements, the greater the risk of gaps or inconsistencies arising. The effort required to keep these collections consistent in the event of changes quickly increases significantly.
Which option you choose depends on the specific situation of your organization. Find a compromise between coherent requirements management and minimal document-related bureaucracy.

At the end of the project start, the project is divided into phases in the process *D.5 Define phases and milestones* and important milestones are planned. The challenge here is that agile sub-projects do not necessarily work in sequential phases, but in iterative sprints, for example. The number of these can vary and therefore cannot necessarily be predicted precisely.
It is helpful to define a certain *change budget* and plan a certain number of sprints based on this. For example, the change budget comprises 20% of the estimated story points and thus takes into account the fact that many user stories are only vaguely specified at the start of the project. The change budget is checked for changes within a sprint. If the change budget has not yet been used up, the changes are implemented. The designation of a change budget promotes the differentiation between important and unimportant changes, but is only compatible with agile values and principles to a limited extent, as changes are limited by the budget.
The *P.1 Work planning* process creates the basis for further detailed planning. In plan-based projects, the project content is usually defined in the work breakdown structure, in agile projects on taskboards or Kanban boards with corresponding user stories.
The process *P.2 Estimation* differs significantly when strictly interpreting plan-based and agile principles:
- In *plan-based* projects, *efforts in staff days* (or similar units, such as person-hours, etc.) are actually estimated.
- In *agile* projects, especially in Scrum, the *size of a task* is estimated. This includes all activities associated with the task. It is estimated relative to a known reference task. Unlike the effort, the size is purely related to the complexity of the task and does not take into account the skills that an employee has to complete it. Story points are usually used as unit for the size.

In hybrid projects, an inconsistent measure has the potential for conflict, as the plans resulting from the estimate (schedules, release plans, etc.) cannot build directly on each other. Nevertheless, there are successful projects that are traditionally planned with efforts in staff days and whose sub-projects work with sizes in story points. Once again, it is important to clearly define the interfaces between the plan-based and agile project parts so that it is clear to everyone involved what is estimated and planned on a plan basis and what is agile.

Example

An electronics company carries out its large product development projects using a company-specific hybrid process model. The projects are planned according to the waterfall model. There is a software group in the development department that creates the software required for the projects as an in-house supplier. This software group works agilely according to Scrum.

The upcoming tasks of the project are estimated on a plan-based principle and the development costs are later calculated from these estimates. The software group, on the other hand, estimates its tasks derived from the customer requirements in story points. The overall project manager does not need to know the estimated user stories in story points for scheduling purposes. The software group's release plan, which includes the planned number of sprints, is sufficient.

The estimated (and actually used) story points are converted into costs for internal allocation and to determine the development costs. This is done by knowing the velocity, the group size and the internal hourly rate for software developers.

Scheduling takes place in the process *P.3 Scheduling*. In *plan-based projects*, network plans or Gantt charts are used here or, in simple projects, pure milestone plans. Thinking in terms of fixed deadlines predominates.

In *agile projects*, deadlines are set either directly via planning on the taskboard or Kanban board, via release planning or the sprint backlog. The focus is on striving for a product that satisfies the customer.

These different plans and focuses can lead to conflicts within a project. The overall project manager often demands fixed delivery dates for parts of the project deliverables from his sub-project managers. Synchronization takes place via milestones or time intervals such as fixed sprint durations. If the agile sub-project working according to Scrum suddenly needs one more sprint, this causes difficulties for the scheduling of the overall project. Change budgets can be used to counteract this.

The planning of the required resources and costs incurred takes place in the processes *P.4 Resource planning* and *P.5 Cost planning*. If different estimates such as efforts in staff days and task size in story points are used, this poses a challenge for the project manager that can be solved in different ways:

- If the agile team works with estimates in story points, these can be converted into staff days using the velocity. If a team of 10 employees usually achieves a velocity of 100 story points in a sprint lasting 20 working days, one story point therefore corresponds to two staff days. The number of employees involved and their hourly rate can be used to specify the costs for each task specified in story points.

- Agile sub-projects often work together on a project for a fixed period of time. Without taking story points and velocity into account, the employees are then simply transferred to the resource plan and their hourly rate is taken into account in the cost plan.
- If an agile sub-project also works for another project at the same time, the costs can be divided proportionally between the projects. For example, if a team consists of 10 employees and works 20 working days (= one sprint) for two projects, the costs are allocated to the two projects in proportion to the story points worked on in each case. For example, if the team costs 40,000 US$ and works on user stories with 40 story points for project A and user stories with 60 story points for project B, 16,000 US$ are allocated to the cost plan for project A and 24,000 US$ to the cost plan for project B.

Synchronization during project realization

During the actual project realization, you are in the control and monitoring phase with the processes

- *M.1 Iterative detail/scenario planning*
- *M.2 Determine project progress*
- *M.3 Analyze project progress* and
- *M.4 Manage progress.*

Here it is important to keep the specified requirements and the planning of all plan-based and agile sub-projects consistent. There is potential for conflict, as changes to requirements are viewed differently in plan-based and agile projects:

- Plan-based projects assume a correct and complete specification and solid plans. *Changes are the exception* and are perceived as a disruption to the project process.
- According to the agile manifesto, reacting to change is more important than following a plan. *Changes help to improve* the project deliverables and increase customer satisfaction. In Scrum, the collection of requirements in the product backlog does not claim to be complete. Changes are therefore usually unavoidable.

Despite these agile ideas, changes in a project consisting of several sub-projects must be coordinated in such a way that the partial results of the sub-projects fit together in the end.

Tip

You can ensure the above-mentioned consistency of the requirements through project-wide change management. How formal this needs to be depends on the complexity of the project and its environment. In a small project, changes can be collected on pinboards or in an electronic list, for example in a spreadsheet. In a large project, there may be a weekly change conference at which change requests are discussed, evaluated and, if necessary, approved.

Different metrics must be combined for the *progress determination* and analysis: Plan-based projects often measure the actual hours worked and compare these with the earned value and the planned value as part of the earned value analysis.
Agile projects like to measure project progress with burndown charts or by monitoring the project flow on the Kanban board.

Of course, you can convert the work performed in agile sub-projects back into staff days by measuring the story points and pending user stories and incorporate the results into your earned value analysis. Conversely, you can also create a burndown chart for the overall project by displaying not only open story points, but also open staff days or pending requirements.

Depending on the complexity of the project, you can also work with different methods and use earned value analysis or a simplified version of it for plan-based sub-projects, while tracking the progress of agile sub-projects using a burndown chart. You can gain insights from both methods that you can use to manage the project. However, it is important that you get a comprehensive overview of the overall project progress.

Synchronization on project completion

All partial results of the plan-based and agile sub-projects must fit together by the end of the project at the latest, so that the project can be formally completed. The organizational framework HyProMM from Figure 5.9 provides the following processes for this:

- *C.1 Deploy and handover*
- *C.2 Analyze project* and
- *C.3 Dissolve project organization.*

During *project handover*, the project deliverables are handed over to the client. In plan-based projects, this usually takes place as part of a formal project handover with an approval protocol. As there are not necessarily integrated feedback cycles in plan-based projects, the client may see the project deliverables for the first time now. However, experienced project managers also involve the client in plan-based projects at an early stage so that there are no completely different views of the project deliverables at the time of handover.

In agile projects, the project is implemented in close coordination with the client anyway. Scrum integrates regular feedback in the sprint review into the project process.

At the end of the project, it is crucial that all partial results fit together and result in the desired project outcome. Ensuring this is the task of project management. If something does not fit, you can now only make corrections to a limited extent or at great expense or delay. There are also clear differences between plan-based and agile projects when it comes to *project analysis.*

In plan-based projects, the project is evaluated at the end as part of the lessons learned, and experiences are saved for future projects. The analysis usually also includes a post-calculation, which also serves to secure knowledge and can also be used as a final calculation.

In agile projects, the striving for efficient and effective processes is one of the core values. Scrum uses the sprint retrospective to analyze the project at the end of each sprint and to secure knowledge. Either the entire project is then reflected on in the retrospective of the last sprint of the project, or an additional cross-sprint retrospective takes place.

As part of the project analysis, you should clarify, for example, whether

- all agreed deliverables of the project have been invoiced
- there are still outstanding claims,
- the calculations prepared before or at the start of the project were correct or need to be adjusted for new projects in the future

- the project processes have worked well or improvements need to be made to the process model
- etc.

Especially in projects with agile sub-projects, the post-calculation helps with the learning process: it provides information on the extent to which changes occurred in the project and whether these changes (which have a positive connotation in agile projects) were cost-effective. These findings can be used to calculate change budgets for future projects. During *organizational dissolution*, the project team and the infrastructure installed for project completion are dissolved. Projects are temporary endeavors. In practice, it is common for a follow-up project to be carried out with at least partially different employees. Agile projects are also designed to be temporary. In some companies, however, the teams work together in the same composition over many projects.

Summary: Parallel application of plan-based and agile process models

- The parallel application of plan-based and agile process models within a project is usually achieved by applying these process models in different sub-projects.
- The different philosophies of plan-based and agile sub-projects can lead to tensions in the project process.
- A success factor for hybrid projects organized in this way is the definition of synchronization points. When defining the hybrid process model, determine where and how these synchronization points look and how the various plan-based and agile methods interlock.

Integrated application of different process models

In the integrated application of plan-based and agile process models in hybrid projects, the plan-based and agile processes, methods and roles cannot be assigned to time phases (as in sequential application) or to sub-projects (as in parallel application). Instead, the use of these processes, methods and roles is appropriate to the situation along the project lifecycle.

Example

- A project manager consistently controls a plan-based project according to the V-model using a Kanban board on which the project tasks are sorted according to the stages to be completed. The team accesses and processes the tasks largely independently.
- In an agile project, progress is determined and analyzed using earned value analysis. Instead of costs or expenses, story points are used as the basis for calculating the earned value.
- A project working according to Scrum uses conventional requirements and efforts in staff days instead of user stories and story points in order to work better with other, non-agile stakeholders. In addition to the product owner, the Scrum master and the developers, there is also a team coordinator. This serves as a kind of "minister of foreign affairs" and is the point of contact for stakeholders outside the project. Unlike the

Scrum master, the team coordinator is part of the developers and is therefore jointly responsible for the project deliverables.

- A company that works plan-based according to the spiral model and has no previous experience with agile project management is gradually moving into the agile world. To this end, agile processes and methods are incorporated into the plan-based spiral model: Customer reviews and retrospectives are integrated into the individual development cycles. The teams are allowed to assign the tasks to be worked on themselves within certain limits, and the weekly project meetings are replaced by short daily stand-up meetings based on the daily Scrum model.

The examples show that the integrated application of plan-based and agile process models takes place in completely different constellations. In many cases, there are **company-specific conditions** that make a textbook implementation of purely plan-based or agile process models difficult. These include, for example:

- regulatory or legal requirements for project realization
- interfaces with clients or suppliers
- the qualification level of the employees
- etc.

Another motivation for the integrated application of plan-based and agile process models is the desire to combine the **advantages of both approaches** and avoid their disadvantages as far as possible. The term "agility" has positive connotations and is associated with dynamism, speed, market focus, customer orientation and innovation. An agile, self-determined approach has a motivating effect on many employees. This is why organizations want to give themselves a modern, innovative look and are developing into agile organizations. Plan-based processes are being broken up and an agile mindset, methods and roles are being introduced.

Warning

Despite all the enthusiasm for agile process models, you should carefully examine and prepare for their use. There are many examples of organizations that use agile processes for their projects. Scrum is usually taken as the model of choice. However, the agile values, principles and role models are not adopted consequently. Instead:

- the line manager is involved in the project
- the team is given neither responsibility nor authority and
- retrospectives and reviews are not carried out or only superficially due to deadline pressure.

In the end, those responsible ask themselves why agile process models do not bring the desired success in their organization. You can find out more about this problem in the practical examples at the end of the chapter.

As part of an *organization-wide* hybrid process model, plan-based and agile process models can be applied in an integrated manner; however, *individual projects* can also benefit from such hybrid approaches. Many project managers use agile approaches within the process models specified by the company in order to motivate their employees, involve customers and manage the project. The delegation of tasks, self-determination, a good feedback culture

and communication as well as transparent progress determination are just some of the key points that stand for an agile approach to plan-based projects.
In practice, it is therefore not enough to design hybrid process models. The understanding of leadership must be adapted to the project management and the people working on it: A modern understanding of leadership places people at the center and focuses on their abilities. In many situations, agile values and principles can form a basis for leadership. However, some employees are overwhelmed by the demand for self-organization. In this case, management must be adapted so that these employees can also successfully complete the project.
The number of possible combinations of plan-based and agile process models is so great that I cannot go into all these possibilities here.

Warning

In the previous section, I described the need to synchronize different project areas and discussed how this synchronization can be achieved. These considerations also apply here in the integrated application of plan-based and agile process models. If you use different plans, methods and roles in your project, you must always ensure that you do not create gaps, contradictions or inconsistencies.

Synchronization within the project tends to become more complex the more plan-based and agile processes and methods you combine. Take a look at Figure 5.20. In principle, it is of course possible to switch between a plan-based and agile approach for every part of the project realization. However, this is unlikely to be useful and efficient in most cases, as the number of interfaces and necessary synchronization points increases and eventually becomes so large that project execution becomes inefficient and error-prone.

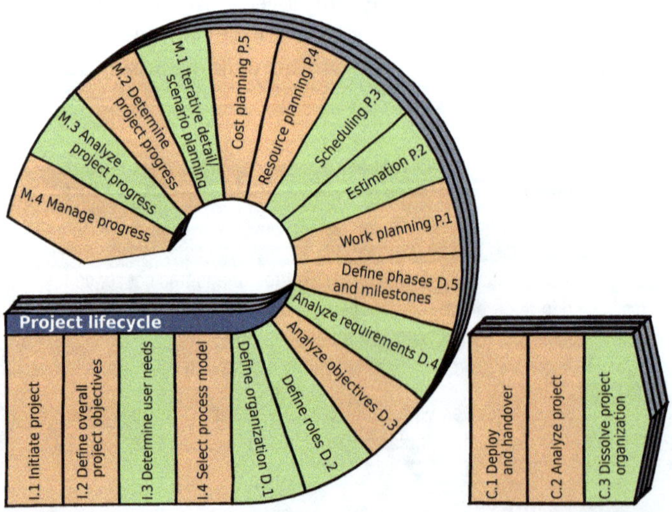

Figure 5.20: Possibly exaggerated combination of plan-based (orange) and agile (green) processes in the regulatory framework for hybrid project management.

Both plan-based and agile process models have shown that projects carried out according to them can be successful and that they are therefore considered legitimate representatives of efficient and effective project management. Hybrid process models attempt to adapt plan-based and agile process models to certain conditions and further optimize them. If plan-based and agile process models are mixed in a wild, unstructured way, projects will not benefit from the advantages of both worlds.

ScrumBan

A now widespread representative of hybrid process models that combines two agile methods is ScrumBan. Even though there is no ScrumBan standard, there are some common features, for example:

- The Scrum taskboard is replaced by the Kanban board.
- The processes and roles from Scrum form the procedural and organizational basis of the process model.
- The Kanban practices are integrated into the processes:
  - the visualization of the work (on the Kanban board)
  - limiting the amount of work started
  - the measurement and optimization of the flow
  - the announcement and enforcement of all agreed rules
  - the consideration of feedback mechanisms and
  - the joint implementation of improvements.

The attractiveness of ScrumBan is certainly also due to the fact that the two process models can be integrated without contradiction as far as possible. The Kanban board is a further development of the taskboard, which not only takes into account the processing status (to do, doing, done) but also the processing stage (specification, implementation, test, etc.)
The limitation of work in progress can be discussed in the daily Scrum, and adequate limits can be set, reviewed and adjusted if necessary. It is also possible to measure and optimize the flow.
Feedback mechanisms and the strive for improvement are already an integral part of Scrum through the daily Scrum, the sprint review and the retrospective.
A significant difference between Scrum and Kanban is the strict timebox principle in Scrum, which does not exist in Kanban. On the contrary: instead of a time limit, Kanban relies on a consistent flow orientation, where time limits are more of a barrier. However, this different approach can usually be solved well in practice. For example, flow orientation within a sprint is not a problem. Rules must then be agreed across sprints on how to proceed here. It is often determined that the rules of Scrum apply.

Summary: Integrated application of different process models

- With the integrated application of plan-based and agile process models, they cannot be assigned to a project phase or sub-project. Plan-based and agile processes, methods, tools and roles are instead selected appropriately according to the situation. This takes place along the entire project lifecycle and in all organizational areas of the project.
- In principle, almost all plan-based and agile processes, methods and roles can be combined with each other. However, not all combinations make sense. In addition, the number of interfaces and synchronization points to be defined increases enormously with the number of combinations.
- Too many interfaces and synchronization points increase the complexity and error rate during project realization. The desired advantage of using the best processes, methods, tools and roles from the plan-based and agile worlds is then turned into the opposite.

Software tools in hybrid projects

What do you associate with project management software? Many people will think of typical programs with which you can create more or less complex-looking Gantt charts. Well-known examples are Project from Microsoft, Redmine and ProjectLibre.

You can use such software to plan processes and put them in a chronological order. They take resources into account and create cost plans. Some of these programs are available as server versions or are connected to a database in order to manage entire portfolios and support company-wide resource management.

Similar programs are also available as web-based applications. Additional functions such as document storage or the creation of templates for standard projects are often offered.

Programs that come from plan-based project management can only be used to a limited extent for agile and therefore hybrid projects. Taskboards or Kanban boards cannot be displayed with them, nor can burndown charts.

Agile projects thrive on the dynamics of small teams that meet regularly on taskboards or Kanban boards to solve problems and analyze progress. In cases where the team works together on site, physical boards such as pinboards are usually the first choice. No (affordable) screen is yet as large, dynamic and, above all, flexible as a pinboard equipped with metaplan cards and structured with markers. However, agile projects can also be successful with distributed teams. In most cases, the pinboard must then be replaced by suitable software.

There is also specialized software for agile projects. It offers the option of creating taskboards and Kanban boards. A simple representative of such web-based software is Trello, the simplest version of which can even be used free of charge. Jira is another representative of agile web-based project management software, for which there is a wealth of extensions, for example for requirements management. It can be used to create product backlogs and manage and develop user stories.

Agile project management software often focuses on functions that facilitate collaboration, which is why some programs are also called collaboration software. For example, there are often options for exchanging files, starting chats or even holding a video conference, including screen sharing. There are now several representatives of mostly cloud-based software solutions that provide agile and plan-based project management functions. These include Wrike, ProjectPlace, Teamwork and Projektron BCS, for example.
Some software solutions now offer an almost seamless transition from plan-based to agile planning. In some cases, Gantt charts can simply be converted into Kanban boards and later transformed back again. This is helpful, for example, if clients expect a Gantt chart for reporting, but the team itself prefers to work with a Kanban board.

Tip

There is a huge range of project management software on offer. Before rushing into an offer, you should determine your requirements and check how you want to work with the software in the future. You should then explore the offers, draw up a shortlist of particularly interesting software solutions and test them as part of a trial. This may sound time-consuming, but it is always better than making a hasty decision in favor of the wrong product. In practice, unsuitable software is often used as an excuse for poor project management.

Practical examples and case studies

Case study 5.1: Criteria and selection of an organization-specific process model

The management of Castona Inc., a medium-sized manufacturer of specialized machinery, is facing increasing economic pressure from competitors. One of the aims of the IMPROVE initiative is to improve project management within the company.
Castona has 10 different series products, including plate lifting clamps, for example. However, many products are developed as individual items on direct customer order. For this purpose, the company has established a kind of modular system of frequently used components.
Two project managers, who have access to a team of 15 engineers and technicians, are responsible for the 10 series products and their further development. For customer orders, a further 30 employees are available, five of whom have received introductory project management training and are allocated accordingly. Most projects have a size of seven to ten employees and last between six and 18 months. The company has purchased licenses for standard project management software for the project managers. The software runs on the project managers' computers without a central database connection. The project managers use the software almost exclusively for scheduling with Gantt charts.
To launch IMPROVE, the management initiated a customer and employee survey, some of the feedback from which is outlined in Figure 5.21.

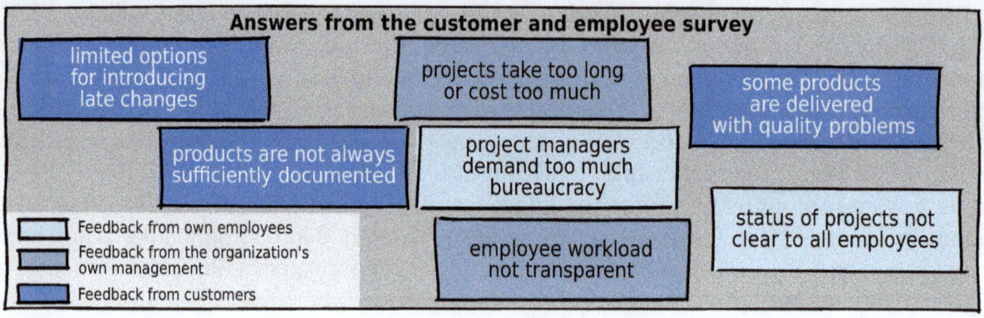

Figure 5.21: Answers from the customer and employee survey as part of the IMPROVE initiative of Castona Inc.

Guiding questions:
- How do you rate the feedback from the customer and employee survey?
- Which process models do you consider suitable as a basis for further individualization, which are less suitable?
- Do you have any initial ideas on how individualization or a fusion of several process models could look for Castona?

Case study 5.2: Situation analysis and target definition

The management of Claire Stutzig Ltd. recognizes deficits in its project management. Recently, a number of tenders have been missed and lost to competitors. The company repeatedly receives feedback from customers that although the quality of the products is good, the development and production times are relatively long. This is particularly true when subsequent changes to requirements need to be integrated. Some competitors have managed to act much more flexibly and quickly here.

This development leads to the management requesting a situation analysis of its own project management. This should then serve as the basis for improvements so that former customers can be won back and new customers acquired. Figure 5.22 shows a summary of important aspects of the situation analysis.

The projects are complex. The good product quality is seen as a strength and there is concern that this could suffer as a result of changes to processes. The corporate culture is seen as suitable for agile working. Even if the teams are highly qualified, they are not used to organizing their work themselves. The only medium stability of the teams is due to the fact that many specialists work in the teams who only support the projects sporadically and are needed in other projects at the same time. As the company only has one site, the distribution of the team is not a problem. Although the customers want to have the opportunity to make changes, they are not interested in investing time in regular feedback loops. The documentation requirements are reasonably high due to various international standards that must be complied with, but the stability of the requirements is very low. The risk potential is rated as medium and the complexity of the project deliverables (see Stacey matrix) as high.

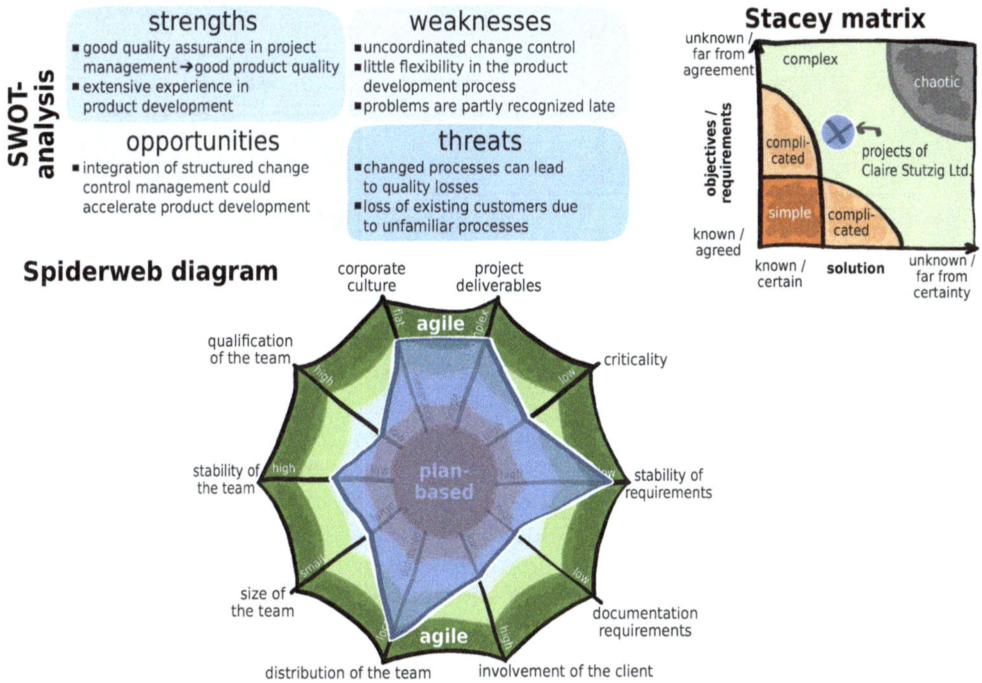

Figure 5.22: Situation analysis of the Claire Stutzig Ltd., consisting of a SWOT analysis, Stacey matrix and spiderweb diagram with criteria for selecting process models.

Guiding question:

- What targets for a new process model would you define on the basis of the situation analysis presented here?

Case study 5.3: Development of an organization-specific process model for successful project management

Casey Holmes is a project manager at Kalgusta Engineering Inc., which carries out projects on behalf of customers to support their development projects in the fields of electrical engineering and software. The project teams at Kalgusta Engineering usually consist of three to eight developers. From time to time, Kalgusta Engineering itself outsources individual services to subcontractors in order to overcome capacity and competence bottlenecks. Most projects last between 6 months and 2 years.

Kalgusta Engineering Inc. employs a total of 60 developers and, together with Casey Holmes, a total of five project managers. For less complex and small projects, the developers themselves sometimes take on the role of project manager. The company's organizational structure is shown in Figure 5.23.

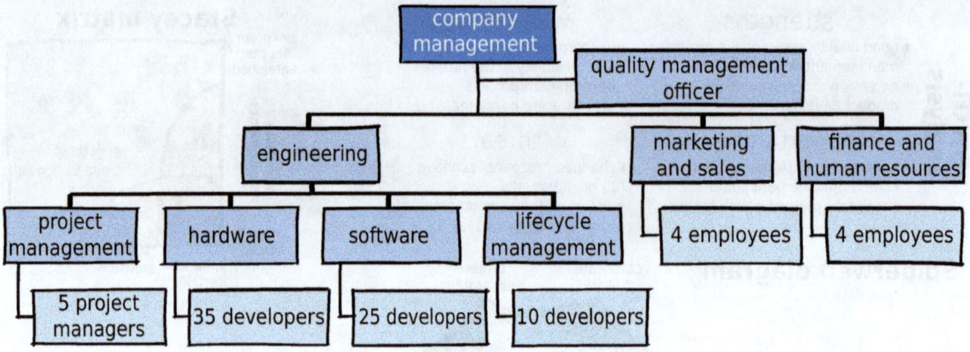

Figure 5.23: Organizational structure of Kalgusta Engineering Inc.

The projects use the waterfall model for realization. Customers contact the company with a list of customer requirements, which is reviewed by a preliminary project team. A sales employee prepares an offer for the customer on the basis of this review. If the offer is accepted, the project manager forms a project team in consultation with the head of the engineering department and the hardware and software group managers, which draws up a the requirements specification. Once this has been approved by the customer, the engineering work is carried out, tested and finally handed over to the customer. The phase transitions during development are handled quite flexibly. There is hardly any coordination with the customer during development.
The company management is not satisfied with the project execution. Although the employees are committed, rework is always necessary and there are misunderstandings with the customers. Together with the head of the engineering department, they decide to form a small team to modernize the project management processes and structures.

Key question 1:

- Who would you appoint to the project team?

The team will be led by Casey Holmes and will include one employee each from the hardware and software group as well as the marketing and finance departments. The company management and the head of the engineering department will form the steering committee together with Ms. Holmes.
Ms. Holmes and her team first gather information on the current status. This shows that interdisciplinary cooperation is valuable, as the marketing and finance departments bring in aspects that she would not have thought of. It becomes clear that the company has an established customer base and that it is important to work well with them in the long term. Customers generally insist on fixed prices and are used to traditional, plan-based process models. The representative of the software group reports that customers find it difficult to define clear requirements for the software. This is where a lot of rework is required because customers are not satisfied with the results. When the representative of the software group

starts to talk about the advantages of Scrum, Casey Holmes interrupts the discussion as she does not want to talk about solutions to the problems at the moment.
Based on the information gathered, objectives are now being defined for a new process model. Among other things, this should:

- enable the customer requirements to be taken into account
- allow the quotation of a fixed price
- involve the customer regularly during development
- ensure a high quality of the project deliverables
- get by with as little bureaucracy as possible for the relatively small project teams and
- report transparently on the progress of the project within the company.

Only now does Ms. Holmes want to discuss possible solutions. As the idea of Scrum is already on the table, this is now taken as the starting point for the first alternative. The processes with iterative development progress and the involvement of the customer are immediately welcomed. A critical view is taken of how the customer requirements should be integrated and on what basis a fixed price can be calculated.
The second alternative being discussed is the conversion of the current waterfall model into a V-model. This would be much easier to integrate into the current processes than Scrum and would increase product quality thanks to verification and validation. In future, the customer could receive a report on the interim status of the project on a regular basis, i.e. approximately every four weeks, together with the opportunity to provide feedback on specific issues.
After comparing the objectives and prioritizing them, the team finally proposes a hybrid process model to the steering committee that is based on Scrum but includes customer requirements, a fixed price, verification and validation.

Guiding question 2:

- Before you read on: How would you integrate the above aspects into a Scrum-based process model?

The steering committee is initially skeptical, as a hybrid working method is new territory for almost everyone involved. It therefore calls for an initial evaluation of a small project to serve as a prototype. Casey Holmes is appointed as a coach for the new project, which she is not supposed to lead.
All team members and Ms. Holmes receive three days of Scrum training. On three further days, the team works with Ms. Holmes to develop the company-specific adaptations to Scrum.
When a new project is due, it is designated as a prototype. The customer has already submitted the customer requirements for the project. These are now analyzed by the team. As there is still no practical experience with Scrum and hybrid process models, the costing is still carried out as usual. The risk of some things taking longer and becoming more expensive as a result of the new processes is taken by the company following an agreement with the steering committee. When the customer agrees to the offer, a large task board is first installed and the team begins to define the customer requirements as user stories, as learned in the training, and to structure them according to user groups. The estimation is still traditionally done in

staff hours, as the team has no experience with story points. The team develops a release plan from the user stories together with the sales employee, who takes on the role of product owner. This plan is sent to the customer for approval together with a tabular overview of the user stories.

At first, the customer reacts with surprise as it misses the familiar requirements specification, but after a discussion with the company management and the sales employee, it agrees to the new form. However, the customer is irritated by the invitation to come to a sprint review every four weeks. This seems to be unnecessary travel time. The customer agrees with the product owner to at least come to the first review and then decide whether further reviews are necessary.

With this in mind, the sprint planning is revised once again and an attempt is made to address all particularly critical risks in the first sprint. The customer's feedback should then enable the risks to be eliminated as far as possible. Although many open questions can be clarified in the first review and the review is therefore successful, the customer does not want to come to the review every four weeks. They agree on two further reviews every eight weeks. However, the team leaves the sprint length at four weeks. Whenever the customer is unable to attend the review, the product owner is asked for feedback.

For internal company reporting, the team uses a burndown chart that is displayed next to the task board for the entire department to see. On this chart, the remaining workload of the user stories is plotted in staff hours over the duration of the project. After each sprint, a remaining effort estimate is made for all open user stories. At the same time, the project manager keeps a list of the expenses so that the finance department can create corresponding budget invoices.

After the last sprint and the handover of the project deliverables to the customer, an additional joint retrospective is held between the development team, the process improvement team and the steering committee. On a positive note, the introduction of the iterative process model worked surprisingly well within the company. The team found the independent task assignment, the transparent overview on the task board and the feedback from the customer to be motivating. At the beginning, some employees found it difficult to get the right tasks from the task board. Many had been used to being assigned to tasks by the project manager for years.

The project manager was pleased with the good project result, but was dissatisfied with his own role. Due to the self-organization of the project team, his work was essentially reduced to administrative tasks such as updating the cost plan and product documentation.

The customer itself was very satisfied with the project result, as no rework was required. In the end, this also compensated for the time invested in participating in the sprint reviews.

The project was a great success for the finance department. Thanks to the fact that no rework was required and the project result was delivered on time, the cost plan was adhered to or even undercut compared to previous similar projects.

In the final meeting between Casey Holmes, the company management and the head of the engineering department, the overall conclusion was positive. However, there are also points that should be considered or changed again when introducing the new processes:

- Customers must be better informed about the new form of cooperation. In future, offers should include participation in a certain number of reviews.
- As customers working on a plan-based basis are used to requirements specification for approval, customers are free to decide whether they want to receive a corresponding requirements specification or alternatively work with the release plan and user stories.

- The scheme for calculating quotations is no longer correct and needs to be adapted to the new way of working. In future, the effort estimate is also to be converted into a size estimate in story points.
- Employees need to be trained in the new processes and the role of the project manager needs to be scrutinized.

All those involved are aware of the risk that the new process model may meet with resistance from some of the workforce once it has been introduced for all projects. The agile approach is viewed particularly critically by long-standing project managers. They fear for their role and their expert status as the know-how carrier of the previous project management. The following case study continues the topic of the introduction of the new process model.

Case study 5.4: Introduction of a new process model

Let's take another look at Kalgusta Engineering Inc. from the previous case study. We will now look at the introduction of the new process model developed in the previous case study. When Casey Holmes was appointed project manager for the process optimization and change project by the company management and the head of the engineering department, she quickly realized that the project had two main focuses. In addition to the substantive work on improving the process model, the focus is on supporting the change process through to the successful and sustainable introduction of the developed process model. She herself and her internal clients are aware of the problems with the previous way of working. However, many employees lack this awareness of the problem. Either they lack the idea of how things could work better, or they have settled for the current situation.

Guiding question 1:

- How can you make the problems visible to Casey Holmes' colleagues?

Her first idea is to collect the costs for rework for every project from previous years and post them on a central bulletin board to convince the employees of the considerable extent of the problems. However, she quickly rejects this idea, as she fears that it will drive the employees too far into a corner and provoke an attitude of justification.

Instead, she has a whiteboard hung up in the company's coffee kitchen. On it, she only illustrates the total costs for rework in recent years, but does not assign them to individual projects. Below the figure, she creates a brainstorming area entitled "board of waste". Together with two other open-minded project managers, she sticks notes on this board. The notes contain examples of the waste and errors in her own projects and the projects of the other two project managers.

In an information event for all employees organized jointly with the head of the engineering department, Casey Holmes explains the background to the whiteboard and encourages her colleagues to fill in the "board of waste" with their own observations so that these can be taken into account in the upcoming process improvements.

Guiding question 2:

- What do you like about the implementation of the "board of waste", and what risks do you see?

There are no new entries on the board. The employees don't understand why they should publicly post problems in their projects or blackball colleagues by citing examples of waste. Casey Holmes recognizes her mistake and thinks about how she can make the situation work in her favor without causing any damage herself. Together with her allied project manager colleagues, she places a few more examples of waste on the board and then invites people to a new info session where she explains the previous examples of waste that led to rework. As expected, there is immediate feedback to justify past actions, which Casey Holmes does not take up. Instead, she explains that on the basis of the examples collected so far and in light of the documented costs for rework, the whiteboard is now being prepared for the next stage, in which ideas for improvements are collected.

The "board of waste" becomes the "improvement board", and indeed some interesting ideas for preventing rework are pinned to the board over the next few days.

In the change phase so far, Casey Holmes was glad to have at least some support from the head of the engineering department and the allied project manager colleagues. Together with the change project team consisting of one employee each from the hardware and software group as well as the marketing and the finance departments, they form the core of the change team. Gradually, however, Casey Holmes would like to find additional supporters outside the project core team. These should carry the ideas for change forward without themselves being suspected of only representing the interests of the change project.

In the days that follow, she uses informal discussions with colleagues to talk about the ideas collected on the improvement board. She quickly notices who is open to change and who is more of a skeptic. However, she listens particularly carefully to the latter in order to understand their concerns and reservations. She hears that some employees believe that the company management is concerned with saving resources. Others, slightly frustrated, point out that they have been drawing attention to certain shortcomings for years, but no one has ever taken an interest in them. A third group blames customers and project managers for the rework: If clients specified more clearly and project managers consistently rejected subsequent changes as not contractually agreed, the problem would not exist.

Casey Holmes realizes that she still needs a lot of convincing. She thinks about how she can communicate her ideas in a credible and convincing way. First, she collects the benefits that employees will gain from the desired change, such as more personal responsibility, closer relationships with customers and less overtime. She then documents how the changes will affect the company. Less rework and more efficient projects will increase margins and make offers more attractive to customers, thus improving the company's economic basis.

Guiding question 3:

- How would you present these arguments to your employees? Who would you involve?

Casey Holmes knows that simply presenting the arguments to employees is not convincing enough. She wants to avoid defensive reactions from employees, which can arise when changes are perceived as being "prescribed from above". She therefore decides not to discuss the changes with the employees first, but to develop a common set of values based on previous discussions within the company. The values should form the foundation for the specific innovations and be supported by the employees.

In several working groups and with the involvement of company management and all employees, common values are developed, such as customer satisfaction with products and cooperation. The values are formulated in such a way that most employees can identify with them. Ms. Holmes and her project team are now linking the ideas for a new, hybrid process model with the values that have been developed.

The company's new vision, the jointly developed values and the new development processes that support them are formally announced at a company meeting. The company management first presents its vision of an innovative, reliable and customer-oriented Kalgusta Engineering Inc. Some of the employees who are positive about the change then present the jointly developed values. Building on this, the project team, led by Casey Holmes, presents the changes to the process model as a consistent implementation of these values in everyday working life.

To signal that customers also want improvements and are open to change, a long-standing customer is selected to report on expectations from the customer's perspective. Ms. Holmes selects a customer who has recently optimized their own project management processes and briefly reports on this.

She would then like to show another successful practical example that illustrates how companies successfully implement projects using hybrid process models. She is able to inspire a former student friend who now works for another company and is willing to talk about his experiences with the introduction of hybrid project management there.

The content of the new hybrid process model has now been worked out. Now it is time to evaluate it on a real project. Before that, however, Casey Holmes identifies potential obstacles to ensure the success of the first project using the new process model. A suitable project that is neither too complex nor too simple is soon found. It is important to Casey Holmes that the project is carried out with a customer with whom cooperation has worked well in the past and who is open to change. The employees involved in the upcoming project are extensively trained in the new approach, and Casey Holmes offers to be available as a project coach herself.

To ensure that all employees benefit from the experience, the project is explained in detail to all colleagues and the project documentation, such as the burndown chart and Kanban board, is displayed for all to see. Casey Holmes invites all employees to a short lunch meeting every Friday at 12:00 noon so as not to disturb the team working on the first project too much, but to keep the other employees regularly informed. She was able to raise a budget for snacks from the company management for this purpose, which provides additional motivation for some employees to find out about the progress of the change project and the first real development project based on the new process model. The customer also has his say and reports after the first sprint review that he is very satisfied with the progress and that the biggest risks have already been eliminated. This meant that the first success stories could be announced long before the first project was fully completed.

The evaluation of the first project according to the new process model shows that it still needs to be further optimized. The employees are also informed transparently about this – in the context of the overall success of the first project. Based on the positive experience, it is decided to use the new process model for all new projects in future. The employees are trained accordingly.
To signal that the improvement of the process model should continue, Casey Holmes leaves the "improvement board" hanging in the coffee kitchen. The costs for rework listed there are replaced by costs for rework per project. This figure is shown once for the projects carried out according to the old process model and once for the new ones. In fact, it can be seen that these costs have been significantly reduced and remain at a low level for the other projects.

Case study 5.5: Agile project management in the construction industry

In the construction industry, a sequential approach based on the plan-based waterfall model is common. Once a building has been constructed, it cannot simply be changed, which is why an incremental approach is generally not possible. Nevertheless, there are many examples where later changes were necessary for technical reasons or to satisfy customer requirements – with correspondingly expensive and time-consuming consequences for the project.
Some companies use hybrid process models specifically to reduce the risk of subsequent changes. The approach can be characterized by two key aspects:

- stabilization of requirements in the planning phase through agile prototype or model construction
- implementation of components whose requirements are still volatile as late as possible.

Innotech Building Construction Inc. has developed its own process model, which is outlined in Figure 5.24. As a rule, planning is carried out as a Scrum sub-project in three sprints, in which the planning object is iteratively detailed. At the end of each sprint, the customer is invited to a review. In order to visualize the planning object as well as possible, models of the later construction are created and evaluated by the customer from the second sprint onwards. Depending on the project, the models are created purely virtually as 3D models on the computer or as a scaled-down model on a model board. If necessary, further sprints can be carried out.

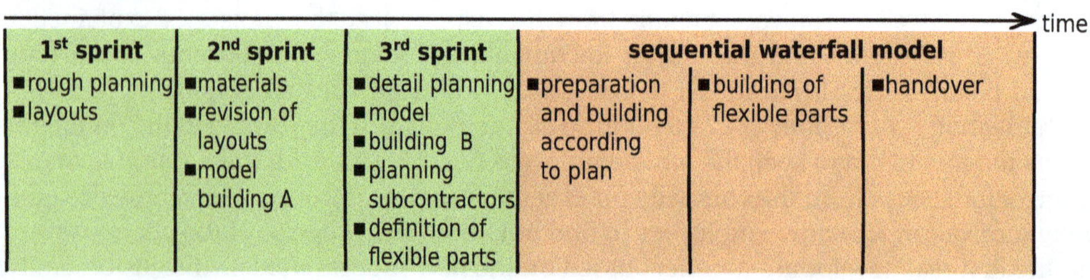

Figure 5.24: Sketch of Innotech Building Construction's process model with agile (green) and plan-based (orange) processes.

This iterative planning was initially unfamiliar to the customers, as they were involved in the planning process earlier and more frequently. At the same time, they appreciated the opportunity to submit their change requests on the basis of concrete models and to be able to approve the plans with a clear conscience at the end of the planning phase.
Together with the customer, the planners defined the components of the construction that should be completed as late as possible so that the customer still had the option of making changes. Depending on the construction project, these might include non-load-bearing walls, materials or window and door cut-outs as well as the position of water and electrical installations. A change budget was agreed with the customer for the calculation, which the customers could dispose of themselves or which was reimbursed to the customers as a bonus if it was not used.

Guiding question:

- How would you rate the degree of agility of the process model described in the example?

Case study 5.6: Plan-based project with Kanban

Keith Rutherford manages a development project at Electrofann Inc., a manufacturer of electronic components. The company's quality management manual contains a product realization process that is based on the plan-based waterfall model, see Figure 5.25.

product realization process				
concept phase	**definition and planning phase**	**design and implementation phase**	**test phase**	**transfer and ramp-up phase**
▪ product concept ▪ customer requirements ▪ proof of concept ▪ profitability analysis ▪ ...	▪ project plan ▪ product configuration ▪ market introduction plan ▪ requirements specification ▪ design specifications ▪ make or buy decision ▪ ...	▪ product design ▪ product catalog ▪ test specification ▪ transfer and ramp-up plan ▪ production instructions ▪ ...	▪ test reports ▪ marketing material ▪ spare parts ready ▪ technical documentation ▪ ...	▪ first samples ▪ validation series production ▪ staff training ▪ transfer report ▪ ...

Figure 5.25: Excerpt from the Electrofann Inc. product realization process.

Hardware components predominate in the products to be developed. To a lesser extent, controller components are integrated with software. Keith Rutherford has gained experience in agile project management through previous work for another employer and would like to bring this to his projects at Electrofann Inc. He appreciates the long-term increase in motivation through the team taking on more responsibility and the direct and short communication channels of agile project management. However, due to the prescribed development process with a sequential approach and established planning documents such as work breakdown structures, Gantt charts and resource and cost plans, he can only contribute agile ideas to a limited extent.

His team consists mainly of qualified engineers and technicians who work carefully but are used to being assigned work packages.

Guiding question 1:

- What possibilities do you see for implementing elements of agile project management in the project?
- What should Keith Rutherford bear in mind when introducing these elements?

Keith Rutherford wants to manage the project using a Kanban board. He uses metaplan cards to depict the work breakdown structure, which has to be created anyway, on a pinboard and places it in one of the frequently used meeting rooms. In order to remain compliant with the specified product realization process, he names the columns of the Kanban board according to the phases of the process, see Figure 5.26. As some work packages are relatively large and therefore not suitable for control, he divides them into several tasks. In the end, each task has a single person responsible and is documented on the Kanban board.

During the preparations, he realizes that the introduction of the Kanban board will not be easy. Many employees are used to waiting for tasks to be assigned. Picking up pending tasks independently will not work straight away, to say the least.

Guiding question 2:

- How could Keith Rutherford change the process to still be able to work with the Kanban board?

Keith Rutherford decides to first convince the sub-project managers to use the Kanban board. They should then distribute the pending tasks documented on the Kanban board within their sub-projects. However, this distribution should take place directly on the Kanban board so that the employees become accustomed to its use. The sub-projects therefore meet at fixed times for short daily meetings directly at the Kanban board. In sub-projects in which this approach works well, employees should then be empowered to take on tasks themselves and work on them according to the basic principles and practices of Kanban. The organizational chart of the project and the Kanban board are illustrated in Figure 5.26.

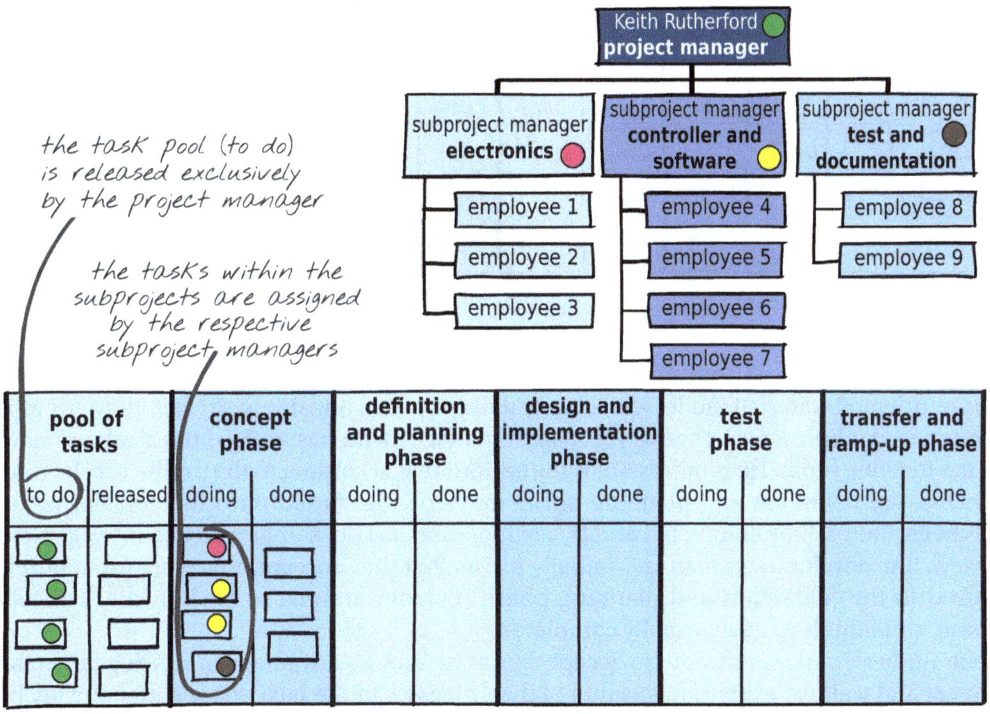

Figure 5.26: Organizational chart of Keith Rutherford's project and Kanban board.

Keith Rutherford reserves the right to release tasks for processing and marks them accordingly in a pool of future tasks. This allows the team to see at any time which tasks are still to come. At the same time, however, they know that these tasks cannot be processed at the moment. Only when the tasks have been released by the project manager and placed in the corresponding column of the task pool can they be processed by the sub-projects.

Guiding question 3:

- What problems do you see for the project when using the Kanban board in a sequential product realization process organized according to the waterfall model?

After a few weeks, the Kanban board looks like the one shown in Figure 5.27. The use of the board by the sub-project managers is going surprisingly well, and the sub-project "controller and software" is planning to delegate responsibility for taking on tasks directly to the sub-project staff in the near future. However, Keith Rutherford recognizes a conflict with the product realization process.

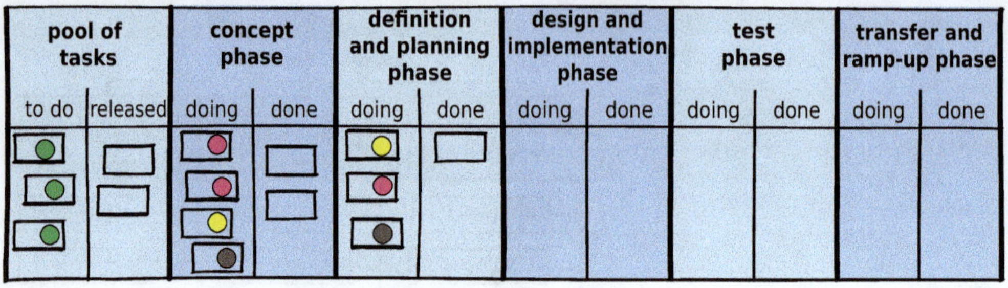

Figure 5.27: Kanban board of the project by Keith Rutherford.

The idea of Kanban, to optimize the flow of tasks through the project, contradicts the phase-oriented waterfall model with clearly defined phase transitions. In fact, Rutherford is approached a little later by the development manager, who has noticed the Kanban board in the meeting room. He reminds Keith Rutherford that according to the product realization process, the definition and planning phase may only be started once the concept phase has been successfully completed and this completion has been approved in a management review. The development manager initially insists that the processes that were prematurely moved to the "definition and planning phase" column are placed back in the "concept phase" until this has been formally completed.
Keith Rutherford does not want to accept this, as he sees no advantage in slowing down the project and waiting, even though some of the processes in the next phase could already be completed. He asks for a longer meeting with the development manager. He would like to introduce him to the concepts and advantages of agile project management and suggest that the project be carried out as a kind of prototype with deviations from the standard procedure. The development manager is torn. On the one hand, he likes Keith Rutherford's arguments and is also interested in involving and motivating the employees more. On the other hand, he feels overwhelmed and presented with a fait accompli, as he only found out about the ideas by chance.

Guiding question 4:

- How would you react as a development manager?

The development manager suggests that Keith Rutherford presents him with a concept explaining how he intends to manage the project to a successful completion. The concept should also include an explanation of how the company's own safety mechanisms for controlled phase transitions can be adhered to and how the progress of the project will be reported to management.

A week later, Keith Rutherford presents the following proposal to the development manager:

- In order to maintain the flexibility of the agile approach and exploit its strengths, he would like to combine the "definition and planning phase", the "design and implementation phase" and the "test phase" into one phase. Only then does the team have the opportunity to optimize the flow over a longer period of time.
- Before the project result is transferred to production, a major management review should be conducted prior to the "transfer and ramp-up phase". This gives management the opportunity to review the result and decide whether it is ready for transfer to (series) production.
- In order to build trust, Keith Rutherford offers to continue recording the usual key figures internally and to present them in the form of an earned value analysis. This is to be supplemented by including snapshots of the Kanban board in the status report and a burndown chart. The aim is to check whether plan-based and agile controlling produce similar conclusions.

The development manager likes the proposals. However, as he still sees risks in the processing, he insists that the usual reporting period be halved from four to two weeks so that he is informed more frequently about the progress of the project. In addition, the combined phases take too long. In order to maintain management's ability to exert influence on the one hand and not slow down the project team on the other, the development manager and Keith Rutherford agree on the following compromise: the project is allowed to go through the phases "definition and planning", "design and implementation" and "test" without a management review. However, as soon as the project is fully defined and planned, i.e. all processes have passed through this phase, a management review takes place.
In the end, the project achieved all the targets set and beat the time target by two weeks and the cost target by 15%. Not all sub-projects dared to enable employees to use the Kanban board themselves. In the sub-project "electronics", the sub-project manager continued to assign the tasks to his three employees until the very end.
Keith Rutherford is appointed to carry out the next project using the modified process model in order to gain even more experience in this area.

Case study 5.7: Agile or hybrid large-scale project?

Carosant Ltd. is a company that develops, produces and distributes medical devices. Regulated processes apply to these. For example, companies are required to structure development in phases, document the development process in detail and make product requirements traceable from the source to verified and validated implementation.
The head of development at Carosant Ltd. previously worked as a software developer and project manager for many years and is very familiar with agile process models. Following the trend towards agile software development, Carosant Ltd. organizes its system development projects in such a way that the software sub-projects can work agilely according to Scrum. The higher-level system development projects follow the V-model, which is widely used in medical device development, see Figure 5.28.

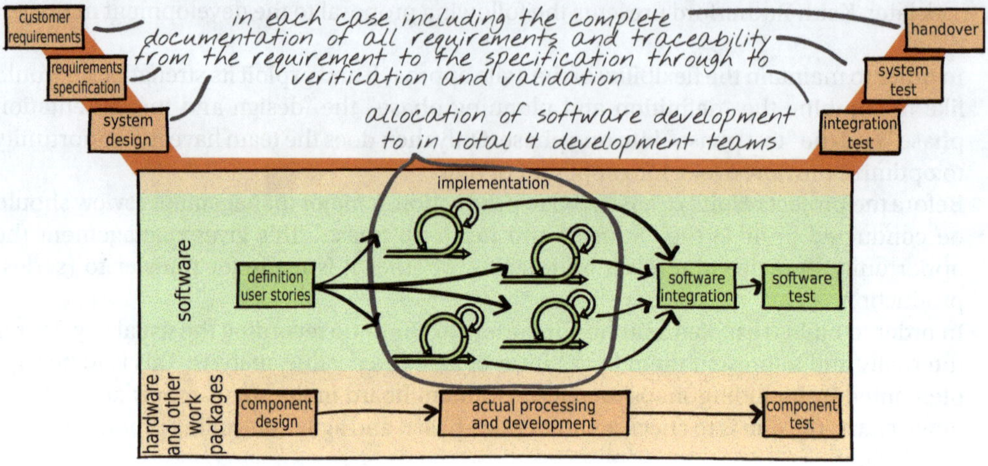

Figure 5.28: Original process model of Carosant Ltd. in which the product development projects use the V-model. Software sub-projects work according to Scrum. These agile sub-projects consist of 2–4 teams, each of which is largely self-organizing.

All non-software-related tasks are planned as work packages and processed according to a plan. The developers of the software sub-project derive user stories from the system design, which they then distribute to usually four development teams of around five developers each. There is neither a sub-project manager for the software sub-project nor people with designated management tasks within the development teams. For coordination purposes, the overall project manager takes on the role of product owner for all software development teams. A standard organizational chart of the development projects is shown in Figure 5.29.

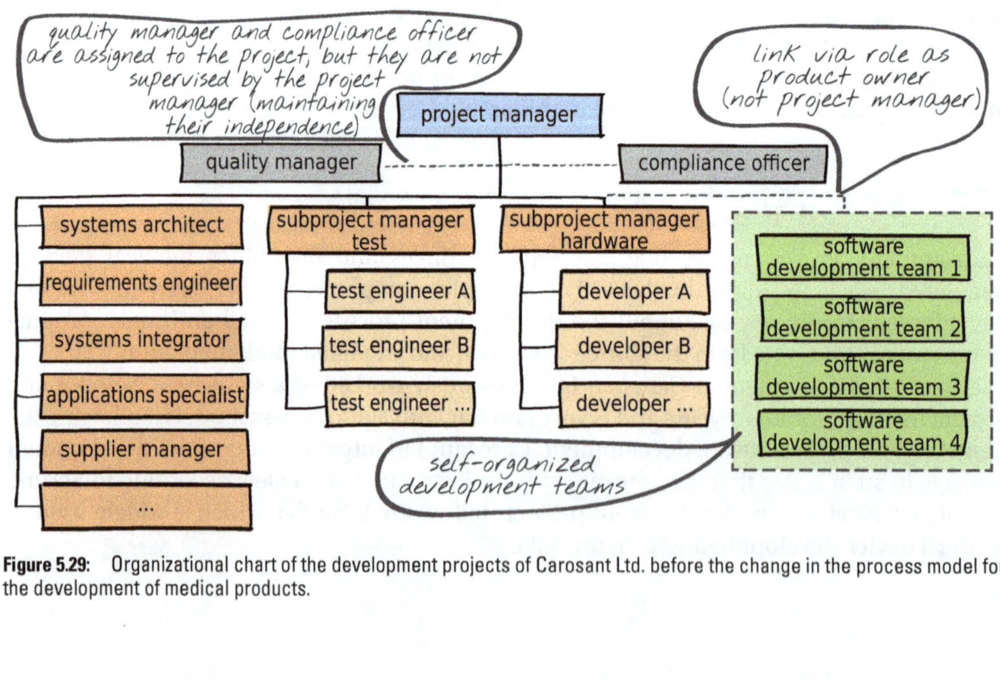

Figure 5.29: Organizational chart of the development projects of Carosant Ltd. before the change in the process model for the development of medical products.

As good experience was gained with the agile approach in the software sub-projects and employee commitment and motivation benefited from the approach, the development manager would like to organize the entire development in an agile manner and at the same time comply with the specifications for medical device development.

The overall projects usually consist of 30 to 40 employees, which is why the development manager divides them into several teams with a maximum of eight employees each. All teams are to work according to Scrum in future. To structure the projects, there are synchronization points at which the assessments required for medical device development must be carried out and important results delivered, see Figure 5.30.

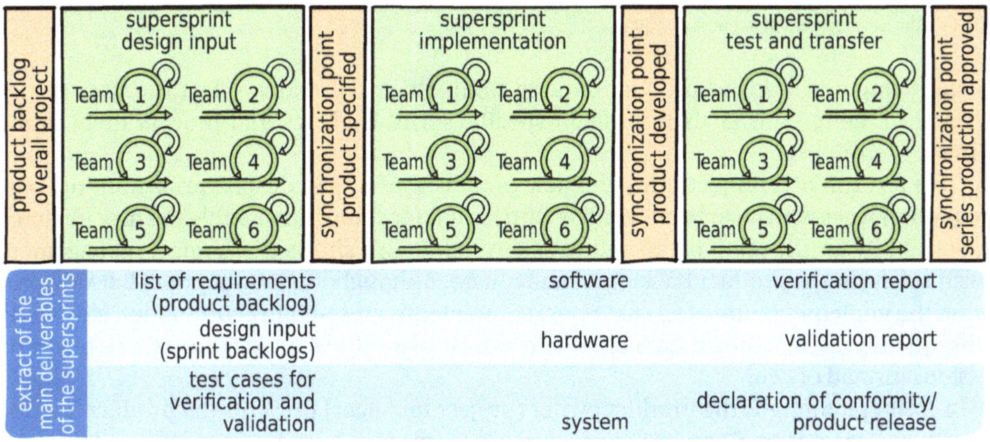

Figure 5.30: New process model at Carosant Ltd. In order to structure the complex system development and take into account the regulatory requirements for medical device development, the agile approach according to Scrum was subdivided into synchronization points. Certain results, usually in the form of documents, must be available at these points in time. These are outlined in excerpts. Depending on the project, a supersprint consists of one or more thirty-day sprints.

Carosant Ltd. refers to the sprints between the synchronization points as supersprints. A supersprint consists of one or more sprints, depending on the project. The sprint length is constant at 30 days. However, the number of sprints per supersprint varies depending on the scope of the project.

The allocation to teams should take the following priorities into account:

- team 1: regulatory documents such as product risk analysis, instructions for use, planning documents, etc.
- team 2: hardware: electrics and electronics
- team 3: hardware: mechanics
- team 4: software: infrastructure
- team 5: software: functionality
- team 6: software: user interface.

The new organization and its structure are outlined in Figure 5.31. Although there are still project managers, they have neither disciplinary nor functional authority over the autonomous, self-organizing agile teams. The project manager acts as the product owner for the teams. There are no sub-project managers within the teams. Overall control is achieved

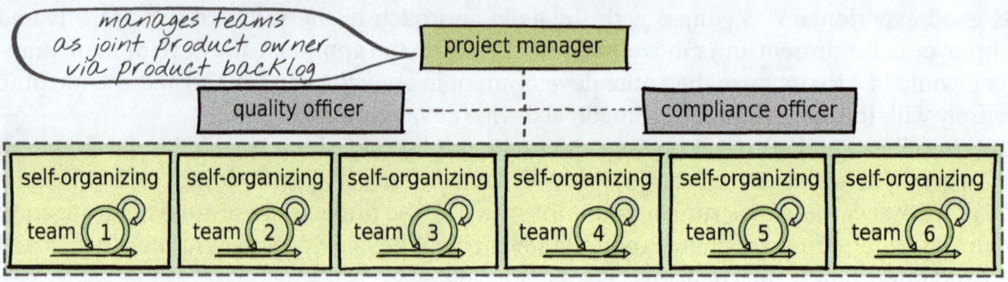

Figure 5.31: New organizational chart for the development projects of Carosant Ltd. following the change in the process model for the development of medical products.

by specifying a prioritized product backlog from which the teams select the user stories relevant to them, transfer them to team-specific sprint backlogs and process them in the teams.

The first test run of a project using the new process shows that, despite prior training, only those employees who have been familiar with Scrum for some time handle the new personal responsibility for the team organization as desired. It also shows that project management based solely on the product backlog is inadequate. Although the joint product backlog promotes the implementation of all customer requirements, cross-team coordination is difficult. Subsequently, some adjustments are made: Sprint planning is divided into three planning sessions instead of two:

- In sprint planning 1, the product owner (project manager) presents the product backlog with its user stories. Representatives of all six teams take part in this meeting. The prioritized user stories are selected together and assigned to the six teams.
- Sprint planning 2 takes place separately in the respective teams. The product owner presents the user stories in detail. The team is given the opportunity to check the allocation of the user stories to the team and correct them if necessary.
- Sprint planning 3 is also carried out separately in the respective teams. The user stories are divided into tasks and the team-internal sprint backlog is compiled.

After planning, the six teams are further synchronized in short daily meetings. In addition to the team-internal daily Scrums of the six teams, a superordinate Scrum of Scrums is introduced, see Figure 5.32. Directly after the team-internal daily Scrums, a representative of each team meets for this Scrum of Scrums. This is where aspects of planning and implementation are discussed that affect cross-team user stories or where team-internal user stories have interfaces with team-external user stories.

An initial test run shows that it makes sense for the project manager to take part in the Scrum of Scrums in his role as product owner in order to coordinate the overall project. The daily Scrums, however, are reserved for the teams and the Scrum master (optionally).

Through further training and the training of several Scrum masters, it was possible to ensure that all teams were able to cope satisfactorily with agile working during the second test run of a development project. The change control management that is required by regulations is coordinated in the Synergy Scrum by the project manager starting from the synchronization point "product specified".

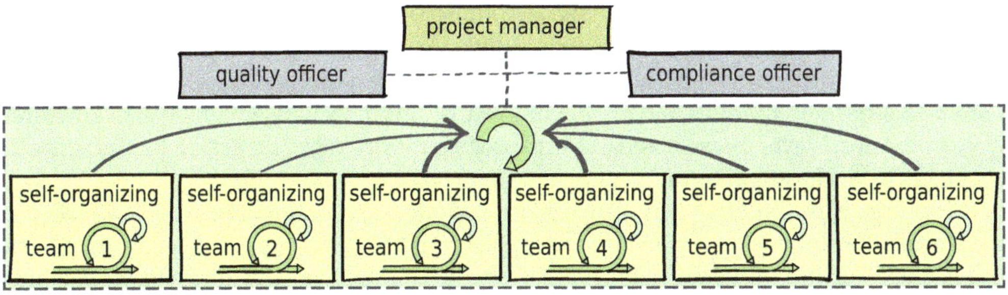

Figure 5.32: Expansion of the new organizational chart to include cross-team coordination as part of a Scrum of Scrums.

Guiding questions:

- How do you rate the original process model consisting of Scrum and V-model and its organization/organigram?
- How do you rate the new process model and its organization/organigram in comparison?
- Assess the degree of agility of the new process model, taking into account the chosen organization, by comparing it with the ideals of the agile manifesto.

Case study 5.8: Failed introduction of a new process model

Verguulst Inc. develops and distributes audio products that are known for their particularly high-quality finish and outstanding sound quality. Recently, however, competitors have launched innovative products on the market that have had a negative impact on sales of the company's own products.

Product development is conducted by a development team consisting of 20 hardware and software developers and the project managers. A typical development project consists of the project manager and four to five hardware and software developers each, as well as other employees from the marketing and sales department, production, testing and the quality department. Often the project teams work together on the further development of an entire product line over the course of many projects. Due to the long period of collaboration, the project managers have a strong position and sometimes take on not only technical but also disciplinary leadership tasks.

As digitalization progresses, products are increasingly incorporating software. The new products recently presented by competitors have innovative user interfaces with touchscreens and voice control as well as connections to smartphone apps. They can be networked with each other and connected to streaming services. The portfolio manager at Verguulst Inc., Mr. Mayer, is concerned that the company's own developments are not being completed quickly enough. In fact, many of the projects working according to the waterfall model are delayed. The sales department often complains about highly technical solutions that are difficult for customers to use. Late changes in the development process may save one or two functions, but they regularly cost additional time and money. Mr. Mayer finds it particularly annoying that sometimes similar software is developed in several projects at the same time and synergies are not used sufficiently.

The company management finds Mr. Mayer's analysis of the project landscape interesting and gives him carte blanche to make improvements. In the opinion of the company management, it is the portfolio manager's task to ensure that project management works.
In order to promote synergies and enable innovative products with a high level of customer involvement, Mr. Mayer has selected a hybrid process model based on Scrum for the software developers to work with in future.
For this purpose, the software sub-projects are completely separated from the projects and the software developers are divided into several Scrum teams with a common product backlog. Mr. Mayer also appoints a higher-level software architect. The software architect becomes the technical supervisor of the Scrum teams, is responsible for their results and promotes cross-team collaboration. The development projects, which are organized on a plan-based basis, can now request the software functions required in their projects from the software architect. The requirements are added to the product backlog, prioritized across projects, assigned to a sprint for implementation and delivered to the product development project after implementation.

Guiding question 1:

- How do you rate the hybrid process model presented?

To ensure that the introduction works, the new process model consisting of plan-based development projects and agile software teams working according to Scrum is announced at an information event. The software developers and the software architect receive a half-day Scrum training course. The software architect, who will lead the Scrum teams in future, will also receive training on the topic of "technical leadership".

Guiding question 2:

Think about the answers to the following guiding questions before reading on. You will then find possible answers directly in the following text.

- How will the project managers rate the new process model – positively or negatively? Give reasons for your assessment!
- What are the advantages and disadvantages of the new process model for software developers?

Most employees are skeptical about the changes. Project managers feel that parts of their team are being taken away, as they no longer have direct access to the software developers but have to request functionality via the software architect. The software developers fear alienation from the product. Whereas they used to be part of a product development team, they now only deliver software to various development projects. They do not necessarily have to deal with the overall product. The company management holds back because it still believes that the portfolio manager will do the right thing. The portfolio manager sticks to his idea of the model and announces that all new projects will work with it in future.
The first projects based on the new process model are finally started. The product backlog is filled with requirements from the development projects and prioritized. The development

projects are committed to certain sprints for the delivery of the required functionality. However, the software teams have taken on too much, which is why there are significant delays in the delivery of the software to the development projects. The project managers use this to complain to their superiors and sometimes directly to the company management about the unpopular process model and its negative effects. The company management and the portfolio manager initially appease the project managers and point to initial difficulties.

Complaints from project managers and delays in software deliveries set alarm bells ringing for the portfolio manager. He fears for the acceptance of his hybrid process model. In his view, the software developers are to blame, as they are clearly unable to work properly in the new constellation. He therefore instructs the software architect to exert more influence on the teams and push them harder. In order to reduce the backlog and be able to handle new projects on time, the software architect orders additional work. At the same time, the delays lead to project managers reprioritizing requirements in order to keep their development projects on track. They prefer to forego less important functionality as long as a certain minimum level of software is delivered on time. As a result, almost all remaining requirements are assigned to the highest priority class. The software team members, in turn, see themselves unfairly criticized and complain to their superiors about the new approach.

Due to the delays in the development projects, Verguulst Inc. is unable to present all the planned products at a major trade fair for the entertainment industry. The company management is shocked and asks the portfolio manager to sit down with the company's line managers and put the portfolio, including the process model, in order.

The line managers disagree with each other. The superior group leaders of the project managers support their demand to reassign the software developers directly to the development projects. Conversely, the superior group leaders of the software developers are happy that their employees are no longer partially assigned to several projects, but that the disciplinary responsibility lies clearly with them. As a compromise, portfolio managers and line managers decide to create a separate product backlog for each project in future. The project managers hope that this will allow them to better control the content and prioritization of the backlog than if they had to coordinate with the other projects on a cross-project basis. The line managers of the software developers are pleased that their employees are at least not directly assigned to projects again. The portfolio manager finds the compromise acceptable, as he at least sees agility in the software area preserved.

In fact, the problems remain even after changing the organization of the product backlog. The coordinating software architect and the Scrum teams do not manage to deliver the projects on time under the pressure of the project managers and the "backlog" in the form of requirements that have not yet been processed. In addition, some deliveries are delayed by an entire sprint, even though only individual requirements have not yet been finalized. The project managers are pushing to make the sprint duration more flexible. They cannot accept having to wait a whole month for the next delivery, even though the missing requirements can be implemented in just two to three working days. Outraged by what they see as massive waste, they turn to the company management and renew their demand to reassign the software developers to their projects.

The company management agrees with the project managers' arguments and instructs the portfolio manager to finally get the portfolio on track for success. Under this pressure, the portfolio and project managers agree on a new compromise: the software developers required to implement the project-specific product backlog are grouped into sub-projects

that report to the project managers. However, these sub-projects should continue to work agilely according to Scrum and coordinate across projects with the software sub-projects of other development projects.
Six months later, after further failures, the portfolio manager realizes that the development projects are actually working according to the old process model. Most project managers have suspended the agile approach in their projects, which was not really pronounced anyway. Compared to the chaotic interim phase during the changeover to the hybrid process model, the projects are now working better, but also with many of the problems from previous years.

Guiding question 3:

- How do you rate the introduction of the new process model?

Case study 5.9: Hybrid product development process

An automotive supplier develops and produces subsystems and components for car manufacturers. For the long-term success of the company, it is important to meet the deadlines for the start of series production in accordance with the specifications of the car manufacturer placing the order. At the same time, the solutions supplied must be innovative. Over the years, the product realization process illustrated in Figure 5.33 has been developed to manage this balancing act of plan-based deadline fulfillment and creative innovation.

Guiding question:

- Consider the individual phases of the product realization process. For each phase, think about why the automotive supplier has chosen the particular (sub-)process model proposed there.

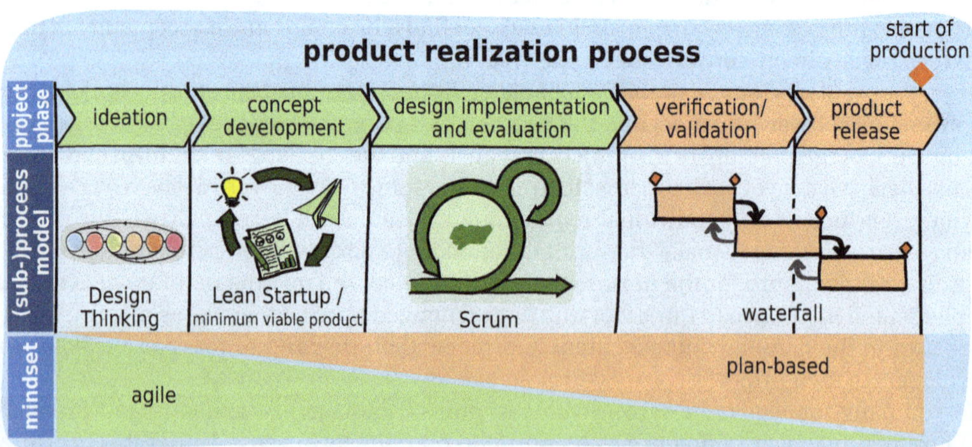

Figure 5.33: Hybrid product realization process.

6 Leadership

In this Chapter ...

- you will learn the basics of traditional leadership concepts for plan-based project management
- you will discover how leadership differs in agile projects and
- you will acquire basic skills for leading project teams in hybrid environments, taking into account special project-specific conditions.

Introduction to leadership and team development

Projects are carried out by people. Each individual member of a project team has their own goals, career ambitions, professional interests and preferred working methods. There are many different motives in a project team that do not necessarily coincide with the project objectives. Project managers must therefore have the appropriate leadership skills to lead a project to success. This is particularly challenging in agile projects without a project manager. Here, the members of the project team need a high level of social skills.

In practice, project managers are often selected based on specialist aspects and not on their leadership qualities. It is therefore hardly surprising that many projects fail not because of technical or content-related difficulties, but because of conflicts and problems on a interpersonal and communicative level.

As a project manager, you cannot rely on your employees automatically being enthusiastic about the project and investing all their energy in working on it. This is especially not the case if they have other line tasks or are working on other projects at the same time. The project manager must therefore first form a team out of many individuals, motivate it and lead it appropriately through the various project phases.

> ***Leadership*** *is the direct or indirect guidance and influencing of others to achieve a goal.*

Leadership can be task-oriented or people-oriented. In **task-oriented leadership**, the aim is to lead employees by defining work packages, responsibilities, authorities andgoals.

In **people-oriented leadership**, leadership is based on appreciation, motivation, conflict management and team composition.

In some texts, leadership and management are used in a similar way. However, the definition of the two terms differs:

> ***Management*** *refers either to the control and organization of something, e.g. a business or an enterprise, or it refers to the people who are responsible for an organization.*

Leadership focuses more on people, behaviors and influencing them as well as on change and change management. Management, on the other hand, describes the control of an organization and the people and roles that are responsible for it. In practice, the distinction is not always easy to make and in this book I do not want to place too much emphasis on differentiating between the two perspectives.

Tip

Project managers have to be able to do both anyway: they lead the people in a project, have to monitor and control the project and are responsible for the project deliverables.
Depending on the project, the people involved and the environment, leadership or management tasks predominate.

Goals of leadership and team development

Project management is a leadership and management task. As a project manager, you must influence the behavior of your employees in such a way that the project goals are achieved. Collaboration in large, complex teams must be organized and managed according to the skills and personalities of the team members.
Team composition and team development play an even greater role than in line organizations, as project teams work together for a limited period of time. Even within the duration of a project, the composition of the team can change, as different skills are usually required at the start of the project than later in the project.
As a result, you have to integrate new employees into the team several times and organize the collaboration.

Leadership concepts and styles

We will first clarify two further terms:

> ***Leadership concepts*** *are general procedures and measures for achieving objectives.*
>
> ***Leadership styles*** *are fundamental behavioral patterns of the person who takes on the leadership task. Leadership styles express the attitude towards employees.*

There is neither the one and only leadership concept nor the one and only leadership style that always fits. Good leadership can act correctly depending on the situation and thus motivate and lead the team in the long term.

Leadership concepts

Well-known concepts are the *management-by* concepts. These include:
- management by objectives
- management by delegation

- management by exception
- management by results.

Figure 6.1 summarizes the key features.

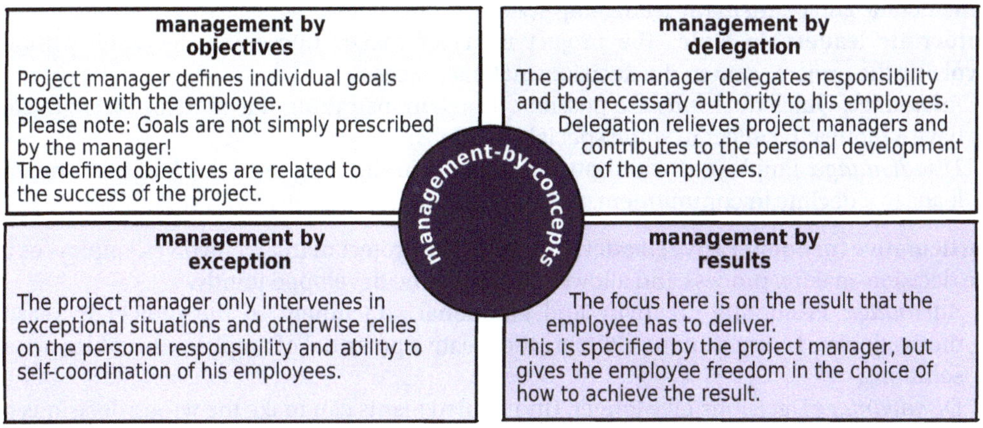

Figure 6.1: Overview of management-by concepts.

Before implementing the concepts, think about what you want to achieve and how you will actually use the concept. Some aspects worth considering:

- Objectives must be achievable and are only accepted and pursued with commitment if the employee stands fully behind them.
- Delegating simple, boring tasks without responsibility and authority is not motivating. It is just as unmotivating if only responsibility for the completion of tasks is delegated, but no the corresponding authority. So give your employees sufficient freedom to shape and make decisions.
- Management by exception requires a sensitive touch: if you intervene too rarely, this can lead to undesirable developments and in some situations be interpreted as a weakness in leadership. If you intervene too frequently and intensively, the desired motivating aspect of personal responsibility is lost. In addition, not all employees can handle the freedom they are given.
- Management by results can exert a great deal of pressure on employees. If the required results are not in line with the employees' personal goals, they will only work on them with limited commitment. Demanding unrealistic results is demotivating.

Another concept is **management by walking around**. This involves the project manager visiting their employees at their workplace. During these visits, the project manager finds out about the processing status, successes and problems from their employees. The fact that the project manager makes an effort to visit, notices good performance, listens to problems and contributes to finding solutions motivates employees. At the same time, project managers gain a deeper insight into their project than if they were to focus solely on formally controlling the progress of the project.

Leadership styles

You can use different leadership styles when implementing the leadership concepts. The best known are the autocratic and participative leadership styles as well as leadership by the members of the group themselves. Other important leadership styles are the task-oriented, bureaucratic and charismatic leadership styles.
Autocratic leadership style: The project manager makes upcoming decisions without involving the project team in the decision-making process.
- *Advantage*: Quick decisions are possible. The demonstration of leadership strength can have a motivating effect in some special situations.
- *Disadvantage*: Employees withdraw emotionally in the long term. Lack of involvement leads to a decline in commitment.

Participative (or cooperative) leadership style: The project manager involves employees in the decision-making process and allows solutions to be developed jointly.
- *Advantage*: Promoting creativity and emotional attachment to the project increases motivation and commitment. Utilizing the creative potential of employees enables better solutions.
- *Disadvantage* Decisions take longer. Unqualified teams can make the wrong decisions or have to be overruled by project managers, which is demotivating. This management style can be interpreted as a leadership weakness in some situations.

Leadership by the member of a group themselves: The project manager takes a back seat as far as possible. Each team member takes responsibility for their own area and manages it. The project manager only intervenes in exceptional cases and for overarching issues.
- *Advantage*: Each team member takes responsibility for their own area. Personal responsibility usually leads to a high level of motivation and commitment among employees. Utilizing the creative potential of employees enables better solutions. The project manager is relieved.
- *Disadvantage*: Higher-level decisions take longer. There is a risk of the team drifting apart. Personal responsibility can overburden employees.

Task-oriented leadership style: The project manager focuses on assigning tasks. Leadership is not focused on people, but on the task at hand.
- *Advantage*: Clear assignment of tasks promotes efficient processing.
- *disadvantage*: Tasks cannot always be clearly assigned. The leadership style neglects personal components of leadership. In the event of conflicts, task orientation does not help, it does not enable the conflict to be resolved.

Bureaucratic leadership style: Cooperation is regulated by function and role descriptions and clearly defined areas of responsibility.
- *Advantage*: The function and role descriptions promote the smooth running of defined processes. Every employee knows their area of responsibility.
- *Disadvantage*: The demarcation of functions and roles leads to solitary thinking, where everyone is only involved within their area of responsibility. The view of the big picture can be lost.

Charismatic leadership style: The project manager leads through personal charisma. In this way, they try to inspire employees and motivate them to achieve their goals.

- *Advantage*: This leadership style can inspire a passion for collaboration.
- *Disadvantage*: The success of this leadership style depends heavily on personal contacts and relationships. If employees and project managers do not have a personal understanding, this leadership style will not work. Charisma is also closely linked to a person's personality and can only be learned to a limited extent.

Laissez-faire leadership style: Personal freedom and the employees' own development dominate. The project manager does not lead, which is why laissez-faire leadership is often not even referred to as a leadership style.

- *Advantage*: Laissez-faire leadership creates space for creativity. Employees feel free. As long as no problems arise, there is usually a good working atmosphere.
- *Disadvantage*: The achievement of goals is not controlled. Laissez-faire management is inefficient as there is no coordination of work. Problems are not solved, but protracted. There is a risk of struggles for authority within the team.

Good managers know how to switch between leadership styles and apply the right leadership style in every situation.

Example

Situational leadership style: In the initial phase of a project, the project manager puts great emphasis on a clear assignment of tasks and responsibilities in order to give structure to the newly formed team. During the definition, planning and control phase, the project manager leads as participatively as possible and thus promotes the joint search for a solution that satisfies the customer.

During the development of the project deliverables, the project runs into a crisis. An important supplier terminates its cooperation. The search for an alternative supplier proves difficult. The project threatens to be significantly delayed and become more expensive. The customer becomes nervous and demands a clear strategy on how the project can be brought to a successful end. The project manager develops a possible scenario with an adapted project plan and presents it to the team. He orders additional work and can ultimately hand over the project to a satisfied customer within the agreed time and cost framework.

Leadership with emotional intelligence

Emotionally intelligent leadership [Goleman, 1996] is based on emotional intelligence.

> ***Emotional intelligence*** *is the ability to correctly perceive, understand and influence one's own and other people's feelings.*

Leadership touches us personally. It would be wrong to believe that leadership is something that can take place purely on a factual level. If you are criticized for your work, it affects you personally and not just your work. Conversely, you are certainly familiar with the effect that you are always particularly committed when you are "on fire" for a topic. The French writer Antoine de Saint-Exupéry (1900–1944) is credited with the following statement in this regard:

> *If you want to build a ship, don't drum up people to collect wood and don't assign them tasks and work, but teach them to long for the endless immensity of the sea.*

Emotionally intelligent leadership would like to embrace the effect of the special engagement that occurs when we are personally touched. Emotions can be positive or negative. Negative emotions can also increase our performance. However, in the medium and long term, this can lead to us becoming numb at work, quitting or looking for another job. In the short term, however, negative emotions can release energy, for example to prevent negative consequences. In the language of emotionally intelligent leadership, positive and negative emotions are called resonance and dissonance.

> ***Resonant leadership** arouses positive feelings, generates enthusiasm and passion and thus strengthens motivation and commitment to the project.*

> *The **dissonant leadership** arouses negative feelings, promotes stress, builds up pressure and discourages employees in the long term. However, employees can be encouraged to show greater commitment in the short term.*

Goleman describes four leadership styles that create resonance (R) and two that create dissonance (D):

Visionary leadership style: (R) The project manager develops a motivating vision of the project and inspires the team to follow this vision with commitment. The goal is defined by the vision, the solutions are developed together.

Coaching leadership style: (R) The project manager supports *individual* team members in analyzing their strengths and weaknesses and in their personal development. The project manager acts as an adviser and manages to achieve a balance between the personal goals of the employees and the project goals.

Affiliative leadership style: (R) The project manager fosters empathy, shows understanding when employees have personal problems and resolves conflicts in a friendly manner. This management style can also be misunderstood by employees as a false offer of friendship if applied incorrectly.

Democratic leadership style: (R) This leadership style largely corresponds to the participative leadership style that you have already become familiar with.

Pacesetting leadership style: (D) The project manager demands maximum performance from the team and consistently exemplifies this through own commitment and dedication. This creates pressure to do the same. The pacesetting leadership style can be successful in the short term and increase the team's performance. In the long term, the effect wears off and demotivates the team.

Commanding leadership style: (D) This leadership style corresponds to the autocratic leadership style that you have already learned about. In the short term, it can give the team support and increase performance. In the long term, it has the same negative effects as the pacesetting leadership style.

The individual styles of emotional leadership must also be applied appropriately in each situation.

Tip

Every type of leadership always arouses emotions in both the person being led and the leader. Project managers who believe they can act purely on a factual level and ignore emotional aspects will rarely be able to realize the full potential of their team.
Agility focuses on people and collaboration between individuals and groups of people. People are characterized by emotions. Agility and the creation of resonance (and dissonance) therefore go well together. I also go into this in the following chapter on leadership in agile and hybrid projects.

Holacracy

The concept of holacracy is based on the ideas of Brian Robertson. It is a way of structuring organizations. Hierarchies should be reduced as far as possible, which leads to a completely different understanding of leadership. The principles of holacracy with a comprehensive set of rules were documented in a separate constitution [Robertson et al., 2021].
Typical hierarchies are replaced by roles and circles, for which the holocratic constitution provides corresponding rules of organization and cooperation. A schematic representation is illustrated in Figure 6.2.

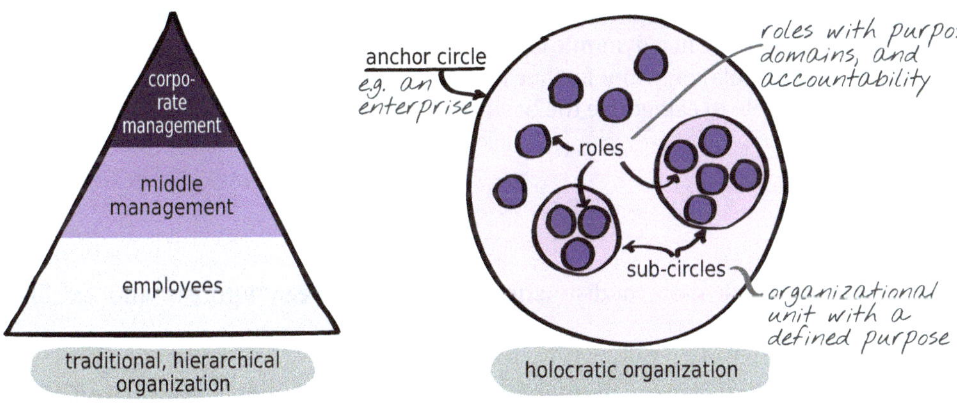

Figure 6.2: A comparison of traditional (left) and holocratic (right) organization.

Employees can take on one or more **roles**. Each role has a specific purpose. Unlike in traditionally organized companies, roles are not necessarily permanent, but can change dynamically. New roles can be created and role descriptions can be changed. Roles that are no longer required are discontinued. Roles are not linked to fixed positions or employment contracts, but are formed from the purpose of the collaboration.
Where several roles follow a common purpose, a **circle** is created. Circles can be compared to specialist departments or projects. The outermost circle, which encompasses the entire organization, is called the *anchor circle*. Within a circle, there can be so-called *sub-circles*. Even though circles work together hierarchically, this hierarchy has a completely different nature

than we know from traditional structures. Circles can be changed much more dynamically and there is a high degree of participatory cooperation at eye level within the circles.
The holocratic constitution defines some **rules of cooperation**. These include:
- the duty of transparency with regard to one's own work and information
- the duty to process messages and requests and
- the obligation to prioritize one's own activities.

The holocratic constitution also stipulates a number of meetings that are an integral part of cooperation and further development. Decisions are made in the circles according to agreed rules without hierarchical structures.
Holocratic organizations avoid interpersonal leadership and a traditional relationship culture. They are replaced by hierarchy-free leadership based on the rules and structures of the holocratic constitution. This requires a willingness to live up to this responsibility. In addition, a certain maturity for constructive, participative cooperation is required.
In fact, there are companies that have implemented the holocratic constitution, at least in part. Typical **successes** reported by these companies are a high level of employee commitment, more innovation and a very flexible organization that can react quickly to changing conditions.
Typical **problems** in holocratic organizations are conflicts between employees who do not adhere to the holocratic constitution and make decisions on their own. In addition, some employees lack certain (traditional) career opportunities. Traditional career paths, such as taking on more (hierarchical) responsibility, greater scope for decision-making and other privileges, are eliminated in holocratically organized companies. Many employees also find it difficult to find their place in a dynamic organization without a hierarchy.
We will not deal with holacracy any further in the following. However, modern managers should know and be able to categorize the basic features explained here.

Motivation

When it comes to motivation, a distinction is made between intrinsic and extrinsic motivation.

> ***Motivation*** *is a state that drives people to engage in target-oriented behavior. It is often understood as a force that explains why we start, continue or end a certain behavior at a certain point in time.*
>
> *With **intrinsic motivation**, the incentive for a certain behavior comes from within us and from the activity itself. We enjoy doing something.*
>
> *With **extrinsic motivation**, the incentive to do something comes from the outside. The incentives can be positive, such as bonuses and rewards, or negative, such as salary cuts, not being considered for promotions and blocking vacations.*

Intrinsic motivation has a more lasting and intensive effect. Project employees who are working on a work package that they enjoy are more committed and have higher quality expectations of their own work. Extrinsic motivation diminishes when the stimulus is removed. Both positive and negative stimuli wear off and must be regularly increased in order to have an effect.

Important types of motivation are:

- **Performance motivation**: Employees want to achieve something and be successful. It is important for them to receive feedback and recognition.
- **Power motivation**: Employees want to demonstrate their own superiority and competence through special commitment.
- **Affiliation motivation**: Employees want to feel that they belong to the group and be recognized by it as a team member.

Knowing these types of motivation helps us to better understand the behavior of our employees. We can derive from this what incentives we can use to motivate our employees.

Warning

It is important for project managers to motivate their team members. However, as every employee has their own preferences, interests and goals, we also have to think about individual framework conditions and incentives with which we can maintain the intrinsic motivation and increase the extrinsic motivation of individual employees. The prerequisite for this is that we have dealt with the inner drive of the employees in advance and derive possibilities for motivation from this.

Example

People with different characters, qualifications and career goals work on a project. Some employees find it motivating when they have the freedom to contribute their ideas and skills when working on a task.

Other employees are afraid of having freedom because they are unsure how to deal with it or tend to get lost. Such employees find it more motivating to be able to work to clear guidelines.

As a project manager, you must always deal with individual needs in order to create conditions that maintain the commitment of your employees.

Motivation due to lack of satisfaction of needs

In the middle of the 20th century, the American psychologist Abraham Maslow published his theory of the hierarchy of needs, which later became known as **Maslow's hierarchy of needs**. According to this theory, there are different evolutionary stages, each of which focuses on different aspects of motivation. The pyramid shown in Figure 6.3 is to be read from bottom to top.

According to this approach, deeper levels of need can no longer motivate us, or only to a limited extent. In other words: If, for example, our financial security is ensured, we are more

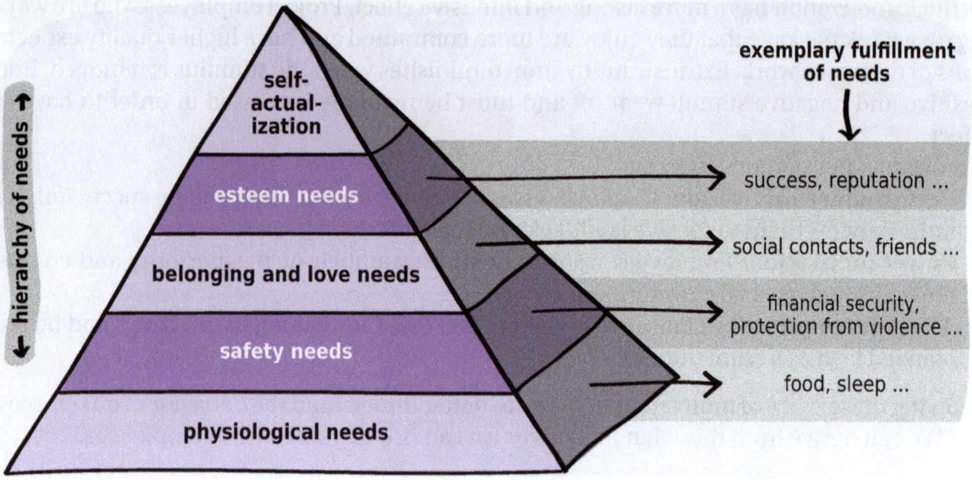

Figure 6.3: Maslow's hierarchy of needs. Physiological needs must first be met before safety aspects motivate us. Once our safety needs are met, we are motivated by the satisfaction of social needs. If these are also satisfied, we can be motivated by individual needs, etc.

likely to strive for the satisfaction of social or individual needs than for a higher financial reward. Conversely, if basic needs are not met, we will not be motivated by incentives at higher hierarchical levels.

The pyramid of needs is of course only a model and should not lead to the false assumption that it can be applied in every situation and to every employee. However, it does show some basic tendencies that are definitely worth noting.

Herzberg's two-factor theory

In contrast to Maslow, Herzberg only distinguishes between two levels of motivation:

- **Hygiene factors** are a prerequisite for motivation, but do not themselves motivate. Hygiene factors include, for example, salary, a pleasant working atmosphere and a comprehensible company policy and vision.
- **Motivators** only have an effect if the hygiene factors are fulfilled. Motivators include personal advancement and development perspectives, opportunities to take on responsibility, recognition and growth.

If the hygiene factors are not met, it is difficult or even impossible to motivate employees. So when managing your employees, ask yourself how you can contribute to meeting the hygiene factors. Examples include realistic time planning, clear assignment of work packages to employees with the necessary skills, clear communication of project objectives and the contributions of individual employees to the achievement of objectives and fair treatment of all team members.

Increasing motivation

Inexperienced project managers sometimes wrongly assume that the project team members are always highly motivated. In reality, however, it is often the case that employees are working to capacity even without the new project and first check for themselves whether the new project is in line with their personal goals.

A **respectful and value-based leadership** that takes these personal goals into account can contribute to a working atmosphere in the project that promotes commitment. Even if you cannot address all personal goals in the project, simply pointing out perspectives and development opportunities can be motivating. Don't be afraid to involve your team's disciplinary superiors in this. However, please also note that the appreciation must be sincere. Nice words alone are not enough; genuine interest in the work of others is also part of appreciation.

The intrinsic motivation mentioned at the beginning cannot be controlled directly, but it can be encouraged: When **assigning work packages**, think about whether the tasks to be assigned match the skills and interests of the respective employee. Of course, not all of your employees' wishes can always be fulfilled. However, if the employee's wishes and the necessity of assigning a work package do not match, you should be aware of this and act accordingly. In this way, you can explain the importance of the work package and its contribution to the success of the project to the employees and offer them the prospect of more interesting work packages. You can maintain the intrinsic motivation of employees through your leadership. However, you can also quickly diminish this through wrong decisions or poor communication. Therefore, when making decisions, always consider how they will be received by your employees.

Delegate not only work, but also **responsibility and authority**.

Introduce a well-working **feedback culture**. Project managers who only appear when things are not going well are not motivating. Show interest in your team's work and recognize good performance and results. Conversely, ask for feedback from your team. You can find out more about feedback later in this chapter in the section on communication.

Be a role model: If you demand extra work from your team, this also applies to you. You can only enforce rules such as punctuality and reliability if you follow them yourself. Some managers evade self-imposed rules by referring to their particularly busy schedule and want to claim a special role for themselves. It goes without saying that this is not perceived positively by the team.

Be mindful: Watch out for motivational problems. Common signs of this are declining commitment, emotional withdrawal or pessimism. Get to the root cause of the observed changes. Declining motivation can be a symptom of conflicts and crises. These need to be resolved.

There are other concepts of motivation theory that you can examine on a case-by-case basis. Examples of these are:

- *job rotation*, i.e. the change between similar activities
- *job enlargement* (horizontal job enrichment), i.e. the expansion of previous activities to include new activities at a similar qualification level with the aim of broadening the personal sphere of influence
- *job enrichment* (vertical job enrichment), i.e. taking on additional activities with a greater scope of responsibility.

Performance and consideration

In return for special achievements, there may be attractive business trips, conference visits and further training or even support from third parties.

Tip

It is important that the reward matches the employee's personality. While an extroverted employee with career ambitions will be happy to make a presentation to the company management, this can be off-putting for employees who simply want to do their jobs. So think carefully about what you want to offer in return.

In some circumstances, you may only be able to agree on benefits in return after consultation with your disciplinary supervisor. There are two points you should definitely bear in mind when motivating through performance and consideration:

- The consideration is only paid if the agreed achievement is provided in full.
- The instrument can wear out quickly and become worthless. You should therefore only use it very occasionally and only for truly outstanding performance.

Relationship level

Motivation can also be fostered through personal relationships. If your employees know that you will stand up for them, take responsibility and, in case of doubt, take responsibility for poor team performance, they will be more willing to stand up for you and complete the tasks you assign to them accordingly.

The **call of duty** of your employees can also promote commitment. You may not achieve top performance, but you will probably at least achieve solid results for your project. Even if not every employee is highly ambitious when they go to work, most of them at least aspire to do their work properly. However, what "properly" means is a matter of interpretation, and this is exactly where you can start: Show your employees, as vividly and clearly as possible, what the consequences are if they do not perform as agreed. In doing so, you should come across as neither bossy nor commanding.

Escalation

If the measures mentioned above do not help to encourage employees to achieve the agreed project objectives, there is probably no way around escalation. Escalation here means reporting an issue to the next level up in the hierarchy. It is important that you really see no other way of getting the employee to achieve the objectives, otherwise you will be accused of having escalated too quickly. Escalation should be based on reliable data. Collect and document information in good time about why you want to escalate and what steps you have taken to prevent the escalation.

The escalation should definitely be pre-announced. You can also set the employee a final deadline within which they can still avert the escalation. It is important that you think about the backing of the higher authority at an early stage. If the authority you want to escalate to does not support you, not only will the escalation fail, but you as a project manager will also be undermined. It can therefore help to reach an agreement with your employees' disciplinary supervisors at the start of the project about what to do in the event of a later escalation. If the conditions for escalation are met, you must escalate consistently and without fail. Otherwise you will lose credibility and lose escalation as a leadership tool.

Team composition and team development

Later conflicts and problems with motivation can already have their roots in the composition of the project team. As a project manager, you should help put together your team or at least analyze its composition and take this into account when leading the team.

Team composition

Teams should not only be put together according to professional requirements. Social skills are at least as important for employees to fit together and work as a team. Högl gives the following **recommendations** [Högl, 1998] for team composition:

- Pay attention to the social skills of employees.
- Consider methodological skills, such as the ability to structure, prioritize and process multiple tasks, etc.
- Pay attention to the preference of employees with regard to work packages to be processed and working styles (teamwork versus individual work).
- Choose a team size that matches the project deliverables - large teams increase complexity, small teams may not cover all the required skills.
- Ensure that the employees have the right level of qualifications. Large differences in commitment and skills can lead to conflicts.

The nine Belbin team roles

Belbin has developed a role model that describes nine team roles, see Figure 6.4.
According to Belbin, each employee has one or more of these nine roles. In a team, roles can be duplicated or missing. But how does this model help the project manager to put together a team?

- If a role is missing, the impact of the shortcoming must be investigated. A research and development project without specialists and plants may have difficulties achieving sufficiently innovative results.
- If roles are duplicated, conflicts can arise that must be prevented. If there are several specialists with different opinions on a particular topic, conflict is inevitable.

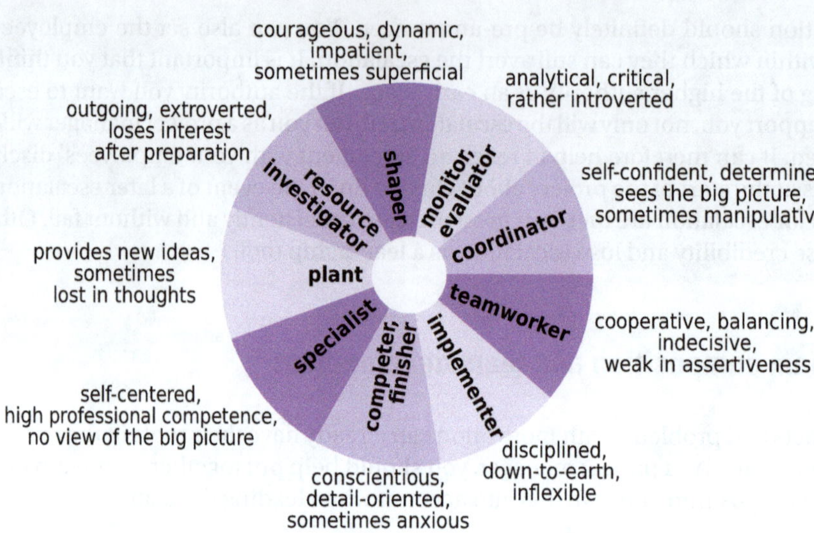

Figure 6.4: The nine Belbin team roles.

Tip

Analyze not only the professional qualifications of your team, but also check the social skills of your employees. Think about what roles the employees play in the team. This includes both formal roles and informal roles. Check whether conflicts could arise from the constellation of roles.
In practice, you will not always have the opportunity to put together your team completely as you wish. However, if you foresee conflicts due to mismatched team members, you should talk to the relevant resource managers and check what alternatives are available when putting the team together.

Team development

You probably know the situation: you are new to a group that has just been formed. The roles are unclear:

- Who can decide what?
- Who has which competencies?
- But also: Who is sympathetic, who can I rely on?

Group dynamics can only be partially controlled. However, with a little basic knowledge, we can prevent undesirable developments and make an important contribution to a successful project team.

Johari window

The Johari window, named after its developers Joseph Luft and Harry Ingham, describes the perception of self and others within a group, see Figure 6.5.

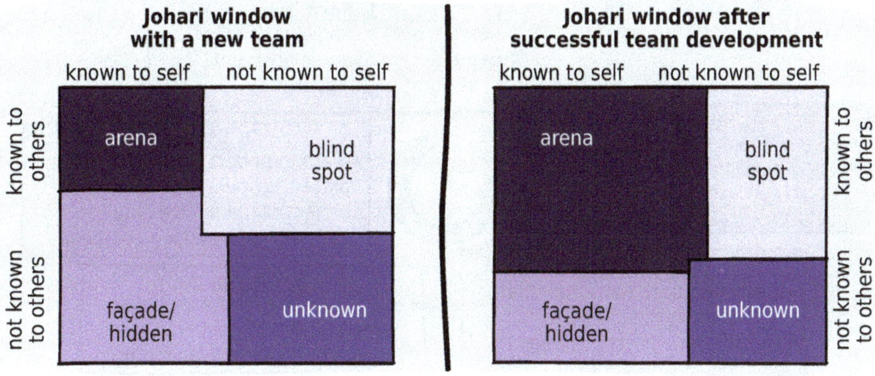

Figure 6.5: Johari window with a new team (left) and after successful team development (right).

The area of the **arena** includes all the things that we and others know about ourselves. In a well-functioning team, this area is quite large. Then we know our strengths and weaknesses and others also know which things we like to do and will do well.

The area **façade/hidden** is the area that we don't want to reveal at first. In a developed team, this area is small – after all, we should work together in a spirit of trust and know our respective strengths and weaknesses, likes and dislikes.

Our **blind spot** is the area that we are not aware of ourselves, but that others know about. In the negative case, such things are the cause of ridicule or conflict, for example when employees constantly stand out as non-constructive critics and don't even realize it themselves. In the positive case, colleagues can help us to discover our own and previously unknown strengths, which we then consistently develop.

The **unknown** area is unknown both to ourselves and to others. Examples can be behaviors whose symptoms we perceive but whose background we do not understand.

As a project manager, you should consider how you can promote the development of your employees in order to recognize and improve strengths and weaknesses. The Johari window can help you to understand correlations and classify employee behavior.

Tuckman's stages of team development

Tuckman's team development model describes the team development process in five stages, see Figure 6.6.

In the **forming stage**, the project manager must integrate the employees. Mutual introductions or team-building events help to get to know each other and build trust. This phase is often neglected as higher management and project managers prefer to devote themselves to the project scope right away due to misprioritized time constraints. The time that we save by omitting the activities that support the forming phase is then used up later in the form of an increased need for conflict management.

In the **storming stage**, employees look for their place in the team. Conflicts are the rule rather than the exception. The project manager cannot and should not suppress this phase, but should use conflict management tools to avoid disputes and help employees find their roles.

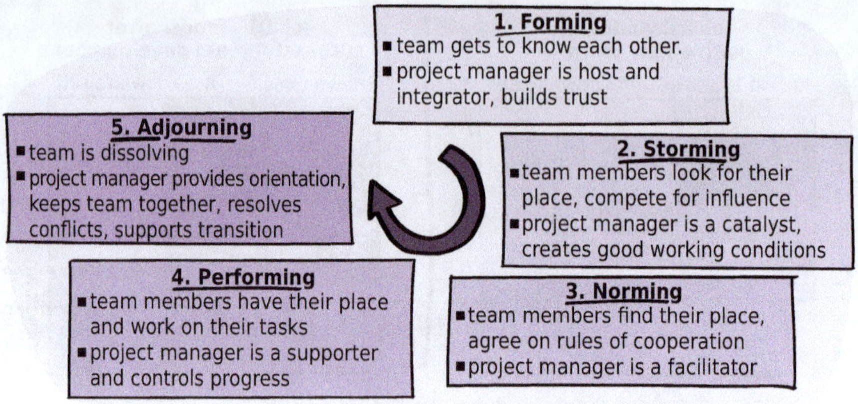

Figure 6.6: Tuckman's team development model with the five stages of forming (orientation phase), storming (conflict or dispute phase), norming (regulation phase), performing (work phase) and adjourning (dissolution phase).

It often helps to establish rules for dealing with each other at an early stage. If these are worked out together, they are more likely to be accepted by everyone involved.
In the **norming stage**, the rules for further collaboration are established. The project manager controls their definition and supports the team members in taking on their respective roles.
The **performing stage** stands for dedicated work on the project deliverables. The project manager only needs to intervene in cases of conflict or problems.
The **adjourning stage** is characterized by the completion of work on the project deliverables and a shrinking project team. The project manager must prevent an uncontrolled break-up of the team, which is not easy. Employees prefer to focus on new tasks and projects rather than on completing a project that is about to end anyway.

Warning

If the team composition changes during the course of the project, for example if new employees join the team or employees leave the project, the team development process is at least partially restarted.

Measures for team development and rules

The opportunities and risks of team development should not be underestimated: A functioning, productive team does not appear out of thin air, but is the result of good team composition and team development. As a project manager, you can promote this development, for example through the following measures:

- Establish rules for cooperation with regard to punctuality, commitment and communication.
- Clearly define competencies, responsibilities and authorities.
- Remain fair – no special treatment of individual employees.
- Communicate clearly and regularly, give feedback and ask for feedback.
- Recognize conflicts and resolve them.

- If individual employees still cannot be integrated into the team, they must be removed from the team so as not to jeopardize the success of the project.

The first item, the establishment of **common rules** for collaboration, will be discussed in more detail here: In many projects, problems and conflicts arise in the course of collaboration. These can manifest themselves, for example, through:

- a team member who is regularly late for project meetings
- employees who do not pass on important information or
- employees who take on work but then do not complete it on time.

In such cases, it is easier to intervene and reorganize the collaboration if rules for collaboration have been jointly established beforehand and compliance with these rules can now be insisted upon. Experienced project managers develop these rules together with the team during the project startup workshop, for example. This joint development and adoption of the rules leads to a high level of acceptance and increases the enforcement of the rules, even by project managers who are not disciplinary supervisors.

Example

As part of the project startup workshop, project manager Clara Smith collects rules for collaboration in a joint brainstorming session with the team. The rules are put up on a pinboard, then grouped and discussed. The team then adopts the rules. Even if not all members completely agree with all the rules, the joint adoption creates a binding moral obligation to adhere to the rules. These are not rules imposed from above, but rules drawn up jointly by the team.

Tip

Comelli [Comelli et al., 2014] and other authors have drawn up typical team rules that you can use as a basis for team development or contribute to brainstorming yourself. Proven rules are:

- Here and now: We live and work in the present and dispense with constant retrospection and arguments from the past.
- Everyone is responsible for themselves. This means that everyone has to take responsibility for their contributions to the project.
- Everyone presents their concerns themselves and does not wait for others to raise them.
- Wishes are expressed. We do not wait for others to act as we expect them to.
- Own information is passed on regularly, required information is requested by ourselves.
- Anyone can say "no" at any time. This means that we do not promise things that we can foresee we will not be able to deliver.
- Problems have priority, i.e., problems are solved first.
- No side conversations.
- I address things clearly, for example "I do" or "you can do" instead of "someone should do".
- I formulate my opinion instead of hiding it behind questions.
- I am open, genuine and honest – as far as tact and timing allow.
- I ensure confidentiality.
- Everyone has the right to make mistakes, everyone has the duty to correct mistakes.

Communication

You already know the important basics of leadership. However, we have not yet talked about an important prerequisite for leadership: Communication between those involved in the project. It determines the way we interact with others and therefore how we implement the aspects of leadership we have learned. First, consider the definition of communication:

> ***Communication*** *is the process of passing information from a sending to a receiving person.*

Communication is not tied to any particular medium; we can communicate via speech, writing, gestures and facial expressions, for example. Communication is also dependent on the context.

Warning

We cannot not communicate! Even what is not said or body language is interpreted as a message by other people.

Shannon–Weaver model of communication

Shannon and Weaver's transmitter/receiver model is a model that explains communication in quite technical terms. Information is sent to a receiver via a transmitter and perceived by a destination. Sources of interference can affect the transmission and distort the information, see Figure 6.7.

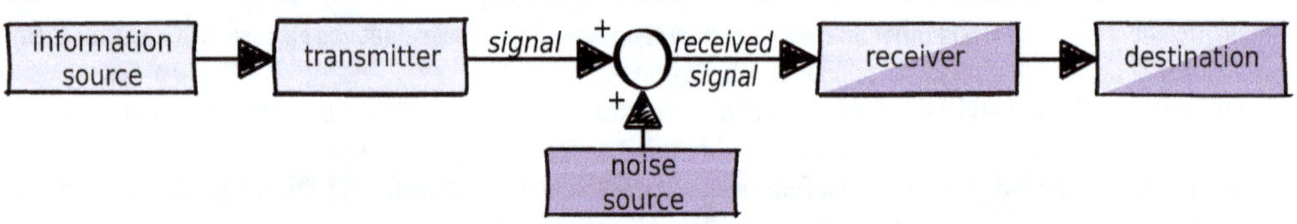

Figure 6.7: Transmitter/receiver model, also known as the Shannon–Weaver model of communication.

What can you learn from this technical model? Fundamentally correct information can be affected by errors at the sender or receiver or by external sources of noise in such a way that it reaches its destination incorrectly.

Warning

Incorrect or faulty communication often causes conflicts. You should therefore not only consider what you want to communicate, but also how you communicate. You are not only responsible for sending the information correctly, but also for ensuring that the information reaches its destination and is understood in the way you intended.

Four-ears model

Let's take a look at Schulz von Thun's communication square, also known as the four-ears model, see Figure 6.8.

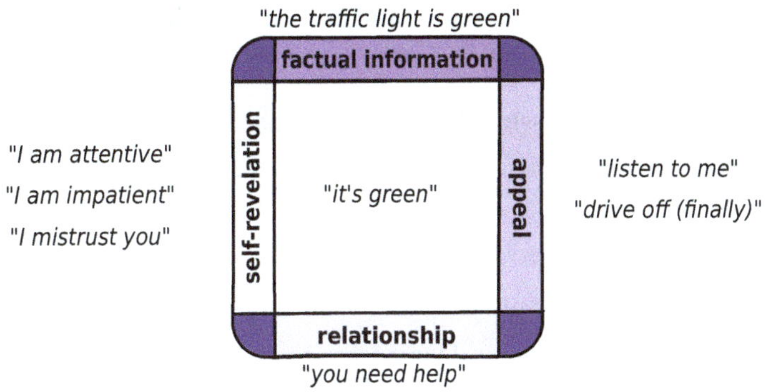

Figure 6.8: Four-ears model according to Schulz von Thun with example (in italics).

According to Schulz von Thun, a message has four sides:

- The **factual side** contains facts and data.
- The **appeal side** is intended to achieve something and contains a request.
- The **relationship side** considers how the sender and recipient relate to each other and how the message can be understood beyond facts and data.
- The **self-revelation side** reveals something about the senders: what do they think, what does the message say about them?

When communicating, we cannot assume that our messages only contain facts and figures on a factual level. The crucial factor is what reaches the recipients and what information they draw from the message.
Of course, not every sentence can be examined with regard to all four sides of the communication square before it is communicated. For important messages, however, you should consider whether the message will reach the recipient the way you want it to.

Successful communication in projects

Communication via media that do not transmit the full spectrum of information can easily lead to misunderstandings. The content of a funny comment, accompanied by a wink and a smile, could come across as rude and impertinent if sent by email. The lack of gestures and facial expressions changes the message of a piece of information, even though the factual content remains the same.
Good communication is characterized by:

- not only being able to talk, but also to listen attentively
- always adapting the message to be communicated to the medium and its strengths and weaknesses

- paying attention to etiquette in written communication – the message is documented and may otherwise reflect negatively on you at some point
- adhering to feedback rules (see below in this chapter).

Success factors for good communication in meetings are:
- Allow sufficient time for final discussions and decision-making.
- Avoid and prevent killer phrases such as "we've never done it this way before" or "you have no idea".
- It's okay to have a tough discussion. However, both parties communicate in an appreciative manner.
- Think first, then speak!

Feedback rules

A good **feedback culture** helps you and your team to improve and avoid conflicts through timely communication. It is important that feedback is given promptly. Praise and criticism are not convincing and do not have the desired effect if they are communicated six months later. The acceptance of negative feedback increases if you also acknowledge positive results. Positive feedback also has a motivational effect. However, you should not only give feedback, but also have the confidence to ask for feedback from your team in return.
A good feedback culture requires feedback givers and feedback receivers to adhere to certain rules. If you give **feedback**, the following applies:
- Don't give feedback without asking – ask whether the recipient wants to hear feedback at all.
- Give feedback as quickly as possible.
- Pay attention to an appropriate framework. Take your time and ensure confidentiality.
- Describe your own perception using the first-person perspective such as: "I thought your presentation today was excellent because ..." instead of "The presentation was good." Your feedback is always subjective and should therefore be formulated in this way.
- Only give feedback on things that the recipient can change.
- Be specific: general phrases do not help the recipient.

If you receive **feedback**, please note:
- The feedback you receive is subjective – your own perception can and may differ.
- Do not justify yourself, but take the feedback on board and take your time to consider what conclusions you can draw from it.
- Thank the feedback giver for taking the time to give you feedback.

To use feedback as an opportunity for improvement and to increase acceptance of the feedback rules, you should communicate these rules to your team right at the start of the project and apply them from the outset.

Conflicts and crises

Even if you as a project manager have done everything right so far, conflicts can arise in the project. This can be due to conflicting expectations and professional or personal problems.

*A **conflict** is a disagreement between two or more parties with different interests.*

***Conflict management** refers to the process of conflict identification, analysis, resolution and sustainable conflict avoidance.*

*A **crisis** is a seemingly hopeless conflict that can lead to a project being stopped or even canceled completely.*

Without conflict management, conflicts can lead to crises and therefore pose a fundamental threat to the project. However, conflicts can also have a positive effect: for example, they can point out shortcomings, lead to better processes, open up new solutions and lead to clearer organizational arrangements in the project.

Causes of conflicts

Conflicts can have **psychosocial** or **factual** causes. Psychosocial conflicts occur when people with incompatible personality patterns come together. This means that a conflict exists for no objective reason due to different personalities, values, social ties or psychological constitutions. However, psychosocial conflicts are often carried out on a factual level without this being apparent to outsiders or the conflict parties themselves.

Example

You notice that two employees regularly hold different perspectives in your project meetings. These are communicated irreconcilably and peppered with personal attacks. Although it is always about a specific issue, the reason for the constant conflict lies elsewhere. The two employees do not understand each other and have already been unable to work together properly in previous projects.

Sometimes, however, there are different points of views that lead to conflicts. Such factual conflicts can arise when

- the objectives of the conflicting parties are not compatible (**factual conflict**)
- the resource allocation is not perceived as favorable (**distribution conflict**)
- an issue is assessed differently (**conflict of judgement**) or
- the respective roles in the project lead to incompatible situations (**role conflict**).

Differentiating between psychosocial and factual conflicts is not always easy and can often only be done in the context of the conflict, by talking to the parties involved and by observing them over a longer period of time. Conversely, as you should not wait unnecessarily long to resolve a conflict, it is advisable to seek dialogue quickly and get to the bottom of the cause of the conflict.

When I discussed emotional leadership, I emphasized the importance of emotions for engagement. You have learned that good leadership does not work on a factual or technical level alone, but that you must always consider the other person as a whole. This also applies if you want to resolve a conflict. You will not be able to resolve psychosocial conflicts on a factual level. However, inexperienced managers often feel more comfortable at the factual level and are reluctant to address the psychological or social causes of a conflict. As a result, conflicts that appear to have been resolved (at the factual level) will flare up again.

Identification of conflicts

Conflicts often begin with minor problems and are not noticed by outsiders. Without conflict resolution measures, the intensity of the conflict increases until the conflict is recognized at some point. Typical indicators that point to a conflict are

- declining commitment and motivation, periods of absence, working to rule
- rumors about other team members
- disturbed communication, sarcasm, verbal attacks, ignoring other people in the team and
- group formation within the team and differentiation between "them" and "us".

Problem awareness

Perhaps you have already made one of the following observations during conversations with employees:

- You have a different opinion to your employee and describe your observation to them. **They deny their opinion** or their behavior. This can happen consciously because it makes them uncomfortable, or unconsciously because they actually perceived the situation under discussion differently.
- Trivializations are very similar: When confronted with a situation that displeases you as a manager, the employee **plays it down, trivializes it**.
- However, it may also have happened to you that the person you are talking to – when asked about a certain situation – blames **other employees or even yourself**.
- It can also happen that **employees react helplessly** or show signs of resignation because they cannot deal with the situation and see no possibility of acting differently than they did.
- Fortunately, there are also **employees who show insight**, gratefully accept your feedback and work to improve the situation under discussion.

As a project manager, you should get into the habit of the following behaviors in order to be able to deal with the situations mentioned:

- **Give feedback promptly** and be prepared by being able to back this up with **facts and very specific examples** to counter possible denial, trivialization and blame.
- **Be constructive** and **offer your support** if you notice that your employee is overwhelmed by the situation. Take time for the employee and show that you value him or her.

The stages of conflict management

The overarching goal of conflict management is to avoid conflict. If conflicts nevertheless arise, the following stages can be followed:

1. Recognizing the conflict, for example by paying attention to the indicators mentioned.
2. Analyzing the root cause of the conflict and the conflict parties; only if you know the cause of the conflict and the people involved in the conflict can you resolve the conflict sustainably.
3. Determine measures for conflict resolution, see following paragraph.
4. Check whether the conflict has been resolved sustainably.
5. Identify and implement preventive measures to avoid similar conflicts in the future.

Tip

Inexperienced project managers sometimes try to resolve only the factual level of a conflict and suppress the personal, also known as psychosocial, level of the conflict. It is correct to separate the factual level and the personal level. This makes it easier for us to have a decent discussion with each other despite a conflict. However, we must also consciously address the personal level and tackle and resolve conflicts here – even if this is more challenging.

Measures to avoid conflicts

As a project manager, you should learn from past conflicts and consider what you need to change in order to avoid new conflicts. Typical measures are:

- clear and unambiguous communication
- building trust in the team and treating all employees fairly
- alignment of project objectives with the personal goals of employees
- establishing a feedback culture
- creating clear organizational structures in the project, establishing rules of cooperation and communicating them
- removing conflict-prone employees from the team.

Measures for conflict resolution

Typical measures for conflict resolution according to Schelle [Schelle et al., 2006] are:

- **adaptation**, i.e. the giving up of a conflict party's own position; this ends the conflict in the short term, but can lead to frustration and a later recurrence of the conflict for the party giving up
- **compromise**, i.e. both parties in conflict give up part of their position in order to resolve the conflict. Depending on the extent to which the position is given up, there is a risk of the conflict breaking out again
- **agreement**, which means that both parties to the conflict agree on a permanent solution that both sides support
- **delegation**, i.e. the involvement of a third party, for example a moderator or even a mediator
- **dispute** or **fight**, i.e. the open resolution of the conflict; in most cases, all parties to the conflict lose reputation and influence
- **use of power**, for example as a company to terminate the conflicting supplier or as a supervisor to relocate the conflicting employee
- **suppression**, i.e. ignoring the conflict; this usually only works for a short time and the conflict comes back to life later.

Lateral leadership

You have now acquired some basic principles of leadership that are valuable for all managers. In this chapter, I will focus specifically on the situation in projects: In most cases, the project manager has a functional leadership role, but is not the disciplinary supervisor of the team members. I will briefly differentiate between these terms:

Disciplinary supervision *relates to compliance by all employees with the standards of conduct and behavior applicable in the company when performing their duties. In the event of negative or positive deviations from these standards, the disciplinary supervisor can discipline or reward employees. The tasks of the disciplinary supervisor include approving vacation, granting salary increases and assigning employees to specific tasks and projects.*

In the case of ***functional supervision****, the supervisors are authorized to specify the manner in which tasks are to be performed. They decide on the modalities of carrying out a certain task.*

Leading in such a constellation requires particularly good leadership skills, as the project manager must lead with conviction, personal authority and other tools but without having disciplinary measures at hand as an aid. Interestingly, these high standards of competence are not always reflected in practice: many project managers receive no training in leadership and have to see how they manage more or less intuitively.

Main leadership activities in projects

Before we look at the possibilities for professional leadership, let's take a look at the activities that characterize it. According to [Rattay, 2013], the main leadership activities within a project include the following aspects:

- the selection of employees, forming a team and agreeing on common goals
- negotiating suitable framework conditions for working on the project
- the planning and distribution of tasks
- promoting communication in the project
- making and demanding decisions and standing up for them
- considering the project environment and actively recognizing and resolving conflicts
- determining the progress of the project and initiating suitable control measures
- recognizing the need for change and implementing the necessary measures
- establishing a feedback and learning culture.

As a project manager, you have to carry out these activities in a special environment:

- Your **team is put together for a limited time** and assigned to you.
- You have to **create new deliverables**.
- **Your team is** mostly **interdisciplinary** and therefore more heterogeneous than employees grouped in line organizations.
- You and your team are subject to **high expectations** regarding the project results.
- You have (mostly) **no disciplinary authority**.

Ideals of a leader and core values

A manager is responsible for achieving the goals of the organizational unit (sub-project, project, program, department, etc.) entrusted to them. To this end, they have employees at their side who have their own personal goals, which generally differ from the goals of an organization, but should at least be consistent with them.

Good leaders are characterized by their ability to inspire and motivate, which is necessary to engage employees and win them over for the joint achievement of organizational goals. A leader who tries to achieve this only through power and disciplinary authority will fail in the long term. With project managers, the question of disciplinary authority does not usually arise. They are typically functional managers and have to lead by other means, which I will discuss later in this chapter. Employees appreciate leaders who:

- have **visions** for the future
- are **role models** and at the same time **appreciate diversity**
- have **leadership skills** and can coordinate
- can **form a strong team**
- **challenge** and **promote** employees in their personal development
- act as **coaches** and **mentors**
- are willing to **develop themselves further** and **are open to new ideas**
- **are credible** and
- **trust others** and **are themselves trustworthy**.

Types of power in leadership

In this chapter, we look at the question of how to lead *without disciplinary* authority or power. Why is the term "disciplinary" emphasized here? Are there other forms of power? You already know that there is disciplinary authority and that leadership that is based solely on disciplinary sanctions leads to a loss of motivation. Managers who rely solely on disciplinary powers undermine their own authority and prevent the creation of a productive working atmosphere. We can distinguish between the following types of power:

Legitimate power

People in a certain (hierarchical) position are given the authority to exercise influence and power. This is the typical basis of disciplinary authority. Superiors with their hierarchical role can issue instructions to their employees.

Power through rewards

Someone who has something to offer that we would like to have can influence our behavior. Rewards can be salary increases or additional vacation days and are therefore closely linked to disciplinary authority. However, a reward can also be to be allowed to present your work results to a certain committee, to be praised or to receive support elsewhere for achieving your own goals. Such rewards can also be given by persons who have no disciplinary authority.

Warning

Rewards are a helpful leadership tool, but they are also counterproductive because employees only perform in return (reward). Rewards should therefore only be given for truly outstanding performance.

Coercive power

Threats and punishments are intended to act as a threat and encourage employees to perform particularly well. Examples include ordering weekend work if certain milestones are not met or assigning unpleasant tasks. Similar to rewards, threats can wear off in the long term and also lead to a poor working atmosphere due to their negative nature.
If a manager decides to use threats and punishments, they will also have to implement them in the announced case so as not to undermine their own credibility. The feasibility of the threatened punishments must therefore be checked before they are announced.

Warning

However, you should think carefully about whether you really want the negative effects of threats and punishments. The working atmosphere in your project will in all likelihood deteriorate and reduce voluntary commitment.

Expert power

Employees with special knowledge or skills can use or refuse to use expert power in order to influence others. Expert power can usually only be built up in conjunction with personal authority. Pure specialists or people with very limited social skills are unable to use their expert status as a leadership tool.
Leaders who have little or no knowledge of the topics their employees are working on are only accepted by them to a limited extent. Project managers can use expert power as a component of their own leadership skills if they have the relevant knowledge and skills. Conversely, however, they must be careful not to neglect their leadership tasks in favor of specialist tasks.

Referent power

Referent power refers to power based on personal charisma. You certainly know people from your environment who, regardless of their legitimate power, have authority and leadership strength. They are recognized as role models and opinion leaders. Referent power is difficult to establish if the personal prerequisites are lacking. However, some aspects that are helpful in gaining referent power can certainly be trained or adopted in one's own behavior:

- friendly, inspiring rhetoric
- consistent and reliable communication
- open way of approaching others
- interest in one's own environment
- trusting interaction and the expression of trust
- standing up for yourself and others
- accepting and justifying responsibility
- constructive problem-solving skills
- taking time and being able to listen
- etc.

Warning

Many of the behaviors mentioned here have to be developed over the long term, but can be lost very quickly. A manager who abuses trust when realizing their own career ambitions or who does not stand up for their team will find it very difficult to have referent power at some point.

Informational power

Those who have exclusive or early access to important information can use this knowledge as an instrument of power. While expert power is about knowledge in the sense of specialist knowledge, power through information focuses on knowledge in the interdisciplinary sense. Such knowledge or information is, for example, knowledge about new business areas, personnel decisions, new projects or upcoming company mergers. Networkers among managers generally find it easier to access such information and can use this information as an advantage.

Lateral leadership of project teams

You have now become familiar with many building blocks and aspects of leadership and, in particular, leadership without disciplinary authority. You also learned that we call this type of leadership lateral leadership. We now want to put everything together like a puzzle and use this to create a leadership concept, which we will discuss using Figure 6.9.

<table>
<tr><th>action</th><td>expand your own power base</td><td>showing authority without being authoritarian</td><td>communicate and act consistently</td><td>value employees</td><td>keep employees motivated</td><td>promote a feedback and error culture</td></tr>
<tr><th>context</th><td colspan="2">empathize with the motives of your employees</td><td colspan="2">agree on rules for collaboration</td><td colspan="2">know escalation instances and escalate appropriately depending on the situation</td></tr>
<tr><th>basis</th><td colspan="2">know your own strengths and weaknesses</td><td colspan="2">know your own impact on others</td><td colspan="2">know your own authority and clarify it with relevant supervisors of your team members</td></tr>
</table>

Figure 6.9: Leadership concept for project managers without disciplinary authority.

The first step is to create a **basis** for leadership in the project: As a project manager, you should know your own strengths and weaknesses. You won't be doing yourself any favors if you repeatedly embarrass yourself in front of your team.

Be aware of your **impact on others**. How are you perceived, what do others see as your strengths and weaknesses? This will also help you to increase the confidence of your presentations and be perceived as a committed and competent project manager.

You should clarify your **own authorities** and coordinate them with the relevant supervisors of your team members so that you don't get into trouble later on by announcing things that you have to take back a short time later. This includes questions such as:

- To what extent can you have access to resources?
- What is your project budget and what flexibility do you have?

- How do the disciplinary supervisors of your employees support (or not support) your project?
- Are you allowed to order overtime – if necessary after consultation with other instances in the company?
- What reward and sanction instruments are available to you?

In other words, you need to know the content of your leadership toolbox! It undermines your authority if you give the impression that you don't know what powers you actually have.
This provides a basis for successful leadership and you can now pay more attention to the **environment of your project**:
Try to put yourself in the **shoes of your employees.**

- What drives them?
- What motivates them?
- Which tasks do they like to do, which do they like less?
- What career ambitions do they have?
- What are their strengths and weaknesses?
- etc.

Only if you know your team members can you lead them in your interests and try to generate enthusiasm.
To ensure that **collaboration** within your environment is successful, you should agree on **rules.** If you succeed in developing the rules together with the stakeholders in your environment, you will increase acceptance and make it easier to enforce the rules. You have learned about typical rules of cooperation in the section on "team development". The rules mentioned there are a good basis, but should always be adapted to the respective situation, the employees and their environment.
For you as a project manager without disciplinary authority, the **escalation instances** are part of the management environment: clarify to whom you can escalate which issues. Try to find out the attitude of these instances to your project, for example how the disciplinary supervisors of your team members think about the project. A failed escalation can damage you as a manager. This is why you should start your project marketing early on and win over important escalation instances for yourself and your project.
In everyday project work, you must now **act and actively lead**. You now know where your strengths and weaknesses lie and how you are perceived by others. You know the different types of **power** and can continuously work on maintaining and **expanding** them. Does that sound too martial? It's not about leading your team in an authoritarian manner and taking all the power. It is more about increasing your personal authority and reputation. This will make your leadership role easier. Remember: you have no disciplinary authority. Your influence on your team can only be as strong as *you* shape it.
Your team should know that you are a project manager who knows how to achieve the project objectives and who also has the necessary **authority** to steer the project even in difficult times. However, wherever possible, you should lead cooperatively and delegate tasks. In certain situations, and especially in agile projects, you can and should hand over responsibility and authority to your team, either in full or in part. Successful project managers are able to switch appropriately between different leadership styles depending on the situation and appear confident at all times.

Communicate consistently. Employees react sensitively if you, as a manager, make different statements depending on the context or communicate contradictory information to employees.

This category also includes communicating and acting decisively, quickly and clearly. If your project gets into difficulties, you should know where to go and be able to communicate an appropriate vision.

Value your **team members**. As a project manager, don't just show your face when something goes wrong, but also reward good performance and show your pleasure when work packages are delivered on time. Signal that willingness to perform stands out and support your employees so that this information also reaches their respective disciplinary supervisors. It is important not to praise things that are taken for granted. On the one hand, this does not come across as particularly credible and, on the other hand, you increase the risk that employees will regard self-evident achievements as a great success.

We have already talked in detail about **motivation** and you know that this can only be done with difficulty and sometimes indirectly. You have learned that it is easier to reduce the intrinsic motivation of employees than to motivate employees who are already demotivated. Consider whether the hygiene factors described by Herzberg are fulfilled. If this prerequisite for further employee motivation exists, you can take appropriate steps. Stöwe and Keromosemito recommend the following stages [Stöwe and Keromosemito, 2012]:

- Try to encourage enthusiasm by **motivating tasks**, taking into account the knowledge of your employees' preferences and skills.
- If this does not work, you can create an incentive for commitment in individual cases by **offering rewards for outstanding performance**.
- As the next level of motivation, you can bring in the **personal relationship** with the employee and suggest a common path that combines your employee's goals and your own or those of the project if possible.
- If the previous steps were not successful, you can at least appeal to the employee's **sense of duty**. This will probably not result in maximum performance, but often ensures that at least the minimum level of quality, time and costs required for the project is achieved.
- In the fifth stage, **escalation** is used as a last measure. Remember to announce this in advance. Provide yourself with evidence-based information so that you have proof of the reason for the escalation. If you have announced the escalation, you must also execute it in order to remain credible.

Try to establish a **good and open feedback culture** in your projects. If feedback is the exception, any feedback to your employees will lead to an exceptional situation. If, on the other hand, you give regular feedback, both positive and negative, openly addressing mistakes and opportunities for improvement is part of everyday life and helps your employees to improve and know where they stand. But also ask for feedback – this is the only way you can improve yourself. Some managers see feedback towards themselves as a leadership weakness. In fact, the opposite is true: if you ask for feedback yourself and handle it confidently, respect for you and trust in you will increase.

Of course, implementing the aspects explained here requires practice and experience. Don't drive yourself crazy in your first big project, but try to take the points mentioned here to heart. Regularly put yourself in the position of a team member and think about how the project manager is perceived. Then step by step apply the leadership concepts you have learned here.

Attitudes of team members

Finally, let me briefly discuss the basic attitude of employees. You will probably find very different personalities with very individual goals in your project. It is important that you engage with your employees and think about what drives and motivates each individual. Good leadership focuses on the individual employee.

Example

Micromanagement is usually understood to mean that the manager takes care of all the details for their employees. The manager sets very detailed instructions and closely monitors compliance with them. For many employees, this is demotivating as it means they lose all freedom to be creative and make their own mark. However, there are employees who appreciate precise instructions for (small) steps, as they can use them as a guide and then simply work through them.
The use of leadership tools therefore always depends on the individual case. What is motivating and supportive for one employee can be frustrating for another.

When analyzing the personalities of your project team, you can use Belbin's role model or other detailed methods to determine the characterizing preferences of your employees. It is important that you realize how you can gain emotional access to your employees and inspire them for your project. However, you should also be aware that you will not turn all employees into outstanding performers.

Dealing with low employee motivation

In almost every company there are employees who struggle to get to work every day, find no fulfillment in their work and have no personal ambitions to improve their situation.
There can be many reasons for this: False promises from supervisors, work processes that are perceived as bureaucratic or even harassing, problems in the private sphere, etc. As a project manager, you will not be able to achieve much with these employees. Try to demand the frequently quoted "call of duty" from these employees and prevent them from negatively influencing other employees or otherwise creating a bad atmosphere in the team. Constant complaining can encourage other dissatisfied employees to join in. Another danger arises when employees who are actually willing to perform complain because they have to take on the tasks of poor performers.

Dealing with average employee motivation

Employees with average motivation have often found a position in the company that they feel comfortable with and are generally happy with. They want to do a good job, but do not necessarily have higher career ambitions. In your leadership activities, make sure that you maintain the existing motivation, for example by:
- showing employees appropriate appreciation for their tasks and
- outlining their contribution to the overall project goals and successes.

Dealing with high employee motivation

Highly motivated employees see your project as a stopover for the next step in their career or are simply interested in the topic. Use their commitment so that these employees see your project as a potential stepping stone and not as a stumbling block. In the former case, employees perform at a high level. They know that if your project is successful, they will be perceived as high performers and get closer to their next career goal. In the second case, employees turn their backs on your project and become low performers because they do not see the benefit for their own career goals.

For example, you can assign them their own areas of responsibility in the form of sub-projects, provide a platform for presenting their own successes or assign particularly demanding tasks with high visibility.

However, you must ensure that putting these employees in the spotlight does not damage the overall structure of your project and demotivates other employees.

Of course, this threefold division of employee motivation is very simplistic and does not do justice to all actual personality patterns. However, you have probably gained an impression of what it means to use the right leadership tools according to the situation and targeted at the individual person.

Leadership in agile environments

You now know the general principles of leadership and the special characteristics of lateral leadership without disciplinary authority in projects. Now I will give you some thoughts on the topic of *leadership in agile and hybrid projects*. To do this, we first recall the values of the Agile Manifesto [Beck et al., 2001], which can be used as a reference for measuring the degree of agility of an organization:

- **Individuals and interactions** are valued more than processes and tools.
- **Working software** is more valued than comprehensive documentation.
- **Customer collaboration** is more valued than contract negotiation.
- **Responding to change** is more valued than following a plan.

The first point is particularly important for the topic of leadership: *People are at the center and are more important than processes and tools*. If you only consider Scrum when you think of agile projects, you may wonder to what extent leadership plays a role there, as the teams organize themselves and the product owner and Scrum master do not have the role of a functional or disciplinary manager. Although the product owner is responsible for the product, the team alone decides on the type of implementation. The Scrum master, on the other hand, takes on the role of process owner, but is more of a coach than a supervisor. However, Scrum does not stand for all agile process models, and even with Scrum the considerations in this chapter can be helpful. This applies all the more when medium-sized and larger projects are carried out with Scrum of Scrums, i.e. when several teams working according to Scrum need to be coordinated.

Another aspect is that the ideals of *hierarchy-free teamwork* are sometimes anything but ideal in practice. Malik, for example, speaks of the myth of teamwork [Malik and Bußmann, 1998] and Korn lists specific points that you should at least be aware of when organizing leadership in your own company [Korn, 2014]:

- Teams are more willing to take risks – the responsibility rests on many shoulders and no one wants to be seen as fearful or a doubter.
- Project teams are temporary teams and are constantly being reassembled in practice. This means they have to rediscover themselves, which takes time.
- Depending on the constellation, teams can level out the individual performance of their members – and thus slow down particularly high-performing, creative and innovative people.
- In practice, non-transparent hierarchical orders can form in officially hierarchy-free teams.
- After an initial euphoria, individual employees can become comfortable within self-organized teams.

For this reason, many companies deviate from the ideal of the hierarchy-free team even in Scrum and install team leaders or similar functions. On the one hand, these should give the team members as much autonomy as possible, but on the other hand, they should signal that negative aspects of teamwork, such as those mentioned above, are not accepted. Even in teams without a formally appointed leader, hierarchies usually emerge. Often eloquent, technically skilled employees take the lead and are followed by the rest of the team.

Warning

Self-organizing teams require that the team members also have skills in self-organization. A technically skilled employee does not necessarily have to be a good organizer. Remember Belbin's role model: do you believe that a perfectionist will purposefully and independently take the next task from the Kanban board as long as the previous task has not been solved perfectly?

You have already learned a lot about leadership in this chapter and know which conditions can motivate employees. If you have a small team with consistently committed members who have the skills to organize themselves, you can let the team lead itself. In any case, check carefully whether the requirements for this are met. If not, many of the leadership principles and tools presented here can also be used in agile and even more so in hybrid projects.

Role of the project manager

Once a project reaches a certain complexity and size, it becomes almost impossible to rely purely on the self-organization of the project team. Process models such as Scrum and Kanban have been optimized for small teams, which – equipped with suitably qualified employees – assign tasks to themselves. They are able to oversee the entire project and deliver the result expected by the customer, including predicted costs and deadlines.

Today, there are many examples that prove that even larger teams can work independently and achieve excellent results. But already the idea of the Scrum of Scrums, i.e. the Scrum approach model scaled to large teams, uses hierarchies. In practice, the individual Scrum teams are often coordinated by team leaders whose role is very close to the *project managers who*

are acting on cooperatively. Depending on the exact shape of the leadership role in your agile projects, you can therefore also apply many of the content already learned in this chapter.
At the end of the chapter, we are now discussing three important leadership factors for successful projects:
- enthusiasm and appreciation
- understanding problems as an opportunity for improvement
- good personal network and project marketing.

First success factor: Enthusiasm and appreciation

Have you ever been particularly committed to a task, then presented the results to the team and only received a brief nod from your manager in response? If so, what effect did this have on you? Unfortunately, situations like this happen again and again and encourage indifference towards our own actions and existence. If what is important to us is not recognized by others, this demotivates us in the long term.
Another situation: An employee is newly assigned to a project that has been running for some time and comes to his first project meeting. The project manager barely acknowledges his presence and does not welcome him as a new team member.
We want to be respected as a person and with our performance. As a good project manager, you know this and act accordingly. Eye contact with presenters, highlighting special achievements, thanking people for their performance and creating platforms for all team members are important elements in giving recognition and attention to each individual. In this way, we tie employees to us and the project and signal that their commitment to the project is recognized and appreciated. Although recognition and attention neither stress the project budget nor jeopardize the schedule, the two situations described at the beginning are widespread.

Warning

Recognition and attention are important to almost everyone. However, we react very sensitively if these are not credible. It is therefore important for you as a project manager to give recognition and attention in a way that fits in with your leadership style and personality. A rule of thumb: show your employee recognition because you are recognizing their performance, not because you think they need your recognition. Employees notice this. You as a leader are not smarter or otherwise superior as a person, you just have a different role in your organization.

Let's look at a third situation: The project team is sitting together and pondering a tricky challenge when working on the project assignment. The project manager urges the team to develop a solution quickly and conveys the message "those up there" don't care what the solution looks like anyway, as long as the budget is met.
Resigned, disillusioned or indifferent project managers are unfortunately not that rare. How can they inspire others with this way of working and motivate them to achieve outstanding results? As a successful manager, it is important that you demonstrate *confidence* and that you have a *vision* of how the project can be successfully completed in the end. Of course, it is also important to be credible. Most things that happen have positive and negative aspects, sometimes more of the one, sometimes more of the other. It is legitimate for you as a manager to make your employees more aware of the positive aspects. To do this, you need to put your own concerns and worries aside.

The leadership aspects described, i.e.:
- to give recognition and attention and
- to demonstrate confidence and a vision and communicate these

are an excellent fit for agile and hybrid projects, as they put people first.

Second success factor: Understanding problems as an opportunity for improvement

An open error culture is important in order to identify and resolve errors at an early stage. In the opposite case, errors and problems are hidden for as long as possible, wasting the opportunity to solve them while they are still small.
As a project manager, you have a special role to play here: deal with problems and risks proactively and with foresight. Successful project managers maintain a list of project risks and discuss these in every project meeting. If you recognize risks (and opportunities) at an early stage, you can work on their probability of occurrence or amount of damage (in the case of opportunities: profit) and initiate measures that are beneficial for the project. The message to the team is important here: a project is a unique endeavor – the project team is a pioneer. As with all things, risks can occur and become problems. These are actively managed.
Make a clear distinction between risk (= potential problem) and problem (= actual damage). Good project risk management will generate several positive effects for your project:
- less time and costs for troubleshooting
- fewer conflicts due to blame being assigned
- strength of action and motivation through the feeling that risks are being actively counteracted
- opportunities to do things better by addressing potential problems at an early stage and developing better solutions.

In addition to good risk management, a positive error culture is one of these success factors. Errors and problems can often be resolved quickly, with little delay and at a manageable cost if they are identified at an early stage. Reporting your own errors and asking for help in correcting them usually takes effort. The willingness to point out errors is made more difficult or even prevented if this is not valued in the project. There are (project) managers who, when they become aware of an error, first want to identify the person to blame (excluding themselves, of course) instead of looking for a solution to eliminate the error. In order to communicate a positive attitude towards errors from the beginning, this basic attitude can be included in the rules for collaboration. In the section on team development, for example, you learned the following rule:
- Everyone has the right to make errors, everyone has the duty to correct errors.

You will not reduce the number of errors in your project by creating a culture of fear of errors. As a leader, you should instead create a culture that:
- enables the early detection of errors and
- collects and disseminates experience in the creation, detection and correction of errors within the team so that lessons are actually learned from mistakes.

Third success factor: good personal network and project marketing

As a project manager, you are on the one hand a leader and responsible for the team and the project result to be delivered. On the other hand, you will need allies because you are subject to certain framework conditions in almost all projects and only have limited room for action of your own.

Carry out a careful stakeholder analysis for yourself and think about who could be useful or dangerous to you or your project. You should build up a good personal network with stakeholders who are particularly beneficial or critical to success.

Example

A project manager is in charge of an internal company development project. The project was commissioned directly by the development manager and has a duration of 3 years. The project manager only reports to the client at the official reporting times. In the reports, the project manager looks at upcoming challenges, but hardly reports on past successes. The development manager, on the other hand, increasingly perceives the project as a headache that is not running smoothly. When he is put under pressure by the company management to open up a new business area and provide development capacities for this, he withdraws them from the "problem project".

Successful project managers not only see themselves as internal managers, but also as "foreign affairs ministers" and "marketing strategists" for their project. Do not assume that everyone knows the purpose and value of your project and will automatically support it, but work on a correspondingly positive perception by project outsiders.

Make a conscious effort to engage with critics. Everyone prefers to talk to people who are well-disposed towards them rather than people who have a negative attitude towards their own work. However, discussions with these critics can provide constructive input for your project or for your project marketing. Moreover, not every critic is negative; some have a positive attitude and see things differently from you.

Other leadership roles in agile and hybrid projects

In addition to the project manager, there are other roles in agile and hybrid projects that (have to) take on management tasks. The client of the project, for example the product owner in a Scrum project, is dependent on the project and its outcome. Without this, they will not be able to achieve their own (economic) goals. They are well advised to be aware of their role, their authority and their responsibilities. At the same time, they need to come to terms with the self-image of the project team acting on its own responsibility.

The concepts of traditional and emotional leadership concepts presented in this chapter and the discussion of lateral leadership without disciplinary authority will help you to select suitable leadership tools.

Tip

One final piece of advice: conflicts are often caused by poor communication and therefore inappropriate leadership styles or miscommunicated decisions. Many conflicts can be avoided if we sit back for a moment and find out in advance the expected reaction of our communication partner to our decision or communication.
However, there are also some conflicts that we deliberately don't want to avoid. In this case, it is important that we consider in advance how the conflict will unfold and how we will resolve it in the end. By the way: when in doubt, friendliness often gets you further. True to the motto: be engaging in tone, firm in substance.

At a glance

- Project teams are made up of individuals. The project manager is a leader and ensures that the team functions properly, promotes motivation and good communication and resolves conflicts and crises sustainably.
- Most project managers have to lead without disciplinary authority.
- There are different types of power that help to influence the project and its employees, including expert and referent power.
- A good leadership concept builds on the foundation (knowing oneself with strengths, weaknesses and powers), the environment (seeing others as individuals and their motives) and concrete actions (demonstrate authority, communicate consistently, show appreciation, motivate and promote a feedback culture).
- When motivating employees, it has proven to be a good idea to start with encouraging tasks. If this is not enough, you can gradually offer benefits in return, utilize personal relationships and call to the employee's sense of duty. As a last option, you can escalate to higher authorities.
- Focus your "leadership energy" on the employees with whom you can make a difference.
- Do not naively assume the ideal of self-organizing teams for agile projects, but check in advance whether the team members have the necessary prerequisites.

Practical examples and case studies

Case study 6.1: Team selection

Henry Schmidt is a project manager tasked with planning the series production of a new product for a medium-sized company. The project is still in its early stages and currently only consists of him and a developer, who is to accompany the transition from product development to production from a development perspective.
In Mr. Schmidt's company, production preparation projects are carried out applying a functional project organization. This means that the project manager receives quite flexible support from employees from different departments and disciplines. At the same time, however, the project manager is neither the functional nor the disciplinary supervisor of the team. That is why it is important to Henry Schmidt that the team can work well together and "works".

The rough planning is almost complete. In the further course of the project, Mr. Schmidt needs a process engineer, a production engineer, a plant technician and a quality assurance employee in addition to the already assigned developer Marcus Zander, see Figure 6.10.

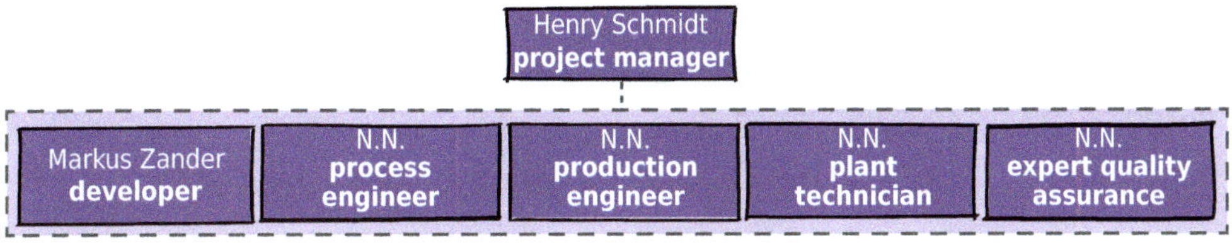

Figure 6.10: Organizational chart of a project to prepare for production with a purely coordinating project manager (functional project organization). So far, only Markus Zander has been appointed as a team member, the other employees have yet to be determined (N. N.).

Guiding question 1

- How should Henry Schmidt proceed with the further staffing of the team and what should he consider?

For most projects in the company, the project manager submits a resource request to the line managers and is then assigned available employees. However, Mr. Schmidt is quite experienced and knows that the success of the project and the efficiency of the processing depend very much on a functioning team. He has therefore gotten into the habit of finding out about potential employees in advance and making a proposal to the line managers. In addition to technical qualifications, he pays particular attention to ensuring that employees can work well together as a team. It is important to him that there are "inventors" or "innovators" in the team. He considers some professional competition to be healthy as long as employees are able to translate their ideas into constructive solutions.

Henry Schmidt holds Mr. Zander in very high professional esteem. Mr. Zander is a conscientious employee who has a wealth of experience and knows the products developed and manufactured by the company inside out. However, he also knows that Mr. Zander can be very stubborn and tends to rely on tried and tested ideas rather than new approaches. In this project, however, Mr. Schmidt has been tasked with significantly reducing production costs, as the product to be produced would otherwise generate too small a margin on the market. The team will therefore have to come up with new solutions.

To his displeasure, Mr. Schmidt does not receive the employees he wants for the open positions in the project. Only Hannah Snyder, a process engineer, is assigned to him as requested. Ms. Snyder works reliably and knows both the current state of production technology and the conditions in the company very well. Carl Timmer is assigned to the project as production engineer. Although he usually has quite good ideas, he is very reserved and will find it very difficult to make his mark. Richard Blum, who is not Mr. Schmidt's first choice but is still acceptable, is assigned as plant engineer. Mr. Blum will carry out the tasks assigned to him, is very familiar with the systems, but has hardly provided any momentum of his own in the past. Ulrike Wagenschoner will be responsible for quality assurance. Ms. Wagenschoner is known in the company as an extremely precise and demanding employee. She brings new ideas to the table and is an energetic advocate for them. However, her approach is not always constructive and cooperative. Mr. Schmidt knows from colleagues' projects that there were considerable conflicts between Ms. Wagenschoner and Mr. Zander in past projects.

Guiding questions 2

- What do you think about the analysis of Mr. Schmidt's employees?
- How should Mr. Schmidt react to the assignment of employees?

Case study 6.2: Communication within the team

Ms. Myers manages an internal company development project and is dissatisfied with the way Mr. Kercado, who has the role of system architect, works. She does not complain about the content of his contributions as such, but criticizes the lack of adherence to deadlines and the quality of the documentation. Ms. Myers was pretty sure that she would be able to work well with Mr. Kercado. They knew each other from previous projects, when Ms. Myers herself coordinated requirements management and had to work closely with the system architect. However, since she was promoted to the position as project manager, Mr. Kercado always reacts very sensitively to her conversations.

For this reason, Ms. Myers proactively provided Mr. Kercado only with work packages that suited him and corresponded to his qualifications. Unpleasant ancillary work was assigned to other people within the project. Ms. Myers does not want to offer Mr. Kercado a deal in return for the work he has done, as she knows that he will not accept it. She would now like to talk to him about further cooperation. Once the system architecture is complete, Ms. Myers expects to have some time-critical design drafts for components of the system. The goal will be to inform Mr. Kercado of her requirements for the quality of the deliverables and to define ways to enable him to deliver that quality on time.

Ms. Myers: "Mr. Kercado, I'm glad you have time for this conversation so quickly. We have been working together for a few years and you know that your and my views on product requirements and system architecture are very similar. I really appreciate your contributions to the project. However, you've probably noticed that we haven't always agreed on their delivery recently."

Mr. Kercado: "What do you mean 'disagreed'? Were the results good or not?"

Ms. Myers: "In terms of the quality of the content, I have nothing to complain about. However, I would have liked you to meet the deadlines agreed in the project plan. I was also surprised that you used an outdated document template for all the documents."

Mr. Kercado: "But you could have said that beforehand. Why didn't you put the current templates on our project drive?"

Guiding question 1

- What is Mr. Kercado's strategy and how should Ms. Myers respond?

Ms. Myers: "I'll be happy to take up the suggestion of how we can ensure that everyone on the team has easier access to the current templates later. But now I would first like to talk about our collaboration and the upcoming work packages for component design. You know that these are time-critical and demanding in terms of content. That's exactly why I wanted you on the team. How can you ensure that we reach the next milestone? Do you need support and, if so, in what form?"
Mr. Kercado: "Why don't you just let me do my job? You know I have a lot to do. Instead of taking care of the really important tasks, you want me to take care of some template documents. That's your job as project manager."

Guiding question 2

- How does the conversation develop? How should Ms. Myers react to this development?

Case study 6.3: Project vision and participation

Timo Jelischek is leading a software development project that he plans and manages in a largely agile manner using lean principles and a Kanban board. The project aims to develop new software for museums that allows museum visitors to call up interactive information as well as audio and video contributions on the museum's exhibits using a QR code. At the same time, customers will be able to give feedback on the museum or specific exhibits. It should also be possible to obtain tickets and travel information via the software.
The development team consists of 12 people, including Mr. Jelischek as project manager, a graphic designer, an online editor, two application specialists, a software architect and six software developers.
It is common practice in the company for teams to work on up to two projects at the same time. In Mr. Jelischek's team, four of the software developers, the system architect and the graphic designer work on another project. The other team members are only assigned to Mr. Jelischek's project and smaller line tasks.
Timo Jelischek attaches great importance to his employees understanding the overall context of the project and being able to categorize their own contributions. He wants to prevent a silo mentality and to regularly remind employees of the project objectives, especially those who are still working on other projects. He has therefore got into the habit of visualizing the company's vision, the project objectives and the most important sub-objectives on a board and discussing them with the team members every two weeks.
These discussions take place after the daily stand-up meetings, but only every two weeks. In addition to the Kanban board, Timo Jelischek then brings up the board of objectives on a separate pinboard during the meeting. The team members check whether the targets are still up to date. Everyone then receives red and green stickers for each objective. If an employee sticks a green dot, this indicates that they are convinced that the goal will be achieved. If an employee sticks a red dot, they believe that the achievement of the objective is at risk. Figure 6.11 shows the target board with voting results.

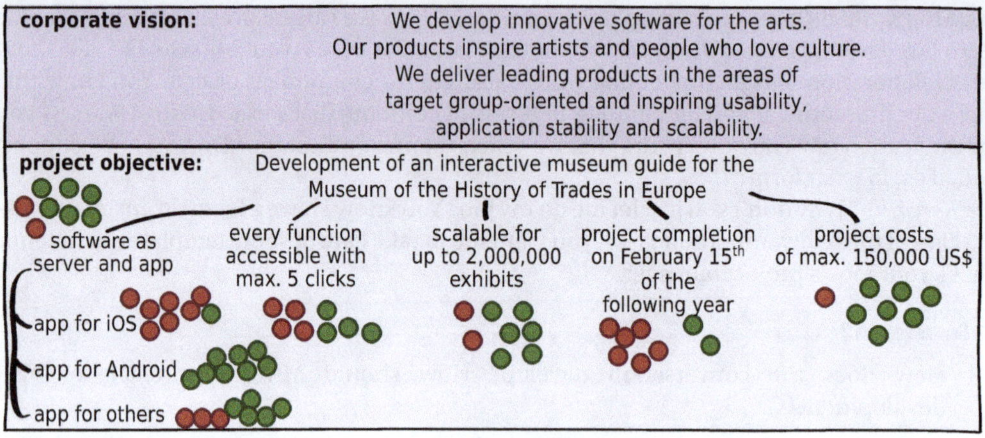

Figure 6.11: Board with company vision (top) and hierarchically structured project objectives (bottom). Red and green markers represent the overall assessment of the project team. Red markers mean that the achievement of objectives is assessed as at risk, green markers mean that the achievement of objectives is assessed as probable.

Guiding questions

- What are the strengths and weaknesses of the target board?
- What should Mr. Jelischek do based on the voting result?

Case study 6.4: Motivation

Helena Kraus is leading an organizational project. The aim of the project is to revise the division-wide procurement process for the development, marketing, sales, service and quality departments. The project is organized as an functional project. Ms. Kraus is a product manager herself and is based in the marketing department. Like all team members, she has to work on the organizational project alongside her line organization tasks and other projects.

The team includes one employee from each of the departments concerned as well as an employee from the data center, who makes any necessary changes to the ERP system. This software is intended to provide company-wide support for (as far as possible) all of a company's business processes, for example for procurement, finance, production, etc. The employees have been with the company for some time and know the procurement process well.

The team is initially making rapid progress: the most important milestones have been planned and an analysis of the current process and its weak points has been carried out. The team is now starting to develop improvements and discuss their feasibility with the data center employee, Mr. Walter Immenbauer.

Many of the ideas are immediately assessed by Mr. Immenbauer as not feasible. Either the ERP system does not offer the desired functionality or the effort required to configure it is too great. Helena Kraus is somewhat annoyed by the way Mr. Immenbauer rejects the ideas, as he does not present any solutions himself and does not seem to check the feasibility of the others' ideas very carefully.

After a project meeting, Ms. Kraus therefore takes Mr. Immenbauer aside and asks him to take some time to talk to her. Unfortunately, Ms. Kraus does not see any opportunities to give Mr. Immenbauer more motivating work packages. She needs his input precisely for the feasibility check and subsequent implementation. Nevertheless, she tries to get Mr. Immenbauer more enthusiastic about these work packages by explaining their importance for the success of the project in more detail:

Kraus: "Mr. Immenbauer, that was once again an exciting project discussion. I would be interested to hear your opinion on the ideas and solutions developed."

Immenbauer: "I realize that the current procurement process is not optimal. But the solutions you have proposed will not solve the problem, as none of them are feasible."

Kraus: "That was my understanding of you so far. Can you clarify that for me again? Do you judge the effort to be very or even too great, do we not have the necessary rights to customize the software, or is it simply not possible?"

Immenbauer: "Some ideas don't work – we would have to purchase a completely new ERP system. You know how expensive that is. Other suggestions could theoretically – and I emphasize theoretically – be implemented. But firstly, we would have to familiarize ourselves with them and then develop new functions ourselves. I think you are underestimating the effort involved."

Kraus: "That's exactly why I have you, the expert, on the team. How big is the effort exactly, and what would we need to be able to implement the ideas?"

Immenbauer: "I can't tell you that off the top of my head. But as the only member of the team, I wouldn't be able to do it within the set time frame. That's why I say: even if the current process is not optimal – it is still running somehow."

Kraus: "I agree with you: the procurement process runs somehow – but if we calculate how much unnecessary capacity we tie up with each procurement process, I am convinced that the effort to be invested would quickly pay for itself with a more fluid process. However, in order to be able to make a final assessment, we need to know this investment, and for this we need your expertise."

Immenbauer: "Just creating the work plan and the estimate based on it is more than I can do at the moment. We are currently migrating the data center to the new operating system. As you can imagine, this has top priority, as it really does affect all employees."

Kraus: "What do you think of the following proposal: I try to find some support for your work on the operating system conversion, and if I succeed, you provide me with the required effort estimate?"

Guiding questions

- How do you assess the course of the conversation so far?
- What strategy does Helena Kraus use at the end?
- What other options does Ms. Kraus have if this strategy fails?

Case study 6.5: Escalation

Mr. Kumar is the project manager of a development project. His team consists of 20 people. The company's projects are organized as a matrix and most employees are assigned to two or three projects. Mr. Kumar is therefore the functional and not the disciplinary supervisor of his team.

For some time now, Mr. Kumar has been having problems with the project's test manager, Mr. Wang. He is supposed to specify the infrastructure for the system test and then assemble it. However, he is significantly behind schedule with his work packages. Mr. Kumar already knows Mr. Wang from previous projects. So far, the cooperation has been satisfactory for the most part. Mr. Kumar characterizes Mr. Wang as a not particularly motivated employee who has no major career ambitions. At the same time, however, he carries out his tasks dutifully. However, Mr. Wang is also considered to be a stubborn employee who sometimes prefers to follow his own plans rather than fit in with the plans of others.

In the current case, the previous discussions between Mr. Kumar and Mr. Wang have not led to a solution. It has not been possible to motivate Mr. Wang, and appealing to his sense of duty has not helped either, as Mr. Wang is questioning the timetable of the project itself. He is of the opinion that it is quite sufficient for his work to be completed at a later date. In Mr. Wang's opinion, this will not delay the project.

Mr. Kumar had actually scheduled the work packages a little earlier than theoretically necessary, as he sees risks in the new test setup that he would like to eliminate as early as possible. In addition, Mr. Wang, who is still working on another project, was assigned to Mr. Kumar's project by his supervisor *now*. Therefore, Mr. Kumar also insists that the work package be completed at this time. Mr. Wang, on the other hand, is of the opinion that the risk will not occur in his estimation and therefore there is enough time to complete the work package later.

After Mr. Kumar has documented all the relevant facts about the case at hand, he announces to Mr. Wang that he will speak to his line manager if the work package is not processed quickly now. As there is still no progress, Mr. Kumar actually approaches Mr. Wang's line manager and describes the situation to him, including the negative consequences for the project. Together they decide to have a meeting with Mr. Wang, which takes place a short time later. However, this goes differently than Mr. Kumar expected, as the line manager suddenly agrees with Mr. Wang's opinion and suggests that the work package should be postponed.

Mr. Kumar is not satisfied with the result and communicates this to the two meeting partners. He explains that the postponement means that a risk-mitigating measure cannot be implemented and that he will therefore inform the internal project client and the interdepartmental steering committee, which he then does.

Guiding questions

- How do you rate the project manager Mr. Kumar's approach so far?
- What do you think will happen in the steering committee?

7 Interfaces of hybrid projects to programs and portfolios

In this Chapter …

- you will learn about the relationship between projects, programs and portfolios
- you will discover the basics of setting up, planning and managing programs and portfolios and
- you will gain an insight into the importance of agile and hybrid approaches for programs and portfolios.

Overview

This chapter is no longer about individual projects. Instead, we look at how projects fit into a company's project landscape.
As soon as there are several projects in a company, these can be part of a program and a portfolio. Interfaces then exist between projects, programs and portfolios. In order to work well with the people responsible for the programs and portfolios, you should therefore be familiar with some of the basics of program and portfolio management (also known as multi-project management):

> *A **program** consists of a series of closely linked projects that pursue a common, over-arching strategic goal.*
>
> *A **project portfolio** comprises a series of projects and/or programs that are not necessarily related to one another. Rather, the projects and programs are grouped together for administrative purposes, for joint controlling or for cross-project resource management.*

The following can be derived from these definitions: Like projects, programs have a duration, i.e. a beginning and an end. In between, they go through phases similar to projects: Programs are defined, planned, implemented and completed. Portfolios are used purely for control and coordination. These are continuous management tasks and therefore designed for the long term.

Corporate strategy, portfolio and programs

The corporate strategy defines the company's long-term goals, such as the business areas in which the company wants to operate, the markets, continents and countries in which it wants to operate and the target groups it wants to address.

The strategic goals are detailed and specific products are defined in the individual business divisions. This also includes decisions on how and where the products are developed and later produced.
Resources are allocated to the individual business areas using methods of portfolio management. It is a strategic task of the corporate management. Implementation within the individual divisions has both strategic components (which products do we want to have on the market when and where?) and operational components (how and where do we develop and produce the products?). These considerations are part of program management within the business units. Figure 7.1 illustrates the interplay between corporate strategy, strategies of the individual divisions and operational implementation.

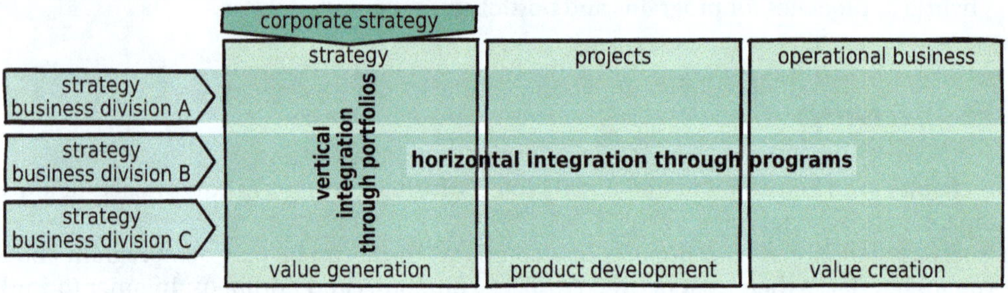

Figure 7.1: Interaction of corporate strategy, portfolio and programs based on Thiry [Thiry, 2015]. The corporate strategy is implemented in the individual business units of the company, which in turn form programs for this purpose.

Roles in program and portfolio management

A typical organizational chart of a project-oriented company is shown in Figure 7.2.

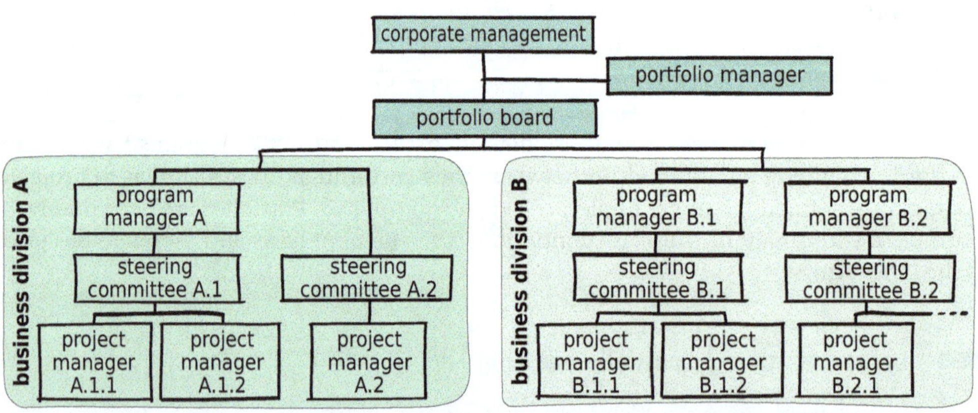

Figure 7.2: Exemplary organizational chart of a project-oriented company with several business units.

The corporate management is responsible for the business divisions. It is supported by a portfolio manager who collects information and prepares management decisions. Alternatively, there may be one (or more) separate portfolio managers for each business division. One or more programs exist within each business unit. A program in turn consists of several projects. Steering committees mediate and coordinate issues within a program or project. They can be responsible for a single project or for several projects simultaneously. The portfolio board acts company-wide. The individual roles and functions from Figure 7.2 are:

- **Portfolio managers** (also known as multi-project managers) coordinate the portfolio. They analyze the composition and allocation of resources, identify dependencies and the resulting risks for programs and projects, collect information about the projects, especially with regard to their progress, and prepare the information for management decisions. Portfolio managers usually do not have their own budget responsibility.
- **Program managers** have leadership responsibility in contrast to portfolio managers. They are usually responsible for the strategic composition of their program and its operational implementation and are responsible for the budget. Program managers are often the functional, sometimes also the disciplinary supervisors of the project managers.
- The **portfolio board** consists of members of the corporate management. It decides on the allocation of resources within the portfolio, decides on projects to be started or discontinued and mediates or decides in the event of conflicts between the programs or business divisions.
- **Steering committees** can be responsible for individual projects or an entire program. They make decisions that exceed the authority of the individual project managers and monitor the progress of the project.

In practice, the terms "portfolio board" and "steering committee" are not always clearly separated. In this respect, the names and responsibilities of the respective committees must be checked within the company.

Methods for project selection

Projects must be selected and prioritized in both program and portfolio management. That is why you will now learn about some methods that you can use in both disciplines.

Tip

In practice, you will come across a surprising number of projects that have not been officially mandated by anyone. As these projects nevertheless consume resources and their contribution to the company's success is unclear, they pose a threat to the projects that are deliberately implemented. Transparent project selection based on clear criteria can reduce this resource-consuming proliferation of projects.

Checklist for preselection of projects

You can use a checklist to filter out projects that do not meet certain minimum criteria. The method is not suitable for prioritizing several projects, as there is only a distinction between promising and non-promising projects. There is no further differentiation within these two categories. Figure 7.3 shows such a checklist approach. The criteria of the review scheme can be adapted to individual needs of a company.

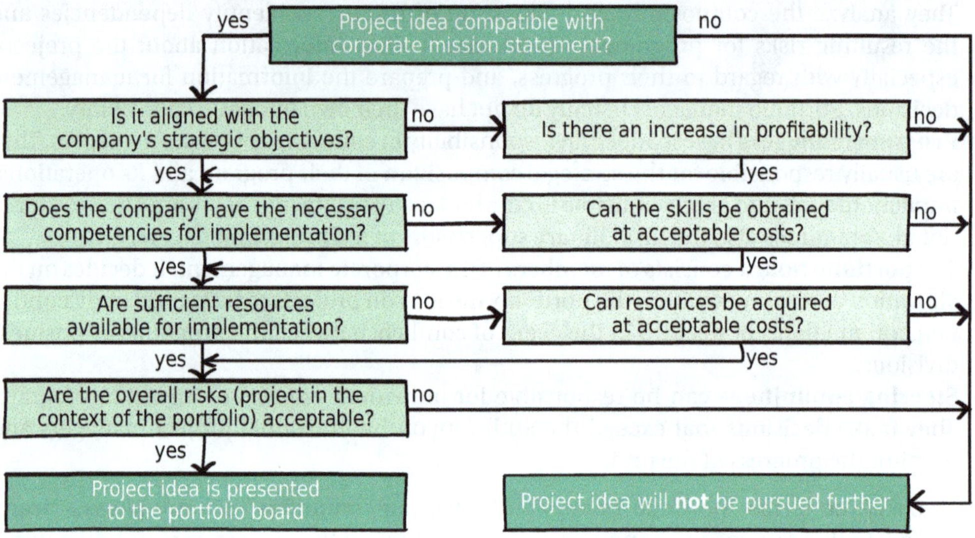

Figure 7.3: Checklist for preselection of projects based on Schelle [Schelle et al., 2006] and Lomnitz [Lomnitz, 2008].

Must, should and could projects

An initial prioritization of the projects that have passed the preselection checklist can take the form of a classification into must, should and could projects.

Must-do projects are projects that have to be conducted due to certain constraints. Such constraints arise, for example, when a supplier discontinues a component in an existing product and the product must therefore be modified. New laws, guidelines and standards can also result in a need for change.

Should-do projects are projects that bring high benefits to the company, for example by opening up new markets or significantly increasing productivity.

Could-do projects can be carried out if sufficient resources are available, but are rated as less important for the company's success than must and should projects.

Figure 7.4 summarizes the differentiation between the three categories in a diagram.

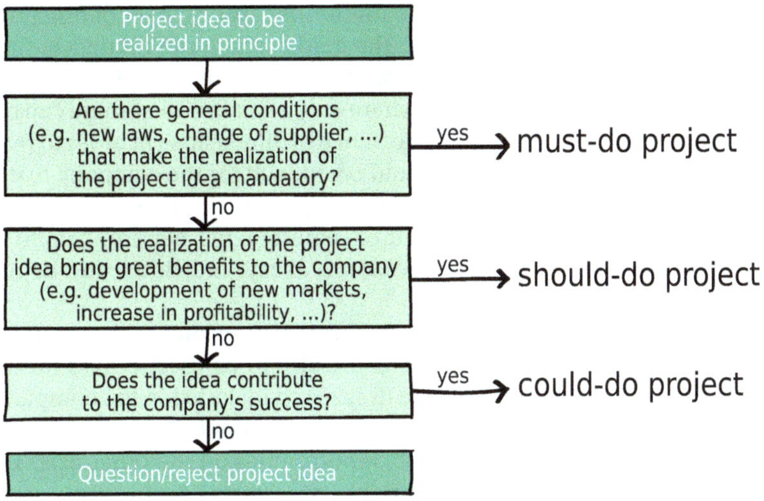

Figure 7.4: Scheme for differentiating between must, should and could projects. The starting point can be projects that have successfully passed the checklist approach and now need to be prioritized more precisely.

Portfolio analysis

In the portfolio analysis, projects are mapped in to a multidimensional diagram. The projects can be compared and then prioritized based on a variety of criteria or dimensions. At the same time, the portfolio presentation provides an overview of the balance of the projects with regard to the selected criteria. Figure 7.5 shows an example of a portfolio representation.

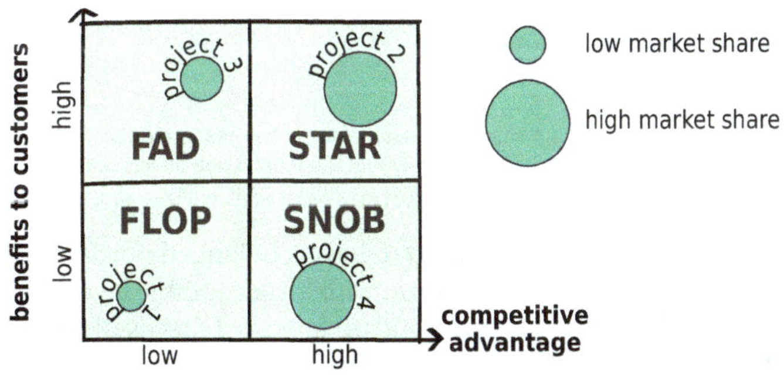

Figure 7.5: Example of a portfolio analysis with three dimensions (competitive advantage, benefits to customers and market share).

Projects with a high competitive advantage and benefits for the customer are also referred to as *stars*. The project result is likely to sell extremely well later on. If the competitive advantage and customer benefit are low, it is referred to as a *flop*, which will only generate a small profit. If the competitive advantage is high and the customer benefit is low, a good marketing campaign can generate customer demand and thus help these *snob* projects to succeed. Projects with high customer benefit and low competitive advantage are referred to as *fads*. Examples are (quite successful) copycat products at lower prices. In a third dimension, for example,

the market share can be shown. In Figure 7.5, this is done using circles of different sizes: The larger the circle, the larger the market share. The market share can influence the discussion that takes place. For example, investing in a snob project with a large market share can be more profitable than realizing a star project where only a small market share can be achieved. It is almost always necessary to consider more than two dimensions in order to reach a sound decision. Without further information, it would be difficult for the company to decide for or against snob or fad projects, for example. Considering other criteria such as revenue, margin, degree of innovation, customer loyalty, resource requirements, etc. can help to identify the projects that best suit the company. The decision is always individual and depends on the respective company situation.
However, it is difficult to visualize more than three dimensions at the same time. If, for example, sales, margin or degree of innovation are also to be used as axes for the portfolio comparison, several portfolio charts must be drawn and included in the comparison.
Another common portfolio representation is the risk-revenue-diagram. Here, the project risks are shown on one axis and the revenue opportunities on the other axis. The resource requirements are often shown as the third variable, see Figure 7.6.

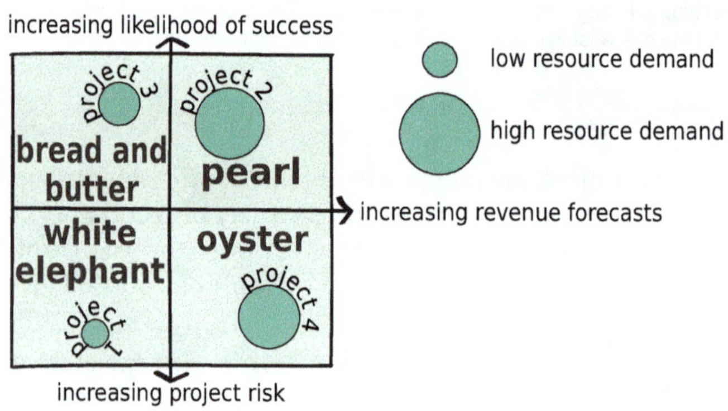

Figure 7.6: Portfolio presentation with the dimensions of risk, revenue and resource demand. The higher up along the vertical axis a project is positioned, the higher the probability that the project can be successfully completed. Projects at the bottom have a high project risk with a correspondingly low probability of success.

It is important for the corporate management to have a balanced portfolio, as this is the only way to ensure sufficient returns in the long term with an acceptable level of risk and use of resources. *Pearls* promise high revenues with good prospects of success. However, the company must check whether it can provide sufficient resources for implementation. The demand for resources is shown in Figure 7.6 as a third dimension in the form of small and large circles. The larger the circle, the more resources are required for implementation. In the case of so-called *oysters*, the company must take significantly higher risks for comparable revenue forecasts. *Bread and butter* projects involve few risks, but promise fewer returns. This also applies to *white elephants* – with the difference that these also entail considerable risks.
Surprisingly, there are still white elephants in many portfolios. These may be necessary, for example if they serve to avoid losing certain customer segments. Nevertheless, caution is advised if the company regularly focuses on risky projects that promise little revenues and invests a lot of resources in them.

The portfolio analysis is perceived as transparent and easy to understand thanks to the graphic presentation. However, there is a risk of focusing on too few criteria and overlooking other important factors. Conversely, if all relevant criteria are taken into account, several portfolio presentations must be drawn. However, this means that clarity suffers.

Utility value analysis

An alternative or supplement to the portfolio approach is the utility value analysis, which you learned about in Chapter 5. There, we used the utility value analysis to compare different process models. In general, you can use the utility value analysis to compare two or more alternatives based on defined criteria.

You can use the dimensions discussed in the portfolio analysis (customer benefit, competitive advantage, revenues etc.) for the project comparison as criteria in the utility value analysis. Due to the tabular presentation, the simultaneous use of many criteria is methodologically no problem. The challenge lies in the correct weighting of the criteria and in the evaluation of the criteria fulfillment.

Warning

Do not overestimate the reliability of clear utility values. They only have the desired significance and objectivity if you have determined the criteria, weighting factors and evaluations carefully and as objectively as possible. Conversely, always question the methodology used when you are presented with the results of a utility value analysis.

In the case studies at the end of the chapter you will find an example of a utility value analysis for prioritizing projects.

Profitability calculations

Ultimately, the products developed in the projects must generate revenues. For this reason, profitability calculations should also be included in the prioritization of projects. Examples are the cost and profit comparison calculation, the amortization calculation, the net present value method (NPV method) and others. As I am focusing on project management in this book and am only providing an initial insight into program and portfolio management, I will not go into these topics in detail here.

Program management

Programs are very similar to projects: They have to be planned, managed and successfully completed. To do this, program objectives must be defined, teams put together and plans drawn up. You can also apply many of the methods you have learned for project management to program management.

Tasks of program management

Even though there are many parallels between programs and projects, program management is more than the management of several projects. The tasks include:

- definition of the program objectives
- selection of the program projects to achieve the objectives
- setting up the internal program infrastructure for the implementation of several projects, e.g. setting up reporting and controlling
- monitoring and controlling the program
- cross-project risk management
- cross-project stakeholder management.

Planning

Programs are part of a business division and must contribute to the company's success. Long-term strategic considerations include determining when the company wants to have which products on the market. The result of this planning is a **product roadmap**, which specifies when which products are to be launched on the market and how long they will remain on the market. Based on these considerations, backward planning can be carried out. This shows when corresponding projects need to be carried out in order to be able to offer the products on the market at the right time. A simple example of a product and project roadmap is shown in Figure 7.7.

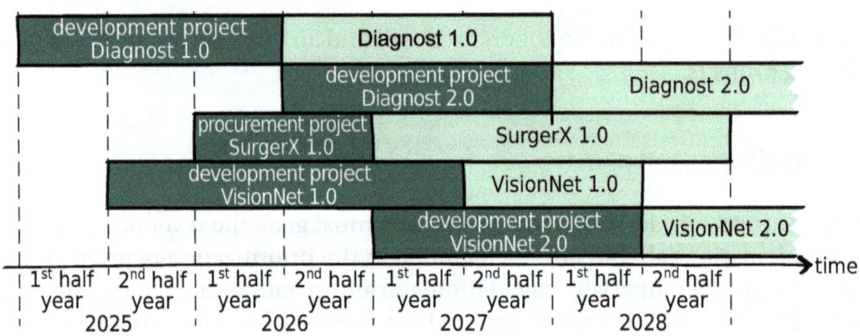

Figure 7.7: Based on a product roadmap (light with black font), specific projects (dark with white font) of the program are planned.

A program can consist of different types of projects:

- research, pre-development and development projects
- procurement projects
- process improvement projects
- maintenance and lifecycle projects during the product life phase.

One of the challenges is to maintain a balance between *short-term* and *long-term* projects in order to ensure the continued competitiveness of the business division. If the program consists only of projects whose results will generate revenues in the next one to two years and at

the same time fails to secure long-term technologies and competencies through appropriate projects, long-term success may be jeopardized.
The planning of the product roadmap and the projects derived from it take into account not only the corporate strategy but also customer needs, requirements from marketing and sales, research, social change and the market and competitive situation.

Control

The program must be managed during the realization of the projects that are part of the program. Particularly important are:
- the monitoring of cross-project risks and
- progress monitoring and control, for example with the help of earned value analysis.

Depending on the constellation, cross-project configuration and change management can also be a relevant component in the control of programs.
As a program manager, you need to set up a *reporting system* to obtain the required information from the project managers. Standardized report forms, which are usually filled out and transmitted electronically, can help to obtain comparable data on the projects.
You can use the control measures known from individual projects to control programs. These allow you to reduce possible deviations from project plans. For example, you can shift resources across projects within your program. However, be sure to analyze the consequences in advance, as knowledge may be lost as a result of such measures, plans may have to be changed and new problems may arise.
Projects that do not deliver the required progress, that are unlikely to succeed or that no longer fit in with the program objectives may have to be terminated early. The resources freed up can then be invested in new projects that fit the program. Please bear in mind, however, that a terminated project will result in the costs invested up to that point being written off.

Portfolio management

A portfolio is used to control and coordinate projects and programs. As a portfolio manager, you must have an overview of all relevant and partially interdependent data across all projects. At the same time, you are the person in the company who is responsible for developing professional project management and continuously improving the associated processes and infrastructure.
The **tasks** of portfolio management are explained below.

Project and program selection

Within a program, the projects are selected by the program manager. However, the selection of programs and individual projects at company level is the responsibility of the portfolio board. To do this, it needs information about the programs and projects. This information is compiled by the portfolio manager and prepared for the portfolio board so that it can make a decision. You have already become familiar with methods for project and program selection. As soon as you know which projects and programs are to be implemented, the question remains:

- Which projects and programs can be implemented given the available resources?

In practice, you will see portfolios whose projects are not progressing because they have been allocated too few resources. This shortage creates bottlenecks. Resources then have to be transferred between projects in order to resolve the bottlenecks. However, the transfer of resources in itself generates effort and therefore costs due to the necessary training of employees and the project plans to be revised. This represents a waste and should therefore be avoided. As a portfolio manager, you need a balanced portfolio-wide resource plan. You can see an example in Figure 7.8.

			quarters/year														
			Q1/ 2025			Q2/ 2025			Q3/ 2025			Q4/ 2025			Q1/ 2026		
priority	project	project manager	GM	SW	HW	GM	SW	HW	GM	SW	HW	GM	SW	HW	GM	SW	HW
1	Arcos	Ms. Schmid	6	10	18	3	10	12	3	3	3						
2	Tokka	Mr. Mayer	3	14	3	3	12	3	3	9	6	3	3	3			
3	Brunt	Mr. Gross	10	6	6	3	3	3									
4	LaSall	Ms. Klein				9	3	3	6	9	9	6	12	15	3	3	6
5	Ramos	Ms. Zeiss				3	3	9	3	6	6	6	3	3			
6	Cesars	Mr. Lorence							3	3	3	3	3	3	6	18	12
7	Zars	Ms. Neff													3	3	3
		required resources	19	30	27	21	31	30	18	30	27	18	21	24	12	24	21
		available resources	20	30	30	20	30	30	20	30	30	20	30	30	20	30	30
		balance	+1		+3	-1	-1		+2		+3	+2	+9	+6	+8	+6	+9

staff months per department

Figure 7.8: Example of a portfolio-wide resource plan. For each project, the resources per department GM (general management support), SW (software) and HW (hardware) are determined for each quarter (Q1, Q2, Q3, Q4). The total resources required are compared with the total resources available. The quarters with too few resources are marked in dark green with white lettering.

It depends on the complexity of the portfolio how detailed the resource plan should be prepared. In the example, we make the simplifying assumption that all employees within a department are interchangeable. Therefore, their capacity can be totaled department-wide. While the resources are quite well booked in the first quarters of the period under consideration, there is room for the start of new projects in future quarters. In addition, resources needed in the near future can be predicted more accurately than resources for projects that will only be needed in a few months or even years. Nevertheless, this forecast is important to avoid scheduling too many projects (think of employee overload and harmful multitasking, which we talked about in Chapter 4 on critical chain project management).

The plan in Figure 7.8 shows a bottleneck in the GM and SW departments in the second quarter of 2025. This is relatively small and can possibly be bridged by short-term overtime. However, this should be investigated further. It may be necessary to change the project plans or provide additional resources to resolve the bottleneck.

You can create a similar plan for the project costs and thus monitor the required financial resources.

Portfolio-wide risk management

Risks are part of every project and result from their unique nature. Good risk management ensures that risks are carefully identified, assessed and, if necessary, reduced. However, if there are dependencies between the projects and their risks, a risk that actually occurs may not only affect one project, but the entire portfolio.
Preventing this is the task of the portfolio manager. In this role, they should be aware of the dependencies between the projects, know the risks and work towards reducing or eliminating them. Dependency diagrams such as the one shown in Figure 7.9 can help to uncover such relationships and subsequently take them into account when assessing portfolio-wide risks.

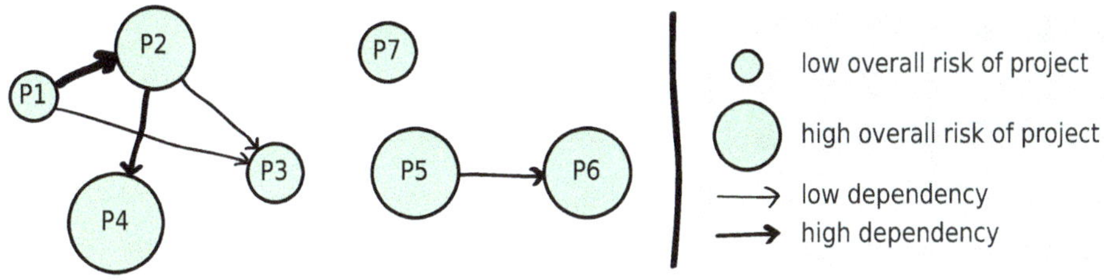

Figure 7.9: Illustration of dependencies between projects P1 to P7 based on Lomnitz [Lomnitz, 2008].

In addition to such dependency diagrams, joint risk lists and action plans can be maintained and monitored.

Project control

Portfolio managers need to collect and evaluate data in order to identify the required project resources, risks and dependencies. Data on project progress, such as the key performance indicators of the earned value analysis, is also required.
Based on this data, you can see whether the projects are being processed according to plan or whether adjustments need to be made. The latter can take the form of additional resources, a larger budget or a change in project prioritization, for example. The controlling data also allows you to identify scope for new projects in the future.
Standardized report forms are recommended for collecting the data. If you find a standardized reporting format that covers the needs of both program management and portfolio management, you will reduce the workload for the reporting project managers.
If the portfolio comprises many projects, it is helpful to use automated report forms with the help of appropriate IT. Software packages with integrated project, program and portfolio management functionality are available for this purpose.

Setting up and developing the infrastructure for successful project management

As a portfolio manager, you should be the person in the company who knows the company's project management best. The establishment and further development of efficient structures and processes for successful project management are therefore also part of the portfolio manager's responsibilities.

Wherever it makes sense, try to create standards that help the people involved to find their way around quickly and carry out projects efficiently and successfully. The considerations for developing company-specific process models that we discussed in Chapter 5 can help you here. You also define key performance indicators to be used in reporting, define templates for plans and report forms as well as checklists.
As a portfolio manager, you are also responsible for selecting suitable software support. The software should offer sufficient functions, because otherwise many things will still have to be planned, documented and reported manually. However, it should also not be too complex, as time-consuming training and maintenance will lead to a lack of user acceptance and more operating errors. Before using new software, you should carry out trial runs and evaluate them carefully.

Characteristics of agile and hybrid program and portfolio management

Introduction

You are now familiar with some basic aspects of program and portfolio management. So far, I have implicitly assumed more or less traditional management approaches. Although a full discussion of agile and hybrid program and portfolio management principles is beyond the scope of this project management book, I will at least explain a few characteristics.

Launch of new projects and programs

Resource management

Let's first look at the initiation of new projects and programs and their launch. In principle, some agile process models demonstrate their strengths particularly well when the project teams work together over a longer period of time. Knowledge of the team's own velocity and a shared understanding of size and effort estimation can then be used effectively across projects. This means that in agile and sometimes hybrid projects, the team composition remains the same across projects and the project duration scales instead. If one and the same team is to take on a large project task, it will take longer than if it is to work on a small project.
In terms of resource management, this means that when putting together teams, care must be taken to ensure that the teams can cover all the work that arises during the course of project realization. The sharing of resources between several projects should be reduced as far as possible and the team composition should be kept constant.

Example

In a plan-based development project, the project start and the definition phase are carried out by a small core team consisting of a project manager, system architect, requirements engineer and application specialist. Gradually, software and hardware developers are involved in the implementation of the requirements. Test engineers, technical writers and production planners are also active in certain project phases, but not necessarily for the entire duration of the project. To optimize workload, employees are often distributed across several projects.
In an agile development project based on Scrum, an interdisciplinary project team is put together that can estimate user stories, create and process tasks and test their implementation. When putting together the team, care is taken to ensure that the team can complete all tasks independently from the beginning to the end of the project. All team members are only active in this one project.

In practice, however, it is not always possible to maintain a constant team composition and avoid assigning employees to several projects at the same time. There are mixed forms between the agile ideal just described and the plan-based resource management of programs and portfolios.
Another aspect to consider in resource management is the **staggering of projects** along the bottleneck. In critical chain project management, we learned about this principle: it jeopardizes a project and the entire portfolio if an employee identified as a bottleneck is working on several projects at the same time. Projects that have the same bottlenecks must therefore not be carried out in parallel, but must be staggered so that the bottleneck does not affect several projects at the same time.

Timing

If your projects use Scrum as a process model in whole or in part, projects can only be started at the beginning of new sprints. Conversely, they are always completed at the end of the sprint. If cross-project synergies are to be used during development, this must be taken into account when scheduling the projects, see Figure 7.10. Such synergies can be helpful, for example, if a supplier delivers to several projects and these then implement part of the development together. The sprints should therefore be the same length across all projects and start at the same time.

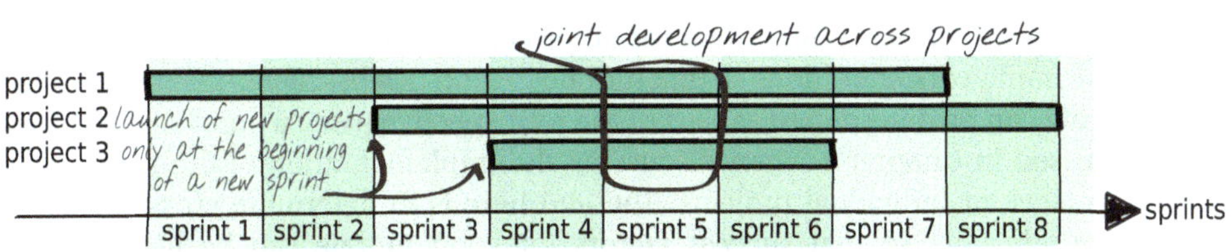

Figure 7.10: Influence of timing in sprints on portfolio planning: Projects can only ever start at the beginning of a new sprint. Cross-project joint work must be scheduled accordingly in a specific sprint.

Definition of content and scope

When planning new projects within a program or portfolio, the ideas of agile project management can be scaled. For example, a general program or portfolio backlog can be created in which requirements for future projects or programs are collected in the form of user stories or epics. If the expected duration of the projects already running is recorded along a timeline, the user stories of the backlog can be grouped and form new future projects, see Figure 7.11. However, this variant of planning new projects also shows the limits of program and portfolio management for agile projects. If additional sprints are required due to changes or velocity changes, the long-term deployment and project planning must be revised accordingly.

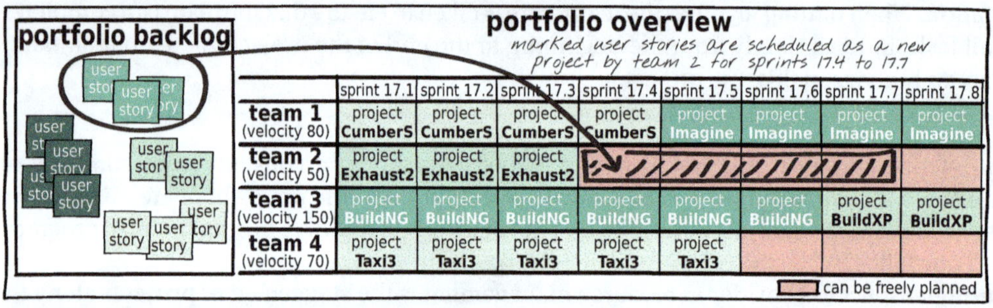

Figure 7.11: Requirements and ideas for future projects are stored in the portfolio backlog in the form of user stories (or epics). The availability of the teams is planned along a timeline. For example, team 2 can be scheduled for new projects from sprint 17.4 onwards. New projects are created by grouping together related user stories. By knowing the velocity of the individual teams, medium- and long-term rough planning can be conducted. In the example, some user stories were grouped together and a new project was planned for team 2 in sprints 17.4 to 17.7.

Control of agile programs and portfolios

You should also use agile approaches when managing agile and hybrid projects, programs and portfolios. When determining the progress of agile and hybrid projects, you will be interested in different key performance indicators than for purely plan-based projects. **Burndown charts** and the **measurement of velocity** also help with programs and portfolios across projects.

Agile, self-organizing teams benefit in particular from the commitment of individual employees. Krebs therefore suggests recording unusual key performance indicators such as team morale in agile projects, evaluating them for all projects in the portfolio and, if necessary, improving the framework conditions so that morale is good [Krebs, 2008]. The measurement of **work morale** can be determined as part of the retrospective using a point scale. This was already discussed in Chapter 4 when discussing the implementation of the retrospective in Scrum projects for individual projects. The aim here is to determine cross-project key performance indicators so that success factors for good morale can be identified and transferred to other projects. Other relevant key figures can be:

- the number of unresolved error messages
- the duration of each troubleshooting
- the throughput and
- others.

Other agile tools can also be used: For example, the Kanban board can help you to identify bottlenecks in agile projects and improve throughput. Each project then corresponds to a card on the program or portfolio Kanban board.
These considerations are intended to give you a brief insight into aspects of plan-based, agile and hybrid program and portfolio management.

At a glance

- Projects can be part of programs.
- Projects and programs belong to a portfolio.
- Programs serve the strategic and operational implementation of a business division's objectives.
- Program managers are responsible for their programs and their projects and have a leadership function.
- The planning and control of programs has some parallels to project planning and control. In addition, there are cross-project planning and control activities to achieve the strategic and economic objectives of the program.
- Projects are grouped into portfolios for control and coordination purposes.
- Portfolio managers monitor the portfolio's projects and prepare information for management decisions. However, portfolio managers are not responsible for the projects.
- Portfolio managers are involved in the selection and planning of the project landscape and further develop the project management infrastructure.
- Agile program and portfolio management must take into account the special features of agile and hybrid projects, for example the fact that these are often not planned in detail over the long term, but proceed iteratively.

Practical examples and case studies

Case study 7.1: Project comparison with the utility value analysis

Resources have been freed up in the product development department of a telecommunications company, enabling the launch of a new project. There are two projects to choose from, which address the success factors relevant to the company's products in different ways. The success factors were determined on the basis of past and current market data. For the company's customers, *intuitive operation* is more important than *product costs, design* and *customer service.* At the same time, a beautiful *design* is more important to them than the *product costs.* However, the *product costs* are more important to most customers than *customer service.* Finally, they rate the *design* and *customer service* as equally important.
The brief description of the two projects available for selection is as follows:
Project High End: A new smartphone is to be launched on the market for which the company holds the exclusive distribution rights. Due to this exclusivity, the costs are significantly higher than for competitor products. Customer service is significantly reduced in order to keep costs down. In the event of problems with the product, a hotline must be called for a

fee. The exclusivity goes hand in hand with an outstanding design. The usability is intuitive, but not outstanding in comparison with competitors.
Project Business: A new smartphone is to be developed for business customers. The design is simple and not spectacular. The operation is geared towards business customers and is very intuitive. The costs are very high due to the robust design of the smartphone. Free customer service is offered for the device, including the provision of a free replacement device in the event of necessary repairs.

Guiding question:

- Use a utility value analysis to decide which project the company should carry out.

Case study 7.2: Hybrid program and portfolio management

Lafroga Inc. manufactures health and medical products for use at home. The products are developed in-house and consist of hardware and software components. Until recently, the company carried out its projects according to the plan-based V-model. The company now has a hybrid company-specific project management standard and uses this in all ongoing projects. The project management itself follows a simplified V-model to ensure compatibility with regulatory requirements. However, the hardware and software sub-projects are carried out in an agile manner. The agile approach follows an iterative mixture of Kanban and Scrum, see Figure 7.12.

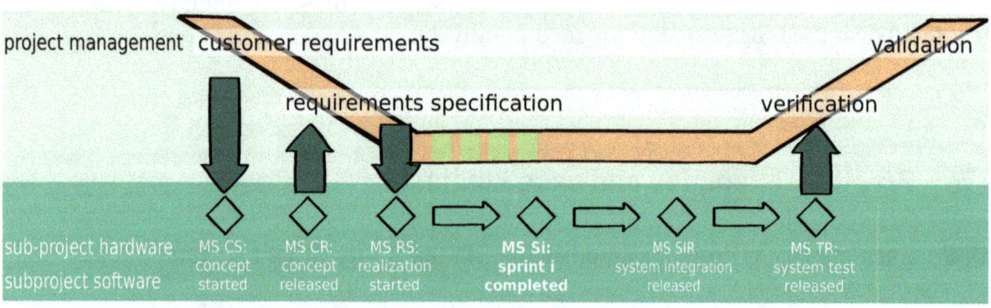

Figure 7.12: Hybrid development model with plan-based project management process and agile, iterative hardware and software sub-projects.

The commissioning marketing department hands over the customer requirements to the project. This creates a corresponding requirements specification. The implementation is done iteratively in i sprints, each lasting one month. To synchronize the sub-projects, each sprint formally ends with a milestone *Si*. Each sprint *i* follows the familiar Scrum process of detailed planning, implementation, review and retrospective. After implementation, the results are integrated into an overall system, verified and validated.
The project is managed using a Kanban board. There are separate lines for the tasks of the overall project or project management, the hardware sub-project and the software sub-project.

The hybrid projects are now running well, and employee commitment and their willingness to take on responsibility have increased noticeably. After the switch to hybrid project management, program and portfolio management was not changed initially. The portfolio manager has so far felt little direct impact of the new process model on his work. However, the teams have been demanding for some time that they are not reorganized for every new project, as this would mean that:

- knowledge about size estimates
- experience gained from retrospectives and
- the social team structure

are at least partially lost.

The program managers in turn require portfolio management to take measures to avoid redundant developments. Within the programs, the program managers ensure that functionalities are not developed more than once. However, this is not done systematically at portfolio level.

Guiding questions:

- Should the composition of the project teams remain constant? What advantages and disadvantages do you see compared to a project-dependent reorganization of the teams?
- How could redundant developments be avoided through portfolio management?

The company decides to continue to rely on dynamically organized teams. However, measures are introduced to safeguard knowledge in order to at least partially address the criticism of the teams:

- In future, size estimates will be cataloged in a database and retrieved from the database for new projects. As part of the retrospectives, the quality of past estimates is regularly reviewed and the values stored in the database are corrected accordingly.
- The results of the retrospectives are divided into project-specific and cross-project optimization potential. The latter are documented in a wiki on the intranet. This means that the results are also available to other teams.

To avoid redundant developments, a portfolio-wide backlog will be created in future, in which all program managers will document the planned functionality in the form of user stories. In future, the user stories will be jointly prioritized and scheduled in a synchronization meeting between the program managers and the portfolio manager. The priority of a user story therefore no longer depends solely on its importance for an individual project, but on its importance for all currently planned or implemented projects.

Guiding questions:

- What prerequisites must be met for the synchronization of functionality to work?
- What problems can arise if the individual projects are to create shippable product increments after each sprint through synchronization, and what are the possible solutions?

For synchronization to work, the start and end times of the sprints were first standardized portfolio-wide, see Figure 7.13. Another prerequisite is that the program managers adhere to the portfolio-wide prioritization of user stories.

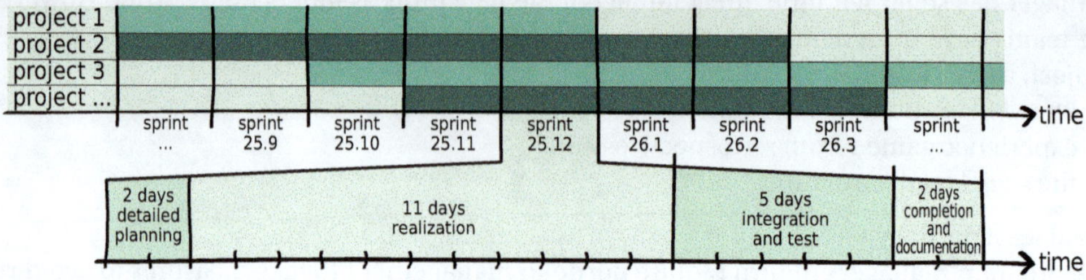

Figure 7.13: Portfolio-wide synchronization of the sprints through uniform start and end dates. In addition, the sprints themselves are divided into detailed planning, implementation, integration and testing as well as completion and documentation in order to facilitate cross-project collaboration.

Cross-project synchronization can result in additional work, as the development results have to be integrated and tested in several products. This reduces efficiency and can lead to a decline in acceptance of portfolio-wide synchronization. The company therefore decides that a shippable product increment will not be provided after each sprint. Instead, a distinction is made between

- internal innovation increments, which are created after each sprint for integration, testing and demonstration purposes and
- the actual product increments, which are of a quality that enables direct delivery as a finished product.

The program managers must plan the sprints after which they require the corresponding product increments for their respective programs. Portfolio management should ensure that the program managers agree on product increments that are as uniform as possible in order to actually reduce the additional work.

8 Toolbox

In this Chapter ...

- you will get to know methods and tools for modern project management and
- you will learn how to fill the framework HyProMM from Chapter 5 with concrete methods and tools.

Introduction

This chapter provides an overview of a variety of methods and tools that can be combined using the HyProMM framework to create a modern process model tailored to your individual needs. The framework already presented in Chapter 5 is illustrated again in Figure 8.1.

The regulatory framework consists of three areas: leadership, project lifecycle and ongoing tasks. There are tasks in each area, such as I.1 Initiate project or P.1 Work planning. For each of these tasks, the framework provides three dimensions, one for the processes describing the tasks, one for the methods and tools to be used and one for the roles involved in completing the task.

This chapter presents excerpts of methods and tools for completing tasks in the HyProMM Framework for the project lifecycle area. The presentation is based on short profiles, the structure of which is explained in Figure 8.2. The profiles are organized thematically, but not chronologically. Not all of the methods and tools presented in this chapter are absolutely necessary when setting up a company-specific process model and the order may vary depending on the project constellation. Conversely, additional methods and tools can of course be sorted into the framework and applied.

Tip

This chapter is intended to fill the framework HyProMM with life and also serve as a reference guide. The methods and tools presented are evaluated in terms of their suitability for plan-based and agile projects and other criteria. This evaluation reflects a general tendency. It can make sense to use certain methods and tools from plan-based process models in otherwise purely agile projects and vice versa.

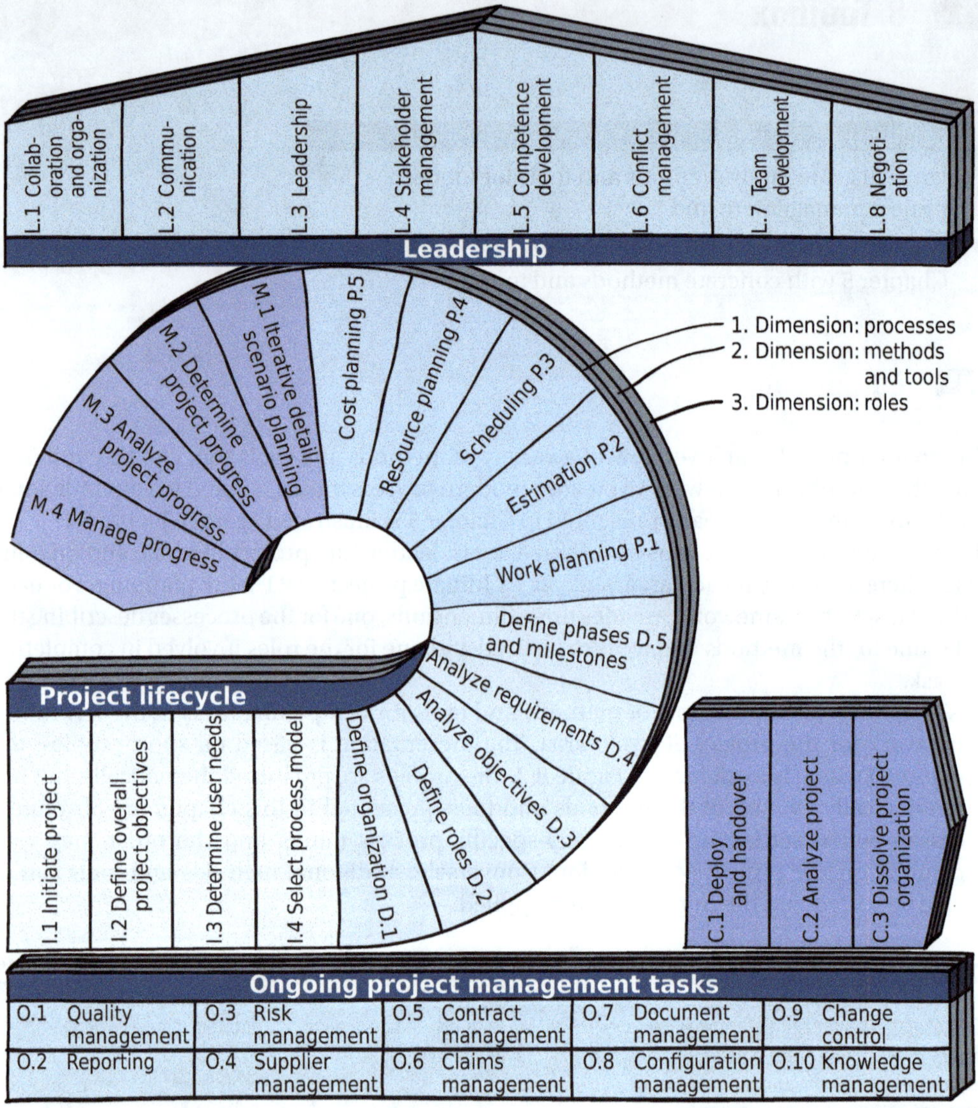

Figure 8.1: HyProMM framework for modern project management with the three areas of leadership, project lifecycle and ongoing tasks.

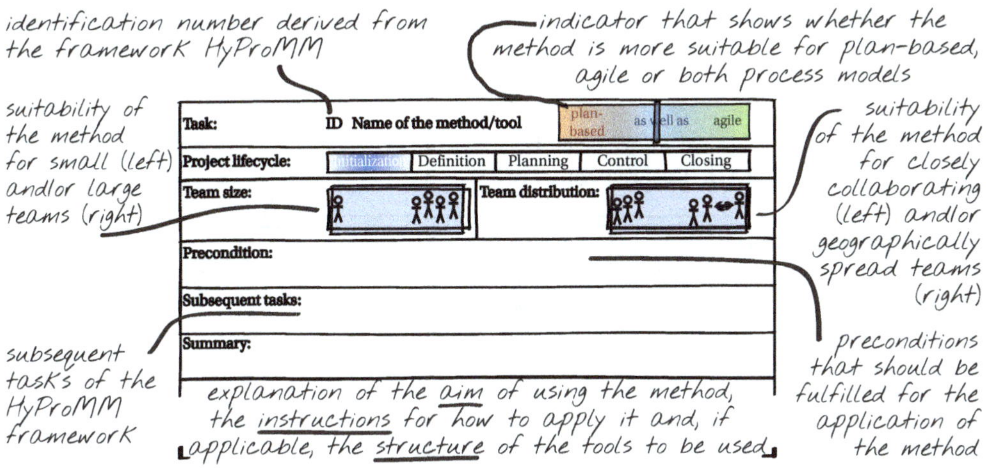

Figure 8.2: Structure of the profiles describing the methods of the framework HyProMM. The team size field helps to assess whether the method presented is more suitable for small teams (bar on the left) or large teams (bar on the right) or for both small and large teams (bar covers the entire team size field). The team distribution field indicates whether the method is suitable for teams that physically work together in one place (bar on the left) or for distributed teams (bar on the right) or for teams of any distribution (bar covers the entire field). These illustrations are to be understood as a rough recommendation. In certain cases, it is possible to deviate from them.

I Initialization

I.1 Initiate project

Task:	**I.1.1 Create project brief** (plan-based as well as agile)
Project lifecycle:	Initialization | Definition | Planning | Control | Closing
Team size:	**Team distribution:**
Precondition:	Recommended: rough idea of the project
Subsequent tasks:	I.2 Define overall project objectives I.3 Determine user needs
Summary:	The project approval is often documented in a project brief, also known as a project one-pager. The project brief can also be used to gradually approve the project from phase to phase. It should contain all information relevant for the company or the client.

Aim:

In the project brief, the planned project is described in short and concise terms. With the signature of the project sponsor or the client, the project is considered approved.

The document can also be used to describe only certain phases of the project and to approve the project gradually from phase to phase. The project brief is then updated and supplemented with information on future phases as it becomes available or new knowledge is gained.

Instruction:

Depending on the situation, the project brief is prepared by a preliminary project team, the project sponsor or the client. All parties involved must sign the final document to approve the project. If some information is missing at this stage, provisional approval may be granted for an initial project phase only. More information can be prepared, e.g. during a → I.1.2 project startup workshop.

Structure:

Project brief		
Name of the project and logo		
Brief description:		
Objectives:		
Start date of project:	End date of project:	
Milestones:	Resources:	Costs:
Risks and mitigation measures:		
Project sponsor / owner:	Client(s):	
Project manager:	Project team:	
Approval for phase: o Initialization	o Definition	o Planning o Control o Closing
Signatures:		

Task:	**I.1.2 Project startup workshop** — plan-based as well as agile
Project lifecycle:	Initialization | Definition | Planning | Control | Closing
Team size:	 **Team distribution:** 
Precondition:	Recommended: rough idea of the project
Subsequent tasks:	I.2 Define overall project objectives I.3 Determine user needs
Summary:	The project startup workshop is a working meeting in which the team collaboratively prepares an initial (and rough) project plan. If the workshop is held very early on, elements of team building can be integrated. The results of the workshop are a completed project brief (or project canvas) and an initial project plans.

Aim:
The aim is to collaboratively prepare an initial plan for the project.

Instruction:
Typically, the project manager invites the core team, the sponsor or client and other key stakeholders.
Depending on the complexity of the project, the meeting lasts between half a day and several days. Preliminary work by the core team can speed up the workshop and is particularly helpful for distributed teams with travel times. If the team does not yet know each other, elements from →I.1.4 team building can be integrated into the workshop.

Structure:
The location should provide sufficient inspiration and prevent disruptions caused by daily business. The agenda should be sent to the participants in advance. A typical agenda might be as follows:

- Welcome and introduction as well as a survey of expectations for the meeting
- Collection of information
- Overview of the planned project management / process model
- Identification of the stakeholders and definition of the overall project objectives
- Collaborative development of cooperation guidelines
- Preparation of an initial work breakdown structure as well as effort and cost estimates
- Structuring the project into project phases and definition of important milestones
- Definition of the project organization including responsibilities for information and communication
- Identification and assessment of project risks
- Definition of next steps and, if necessary, detailed planning of the next phase
- Feedback

Task:	I.1.3 Kick-off meeting	plan-based as well as agile
Project lifecycle:	Initialization | Definition | Planning | Control | Closing	
Team size:		Team distribution:
Precondition:	Approved project brief or project canvas	
Subsequent tasks:	I.2 Define overall project objectives I.3 Determine user needs	
Summary:	The kick-off meeting is used to inform stakeholders. They are to be informed about the new project. The exact content of the meeting is based on the information already available.	

Aim:
Important stakeholders of the project should be informed about its start and relevant project-specific facts.

Instruction:
The project manager invites the core team, the sponsor, the client and other relevant stakeholders. The duration of the meeting is relatively short and, depending on the complexity of the project and the stakeholder structure, lasts from less than an hour to a few hours.
In contrast to the → I.1.2 project startup workshop, the kick-off meeting is primarily for information purposes and is not a working meeting.
As the project has only just begun, the focus should be on motivation and explaining the scope and boundary conditions. The timing of the kick-off meeting depends on how much information is expected from the participants. This can be directly after project approval or after the initial planning.

Structure:
The location should be appropriate for an information event and contribute to a positive perception of the project. The agenda should be sent to the participants in advance.

A typical agenda might be:

- Welcome and introduction
- Presentation of the overall project objectives
- Introduction of the core team
- Explanation of known boundary conditions
- Next steps

Task:	I.1.4 **Foster team building**		plan-based as well as agile
Project lifecycle:	Initialization / Definition / Planning / Control / Closing		
Team size:		**Team distribution:**	
Precondition:	Recommended: defined core team of the project		
Subsequent tasks:	I.2 Define overall project objectives I.3 Determine user needs		
Summary:	Team building supports team development and should help to reach the performing phase (according to Tuckman) as quickly and conflict-free as possible and to facilitate communication.		

Aim:
The aim is to improve communication and cooperation within the project team through team development measures. If there is a lack of team development and/or if teams are spread across several locations, cooperation can suffer as a result of poor communication, incorrect mutual understanding and other disruptions.

Instruction:
The project manager invites the project team to a joint team building activity. Particularly in the case of distributed teams, this promotes subsequent collaboration, as all team members will have gotten to know each other personally. Typical team-building measures include joint activities such as outdoor events, canoe trips or moderated workshops followed by a joint dinner. The event should provide space for

- all members getting to know each other,
- clarifying roles in the project,
- the joint definition of rules of cooperation, and
- defining further measures to continuously improve cooperation.

It may be necessary to repeat team building measures several times, for example if conflicts arise in the team or the composition of the team changes.
For teams that have been working together for some time, the team building activity can include a reflection or situation analysis.

This is then used to work out which problems in cooperation or obstacles in the processing of tasks currently exist. Based on this, solutions are then developed.

Task:	**I.1.5 Project canvas**		plan-based as well as agile
Project lifecycle:	Initialization | Definition | Planning | Control | Closing		
Team size:		**Team distribution:**	
Precondition:	Recommended: rough idea of the project		
Subsequent tasks:	D Define Project		
Summary:	The project canvas is used for the structured description, visualization and assessment of a project in a "language" that is understandable for all stakeholders. It can be seen as an advanced version of the → I.1.1 project brief.		

Aim:
The project canvas is used to describe, visualize and assess a project.

Instruction:
The project canvas is developed in a team with important stakeholders, for example as part of a → I.1.2 project startup workshop, for example. The structure of the canvas helps with the complete and structured description of the project. In the control phase of the project, it can be used for continuous status recording and reporting.

Structure:
The exact structure depends on the requirements of the project and the project environment. Frequently used entries are the objectives and purpose, budget, team, boundary conditions, milestones, project scope and understanding of quality, resources, risks and opportunities, expected results and important stakeholders (client, customer etc.)

Project canvas for project:			**Author:** **Last updated:**
Objectives Defining the project vision or purpose and the most important objectives	Scope Description of the purpose of the project (what is the project about?)	Budget Determination of the available budget for the project	Milestones Definition of important internal and external milestones of the project
Boundary conditions/limitations Identification of mandatory framework conditions (e.g. standards, interfaces)	Quality assurance Definition of how the quality of the project results is to be achieved	Required resources Identification of the resources required for project realization	Important stakeholders (sponsor, client etc.) Identification of important stakeholders and their relationship to the project
Risks and mitigation Identification of relevant risks and opportunities and definition of initial measures to manage them	Deliverables Identification of the most important results of the project (exclusions can also be defined here)	Team/organization Rough definition of the project organization and the (core) team	

Task:	I.1.6 Prepare business case	plan-based as well as agile
Project lifecycle:	Initialization | Definition | Planning | Control | Closing	
Team size:		Team distribution:
Precondition:	Recommended: rough idea of the project	
Subsequent tasks:	D Define Project	
Summary:	A business case justifies a project and contains as quantitative a comparison as possible of a project's revenues and the expenses required for a project.	

Aim:
The business case provides the most quantitative justification possible for the project. During the project lifecycle, it serves as a navigator and helps to continuously review the project.

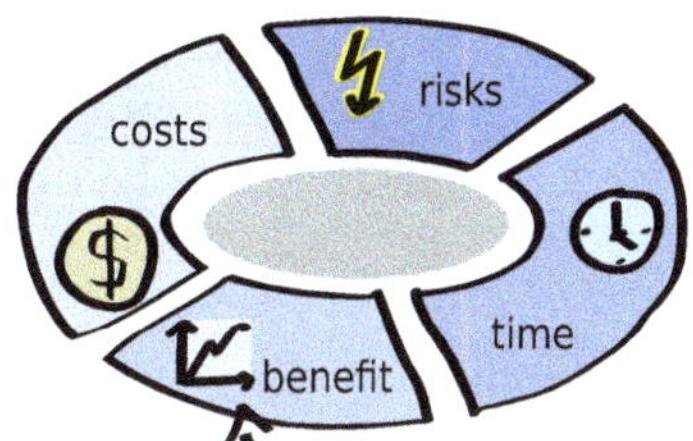

Instruction:
Comprehensive data is required for the creation of the business case, which must be collected, aggregated, structured and finally analyzed and evaluated.

Structure:
The typical contents of a business case include data on the benefits of the project deliverables, the timeline, possible risks and costs. The entire lifecycle of the project results should be considered. In addition to the project, this also includes the subsequent use and maintenance phase and the end of life. Examples of life cycle phases to be considered are the development, production, introduction of a new product, the use phase and the end of life of the product.
Typical elements of a business case are

- a justification of the relevance of the project,
- a description of the consistency with the overarching strategy,
- an initial overview of possible risks
- the expected expenses / costs,
- the achievable revenues, and
- a rough estimate of the project duration.

Other content may include comparisons with competitors, feasibility studies and economic indicators.

Task:	I.1.7 Assess success factors		plan-based as well as agile
Project lifecycle:	Initialization | Definition | Planning | Control | Closing		
Team size:		Team distribution:	
Precondition:	Recommended: rough idea of the project		
Subsequent tasks:	D Define Project		
Summary:	Success factors can promote the success of a project. At the start of the project, it should be checked which factors are fulfilled. Missing factors should be checked for their relevance to the specific project and addressed in risk or stakeholder management if necessary.		

Aim:
Factors for success and failure should be identified and evaluated at the start of the project. Success factors promote the success of the project and should be present; missing success factors or failure factors can jeopardize the success of the project.

Instruction:
Check the success factors listed below. The list can be supplemented with your own experience. If success factors are missing or if failure factors exist, these circumstances should be examined more closely in risk or stakeholder management. Some factors can be influenced in favor of the project through suitable measures, for example through good stakeholder communication.

List of typical success factors:

- high level of project management maturity
- selection and tailoring of a suitable process model
- clear goals and requirements
- appropriate project governance
- fast and transparent decision paths
- support from (top) management
- qualification of project participants
- good communication in project team
- sufficient resources
- good planning and control
- good communication with people of the project environment
- good risk management
- good error culture and ability to learn from mistakes

Task:	I.1.8 Characterize type of project	plan-based as well as agile
Project lifecycle:	Initialization | Definition | Planning | Control | Closing	
Team size:		Team distribution:
Precondition:	Recommended: rough idea of the project	
Subsequent tasks:	D Define Project	
Summary:	The characterization of the project helps to determine appropriate project management.	

Aim:
A new project can be characterized on the basis of known categories and the appropriate project management can be selected on this basis.

Instruction:
The (roughly) planned project and the project environment are used for characterization.

Typical types of projects:

- Client: external client; internal client
- Scope: research and development project; organizational project; investment project
- Project organization: functional project organization; matrix project organization; pure project organization
- Complexity and environment

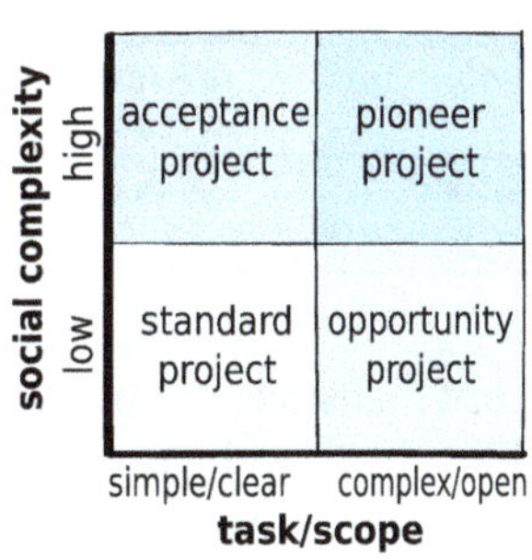

I.2 Define overall project objectives

Task:	**I.2.1 Define SMART targets** — plan-based as well as agile
Project lifecycle:	Initialization | Definition | Planning | Control | Closing
Team size:	**Team distribution:**
Precondition:	Recommended: rough idea of the project
Subsequent tasks:	I.3 Determine user needs I.4 Select process model
Summary:	The project objectives must be defined in the initialization phase. These lay the foundation for the later concretization of the project deliverables with the help of requirements. The objectives are often developed as part of the → I.1.2 project startup workshop.

Aim:
Defining objectives using the SMART rule serves to fully capture the project targets and to ensure their measurable definition.

Instruction:
The project manager collects and formulates together with the core team of the project and in coordination with the client and other relevant stakeholders, the project targets. Following the SMART rule has proven to be a good idea. SMART stands for the terms specific, measurable, achievable, relevant, and time-bound:

S	specific	simple, comprehensible and precise
M	measurable	clear criteria and targets for checking fulfillment
A	achievable	achievable under realistic conditions (but also quite demanding)
R	relevant	relevant to the project and in relation to the overall target
T	time-bound	defined and plannable in terms of time

Targets should be ambitious but realistically achievable. Without measurability, checking for target achievement is difficult or ambiguous and therefore prone to conflict. The project should not be overloaded with irrelevant objectives. Not every individual target of the project does not necessarily have to cover all SMART aspects. Overall, however, the project targets should be formulated SMART.

Task:	**I.2.2 Iron triangle**	plan-based as well as agile
Project lifecycle:	Initialization | Definition | Planning | Control | Closing	
Team size:		**Team distribution:**
Precondition:	Recommended: rough idea of the project	
Subsequent tasks:	I.3 Determine user needs I.4 Select process model	
Summary:	In the initialization phase, the project objectives must be defined as comprehensively as possible. The iron triangle helps to consider basic objectives and at the same time shows the tension involved in achieving them. The objectives are often defined during the → I.1.2 project startup workshop.	

Aim:
The aim is to fully capture the project objectives.

Instruction:
The project manager collects and defines the project objectives together with the project's core team and in consultation with the client and other relevant stakeholders. The categories of objectives of the magic triangle are taken into account, see the illustration on the right.

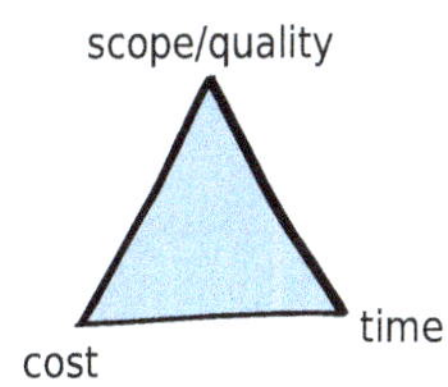

The scope and quality objectives define the content and quality of the project. The cost objectives take into account project-related costs, such as total project costs, and cost targets related to the project deliverables, such as production or sales costs. This also applies to the time-related objectives, which define both project-specific deadlines, such as project completion, and deadline objectives for the project deliverables, such as the maximum production time for a series product to be developed.
The iron triangle illustrates the tension between the categories. If lower project costs are to be achieved, this usually has an impact on the extent to which the scope/quality and time-related objectives can be achieved. Conversely, this also applies to the other categories.
In addition to the aforementioned categories of objectives, social objectives can be defined to satisfy stakeholder interests and non-objectives to demarcate the project. Objectives can also be divided into result-oriented objectives and process-oriented objectives. The completeness and achievability of the objectives are examined later in → D.3 analysis of objectives.

I.3 Determine user needs

Task:	**I.3.1 Collect requirements**	plan-based as well as agile	
Project lifecycle:	Initialization | Definition | Planning | Control | Closing		
Team size:		**Team distribution:**	
Precondition:	Helpful: Known project objectives and stakeholders		
Subsequent tasks:	I.4 Select process model D.4 Analyze requirements		
Summary:	The project objectives must be further refined in order to work on the project deliverables. First, user needs, i.e. customer requirements, are collected and later converted into specifications.		

Aim:
The aim is to collect all customer requirements for the subsequent project definition.

Instruction:
The project manager or an appointed employee collects requirements for the project from all relevant project stakeholders. These are often submitted directly by the client to the project in the form of a customer requirements specification (document).
Requirements are documented representations

- of a condition or capability that contributes to solving a problem or achieving a goal, or
- of a condition or capability of a system to fulfill a contract, a standard, a specification or other documents.

Like goals, requirements should fulfill the SMART criteria. The template according to Rupp[1)] helps to formulate good requirements:

[<when?> <under which condition?>]	the system <name of the system>	must / should / will	<verb/ process word>	- / <whom?>	[<object and/or supplement>]
Examples:					
If the toner is empty,	the copy machine	must	inform	the user	via a message on the display.
If the event is canceled,	registered persons	should	have	-	the option of rebooking their ticket.

Requirements should be formulated individually (only one requirement per sentence), in complete sentences and using standardized terminology. Requirements must be measurable and testable.
The principles presented here can also be applied in agile and hybrid projects. If it is already foreseeable at this stage that the project is to be carried out in an agile manner, the requirements can also be formulated as ➔ I.3.3 user stories.

1) Rupp et al.: Requirements-Engineering und -Management. Carl Hanser Verlag. 2014

Task:	I.3.2 Prepare customer requirements specification		plan-based as well as agile
Project lifecycle:	Initialization | Definition | Planning | Control | Closing		
Team size:		Team distribution:	
Precondition:	Helpful: Known project objectives and stakeholders		
Subsequent tasks:	I.4 Select process model D.4 Analyze requirements		
Summary:	All requirements of the client are usually collected in the customer requirements specification. The requirements can be used for tenders and the definition of the project scope.		

Aim:
The aim is to collect a complete set of requirements from the client's perspective for a project and its deliverables.

Instruction:
The customer requirements specifications are usually drawn up by the client of the project and contain the requirements from the client's perspective. The client can use the specifications to obtain offers from several potential contractors. The requirements should define WHAT needs to be done. HOW and WITH WHAT something is to be done should only be defined in the → D.4.5 requirements specification to be created later, as otherwise the solution space will be restricted unnecessarily early on.

Structure:
In order not to forget any requirements, it helps to give the customer requirements a structure and to collect requirements for different categories.
To ensure → D.4.6 traceability, all requirements should be given a uniform identification number.

1. Introduction
   - What is it about?
   - Description of actual and target status
2. Objectives from the client's point of view
3. Product use
   - Description of the purpose and use of the project deliverables
4. Functional requirements
   - Which functions are expected by the client?
5. Non-functional requirements
   - Requirements for efficiency, scalability, reliability, operation, maintenance and transportation, etc.
6. Interfaces
7. Testing and acceptance
8. Documentation
9. Boundary conditions
   - Laws, standards, etc. to comply with
10. Miscellaneous

Task:	**I.3.3 Collect user stories** — plan-based as well as agile
Project lifecycle:	Initialization | Definition | Planning | Control | Closing
Team size:	**Team distribution:**
Precondition:	Helpful: Known project objectives and stakeholders
Subsequent tasks:	I.4 Select process model D.4 Analyze requirements
Summary:	The project objectives must be further specified in order to work on the project deliverables. In agile projects, this is done by collecting and defining user stories. A user story is a requirement that is written from the perspective of a specific role, often that of the user.

Aim:
The aim is to collect all requirements regarding the project deliverables to be completed.

Instruction:
The product owner collects requirements regarding the project deliverables from all relevant stakeholders of the project. These are defined from the perspective of a specific role. A formulation template can look like this:

AS A – <user role> – I WOULD LIKE – <the objective> – [, IN ORDER TO <reason>]

Typical roles are:

- team member
- user
- service technician
- observer
- production staff
- supplier
- customer
- corporate management
- marketing manager
- logistics expert
- sales representative
- etc.

User stories, like requirements, should not unnecessarily restrict the solution domain. They should be detailed enough to be estimated in terms of their size. User stories should follow the INVEST principle, i.e. they should be independent, negotiable, valuable, estimable, small, and testable.

If a user story cannot yet be formulated due to a lack of information, a so-called epic is formulated. The epic helps to ensure that nothing is forgotten. For example, an epic can simply be "invoicing". It is then specified in more detail later as soon as further information about the exact request of the associated stakeholder is available.

Task:	**I.3.4 Prepare initial product backlog** plan-based as well as agile
Project lifecycle:	Initialization / Definition / Planning / Control / Closing
Team size:	**Team distribution:**
Precondition:	Helpful: I.3.3 Collect user stories
Subsequent tasks:	I.4 Select process model D.4 Analyze requirements
Summary:	The project objectives must be further specified in order to work on the project deliverables. In agile projects, this is done by formulating and collecting user stories. All user stories are compiled in prioritized order in the product backlog. It serves as the basis for subsequent detailed planning.

Aim:

In projects based on Scrum, the product backlog serves as the basis for further planning. In the initialization phase, the existing requirements can be collected, estimated and prioritized in an initial product backlog.

Instruction:

The product owner collects the user stories and asks the team to estimate their size, for example with the help of → P.2.7 planning poker. The product owner prioritizes all user stories and assigns them a unique ID.

An example of a product backlog is shown below.

	A	B	C	D	E	F
1	ID	User Story	Size [Story Points]	Priority	Comments	
2	U5	As a user I would like ...	13	100	-	
3	M3	As a manager I would like ...	20	115	-	
4	P2	As a manufacturer I would ...	8	150	see also Doc. D13	
5	U2	As a user I would like ...	100	200	-	
6	U1	As a user I would like ...	3	205	-	
7	P1	As a manufacturer I would ...	40	210	-	
8	M2	As a manager I would like ...	8	213	-	
9	U4	As a user I would like ...	5	250	align with product T20	
10	M1	As a manager I would like ...	1	260	-	
11	S1	As a service technicians I ...	40	270	-	
12	...	...	...	...	...	

Prioritization can be done in order, by points or in categories according to the MuSCoW principle. The product backlog does not yet provide any information about when the user stories will be implemented and by whom.

Task:	I.3.5 Transfer customer requirements in/from product backlog — plan-based as well as agile		
Project lifecycle:	Initialization · Definition · Planning · Control · Closing		
Team size:		Team distribution:	
Precondition:	I.3.2 Customer requirements specification or I..3.4 Product Backlog		
Subsequent tasks:	I.4 Select process model D.4 Analyze requirements		
Summary:	In hybrid projects with Scrum and plan-based process models, there are different collections of requirements in the form of product backlog and customer requirements specification. Depending on the structure of the hybrid project, it may be necessary to compare and align these two documents.		

Aim:
The aim is to use the requirements documented in the product backlog and/or the customer requirements specification in hybrid projects.

Instruction:
Product backlogs and customer requirements specification have different structures and objectives. In the product backlog, all requirements are collected as user stories, estimated in terms of their size and prioritized. In the customer requirements specification, requirements are usually structured according to certain categories. The requirements are not always prioritized and rarely estimated. Another difference is that user stories are usually formulated from the perspective of a specific role, whereas the entries in the customer requirements specification describe requirements for a (often technical) system. In a hybrid project, it must be ensured that these differences do not hinder the project and/or lead to misunderstandings. This can be achieved as follows:

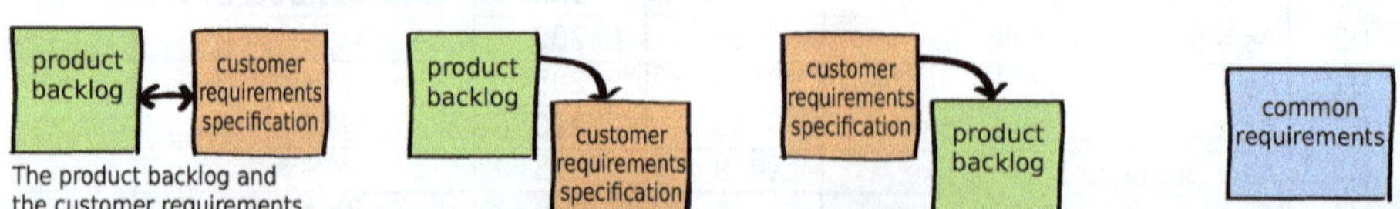

The product backlog and the customer requirements specification are collections of requirements with equal status and are kept synchronized throughout the project.

The product backlog is the master documentation from which the customer requirements specifications are derived.

The customer requirements specification is the master documentation from which the product backlog entries are derived.

There is only one common list of requirements, which serves as both, the customer requirements specification and the product backlog.

In the further course of the project, other documents such as the sprint backlog and the requirements specification must be aligned with the product backlog and customer requirements specification.

Task:	**I.3.6 Prepare traceability**	plan-based / as well as / agile
Project lifecycle:	Initialization | Definition | Planning | Control | Closing	
Team size:		**Team distribution:** 
Precondition:	Helpful: Known concept for requirements management	
Subsequent tasks:	I.4 Select process model D.4 Analyze requirements	
Summary:	In some industries/projects and with some process models, it is important to be able to trace requirements from the idea to the (technical) specification through to verification, validation and start of production. A concept should be developed for this as part of the overall requirements management.	

Aim:
The aim is to create a concept for the traceability of requirements.

Instruction:
The project manager determines the requirements and general conditions for traceability and prepares a corresponding concept.

Structure:
Once a project reaches a certain level of complexity, traceability is usually ensured using suitable software. Requirements, specifications, design drafts, prototypes, parts lists, test specifications, test reports, etc. are linked together in this software. If something is changed during the project, it is possible to immediately trace which components of the project deliverables are affected by the change.

Up to a certain level of complexity, traceability can also be established using lists and tables in a spreadsheet or mind-mapping software.

The concept should include the following aspects
- Recording of all elements in which requirements must be traceable, e.g. requirements specification, functional specification, design, test specification, test report
- Definition of a coding system that can be used to identify the requirements, for example via common numbers: Customer requirements specification C.143 belongs to requirements specification R.143, and this belongs to test specification T.143, etc.
- Definition of the infrastructure for traceability (software, spreadsheet, mind-map, etc.)

I.4 Select process model

Task:	I.4.1 Select plan-based process model	plan-based as well as agile
Project lifecycle:	Initialization / Definition / Planning / Control / Closing	
Team size:		Team distribution:
Precondition:	Project started and project objectives at least roughly known	
Subsequent tasks:	D Definition phase	
Summary:	At the end of the initialization phase at the latest, it should be determined whether the project will follow a plan-based, agile or hybrid process model.	

Aim:
The aim is to define a plan-based process model that fits the project.

Instruction:
The project manager determines parameters that characterize the project and selects the appropriate process model for the project, taking company-specific requirements into account. Plan-based process models are suitable for projects whose → I.3 user needs can be easily determined and for which the solution path is known and associated with low risks. In principle, plan-based process models can be divided into sequential (with or without phase overlap) and incremental process models.

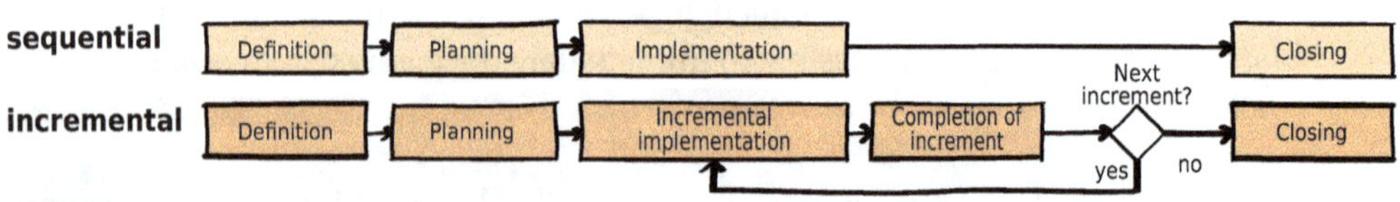

Structure:

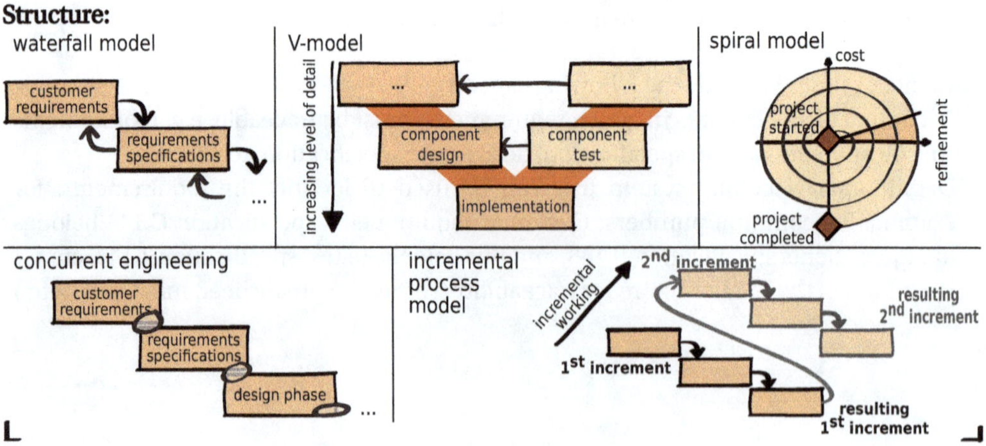

Task:	**I.4.2 Select agile process model**	plan-based as well as agile
Project lifecycle:	Initialization | Definition | Planning | Control | Closing	
Team size:		**Team distribution:**
Precondition:	Project started and project objectives at least roughly known	
Subsequent tasks:	D Definition phase	
Summary:	At the end of the initialization phase at the latest, it should be determined whether the project will follow a plan-based, agile or hybrid process model.	

Aim:
The aim is to define an agile process model that fits the project.

Instruction:
The project manager determines parameters that characterize the project and selects the appropriate process model for the project, taking company-specific requirements into account. Agile process models show their strengths particularly in small and non-distributed teams. They are suitable for projects whose ➔ I.3 user needs or solutions are subject to major uncertainties:

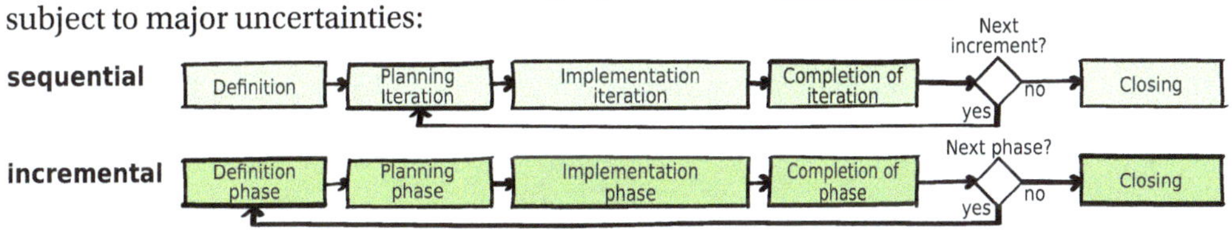

Structure:

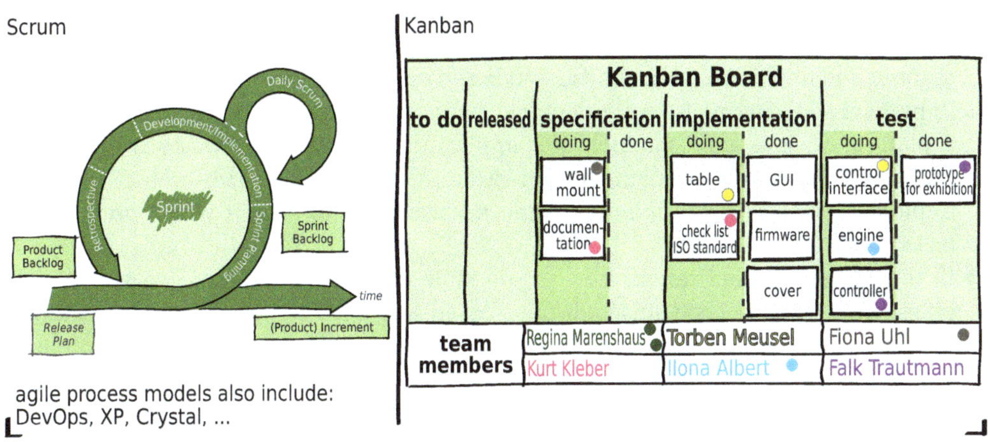

Task:	**I.4.3 Select hybrid process model** (plan-based / as well as / agile)
Project lifecycle:	Initialization | Definition | Planning | Control | Closing
Team size:	**Team distribution:**
Precondition:	Project started and project objectives at least roughly known
Subsequent tasks:	D Definition phase
Summary:	At the end of the initialization phase at the latest, it should be determined whether the project will follow a plan-based, agile or hybrid process model.

Aim:
The aim is to define a hybrid process model that fits the project.

Instruction:
The project manager determines project-characterizing parameters and key performance indicators and selects the appropriate process model for the project, taking company-specific requirements into account.
Plan-based process models show their strengths in large projects with distributed teams and clear objectives and solutions.
Agile process models show their strengths particularly in small and undistributed teams. They are suitable for projects whose objectives or solutions are subject to greater uncertainty.
Hybrid process models combine plan-based and agile processes, methods, tools and roles. The combination can be achieved in various ways:

- In **sequential application**, a plan-based and agile approach is used in successive, self-contained phases. An example of this is the water-Scrum-fall model, in which requirements are first determined based on a plan, then implemented in an agile manner and finally integrated and tested based on a plan.
- In **parallel application**, there are parallel tasks within a project that are carried out in a plan-based and agile manner. The application usually takes place within sub-projects carried out in parallel. For example, a software sub-project can be implemented in an agile manner, while a hardware sub-project works on a plan-based basis.
- In the **integrated application**, it is not possible to differentiate between sequential phases and parallel tasks. Instead, plan-based and agile processes, methods, tools and roles are interlinked within a project. One example is the consistent project management with the Kanban board in a project otherwise carried out according to the V-model.

Task:	**I.4.4 Tailor the process model**	plan-based · as well as · agile
Project lifecycle:	Initialization · Definition · Planning · Control · Closing	
Team size:		**Team distribution:**
Precondition:	Project started and project objectives at least roughly known	
Subsequent tasks:	D Definition phase	
Summary:	At the end of the initialization phase at the latest, it should be determined which process model the project will follow. Once this is known, the selected process model should be adapted to the project-specific characteristics as far as possible.	

Aim:
The aim is to adapt the process model to individual project characteristics.

Instruction:
In many companies, project management is defined by company-wide process models. The associated standardization of project processes simplifies collaboration, supports knowledge management and promotes the continuous development and improvement of project management through the comparability of projects resulting from standardization.

However, if the projects are actually of varying complexity and size or have to interact with very different stakeholders, rigid standardization can impair project efficiency and effectiveness. Many process models therefore allow tailoring, i.e. the adaptation of the process model within certain limits.

Project managers should be aware of these limits and check which possibilities for adaptation exist and make sense. Typical adaptations are, for example

- breaking down the project into suitable sub-projects
- omit phases, tasks and roles that are not required
- adding additional phases, tasks and roles
- adding additional milestones to structure the project over time and for synchronization within the project as well as with external partners
- application of agile processes, methods, tools and roles within an otherwise traditionally organized project

Every adjustment should be well justified and clearly communicated. Adjustments should not be misunderstood as an invitation to work together in an unstructured and chaotic way.

D Definition

D.1 Define organization

Task:	D.1.1 Define organizational form (plan-based as well as agile)
Project lifecycle:	Initialization / **Definition** / Planning / Control / Closing
Team size:	Team distribution:
Precondition:	The project should be approved
Subsequent tasks:	D.1.2 Define organizational chart
Summary:	The organizational form determines how the employees are assigned to the project. Often, no project-specific organizational form is possible, i.e. all projects within a division are carried out either within an functional, matrix or pure project organization.

Aim:
The aim is to determine the organizational form of a project.

Instruction:
In most cases, the organizational form for a business unit is defined in a standardized manner. For larger projects, a separate project entity (joint venture, consortium, etc.) can be established in addition to the organizational forms listed below.

Structure:

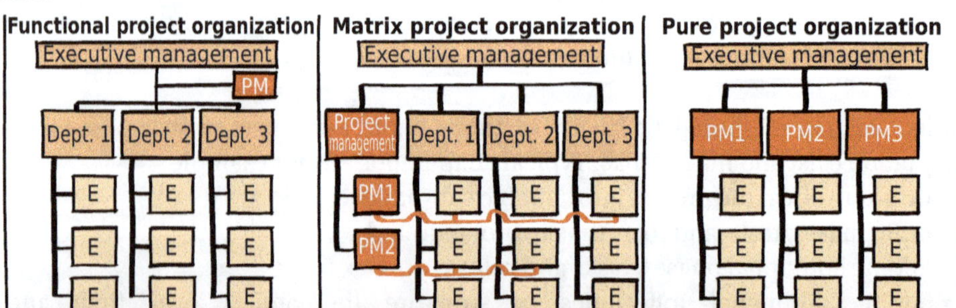

	Functional project organization	Matrix project organization	Pure project organization
Characteristics	employees remain in line, project manager has weak authority	line manager is disciplinary, project manager is functional manager	employees report to project manager in functional and disciplinary terms
Typical project types	small projects, organizational projects	medium to large projects, research and development projects	large projects, investment projects
Strengths	simple structure, no organizational changes required, good workload balance, good exchange of knowledge	good workload balance, good exchange of knowledge	clear reporting lines and responsibilities, low conflict potential, high level of identification with the project
Weaknesses	no managerial authority, coordination difficult, increase in workload in line departments	potential for conflict due to shared responsibilities between project and line, risk of employees being overloaded by project and line tasks	complex changes of project organization, difficult workload management, additional disciplinary tasks, difficult knowledge sharing

PM: project manager Dept.: Department E: Employees

Task:	D.1.2 Make team composition	plan-based as well as agile
Project lifecycle:	Initialization / **Definition** / Planning / Control / Closing	
Team size:		Team distribution:
Precondition:	The core team for the start of the project should be known	
Subsequent tasks:	D.1 Continue definition of organization	
Summary:	Even if there is no final resource plan at this stage, the project manager must put together a core team that will complete the next tasks.	

Aim:
The aim is to put together the (core) team for the project.

Instruction:
A complete resource plan does not yet exist in the project definition phase. Nevertheless, the project manager still needs a small team, which will grow later, to define and subsequently plan the project. Project managers cannot always choose their own team. However, they should always check the composition of the team and work towards changing it if it is not suitable. When selecting the members of this core team, project managers should consider both professional and personal aspects:

- The first step is to identify employees for all upcoming tasks who are basically capable of carrying out these tasks.
- It must be checked whether the employees cover all the necessary formal roles, such as requirements engineer or analyst.
- Additionally, it should be evaluated whether the employees are capable of working in a team and whether there are any existing conflicts between designated team members.
- It is important to check whether the employees also cover all the necessary informal roles. Character traits that may be necessary to lead a project to success play a role here. For example, an innovation project requires employees who can work creatively, challenge the status quo and express their opinions to other employees and managers. Every project needs people who can work as coordinators or integrators and hold a team together. Models such as the Belbin role model outlined on the right can help to verify the existence of relevant informal roles.

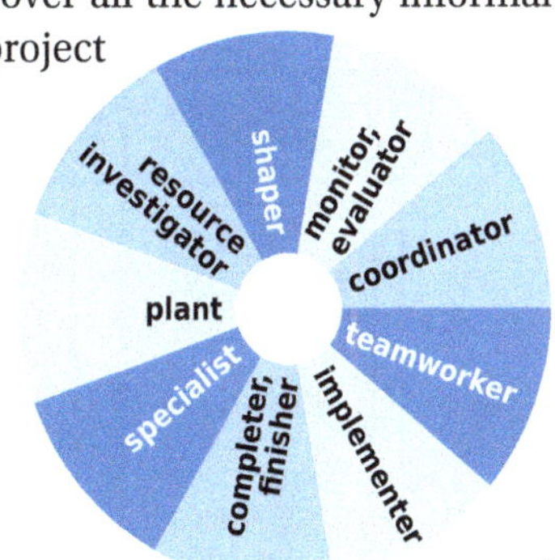

Task:	D.1.3 Define organizational chart	plan-based as well as agile
Project lifecycle:	Initialization / **Definition** / Planning / Control / Closing	
Team size:		Team distribution: 
Precondition:	D.1.2 Make team composition	
Subsequent tasks:	D.1 Continue definition of organization	
Summary:	The organizational chart visualizes the organizational structure of the project, structures organizational units (e.g. sub-projects) and depicts relationships between them.	

Aim:
The aim is to define the organizational structure of the project and its personnel composition.

Instruction:
A complete resource plan does not yet exist in the project definition phase. Nevertheless, the project manager is usually supported by other people and experts. The internal project responsibilities can be defined in an organizational chart. In purely agile projects, the organizational chart is generally not used for the (non-existent) hierarchical structuring of the project, but to define the collaboration between several agile teams, including the product owner and clients/customers.

Structure:
The organizational structure of the project should simultaneously promote and streamline communication. It should also identify clear competencies and areas of responsibility. The following illustration is an example of an organizational chart:

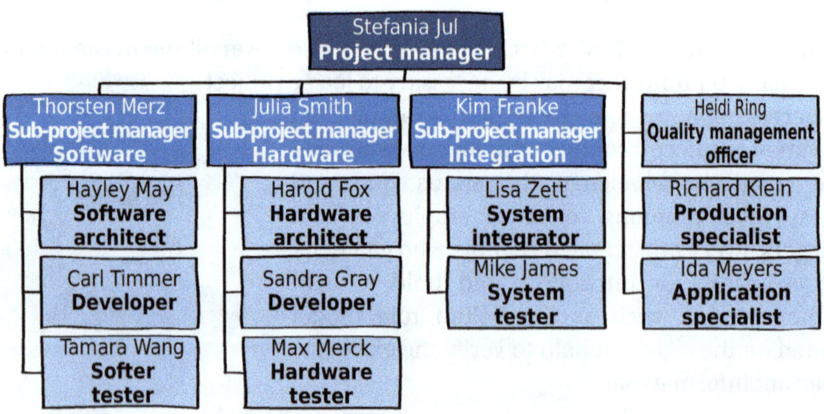

Task:	D.1.4 Define organization and interfaces	plan-based as well as agile
Project lifecycle:	Initialization / **Definition** / Planning / Control / Closing	
Team size:		Team distribution:
Precondition:	D.1.2 Make team composition	
Subsequent tasks:	D.1 Continue definition of organization	
Summary:	The project team must be embedded in its external project environment. Interfaces to the project management office, the client, the steering committee, etc. must be defined.	

Aim:
The aim is to define interfaces to external organizational units.

Instruction:
The project and its organizational environment must be defined. The project owner must be determined, if not already done, a steering committee must be appointed as a steering and escalation committee, and possible suppliers and other stakeholders must be identified.
Administrative support from a project office should be requested, and a framework conditions from a project management office should be adhered to. In a further step, the exact escalation paths should be defined and reporting deadlines, media and scope should be specified.

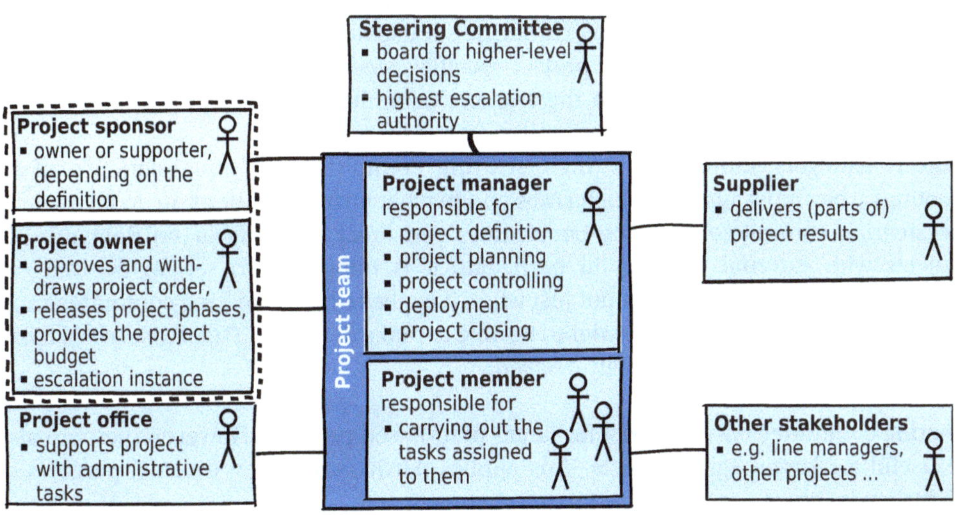

Task:	**D.1.5 Establish steering committee** (plan-based as well as agile)
Project lifecycle:	Initialization · **Definition** · Planning · Control · Closing
Team size:	**Team distribution:**
Precondition:	The project organization should be known
Subsequent tasks:	D.1 Continue definition of organization
Summary:	The steering committee is a higher-level decision-making and escalation committee for the project, made up of representatives of the project owner or client and the contractor (project). At the start of the project, it must be determined who belongs to the committee, how often it meets and what the communication and escalation interfaces look like.

Aim:
The aim is to define the personnel composition, responsibilities, authorities and communication interfaces of the steering committee.

Instruction and Structure:
The steering committee is the decision-making and escalation body. It makes decisions that exceed the authority of the project manager or if the project manager cannot reach an agreement with the client. In companies with a large number of projects, the steering committee can be responsible for all projects in the portfolio.
It can also be the reporting instance for the projects and prioritize the projects within the company. In this case, the steering committee is usually called a portfolio board. Then there are usually additional project-specific steering committees, which are hierarchically positioned between the steering committee and the project and are the first escalation instance.
Project managers should know their steering committee and the frequency of its meetings. They must agree in which cases and in what form escalations are to be made to the steering committee. Steering committees can meet regularly or on demand. For projects with external partners in particular, it is important to set up the steering committee at an early stage and not just when it is needed. The steering committee must have high-ranking members in order to be able to make decisions. Otherwise, escalations and decisions will be unnecessarily delayed.

Steering committees usually exist for larger plan-based projects. However, they can also be useful for larger, agile projects. This applies particularly when external partners or clients are involved.

Task:	D.1.6 Establish project office	plan-based as well as agile
Project lifecycle:	Initialization | **Definition** | Planning | Control | Closing	
Team size:		Team distribution:
Precondition:	D.1.4 Define organization and interfaces	
Subsequent tasks:	D.1 Continue definition of organization	
Summary:	A project office handles administrative, commercial, reporting and controlling-related tasks for a project.	

Aim:
The aim is to set up a project office to provide administrative support for the project manager and the project team.

Instruction:
Project offices are used in larger projects to assist the project manager and other team members with administrative and recurring tasks. These include for example

- support in the creation of standardized plans and documents,
- creation and archiving of meeting minutes,
- recording controlling-relevant key performance indicators such as working hours, costs etc.
- the utilization of knowledge and experience from other projects,
- providing knowledge and experience for other projects.

Project offices can be set up across projects in order to utilize synergies and experiences or to be able to pass them on more easily. Unlike the project management office, however, they usually take on project-specific tasks. When setting up a project office, at least the following questions must be clarified:

- allocation of tasks between project manager, team and office
- reporting channels, responsibilities and authorities
- understanding the role of the project office
- time availability of the project office and scope of work performance
- billing modalities (by the hour, flat rate, etc.)

A note on the project management office (PMO for short): This is active across all projects and is responsible for the entire portfolio. It has no administrative tasks in the individual projects, but serves to support the portfolio manager and the further development of project management. Project offices are often either attached to the project management office or located directly in the projects.

Task:	**D.1.7 Attract mentors** — plan-based as well as agile
Project lifecycle:	Initialization | Definition | Planning | Control | Closing
Team size:	 **Team distribution:** 
Precondition:	The project organization should be known
Subsequent tasks:	D.1 Continue definition of organization
Summary:	The initial project euphoria can quickly evaporate and your lose the interest of the management. Mentors and supporters can be an important building block for the success of the project and should therefore be involved in the project at an early stage.

Aim:
The aim is to identify and recruit suitable mentors and supporters for the project.

Instruction:
Before and after the project is freshly launched, the new project receives the full attention of the management. If there are no quick, outstanding successes or, on the contrary, there are even problems with the project, it can quickly fall into disrepute. Project managers should prevent this by securing the appropriate support from management. Mentors and other supporters can be a valuable building block here:

- Project managers can look directly for **mentors** in senior management who support them and their project, reflect on it and provide feedback. The project-external view helps to revise and improve management's perception of the project. With appropriate feedback from the mentor, project managers can improve their personal management skills and may find a supporter beyond the duration of the project.
- Less personal **supporters** can promote the project in general. These must be regularly informed about the progress of the project and reminded of its importance and successes.
- If the course of the project indicates difficulties, it can be helpful to consult a **coach**. Their external view of the project and the associated (required) critical reflection can provide important inspiration for project management.

A typical mistake made by project managers is to take management support for granted. The measures outlined here should be flanked by further project marketing activities so that the project is not jeopardized by a lack of or incorrect perception.

Task:	D.1.8 **Organize meetings and communication**	plan-based as well as agile
Project lifecycle:	Initialization / **Definition** / Planning / Control / Closing	
Team size:		**Team distribution:**
Precondition:	The project organization should be known	
Subsequent tasks:	D.2 Define roles	
Summary:	Meetings are important and expensive at the same time. Efficient preparation, organization and safeguarding knowledge are therefore important.	

Aim:
The aim is to use meetings efficiently and effectively.

Instruction:
The organizational chart and the project definition can be used to determine which employees need to coordinate on which topics. The number of participants in a meeting should be kept as small as possible and as large as necessary. If there are no decision-makers, the meeting is worthless. If people take part who have nothing to contribute, the meeting will be unnecessarily expensive. The following aspects should be taken into account:

- Balance project meetings of the entire team against often more efficient meetings of sub-projects
- Compare the team's need for information with the costs incurred
- Careful preparation of meetings with a defined meeting location, date, duration and participants
- Timely distribution of an agenda with the planned duration of the contributions and the name of the person responsible for preparation
- Allow sufficient time for the final discussion and resolution
- Team participation, for example through the use of a Kanban board or other tools
- Decisions and action points from a meeting should be recorded in the meeting minutes and made permanently available to all participants

Meetings are important and can help to keep employees motivated if they are conducted in an informative and respectful manner. In many projects, short meetings held regularly and in a standing position are more effective than infrequent and lengthy meetings. The former emphasize the urgency and help to identify and solve problems at an early stage.

D.2 Define roles

Task:	D.2.1 Define roles	plan-based as well as agile
Project lifecycle:	Initialization | Definition | Planning | Control | Closing	
Team size:		Team distribution:
Precondition:	D.1 Define organization	
Subsequent tasks:	D.2 Continue definition of roles	
Summary:	Roles describe the tasks, responsibilities and authorizations that a person takes on in the project. Roles can be formal or informal. Each role also includes the expectations of stakeholders regarding the actions and the way the role works.	

Aim:
The aim is to define the roles involved in the project, their responsibilities and authorities.

Instruction:
The project is already structured in the organizational chart. For larger projects, it makes sense to define the tasks with the associated responsibilities and authorities for each role in order to avoid misunderstandings. Typical roles are project manager, sub-project manager, client, system architect, product owner, Scrum master, etc.
Role descriptions should not be too detailed. Otherwise, conflicts of responsibility will quickly arise, as either several employees feel responsible or none at all. In practice, it will rarely be possible or necessary to describe all tasks within a (unique) project in detail and assign the necessary authorities. The definition of roles originates from plan-based project management, but can also be useful in agile and hybrid projects. You can find out more about this in ➔D.2.2 define plan-based roles, ➔ D.2.3 define agile roles and ➔ D.2.4 define hybrid roles.

Structure:
A role description can be structured as follows:

role	tasks	responsibilities	authorities
project manager	operative management of a project, planning and controlling, reporting	operational responsibility for achieving the project objectives and providing the project deliverables	functional supervisor (lateral leadership) of the project team, authorization to sign up to 20,000 US$

Alternatively, the column division into tasks, competence and responsibility is also widely used and sometimes referred to as **TCR-matrix**.
Note: In organizational theory, the term function is sometimes used instead of role. The latter focuses on the formal aspects of a position. In common project management standards, however, the term role is mostly used. This can be formal (like the function) and informal. The informal roles such as perfectionist, quarrelsome etc. should be known. However, these roles are not documented, but taken into account in communication and leadership.

Task:	D.2.2 Define plan-based roles	plan-based / as well as / agile
Project lifecycle:	Initialization / **Definition** / Planning / Control / Closing	
Team size:		Team distribution:
Precondition:	D.1 Define organization	
Subsequent tasks:	D.3 Analyze objectives	
Summary:	There are various formal roles within a project, whose characteristics should be known. Not in every project all roles are required.	

Aim:
The aim is to define the relevant roles for working on the project deliverables in plan-based projects.

Instruction:
The roles presented in the following overview perform various tasks with corresponding responsibilities and authorities. The exact interpretation may vary depending on the company and project. It is important to create clear responsibilities, keep the project organization as lean and simple as possible and define clear decision-making paths.

	Project sponsor	Project owner/client	Steering committee	Project manager	Team member	Project office	Supplier
Responsibility	can be identical with project owner/client supports project financially, organization-ally or ideally	secures funding and provides the project budget escalation authority in the event of problems	reports to executive management monitoring the progress of the project escalation instance	reports to steering committee planning, execution and control of the project project closure	working on the assigned tasks reporting to project manager	usually administrative support for the project manager	delivery of the agreed deliverables in accordance with the contract
Authority	depending on the arrangement, identical to the client or authority as coach or mentor	issues and withdraws the project order releases the subsequent project phase	decides if the limit of the project manager's authority has been reached decides on major changes balancing the interests of the client and the assigned project	participates in project definition and resource allocation depending on organization, coordinating, functional or disciplinary leadership decides on tasks in the project	within the scope of the assigned tasks preparing and, if necessary, participating in project decisions	requesting data in coordination with the project manager	decisions within the agreed framework in accordance with the contract

Task:	D.2.3 Define agile roles	plan-based as well as agile
Project lifecycle:	Initialization / Definition / Planning / Control / Closing	
Team size:		Team distribution:
Precondition:	D.1 Define organization	
Subsequent tasks:	D.3 Analyze objectives	
Summary:	There are various formal roles within a project, the characteristics of which should be known. Not in every project all roles are required.	

Aim:
The aim is to define the relevant roles for working on the project deliverables in agile projects.

Instruction:
The roles presented in the following overview take on various tasks with corresponding responsibilities and authorities in an agile project according to Scrum. Scrum consistently avoids a designated project manager and instead relies on self-organizing teams. Other agile process models such as Kanban allow the assignment of a project manager. This manager should apply agile leadership principles and allow the team to organize itself if possible. Other roles such as clients, suppliers etc. are also used in practice. In agile projects, the client is involved in the project process earlier and more regularly than in plan-based projects. Nevertheless, the roles described in → D.2.2 define plan-based roles can also be relevant in agile projects in practice.

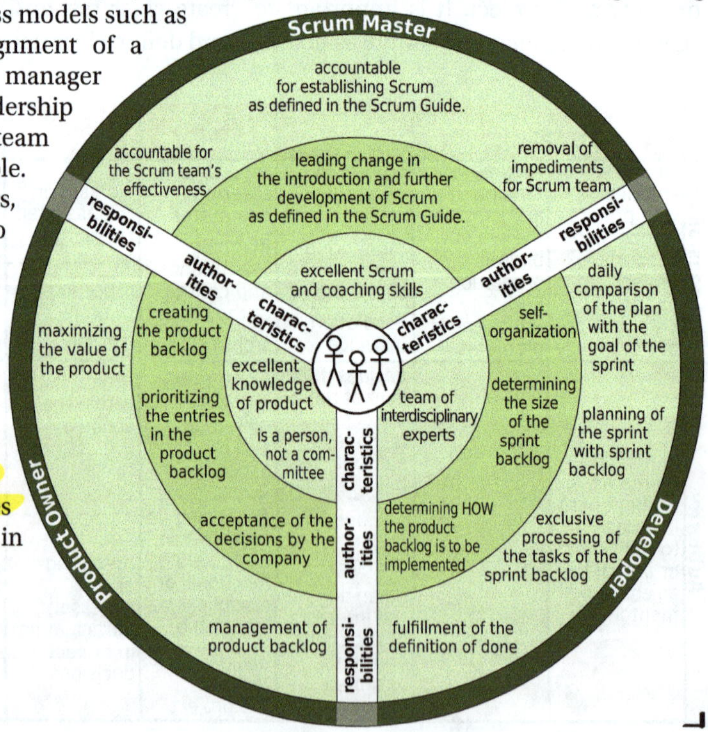

Task:	D.2.4 Define hybrid roles	plan-based as well as agile
Project lifecycle:	Initialization / **Definition** / Planning / Control / Closing	
Team size:		Team distribution:
Precondition:	D.1 Define organization	
Subsequent tasks:	D.3 Analyze objectives	
Summary:	There are various formal roles within a project, the characteristics of which should be known. Not in every project all roles are required.	

Aim:
The aim is to define the relevant roles for working on the project deliverables in hybrid projects.

Instruction:
Depending on the process model and the methods and tools to be used, suitable roles must be appointed in the project. The following considerations are intended to help you select the right roles. An in-depth discussion is not possible due to the many possible combinations of hybrid models.

- In a hybrid project, which is predominantly agile according to Scrum, but which uses traditional requirements documents such as customer requirements and functional specifications in order to be able to collaborate with other projects, similar roles are probably used as in a pure Scrum project, i.e. Scrum master, product owner and developers.
- In a hybrid project, which is predominantly plan-based but includes an agile sub-project, the overall project is managed by a project manager. This project manager interfaces with the client and the steering committee. Depending on the nature of the agile approach and the qualifications of the team, the agile sub-project is assigned a coordinating sub-project manager or works in a self-organizing manner with the typical Scrum roles. The project manager of the overall project acts as the product owner for the agile sub-project.

Structure:
In general, the design of the roles should match the qualification level of the team, the process model, the governance requirements of the project and the interfaces to external project units. Rigid hierarchical structures prevent many of the advantages of agile process models, such as participation, maintaining motivation and meeting customer requirements. Conversely, a lack of structures and hierarchies in a large hybrid project can lead to chaos and jeopardize the success of the project.

Task:	**D.2.5 Define responsibilities with the RACI matrix** — plan-based as well as agile
Project lifecycle:	Initialization | Definition | Planning | Control | Closing
Team size:	**Team distribution:**
Precondition:	D.1 Define organization
Subsequent tasks:	D.3 Analyze objectives
Summary:	Responsibilities, consultation and reporting obligations are clearly defined in the RACI matrix. It can be a supplement to the role descriptions (see → D.2.1 define roles) or replace them in smaller projects.

Aim:
The aim is to define and document responsibilities, consultation and reporting obligations in the project.

Instruction:
RACI stands for **r**esponsible, **a**ccountable, **c**onsulted and **i**nformed. The RACI matrix specifies who is responsible for implementation (responsible), who is accountable for approval or has the budget required for completion (accountable), who must be consulted for advice (consulted) and who must be informed (informed).
For complex projects, the RACI matrix can quickly become confusing, as in principle a row or column can be set up for each stakeholder and each task in the project. Project managers should therefore focus on the particularly critical stakeholders and tasks. Once the RACI matrix has been created, it should be made known to all stakeholders named in it.

Structure:
The relevant stakeholders are usually arranged in columns and the pending tasks in rows. In the fields connecting columns and rows, an R, A, C and/or I is used to document how the respective stakeholder is to be involved in the task. An example of a RACI matrix is shown on the right.

	Program manager	Project manager	Requirements engineer	System architect	Test manager	Developer	...
Project meeting	I	A/R	C	C	C	C	
Project brief	A	R	C	C	C	C	
Requirements Specifications	C	A	R	I			
System design		A	I	R		C	
Test planning		A	C		R	I	
...							

Task:	**D.2.6 Define escalation paths**	plan-based | as well as | agile
Project lifecycle:	Initialization | **Definition** | Planning | Control | Closing	
Team size:		**Team distribution:**
Precondition:	D.1 Define organization	
Subsequent tasks:	D.3 Analyze objectives	
Summary:	Escalation means the orderly forwarding of an issue to the next higher hierarchy level if it cannot be resolved at the current hierarchy level. Escalation paths determine who escalates to whom in which cases.	

Aim:
The aim is to define and document the reporting paths in the event of escalating issues.

Instruction:
In the event of a conflict, a fact-based, solution-oriented approach is often superimposed by personal attacks and blockades. It is therefore advisable to develop escalation paths when setting up the project organization and to decide on these with the project team and other bodies involved.
At the start of the project and as long as no escalating situation has occurred, it is usually possible to determine who is responsible for decisions in the event of a conflict by mutual agreement and with objective justification. The escalation path should include project team members, sub-project managers and the project manager. In the event that the project manager's authority is exceeded, it should also include stakeholders external to the project, such as the higher management, the client or the steering committee.
Before an actual escalation occurs, you should have exhausted your own room for maneuver. It is helpful to find supporters for your own cause. Escalations should generally be announced.

Structure:
Escalation paths can be defined as a flowchart, table or in text form. The following figure shows a simple example.

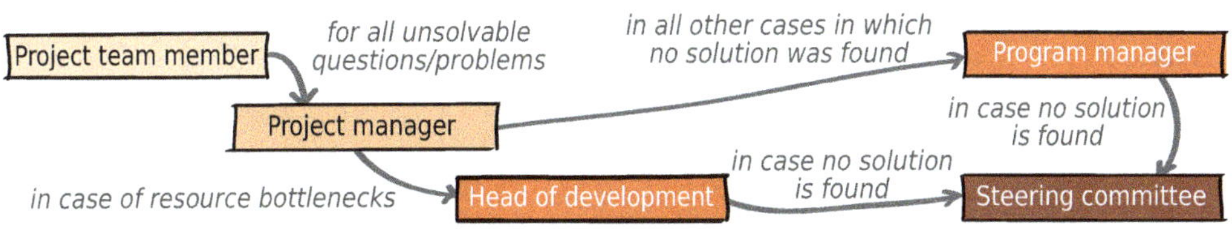

D.3 Analyze objectives

Task:	**D.3.1 Create hierarchy of objectives** — plan-based as well as agile
Project lifecycle:	Initialization / **Definition** / Planning / Control / Closing
Team size:	**Team distribution:**
Precondition:	The project objectives should at least be roughly known.
Subsequent tasks:	D.3 Continue analyzation of objectives
Summary:	At the beginning of the project, the objectives should be defined and checked for completeness and consistency. In order to reduce conflicts of objectives, the objectives should also be prioritized. The hierarchy of objectives serves to check the completeness and measurability of the objectives.

Aim:
The aim is to present the project objectives in a structured manner, thereby facilitating the completeness check and ensuring the measurability of all project objectives.

Instruction:
The project objectives are first collected and then grouped and classified. Each objective must be broken down to such an extent that it can be evaluated using a concrete measure. Due to the plan-based approach of traditional process models, the hierarchy of objectives is much more important in these than in agile process models. However, it can also be used there.

Structure:
The following figure shows an example of a hierarchy of objectives. The overall objective is divided into categories, subcategories, objectives, and subobjectives. Evaluation criteria and the degree of achievement enable later evaluation. Social objectives serve to ensure stakeholder satisfaction. Exclusions delimit the project and help to avoid misunderstandings.

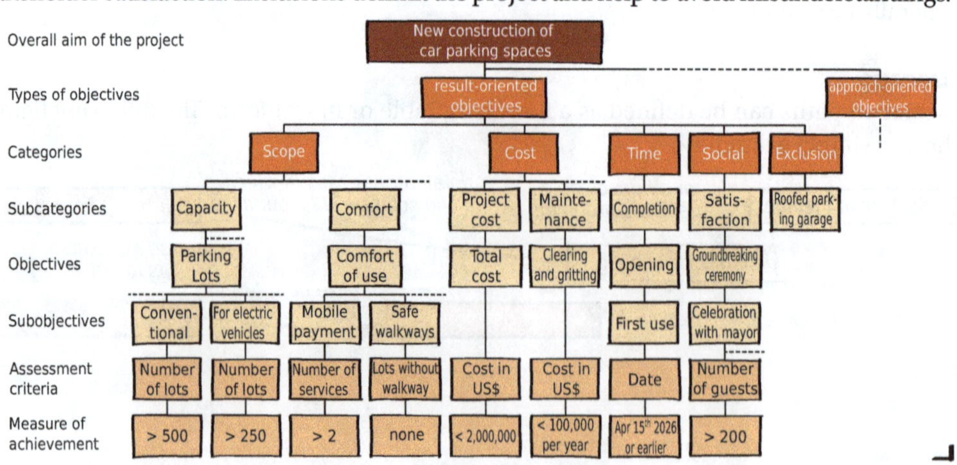

Task:	**D.3.2 Create matrix of objectives**	plan-based as well as agile	
Project lifecycle:	Initialization | **Definition** | Planning | Control | Closing		
Team size:		**Team distribution:**	
Precondition:	The project objectives should at least be roughly known.		
Subsequent tasks:	D.3 Continue analyzation of objectives		
Summary:	At the beginning of the project, the objectives should be defined and checked for completeness and consistency. In order to reduce potential conflicts of objectives, the objectives should also be prioritized. The matrix of objectives is used to identify inconsistencies and conflicts of objectives.		

Aim:
The aim is to identify redundant and conflicting objectives that jeopardize project success.

Instruction:
The project objectives are collected and then compared to each other. The comparison examines whether objectives are mutually exclusive (objective antinomy), whether they are in competition or neutral to each other, whether they support each other or whether they are identical. Competing objectives must be prioritized in order to be able to decide which objective to pursue in the event of a conflict. Antinomies of objectives must be avoided.

Structure:
The figure on the right shows an example of a matrix of objectives.

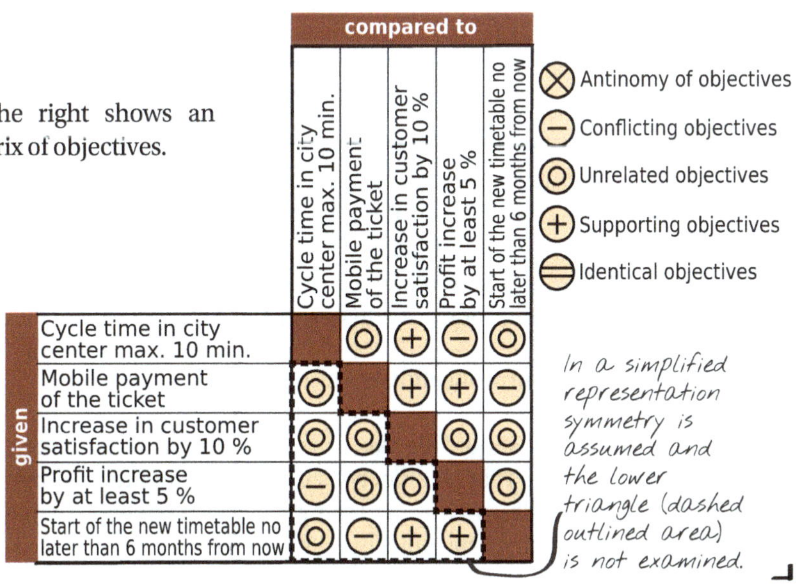

Task:	**D.3.3 Prioritize objectives** — plan-based | as well as | agile
Project lifecycle:	Initialization | **Definition** | Planning | Control | Closing
Team size:	
Team distribution:	
Precondition:	The project objectives must be known .
Subsequent tasks:	D.4 Analyze requirements
Summary:	At the beginning of the project, the objectives should be defined and checked for completeness and consistency. In order to reduce potential conflicts of objectives, the objectives should also be prioritized. The prioritization of objectives determines which objective should be addressed primarily.

Aim:
The aim is to prioritize the project objectives. In the event of a conflict, i.e. if two competing objectives cannot be achieved at the same time, prioritization determines which objective to be followed with priority.

Instruction:
The prioritization of objectives is common in plan-based as well as agile and hybrid projects. Prioritization must be carried out in consultation with the project owner. For example, it can be determined whether the schedule, the cost target or a specific performance objective is to be met in the event of a conflict.
Prioritization can be carried out as follows:

- Determination of a ranking from the most important (priority 1) to the least important objective (priority N for N objectives).
- Prioritization in categories, for example in **mu**st have, **s**hould have, **co**uld have and **w**on't have goals. This categorization is also called **MuSCoW** prioritization based on the initial letters.
- Weighted prioritization in percent, e.g. schedule objective 40 percent, cost objective 20 percent, performance objective 40 percent.

Of course, other types of prioritization are also possible.

When categorizing, care should be taken to ensure that a real distinction is made. In practice, it often happens that all objectives to be implemented are rated as must-haves. In this case, it is not possible to assess the importance of the objectives within the project and therefore no control over the objectives is achieved.

Task:	**D.3.4 Create balanced scorecard**	plan-based | as well as | agile
Project lifecycle:	Initialization | **Definition** | Planning | Control | Closing	
Team size:		**Team distribution:**
Precondition:	The project objectives must be known.	
Subsequent tasks:	D.4 Analyze requirements	
Summary:	At the beginning of the project, the objectives should be defined, checked for completeness and consistency and prioritized. A balanced scorecard can help to make objectives measurable and use the metrics to manage the project.	

Aim:
The aim is to ensure balanced and measurable project objectives. To this end, cause-and-effect chains are considered and analyzed from different perspectives.

Instruction:
The first step is to identify the perspectives relevant to the project. The perspectives of finance, customers, processes and employees are often used as standard. However, these can be replaced or supplemented. Objectives to be achieved are then set for each perspective and measures and target values are defined to monitor the achievement of objectives. Based on this, concrete initiatives are then planned to achieve the objectives.

Structure:
The following figure outlines an example of a balanced scorecard. The perspectives, objectives and measures must be selected in a balanced manner. Care must be taken to ensure that they represent the project in its entirety.

perspective	objectives	measures	target values	initiatives
financial	increasing the margin of the product	production cost	< 500 US$	use of standard components
...	...	...	...	...
customer	increase in customer satisfaction	net promoter score	increase of > 20 percent	intensification of quality planning and assurance; redesign of customer center
...	...	...	...	...
processes	...	...	...	...
...	...	...	...	...
employees	...	...	...	...
...	...	...	...	...

D.4 Analyze requirements

Task:	**D.4.1 Apply structured problem solving**	plan-based as well as agile
Project lifecycle:	Initialization / Definition / Planning / Control / Closing	
Team size:		**Team distribution:**
Precondition:	D.3 Analyze objectives	
Subsequent tasks:	D.4 Continue analysis of requirements	
Summary:	The structured problem-solving cycle helps to understand the problem to be solved by the project, analyze it, develop alternative solutions and select the best solution. It counteracts the temptation to implement the first idea that comes along.	

Aim:
The aim is a structured search for a solution that fits the project and the problem at hand.

Instruction:
The problem-solving cycle is outlined in the figure on the right. It can be roughly divided into the search for a target, the search for a solution and the selection of the best solution. Project managers should ensure that the development of a concept for the project deliverables and the collection of requirements are initially focused on the problem. The solution space should not be restricted too early. Only when the problem to be solved has been analyzed is it possible to define appropriate solutions. Various alternative solutions are then developed and analyzed. The most suitable solution is selected and the decision is made and documented in a comprehensible manner in order to reduce the risk of the solution being called into question later.

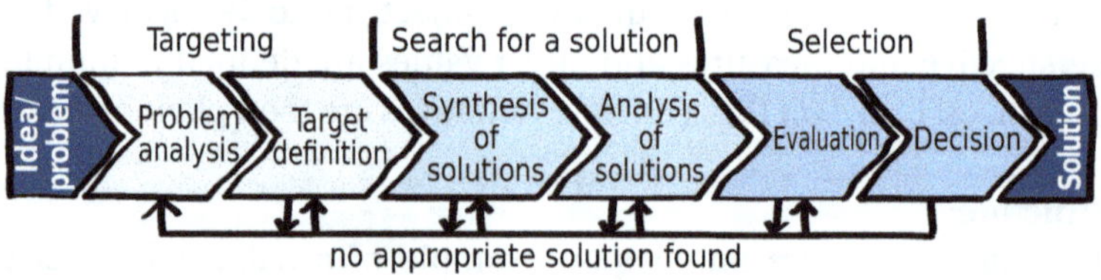

Project managers can and should be familiar with the problem-solving cycle and apply it whenever there is a problem to be solved or when there is a risk of hastily committing to a solution without having understood the problem at hand. The problem-solving cycle originates from systems engineering. It can be used in plan-based, agile and hybrid projects.

Task:	D.4.2 Analyze customer requirements specification	plan-based as well as agile
Project lifecycle:	Initialization / **Definition** / Planning / Control / Closing	
Team size:		Team distribution:
Precondition:	I.3.2 Prepare customer requirements specification	
Subsequent tasks:	D.4 Continue analysis of requirements	
Summary:	Depending on the course of the project, the client may have handed over the customer requirements specification to the project during the initialization phase. This must now be analyzed and turned into the → D.4.5 requirements specification. If there is no customer requirements specification yet, this must now be created or requested from the client.	

Aim:
The aim is to analyze the client's requirements documented in the customer requirements specification with regard to comprehensibility, completeness, consistency, feasibility and traceability.

Instruction:
If the client's customer requirements are not yet available, these must be requested or drawn up by now at the latest.
If the requirements are available (usually in the form of a customer requirements specification document, see also → I.3.2 prepare customer requirements specification), they are now checked by experts for comprehensibility, completeness and consistency (for example with the help of a matrix of objectives adapted to the requirements). It should also be checked whether the requirements can be implemented. For projects with external clients or if this is required by regulations, precautions must be taken to ensure that the requirements can be traced at a later date. For this purpose, the requirements must be marked with a uniform and clear coding. If there are a large number of requirements, these should be transferred to a software application in which the requirements and the specifications, design drafts, test protocols etc. derived from them can be archived in a traceable manner.
To check completeness, the structure presented in → I.3.2 prepare customer requirements specification can be used as a reference and supplemented if necessary. If requirements are missing, these must be requested or developed together with the client. The customer requirement specification document is usually used in plan-based projects. It can also be used in hybrid projects. This is particularly useful if there are interfaces to traditional clients or suppliers.

Task:	**D.4.3 Use house of quality** — plan-based as well as agile
Project lifecycle:	Initialization / **Definition** / Planning / Control / Closing
Team size:	**Team distribution:**
Precondition:	I.3 Determine user needs
Subsequent tasks:	D.4.5 Create requirements specification P.1 Work planning
Summary:	The house of quality is used to synchronize customer requirements and their technical implementation.

Aim:
The aim is to derive and evaluate (technical) specifications from customer requirements.

Instruction and structure:
The approach and structure of the house of quality is explained using the following example. The basic scheme is shown on the left and a specific example on the right. The numbers in the diagram indicate the order of processing.

1. List of collected customer requirements
2. Comparison of customer requirements with products from competitors
3. Listing of the relevant design requirements and technical specifications
4. Measurable target value of the design requirements
5. Investigation of the interdependence of design requirements
6. Comparison of design requirements with competitor products
7. Comparison of the design requirements with customer requirements (high score = good match)
8. Column-by-column sum of the relationship matrix (high score = design requirement that contributes significantly to fulfilling customer requirements) and final prioritization of the design requirements based on the score

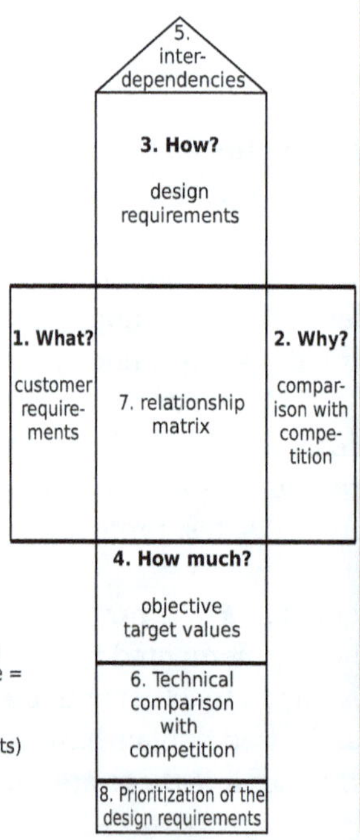

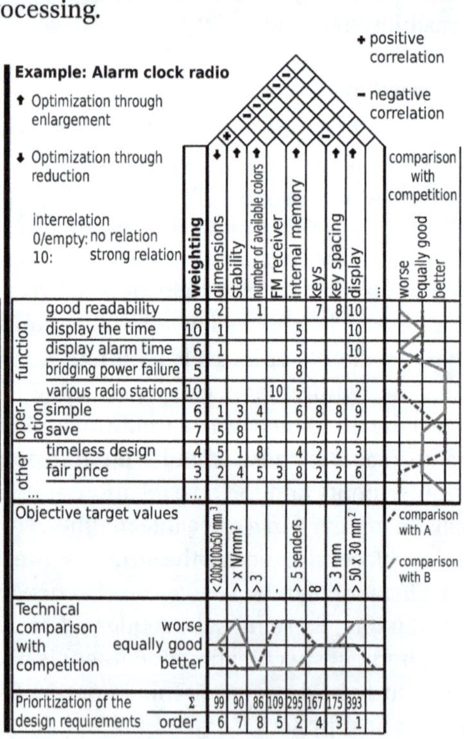

Task:	**D.4.4 Examine use case**	plan-based as well as agile
Project lifecycle:	Initialization / Definition / Planning / Control / Closing	
Team size:		**Team distribution:**
Precondition:	I.3 Determine user needs	
Subsequent tasks:	D.4.5 Create requirements specification P.3.5 Work planning	
Summary:	Use cases model and describe the behavior of a specific role or system when interacting with other participating roles or (sub)systems to achieve a specific target. Depending on the level of detail of the model, use cases help to concretize the requirements of the requirements specification or to derive (technical) specifications for the functional specification.	

Aim:
The aim is to describe and model the behavior of users and system components to achieve a specific target.

Instruction:
Together with the application engineer, the requirements engineer collects all relevant use cases and models and describes them, for example using a suitable modeling language such as UML 2.x.

Structure:
Below is an example of the use case "Pick-up pre-ordered ticket" for a ticket vending machine at a theater box office as a diagram (left) and in text form (right):

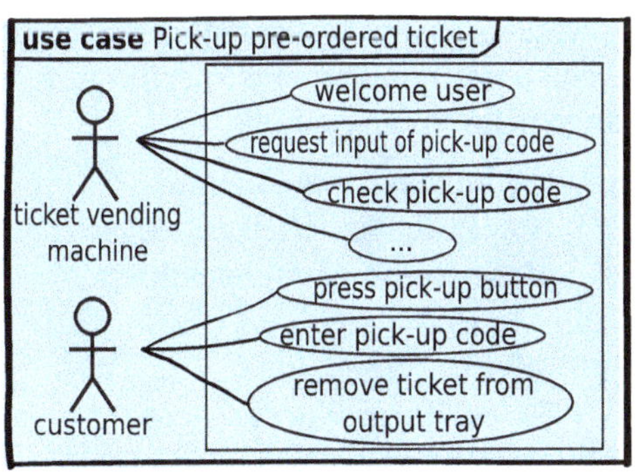

use case:	pick-up pre-ordered ticket
description:	▪ customer presses "pick-up ticket" ▪ system asks for pick-up code ▪ customer enters pick-up code ▪ system checks pick-up code ▪ if valid, ticket is printed ▪ if invalid, customer is informed ▪ ...
exception:	▪ previous pick-up not yet completed
precondition:	▪ pick-up only possible from 30 min. to 120 min. before event starts
postcondition:	▪ process completed and welcome screen displayed

In addition to use cases, there are other tools for detailing requirements, such as structure, sequence, activity and class diagrams.

Task:	**D.4.5 Create requirements specification**		plan-based as well as agile		
Project lifecycle:	Initialization	Definition	Planning	Control	Closing
Team size:			**Team distribution:**		
Precondition:	D.4.2 Analyze customer requirements specification				
Subsequent tasks:	D.4.6 Ensure traceability D.4.7 Prioritize requirements				
Summary:	In the requirements specification, the contractor documents how the client's requirements are to be implemented. This provides the solution path with precise specifications. Once the specifications have been approved by the client, the planning of the implementation can start.				

Aim:
The aim is to create specifications that document how the customer's requirements documented in the customer requirements specification are to be implemented.

Instruction:
Like the customer requirements specification, the requirements specification should be complete. For each requirement in the customer requirements specification, there is at least one specification documenting how the requirement is to be implemented. The specifications must be formulated in a measurable way so that their fulfillment can be checked later. Purely agile projects rarely work with a requirements specification. Instead, the requirements to be implemented are transferred directly into more detailed user stories or tasks derived from them.

Structure:
A standardized structure of the requirements specification helps to ensure that nothing is forgotten. Due to the correspondence with the customer requirements specification, a similar structure is usually chosen, which is outlined below as an example.

1. Introduction
   - How does implementation work?
   - Description of actual and target status
2. Objectives of the project
3. Product description
   - Description of the product structure and relevant components
4. Functional requirements
   - How are the functional requirements of the customer implemented?
5. Non-functional requirements
   - How are the non-functional requirements of the customer implemented?
6. Interfaces
7. Testing and acceptance
   - How is the product delivered?
8. Documentation
9. Boundary conditions
   - Laws, standards, etc. to comply with
10. Miscellaneous

Task:	**D.4.6 Ensure traceability**	plan-based | as well as | agile
Project lifecycle:	Initialization | **Definition** | Planning | Control | Closing	
Team size:		**Team distribution:**
Precondition:	I.3.6 Prepare traceability D.4.2 Analyze customer requirements specification D.4.5 Create requirements specification	
Subsequent tasks:	D.4.7 Prioritize requirements	
Summary:	The traceability of requirements from customer needs, design drafts, development results, test plans and protocols through to the transfer to production helps to check whether everything has been implemented completely and correctly. In the event of changes, the effects on requirements and results already achieved can be quickly assessed and evaluated.	

Aim:

The aim is to establish traceability from the customer requirement to the finished product or project result. Traceability helps to determine the scope of changes, for example. If, for instance, the test of a function fails during the course of the project, the traceability can be used to determine which design drafts, specifications and customer requirements are affected and may need to be revised.

Instruction:

In → I.3.6 prepare traceability, the foundations were laid for the traceability that is now to be established. In small projects, traceability can be achieved using correlation tables or hierarchical tree structures, for example. These link customer requirements and functional specifications as well as, for example, design drafts and test protocols.

Even with a moderately complex project, a large number of customer requirements, specifications, design drafts etc. need to be linked together. Linking these by hand is tedious and error-prone. Special software applications offer solutions for this and allow requirements to be entered, managed and linked to specifications, design documents, etc.

Traceability is particularly relevant for plan-based projects that are carried out according to the V-model. However, agile projects that are subject to certain regulatory frameworks, such as the development of medical devices, must also ensure traceability.

In general, ensuring traceability is time-consuming, but it helps to provide complete proof of compliance with requirements and to manage changes to complex products.

Task:	D.4.7 Prioritize requirements		plan-based as well as agile
Project lifecycle:	Initialization | Definition | Planning | Control | Closing		
Team size:		Team distribution:	
Precondition:	The requirements must be known		
Subsequent tasks:	D.5 Define phases and milestones		
Summary:	In the course of the project, conflicts may arise with regard to content or resources. Finding a solution may mean that not all requirements can be implemented. In order to decide which requirements are particularly important, they must be prioritized.		

Aim:
The aim is to prioritize the requirements. This helps with competing requirements that cannot be implemented at the same time or in combination. Reasons for incompatible requirements can be, for example, conflicting demands in terms of content, resource bottlenecks or time constraints.

Instruction:
The prioritization of requirements is common in traditional, agile and hybrid projects. Prioritization must be carried out in consultation with the client. Prioritization can be carried out using the methods known from → D.3.3 prioritize objectives.

When categorizing, care should be taken to ensure that a real distinction is possible at the end. In fact, it often happens that all requirements to be implemented in the project are rated as must-haves. In the event of a conflict, it is then not possible to decide which requirements must be implemented as a priority.

In practice, it is difficult to prioritize customer and technical, infrastructure or architecture-related requirements. The client will usually prioritize visible and usable functionality higher than a good underlying system architecture. Although this is important for long-term maintainability, it cannot usually be advertised directly as a product feature. One solution may be to keep two catalogs of requirements and prioritize them separately. One catalog contains requirements that delight the customer and the other catalog contains technical or infrastructure and architecture-related requirements. In each project, a specific budget is then set aside for these technical requirements in order to develop a successful product in the long term.

Task:	D.4.8 Prepare product backlog	plan-based / as well as / **agile**
Project lifecycle:	Initialization / **Definition** / Planning / Control / Closing	
Team size:	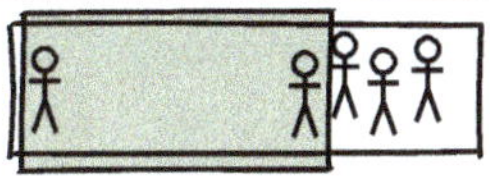	Team distribution: 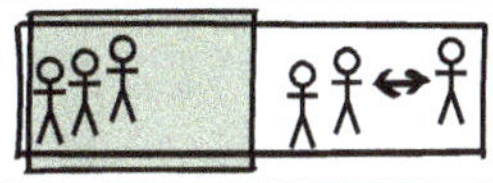
Precondition:	D.3 Analyze objectives	
Subsequent tasks:	D.5.3 Prepare release plan P.1 Work planning	
Summary:	If not already done, the needs and requirements of the users must be collected, estimated and prioritized by now at the latest. In agile projects, this often takes the form of user stories, which are collected in a product backlog.	

Aim:
The product backlog serves as the basis for further planning in Scrum projects. In the de⊠nition phase, the requirements must be collected, estimated and prioritized.

Instruction:
The product owner collects the → I.3.3 user stories and has the team estimate their size, for example using the → P.2.7 planning poker method. The product owner prioritizes all user stories and assigns them a unique identifier. An example of a product backlog is shown below.

	A	B	C	D	E	F
1	**identifier**	**user story**	**size [story points]**	**priority**	**comments**	
2	A5	As an operator, I want to ...	100	100		
3	M3	As a manager, I must be able to ...	20	115		
4	H2	As manufacturer, I expect that ...	8	150	see also Doc. D13	
5	S1	As service technician, I want to ...	13	200		
6	...	...	...	...		

Prioritization can be done in order, by points or in categories (must have, should have, could have, won't have). The product backlog does not yet provide any information about when, by whom and how exactly the user stories are to be implemented. In contrast to the requirements and functional specifications, which must be complete at the start of the project in order to enable the traditional plan-based approach, the product backlog does not claim to be complete. It can contain epics (see → I.3.3 collect user stories) and can be supplemented and corrected during the course of the project.

Task:	D.4.9 Conduct design thinking macrocycle	plan-based / as well as / agile
Project lifecycle:	Initialization / **Definition** / Planning / Control / Closing	
Team size:		Team distribution:
Precondition:	D.3 Analyze objectives	
Subsequent tasks:	D.4 Continue analysis of requirements	
Summary:	Design thinking is a process model for solving problems and for finding and evaluating ideas. Even though design thinking is often associated with agile project management, it can also be used in plan-based process models.	

Aim:
The aim is to solve problems creatively using a structured process.

Instruction:
Design Thinking is divided into a macrocycle and a →D.4.10 microcycle. It is assumed that an interdisciplinary team can come up with good ideas in a creative environment. To develop new solutions beyond established thought patterns, the team is deliberately encouraged to think through even unrealistic ideas and not to consider feasibility too early.

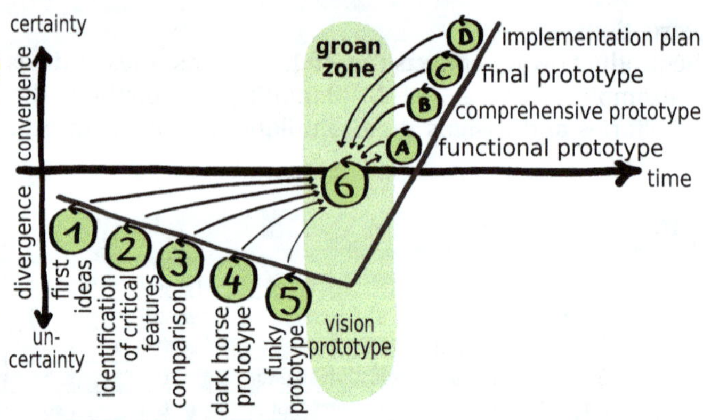

Based on initial ideas (step (1) in the figure above), critical functionalities are first identified (2) and comparisons with established solutions are drawn up (3). A dark horse prototype is then designed in a divergent process (4). This should be surprising and creative. The funky prototype (5) should be even more innovative. It is not subject to any feasibility or profitability limits. After each of the steps (1) to (5), but at the latest after step (5), the team enters the groan zone, i.e. the transition from divergence to convergence. This is where the so-called vision prototype (6) is created, taking into account previous experiences. The prototype should be so concrete that it can be evaluated by potential users. Depending on the evaluation results, a step is taken back to (1) to (5) or convergence is achieved via steps (A) to (D), i.e. via increasingly realistic prototypes. For the final prototype (C), an implementation plan is then drawn up for series production (D).

Task:	D.4.10 Conduct design thinking microcycle	plan-based / as well as / agile
Project lifecycle:	Initialization / **Definition** / Planning / Control / Closing	
Team size:		Team distribution:
Precondition:	D.3 Analyze objectives	
Subsequent tasks:	D.5 Define phases and milestones	
Summary:	Design thinking is a process model for solving problems and for finding and evaluating ideas. In several steps of the so-called → D.4.9 design thinking macrocycle, creative solutions are collected in a divergent process and consolidated in a converging process. Each step goes through a so-called microcycle. This consists of the sub-steps "understand", "observe", "define point of view", "ideate", "prototype" and "test".	

Aim:
The aim is to solve problems creatively using a structured process.

Instruction:
The microcycle consists of six sub-steps, which are run through one after the other, but which also allow steps backwards as shown in the adjacent figure. The basis for creative solutions is that the problem to be solved and its causes are fully understood.

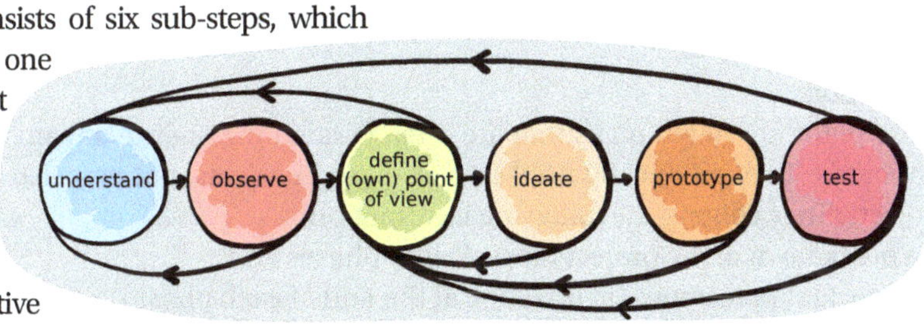

This also includes the identification of relevant stakeholder groups, e.g. customers. The next step is to observe those involved in solving the problem. The problem description should be supplemented by their own impressions. This helps to define your own point of view and use it for further work. The insights gained are collected, analyzed and prioritized. The next step is to find different ideas for solving the problem. Good ideas are developed into prototypes. Prototypes can be, for example, simple handwritten drawings, photorealistic representations, videos and graphic or physical 3D models. The aim is to make the ideas as vivid as possible. This facilitates the final step of the microcycle, i.e. testing. Through regular testing, frequent and meaningful feedback can be collected and applied. Each sub-step allows you to go back, for example if an idea proves to be unsuccessful.

D.5 Define phases and milestones

Task:	D.5.1 Create phase plan	plan-based as well as agile
Project lifecycle:	Initialization | **Definition** | Planning | Control | Closing	
Team size:		Team distribution:
Precondition:	D.3 Analyze objectives D.4 Analyze requirements	
Subsequent tasks:	P.1 Work planning	
Summary:	Phase plans are rough project plans that structure the project into phases and for each phase roughly define the tasks to be completed, resources required, costs and deadlines. The phases are usually divided up using a process model. The phases of the process model are then used as the starting point for phase planning.	

Aim:
The aim is to draw up an initial rough plan of what is to be included into the project, deadlines and costs. It allows the plausibility of the project objectives to be checked at the end of the definition phase.

Instruction:
Phase plans are rough plans that are often created together in a team on whiteboards or pinboards. They are mostly used in plan-based projects, but can also be used in hybrid projects. They contain the following information:

- a breakdown of the project lifecycle into phases
- important milestones (at least one at the end of each phase)
- a rough list of tasks to be completed per phase
- an estimate of the resources, deadlines and costs required for these tasks
- the total costs and total duration of the project

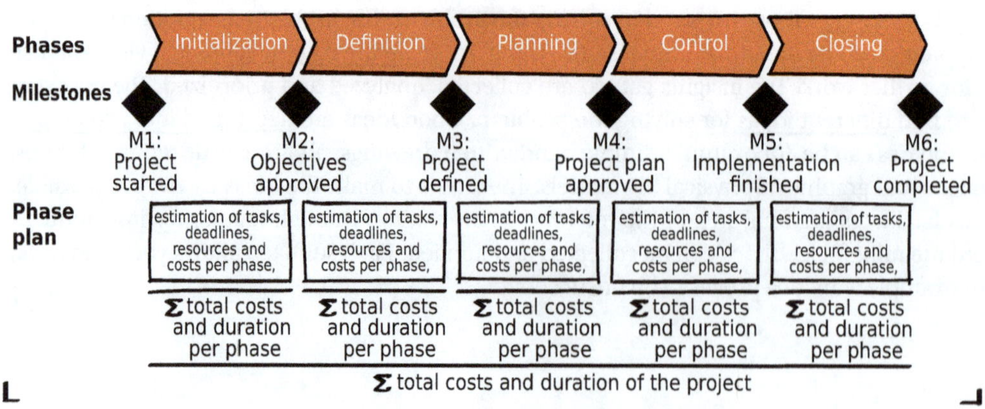

Task:	D.5.2 Create milestone plan	plan-based as well as agile
Project lifecycle:	Initialization / **Definition** / Planning / Control / Closing	
Team size:		Team distribution:
Precondition:	D.3 Analyze objectives D.4 Analyze requirements	
Subsequent tasks:	P.1 Work planning	
Summary:	Milestone plans are rough project plans that link important project dates to specific events (= milestones). They are often used to provide an overview of the roughly planned course of the project.	

Aim:
The aim is to create an initial rough plan of all particularly important dates in the project.

Instruction:
Important milestones of the project are collected and sorted chronologically. Typical milestones are

- project started
- concept created
- planning completed
- implementation completed
- result/product released
- project completed

In order to immediately recognize a milestone as such, i.e. as an event that has occurred, milestones are usually formulated using the noun with a following past participle.

Structure:
Milestone plans are created by simply lining up the milestones chronologically with the planned dates. Due to their simple structure, milestone plans are particularly popular with clients and managers.
Milestones can also be used for controlling projects with the help of the ➔ S.3.3 milestone trend analysis.
One disadvantage of pure milestone plans is that the tasks to be carried out between the milestones, including the resources required and the costs incurred, are not listed and therefore not taken into account. There is therefore a risk of naming purely desired deadlines that cannot be met on closer inspection.

Task:	**D.5.3 Create release plan** — plan-based as well as agile
Project lifecycle:	Initialization · Definition · Planning · Control · Closing
Team size:	**Team distribution:**
Precondition:	D.4.8 Prepare product backlog
Subsequent tasks:	P.1 Work planning
Summary:	The release plan is an overview plan in Scrum that shows which user stories are to be implemented in which sprint. As the product backlog usually still contains epics at the beginning and is updated and supplemented as the project progresses the release plan must also be updated regularly.

Aim:
The aim of the release plan is to roughly plan which user stories are to be implemented in which sprint of a Scrum project. In addition, it can be specified after which sprints purely internal product increments are planned for evaluation and after which sprints official product increments are planned for release.

Instruction:
The following must be taken into account when drawing up the release plan:
- the priority of the user stories,
- their estimated size,
- the velocity of the team and
- the strategic and economic conditions of the product

The product owner needs an idea of how many sprints it will take to implement all relevant user stories. Otherwise, there is a risk that development will take place without taking economic aspects into account. Every time the product backlog is updated, the release plan must also be updated.

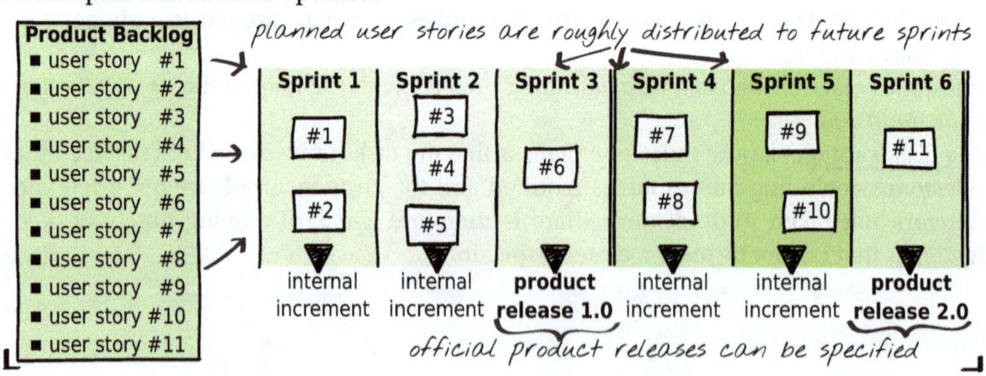

P Planning

P.1 Work planning

Task:	**P.1.1 Create work breakdown structure**	plan-based as well as agile
Project lifecycle:	Initialization | Definition | **Planning** | Control | Closing	
Team size:		**Team distribution:**
Precondition:	D Define project	
Subsequent tasks:	P.2 Estimation P.3 Scheduling	
Summary:	The work breakdown structure defines **what** needs to be done in the project. It contains **all the tasks** of the project. This claim to completeness means that it forms the basis for all other plans (schedule, resources, costs).	

Aim:
The work breakdown structure defines what needs to be done in the project and is intended to be complete. This means that it contains all upcoming tasks of the project including all tasks for project management and project completion.

Instruction:
The work breakdown structure structures the project into (sub)tasks and work packages. A **task** is a piece of work under defined conditions and using certain resources. Depending on the scope and structure, a task can be an entire program, a project, a sub-project, a work package or an activity or an even smaller unit.
A **work package** is a self-contained task that is not further subdivided in the work breakdown structure. The sum of all work packages in the work breakdown structure is everything that needs to be completed in the project. A work package should always be the responsibility of just one person.
A work breakdown structure can be displayed as a hierarchical diagram (left) or in tabular form (right):

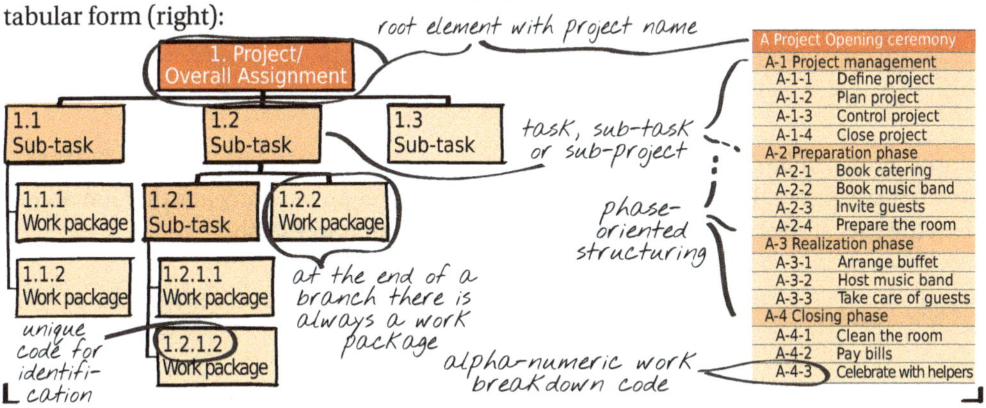

Task:	**P.1.2 Update product backlog**	plan-based as well as agile
Project lifecycle:	Initialization | Definition | Planning | Control | Closing	
Team size:		**Team distribution:**
Precondition:	D Define project	
Subsequent tasks:	P.2 Estimation	
Summary:	The product backlog must be regularly updated. In contrast to plan-based projects, in which the project is planned through to the end on the basis of initially defined requirements, agile projects are implemented iteratively with regularly updated requirements.	

Aim:

In Scrum projects, the product backlog serves as the basis for the detailed planning of the individual sprints. The user stories and epics documented in the product backlog can be changed or made more specific as a result of new findings gathered during project work and sprint reviews. If user stories of a sprint cannot be completed as planned, they must be reprioritized and estimated. These changes are incorporated into the product backlog and thus into the planning for the next sprint.

Instruction and structure:

The product owner collects changes to the user stories, concretizations of epics and changing prioritizations of user stories. He or she initiates the update of the size estimate and provides an updated product backlog. The update must be made in good time before the next sprint planning. The structure of the product backlog is shown in the following diagram.

	A	B	C	D	E	F
1	ID	User Story	Size [Story Points]	Priority	Comments	
2	U5	As a user I would like ...	100	100	-	
3	M3	As a manager I would like ...	20	115	-	
4	P2	As a manufacturer I would ...	8	150	see also Doc. D13	
5	S1	As a service technicians I ...	13	200	-	
...	...	...	...	...	...	

Prioritization can be done in the form of ranking, by scoring or in categories according to the MuSCoW principle. The product backlog does not yet provide any information about when the user stories will be implemented, by whom and how exactly.

Task:	**P.1.3 Informal task planning**	plan-based as well as agile
Project lifecycle:	Initialization | Definition | **Planning** | Control | Closing	
Team size:		**Team distribution:**
Precondition:	D Define project	
Subsequent tasks:	P.2 Estimation	
Summary:	In small projects, upcoming tasks are often collected informally and documented in a spreadsheet or on a simple Kanban board, for example.	

Aim:
The aim is the informal and simple planning of upcoming tasks in small projects.

Instruction:
In small projects, formal work breakdown structures or a product backlog can seem too bureaucratic. Instead, upcoming tasks are collected in spreadsheets or on the cards of a Kanban board.

Despite the informal approach, project managers should ensure that all upcoming tasks are identified, consistent and assigned to a responsible person. An effort estimate should also be made in order to be able to roughly plan the further course of the project. Otherwise, there is a risk that informal task planning will lead to uncoordinated or even chaotic processes and the project cannot be carried out successfully.

Tasks can be collected as part of a team brainstorming session. For recurring similar projects, checklists can help to ensure that no important tasks are forgotten. The checklists are reviewed at the end of the project and any new tasks that have not yet been documented are added to the checklist.

Structure:
The tasks can be collected and documented in a spreadsheet, on cards or using other media. The minimum content is a meaningful name for the task and a responsible person. More complex tasks can be described in detail. Milestones with dates and the estimated time required for the task help with later progress monitoring.

Task:	P.1.4 Plan tests	plan-based as well as agile
Project lifecycle:	Initialization / Definition / **Planning** / Control / Closing	
Team size:		Team distribution:
Precondition:	D Define project	
Subsequent tasks:	P.2 Estimation	
Summary:	In many projects, the project object must be tested before delivery. Defining a test strategy and planning the operational test implementation helps to ensure efficient execution of the tests.	

Aim:

The aim is to define a test strategy and plan its operational implementation.

Instruction:

One of the fundamental questions that the test planning should answer is whether testing should take place once at the end of the project ...

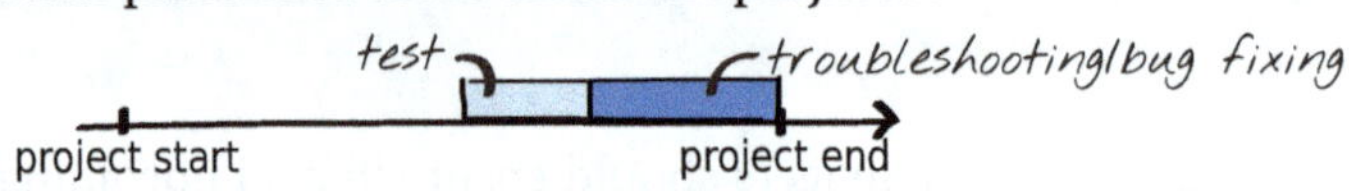

or iteratively during the course of the project.

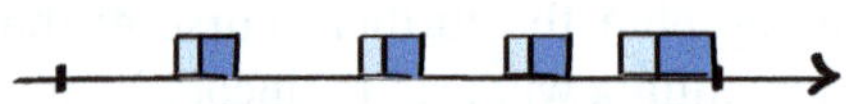

Careful consideration helps to assess which option will incur less expense. Typical questions and arguments when weighing up the options are:

- Is iterative testing at all possible as long as the project deliverables are not yet complete and no overall system is available?
- Can the defects identified during the iterative tests be processed and resolved at an early stage?

A comparison of one-time test execution at the end of the project and iterative test execution over the entire project duration shows which variant is better:

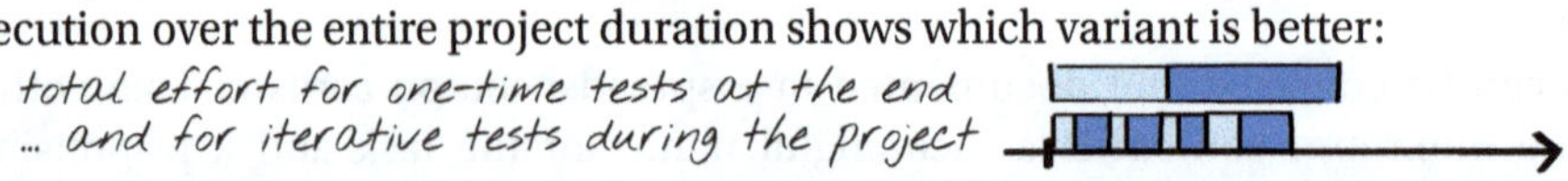

In many, but not all projects, early tests help to reduce the testing effort. Other aspects of test planning are:

- Is it possible to automate tests, including the use of continuous tests overnight and over weekends?
- Are the errors found during the tests prioritized correctly? The team should be prevented from losing time when eliminating errors. Some projects therefore only focus on the 20 highest-priority bugs. Only these may be resolved by the team.

P.2 Estimation

Task:	P.2.1 Perform expert estimation	plan-based as well as agile
Project lifecycle:	Initialization / Definition / **Planning** / Control / Closing	
Team size:		Team distribution:
Precondition:	P.1 Work planning	
Subsequent tasks:	P.3 Scheduling	
Summary:	In an expert survey, one or more experts are asked to estimate the effort of a task. The expert estimation is easy to perform, but the individual estimation is very subjective and prone to error if the wrong experts are chosen.	

Aim:
The aim is to determine the effort involved in a task by consulting one or more experts.

Instruction:
The project manager presents the task to be estimated and asks one (= **single estimate**) or more (= **multiple estimate**) experts to provide an estimate of the effort required. If necessary, the task can be broken down into several subtasks or further experts can be consulted in order to increase the accuracy of the estimate. The assumptions on which the estimate is based should be noted and estimation uncertainties considered in risk management.

The **Delphi method** is a structured form of multiple estimates, the process of which is illustrated on the right. The estimation of several experts is anonymous. The project manager determines the mean value of the estimates submitted and communicates this to the experts. If the experts' estimates differ significantly, a new round of estimates is carried out. In the **broadband Delphi method**, the estimates are discussed (while trying to maintain the anonymity of the estimators). The discussion usually focuses on the on the largest and smallest estimates and the underlying assumptions about the general conditions.

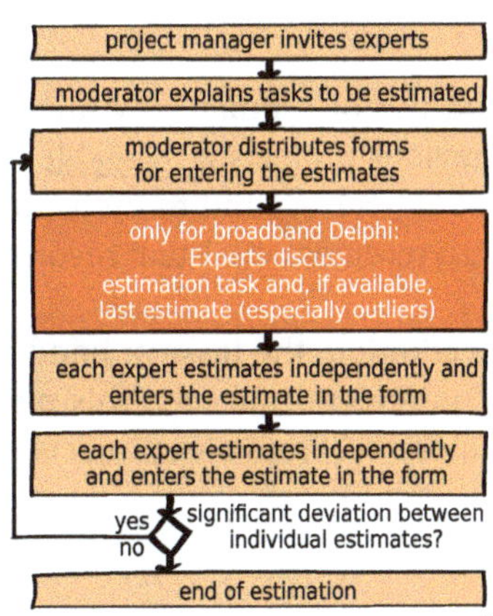

Several experts also estimate in an **estimation workshop**. However, the estimate is made openly and jointly as part of a discussion.

Task:	P.2.2 Estimate by analogy	plan-based as well as agile
Project lifecycle:	Initialization / Definition / **Planning** / Control / Closing	
Team size:		**Team distribution:**
Precondition:	P.1 Work planning	
Subsequent tasks:	P.3 Scheduling	
Summary:	When estimating by analogy, the effort required for upcoming tasks is estimated by comparing them with similar tasks from previous projects.	

Aim:
The aim is to determine the effort of tasks by comparing them with similar tasks of previous projects.

Instruction:
The project manager identifies tasks of the currently planned project that already existed in a similar form in previous projects. The project manager determines the effort required for the tasks in previous projects and applies this to the current tasks.

The effort assumed can be increased or reduced if

- the tasks at that time were somewhat simpler or more complex,
- there are now simpler ways of working (process improvements) or better tools for completing the tasks,
- the employees assigned to the task have a different level of qualification and can probably complete the task faster or slower,
- etc.

Changes to expenses from the past should be well justified. Merely wishing to work faster this time does not justify a reduction in the effort estimation.

The use of estimates by analogy requires that information on the effort of tasks from previous projects is available and easy to find. Where possible, the values from the past should correspond to the actual effort required and not to the estimated effort. This in turn requires consistent recording of actual efforts.

In addition, the tasks at hand and the previous tasks must be truly comparable. If these requirements are met, this method provides very good and objective estimates, as real experience from past projects is used.

Task:	P.2.3 Perform three-point estimation (PERT) — plan-based as well as agile		
Project lifecycle:	Initialization / Definition / **Planning** / Control / Closing		
Team size:		Team distribution:	
Precondition:	P.1 Work planning		
Subsequent tasks:	P.3 Scheduling		
Summary:	The three-point estimation determines an optimistic, a pessimistic and a probable estimate. This allows statements to be made about the uncertainty of the estimate and reduces the risk of non-transparent safety buffers. The method can be combined with almost all other estimation methods. The method is also known as the Program Evaluation and Review Technique (PERT).		

Aim:
The aim is to determine the effort of a task, including its associated uncertainties.

Instruction:
The project manager decides on an effort estimation method. For example the project manager applies the → P.2.1 perform expert estimation. Then not only a single estimate is determined for each task, but instead an optimistic, a pessimistic and a probable estimate. The more these values differ, the more risks are associated with the task and the greater the estimation uncertainty. A useful effect is that employees tend to integrate personal safety buffers into the pessimistic estimate. This gives the project manager a more objective estimate. The difference between the pessimistic and probable value can be used for buffer planning in critical chain project management and the risks behind the difference can be incorporated into risk management. For further planning, the project manager either uses the probable estimate or applies the formula outlined in the figure below to calculate the effort.

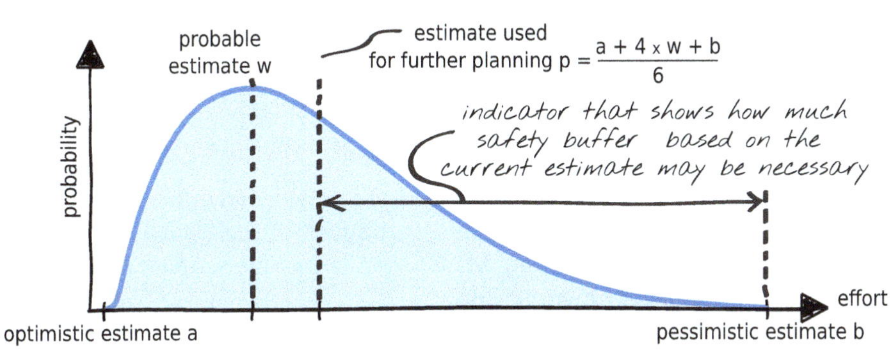

Task:	**P.2.4 Perform parameter-based estimation** (plan-based as well as agile)
Project lifecycle:	Initialization / Definition / **Planning** / Control / Closing
Team size:	**Team distribution:**
Precondition:	P.1 Work planning
Subsequent tasks:	P.3 Scheduling
Summary:	Parametric estimation involves calculating the effort using parameterized equations. The equations are derived from theoretical considerations or models based on experience gained from past projects.

Aim:
The aim is to determine the effort using parameterized equations.

Instruction:
A prerequisite for the application of this method is that appropriate parameterized equations are available. The values of the parameters relevant to the task to be estimated are then determined and the equation is calculated. A simple example illustrates this: It is known from past projects that the effort for the creation of a requirements specification can be calculated from the following parameters:

- On average, 3 staff hours are required to create the document, maintain it and obtain the final approval.
- For the transfer of each user need into the requirements specification 1/2 staff hours are required.

This results in the equation for calculating the effort involved in creating the requirements specification for N associated user needs:

$$[\text{effort}] = 1/2 \text{ staff hours} \times N + 3 \text{ staff hours}$$

COCOMO II (Constructive Cost Model) is well known in software development. The cost in staff months is calculated as follows:

$$[\text{effort}] = m \times \text{KSLOC}^{n}$$

The factor m takes into account the complexity of the project and influencing factors such as productivity. The exponent n represents economies of scale such as development flexibility and the degree of innovation. KSLOC stands for Kilo-Source-Line-of-Code and thus for the expected lines of program code.

Task:	P.2.5 Apply function point method	plan-based as well as agile
Project lifecycle:	Initialization / Definition / **Planning** / Control / Closing	
Team size:		Team distribution:
Precondition:	P.1 Work planning	
Subsequent tasks:	P.3 Scheduling	
Summary:	The function point method originates from software development and is standardized in ISO/IEC 20926. Function points and therefore effort are estimated using component types to be defined, their complexity and influencing factors.	

Aim:

The aim is to determine the effort required to complete tasks.

Instruction:

The procedure is illustrated by steps 1 to 6 in the following figure.

① The task is divided into input, output, user and reference data as well as queries. The degree of complexity (simple, medium, complex) is defined for all types with a value for the weight G = 3 to 15.

② The weights are multiplied by the number A of components to be developed to form the value P

component type	complexity	weight G	amount A	product P=GxA
input data	simple	3	10	30
	medium	4	10	40
	complex	6	7	42
output data	simple	4	20	80
	medium	5	15	75
	complex	7	15	105
user data	simple	7	20	140
	medium	10	25	250
	complex	15	12	180
reference data	simple	5	5	25
	medium	7	5	35
	complex	10	15	150
queries	simple	3	2	6
	medium	4	5	20
	complex	6	8	48
			sum S1:	1,226

③ The sum of the products P gives the value S1

④ Influencing factors are defined and rated with the values 0 (no influence) to 5 (strong influence). The ratings are added up to S2 and converted to S3 using the indicated formula.

influencing factors	rating
Integration with other systems	5
Decentralized processing and data storage	4
Transaction rate and response time behavior	3
Difficulty/complexity of computing	4
Scope of control procedures (data consistency)	1
Number of exceptions	3
Difficulty and complexity of the logic	5
Reusability	5
Conversions of existing databases	2
Ease of use and changeability	4
sum S2:	36
factor of rating the influences S3 = (0.70 + 0.01xS2):	**1.06**

⑤ The function points are calculated by multiplying S1 and S3 and converted into effort using experiences from the past.

S1 x S3 = 1,226 x 1.06 = 1,300

function points	effort in staff months
...	...
1,100	76
1,200	82
1,300	89
1,400	96

⑥ Company-specific experience table from past projects for conversion of function points into effort in staff months

based on Jenny: Project Management. vdf Hochschulverlag. 3rd edition 2014

Task:	**P.2.6 Estimate user story**	plan-based as well as agile
Project lifecycle:	Initialization / Definition / **Planning** / Control / Closing	
Team size:		**Team distribution:**
Precondition:	D.4.8 Prepare Product Backlog	
Subsequent tasks:	P.3 Scheduling	
Summary:	In agile projects, it is usually not the effort of a task that is estimated, but its size. The size reflects the complexity of the task regardless of the resources required to complete it. The size is often specified in the unit story points.	

Aim:
The aim is to estimate the size of upcoming tasks (user stories) in story points.

Instruction:
For the actual estimation, the methods known from plan-based projects can be used. Examples are → P.2.1 perform expert estimation or → P.2.3 estimate by analogy. Alternatively, → P.2.7 planning poker, which is widely used in agile projects, can be applied. In agile projects, it is usually not the effort of the task in staff hours that is estimated, but the size of the tasks. The size of a task depends on its complexity; for example, how does the function to be implemented interact with different components of a system?
The size is independent of whether the requirement is implemented by an experienced or inexperienced developer. This type of estimation decouples the underlying complexity of a requirement from the ability of the developer. This in turn is the prerequisite for the team only being able to decide during the sprint who will take on the task. It also allows the team to measure its own performance development. If more story points were implemented per time unit, this means that the team was able to work more efficiently. This measurability is particularly important for securing experience and optimization as part of the sprint retrospectives in Scrum.
As the complexity and therefore the size of a task is not an absolute value, it is estimated relative to a known task. As part of the size estimation, a sample task is often first defined that can be completed by one person within one day.
A reference value, for example 1 story point, is defined for this. All other tasks are then compared with this exemplary task in terms of their complexity. This means that the size of story points can represent different levels of complexity from team to team. Estimates of the size of different teams are therefore not comparable.

Task:	P.2.7 Play planning poker	plan-based / as well as / agile
Project lifecycle:	Initialization / Definition / Planning / Control / Closing	
Team size:		Team distribution:
Precondition:	D.4.8 Prepare Product Backlog	
Subsequent tasks:	P.3 Scheduling	
Summary:	Planning poker is a method for estimating the size of user stories. The method is similar to the Delphi method. The estimation is carried out using special cards on which the sizes to be estimated are defined.	

Aim:

Planning poker is used to estimate the size of user stories in story points.

Instruction:

The method is usually used in Scrum. The developers meet with the product owner and the Scrum master who moderates the meeting. Each team member receives a set of planning poker cards, see the illustration below. The points noted there represent story points and are based on the Fibonacci sequence. Each user story is presented by the product owner. The developers can ask questions and discuss the story. Each member then places one of the cards face down on the table. The cards are revealed simultaneously. If the estimates differ greatly, the highest and lowest estimates are discussed. A new round of estimations then begins. This is repeated until everyone in the team estimates the same value or the differences between the individual estimates are at least very small. There is a question mark card for ambiguities and a coffee cup card for breaks. A reference value, for example 1 story point, is defined for this. All other tasks are then compared with this exemplary task in terms of their complexity. This means that the size of story points can represent different levels of complexity from team to team. Estimates of the size of different teams are therefore not comparable.

Hint:

Story points indicate the size for the complete completion of a user story (including testing and documentation if required). Story points are a relative size. It can be helpful to explain at the beginning of the planning poker how many story points were completed in the last sprint (= velocity) and then start the estimation round with small user stories. The larger user stories are then estimated relative to these.

P.3 Scheduling

Task:	P.3.1 Create schedule as network diagram	Plan-based as well as agile
Project lifecycle:	Initialization / Definition / **Planning** / Control / Closing	
Team size:		Team distribution:
Precondition:	P.1 Work planning	
Subsequent tasks:	P.4 Resource planning	
Summary:	Network diagrams are used as schedules in medium-sized and large projects. They contain a lot of information, but are more difficult to read than ➔ P.3.2 Gantt charts.	

Aim:
The aim is to create a schedule as a network diagram. The schedule should provide information on when the content planned in the work breakdown structure must be completed. It forms the basis for subsequent resource and cost planning.

Instruction:
The work packages of the ➔ P.1.1 work breakdown structure are first subdivided into tasks and milestones, see figure on the right. A work package can be converted into one or more activities and additional milestones. All processes and milestones are then chronologically organized and combined into a schedule by adding lead times.

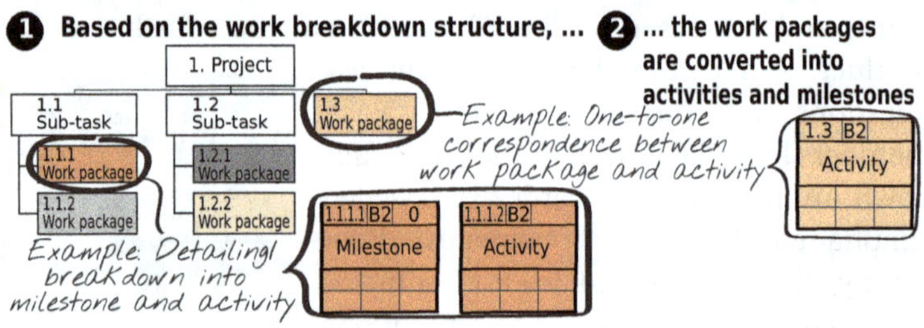

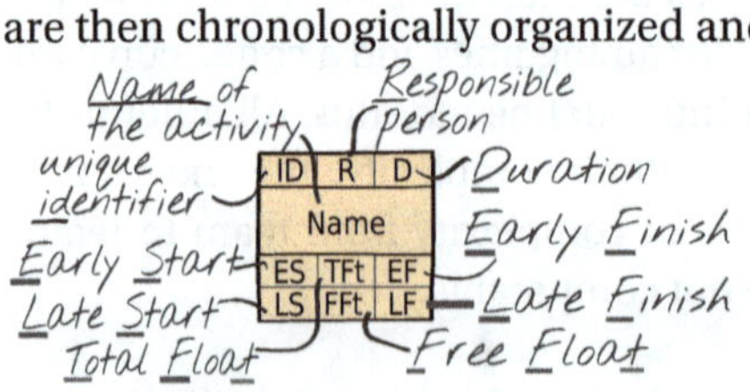

The nodes of the network diagram are structured as shown in the illustration on the left. A simple example of a schedule as a network diagram is shown below. The arrows connecting the nodes are called relationships. Unless otherwise specified, this is always a finish to start relationship. In this case, the respective successor starts immediately after the end of the predecessor node. Other relationships are the start to start relationship (the nodes start at the same time), the finish to finish relationship (the nodes end at the same time), and the start to finish relationship (reversed order of the nodes).

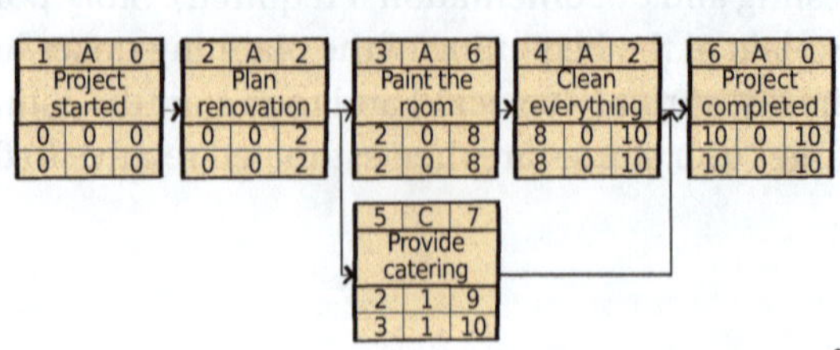

Task:	P.3.2 Create Gantt chart	plan-based as well as agile	
Project lifecycle:	Initialization / Definition / **Planning** / Control / Closing		
Team size:		Team distribution:	
Precondition:	P.1 Work planning		
Subsequent tasks:	P.4 Resource planning		
Summary:	The Gantt chart is probably the best-known project plan of all. It shows the activities to be accomplished in chronological order using bars along a time axis.		

Aim:
The aim is to create a schedule in the form of a Gantt chart. The schedule should provide information on when the content planned in the work breakdown structure must be completed. It forms the basis for subsequent resource and cost planning.

Instruction:
Gantt charts can either be derived from → P.3.1 network diagrams or created directly from the → P.1.1 work breakdown structure. Most project management software can create and display Gantt charts.

Structure:
In Gantt charts, the earliest and latest positions and buffers can be calculated for each activity and milestone and various relationships such as finish to start (FS), start to start (SS), finish to finish (FF), and start to finish (SF) can be modeled. The following figure shows an example of a Gantt chart:

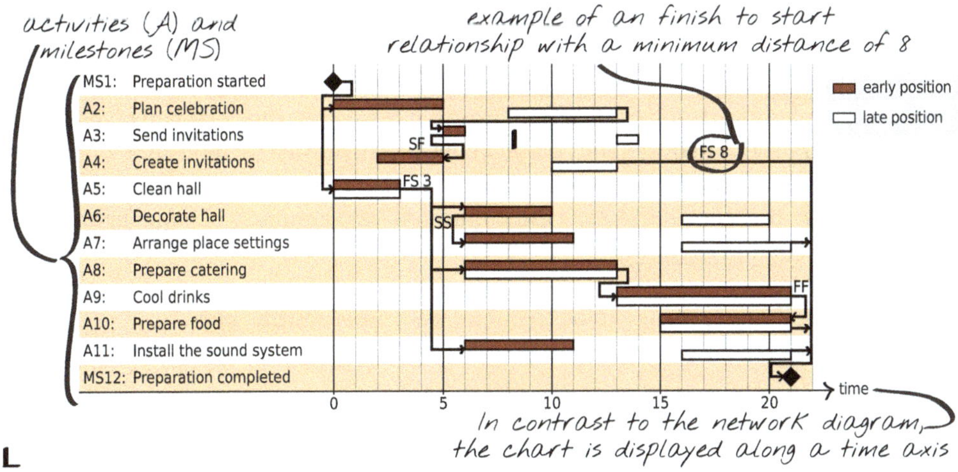

Task:	P.3.3 Create milestone plan	plan-based as well as agile
Project lifecycle:	Initialization / Definition / **Planning** / Control / Closing	
Team size:		Team distribution:
Precondition:	P.1 Work planning	
Subsequent tasks:	P.4 Resource planning	
Summary:	Milestone plans are used in medium-sized and large projects as condensed schedules for the client and the steering committee. In smaller projects, milestone plans can be used as stand-alone schedules. .	

Aim:
The aim is to create a simple schedule or to condense a comprehensive schedule to the most important milestones.

Instruction:
In contrast to the ➔ D.5.2 milestone plan in the definition phase, which served as an rough initial plan, the work of the project has now been planned. The milestone plan to be created now can be used as the sole schedule for small and simple projects. When setting the dates, the work planning and, if available, the effort estimate should be used to date the milestones realistically. The number of milestones defined should be carefully considered. Even too many milestones can be helpful in project management, as the achievement of milestones is an indicator of project progress. Many milestones often provide feedback on this progress and allow early intervention.
On the other hand, too many milestones can also confuse clients and other stakeholders and lead to hectic control measures in the project.

For medium-sized, large and generally complex projects, a milestone plan is generally not sufficient to plan the deadlines in such a way that resources and costs can be planned on this basis and used for subsequent project control. Nevertheless, milestone plans are also justified in these projects: they are used to present the detailed schedules, which are often difficult for the client, the steering committee or members of management to understand, in a condensed form.

The structure of milestone plans can be found in ➔ D.5.2 milestone plan.

Task:	**P.3.4 Update release plan**		plan-based as well as agile
Project lifecycle:	Initialization | Definition | Planning | Control | Closing		
Team size:		**Team distribution:**	
Precondition:	P.1.2 Update product backlog		
Subsequent tasks:	P.3.5 Create sprint backlog		
Summary:	The release plan is an overview plan that shows which user stories are to be implemented in which sprint. As the product backlog usually still contains epics at the beginning and is updated and supplemented as the project progresses, the release plan must also be updated regularly.		

Aim:
The aim of the release plan is the rough planning of which user stories are to be implemented in which sprint. The planning must be updated regularly, especially if not all planned user stories can be completed in a sprint (definition of done).

Instruction:
Based on the current product backlog with prioritized user stories estimated in terms of size, a global target is defined for each sprint. The user stories that are still to be completed are then assigned to the individual sprints. The current velocity of the team and other strategic and economic framework conditions of the product are taken into account.

Structure:
The following figure illustrates the procedure and structure of release planning.

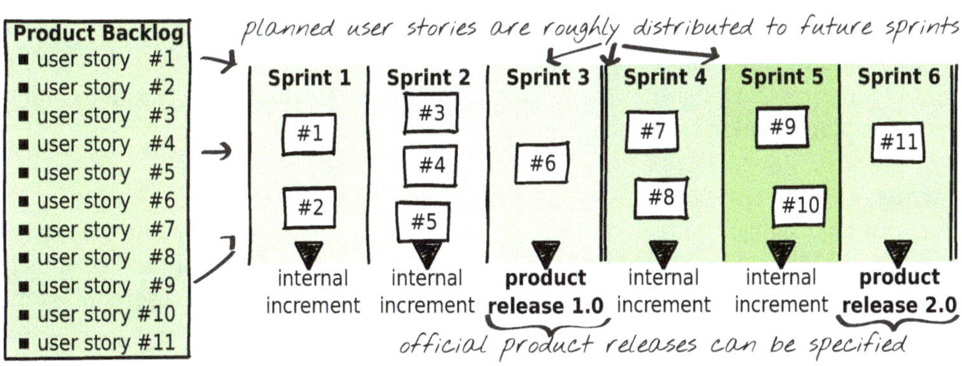

Task:	P.3.5 Create sprint backlog	plan-based as well as agile
Project lifecycle:	Initialization · Definition · **Planning** · Control · Closing	
Team size:		Team distribution:
Precondition:	P.1.2 Update product backlog	
Subsequent tasks:	P.3.6 Prepare taskboard	
Summary:	The sprint backlog contains the user stories to be implemented in the current sprint and the tasks to be completed during implementation. It is created and maintained by the developers.	

Aim:
The aim of creating the sprint backlog is to determine which of the prioritized and estimated user stories collected in the product backlog are to be implemented in the current sprint and how they are to be implemented. For this purpose, the user stories are broken down into tasks.

Instruction:
The sprint backlog is usually created in a sprint planning in two stages, see also the figure on the right. During the first stage, the product owner presents the user stories and works out the target for the next sprint with the developers. The developers then determine the scope of the user stories to be implemented. In the second sprint planning, the user stories are broken down into individual tasks; it is determined how the implementation should be done.

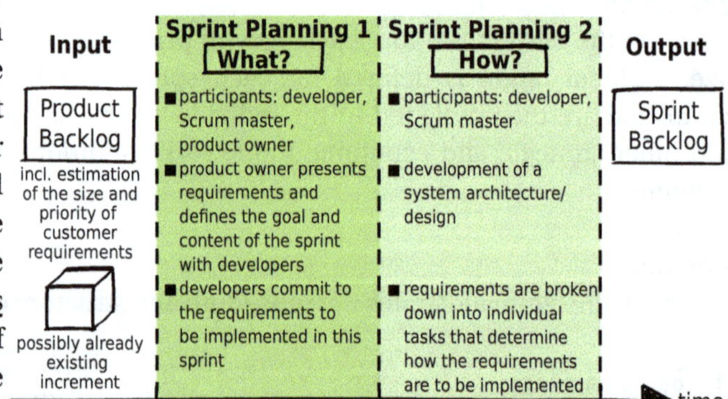

Structure:
The sprint backlog can be maintained as a separate list in a spreadsheet or transferred directly to a → P.3.6 taskboard and tracked there. At the end of the sprint, the status of the sprint backlog is used to update the product backlog. Unfinished user stories are transferred to the product backlog, reprioritized and estimated.

Task:	**P.3.6 Prepare taskboard**	plan-based / as well as / agile	
Project lifecycle:	Initialization / Definition / Planning / Control / Closing		
Team size:		**Team distribution:**	
Precondition:	P.3.5 Create sprint backlog		
Subsequent tasks:	P.3.8 Prepare burndown chart		
Summary:	The taskboard provides a comprehensive overview of the tasks identified in sprint planning. A processing status is assigned to each task. The taskboard is used for transparent status monitoring and project controlling.		

Aim:

The aim is to prepare the task board, which is used during the control phase of the project in order to monitor the project status and control the project. In Scrum projects, the project team meets for the daily Scrum at the task board.

Instruction and Structure:

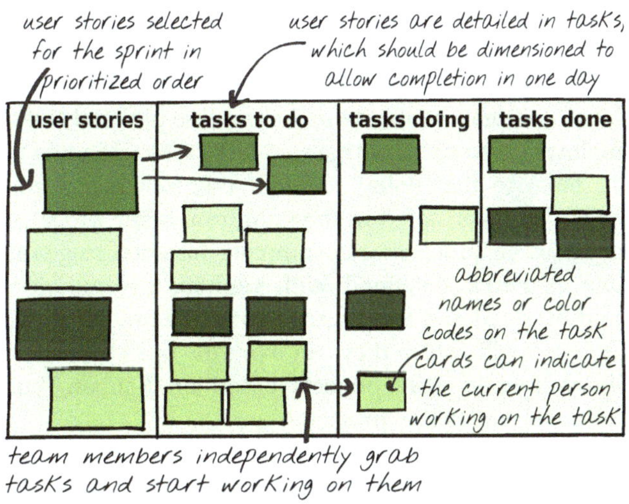

First, relevant statuses for the project are identified as columns on the taskboard, see the figure on the right. In most cases, one state represents the waiting status (to do) before processing begins, one state represents the current implementation (doing) and one state represents the successful completion (done) according to the definition of done.

The user stories selected in sprint planning for the current sprint are placed on the taskboard together with the tasks derived from them. In addition, markers can be prepared that allow the person working on the task to be identified, for example in the form of colored sticky dots. This can be used in the later control phase to indicate who is currently working on a task. Taskboards are widely used in agile projects, but can also be used in traditional and hybrid projects. In this case, work packages and activities are visualized instead of user stories and tasks.

Task:	**P.3.7 Prepare Kanban board**	plan-based as well as agile
Project lifecycle:	Initialization | Definition | Planning | Control | Closing	
Team size:		**Team distribution:**
Precondition:	P.1 Work planning	
Subsequent tasks:	P.4 Resource planning	
Summary:	The Kanban board visualizes the tasks to be worked on in the project along the stages/phases to be completed. Unlike the → P.3.6 taskboard, the flow of work through the project (i.e. through multiple stages is visualized rather than the processing status.	

Aim:

The aim is to prepare the Kanban board for the later monitoring and controlling phase of the project.

Instruction and Structure:

First, the stages relevant to the project are identified, for example specification, implementation, integration, and test. Each stage is assigned a "doing" column and a "done" column, see the figure below. The planned tasks are placed in a "pool" column. Employees who currently have no tasks select pending tasks from the Kanban board and add them to the "doing" column themselves. An additional "released" column can prevent all tasks stored in the pool from being processed at random. It is important to define, when processing is complete for each stage. A stage-dependent "definition of done" is therefore defined. With Kanban, the amount of work in progress is limited to avoid overload and blockages. A limited number of markers can be prepared for each employee, with which they can mark the tasks they have taken on. The Kanban board is mostly used in agile projects based on Kanban, but can also be used in Scrum, traditional, and hybrid projects.

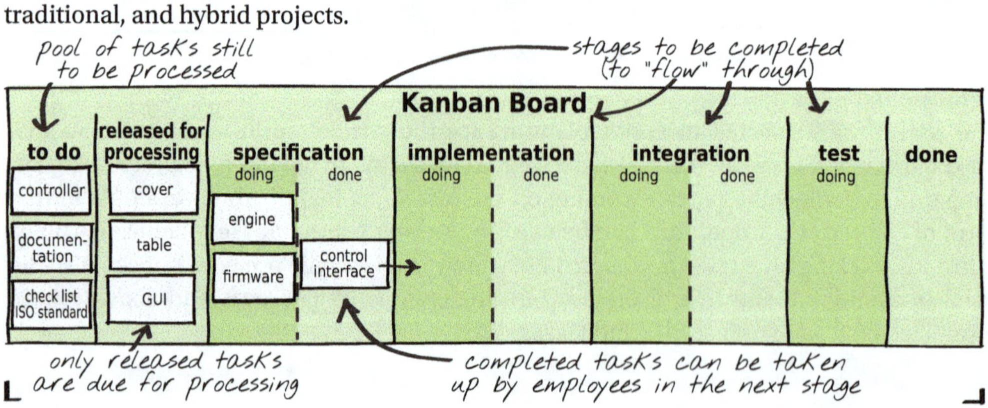

Task:	**P.3.8 Prepare burndown chart**	plan-based as well as agile
Project lifecycle:	Initialization · Definition · Planning · Control · Closing	
Team size:		**Team distribution:**
Precondition:	P.3 Scheduling	
Subsequent tasks:	P.4 Resource planning	
Summary:	The burndown chart is used to determine and visualize the progress of the project. It quantifies open work up to a certain deadline.	

Aim:
The aim is to prepare the burndown chart for the later monitoring and controlling phase of the project.

Instruction:
The prerequisite for creating the burndown chart is that a deadline has been set by which defined tasks are to be completed. In agile projects according to Scrum, the deadline is the upcoming end of a sprint. The (still unfinished) tasks up to this deadline are then quantified. The quantification can take the form of form of

- number of open tasks,
- sum of the story points of the still open user stories, and
- effort of open tasks.

Structure:
The structure of a burndown chart is illustrated in the following figure. Ideally, the number of open tasks decreases linearly over time and is zero by the planned deadline. The measured progression shown in the figure is only determined and entered during the control phase of the project. The project status can be determined by comparing it with the idealized progression.

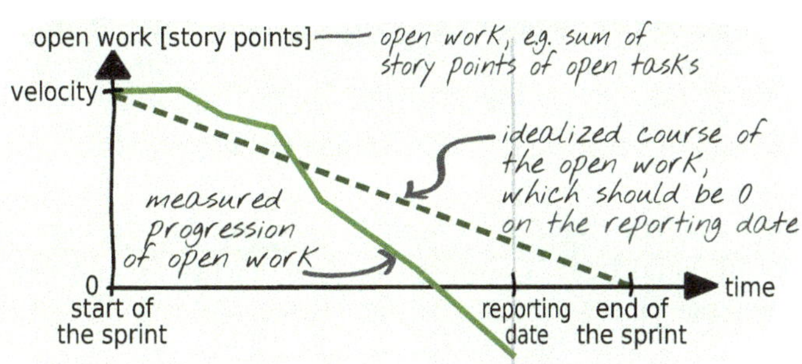

Task:	**P.3.9 Plan (time-based) contingency and management reserve**	plan-based as well as agile
Project lifecycle:	Initialization / Definition / **Planning** / Control / Closing	
Team size:		**Team distribution:**
Precondition:	P.3 Scheduling O.3 Risk management	
Subsequent tasks:	P.4 Resource planning	
Summary:	Reserves (for example time buffers) can protect the project against delays in the event of risks occurring. However, they must not be set arbitrarily, but must be planned and monitored transparently depending on the project risks.	

Aim:
The aim is to plan reserves (e.g. time buffers) to safeguard the progress of the project in the event of risks occurring.

Instruction:
The project manager decides on an effort estimation method. For example the project A distinction is usually made between two different types of buffer:
- The contingency reserve protects the project against known risks.
- The management reserve protects the project against unknown risks.

To determine the **contingency reserve**, the identified risks are analyzed with regard to their potential to delay the project. A time buffer is planned depending on this potential delay. This buffer can be placed with the respective tasks that carry the associated risk. Alternatively, it can be placed at the end of the project or at other neuralgic stages of the project, for example to secure the fulfillment of important milestones in the project.
Under no circumstances, however, should the buffer be distributed randomly across several tasks. According to Parkinson's law (work expands to the extent that there is time to complete it), the buffer is otherwise used without need and unnecessarily delays the project.
The **management reserve** represents a time buffer for delays due to unforeseen risks. As the risks are therefore unknown, the size of the management reserve is based on experience from past projects and risk awareness within the company. The management reserve is usually planned at the end of the project or before critical points in time (e.g. important milestones).
Contingency and management reserve should be monitored and reported separately. If a risk occurs and additional time is required as a result, this requirement is subtracted from the respective reserve.

P.4 Resource planning

Task:	**P.4.1 Create static resource plan**	plan-based as well as agile
Project lifecycle:	Initialization / Definition / **Planning** / Control / Closing	
Team size:		**Team distribution:**
Precondition:	P.3 Scheduling	
Subsequent tasks:	P.5 Cost planning	
Summary:	A static resource plan is a simple list of the resources required in the project.	

Aim:
The aim is to record the resources required in the project in order to ensure their availability. Based on this, the costs incurred for the resources can later be determined.

Instruction:
The tasks planned for working on the project deliverables are usually planned as work packages and activities in plan-based projects and often as user stories and tasks in agile projects. During resource planning, the resources required to complete each task are determined, for example

- employees with certain qualifications,
- machinery and equipment,
- infrastructure (premises, information technology, etc.),
- services, and
- licenses.

Project managers should ensure that all resources are known and their availability has been checked. Otherwise there will be delays in the further course of the project due to a lack of resources.

Structure:
In small and simple medium-sized projects, it is sufficient to record and check the resources using a list or table. The resources and their scope in staff hours, machine days and quantities can be recorded as illustrated in the following figure.

	A	B	C	D	E
1	**ID**	**work package (or user story)**	**resources**	**scope**	**availability approved**
2	1.2.1	design system architecture	Mr. Myers (system architect)	10 staff hours	yes
3	1.2.1	design system architecture	software ArchiDesigner XT)	1 licence	yes
4	1.2.2	...	...	...	...

Task:	P.4.2 Create time-resolved resource plan	plan-based as well as agile
Project lifecycle:	Initialization / Definition / **Planning** / Control / Closing	
Team size:		Team distribution:
Precondition:	P.3 Scheduling	
Subsequent tasks:	P.5 Cost planning	
Summary:	A time-resolved resource plan lists the required resources depending on the schedule of the project.	

Aim:
The aim is to schedule the resources required for the project in order to ensure their availability. Based on this, the associated costs can then be determined later.

Instruction:
Based on the schedule, it is determined for each activity which resources are required and when. These resources (employees, materials, services, etc.) are listed along a time axis and aggregated.

Structure:
The figure on the right shows the structure of a time-resolved resource plan. The scaling should be chosen appropriately. If planning is too detailed, it reduces the readability of the plan. In the event of time-related delays or changes in general, it also leads to a high need for adjustment during project control.

Resource plan (required resources in hours [h] or quantity [#])

		CW 4	CW 5	CW 6	CW 7	CW 8	CW 9	CW 10	CW 11
Activity 1:	Requirements engineer [h]	30	30						
Activity 2:	SW developer [h]			120	120	120	120	120	
Activity 3:	SW developer [h]						30	30	
	Test engineer [h]						60	60	
	Test system (prototype) [#]						1	1	
Activity 4:	Requirements engineer [h]								30
	Technical writer [h]								60
Demand (total)	Requirements engineer [h]	30	30						30
	SW developer [h]			120	120	120	150	150	
	Test engineer [h]						60	60	
	Technical writer [h]								60
	Test system (prototype) [#]						1	1	
Balancing with capacity									
Capacity:	Requirements engineer [h]	30	30	30	30	30	30	30	30
	SW developer [h]	120	120	120	120	120	120 (overload)	120 (overload)	120
	Test engineer [h]					50	50 (overload)	50 (overload)	50
	Technical writer [h]					60	60	60	60
	Test system (prototype) [#]						(overload)	1	1

Overload (shortage)

time

Task:	P.4.3 Create resource plan in agile projects	plan-based as well as agile
Project lifecycle:	Initialization / Definition / Planning / Control / Closing	
Team size:		Team distribution:
Precondition:	P.3 Scheduling	
Subsequent tasks:	P.5 Cost planning	
Summary:	Agile projects generally rely on stable teams. Nevertheless, resources must also be planned here and checked for availability.	

Aim:
The aim is to plan the resources required and to check the availability of these resources for the project.

Instruction:
In the theory of agile project management, the topic of resource planning plays a minor role. Scrum teams like to rely on stable collaboration across several projects. Experience from the → M.3.10 sprint retrospectives and when estimating the size of user stories in story points can then be transferred from one project to the next. In practice, however, very few projects work without resource planning. Resources include, for example,

- employees with certain qualifications
- machinery and equipment,
- infrastructure (facilities, information technology, etc.),
- services, and
- licenses.

In any case, the agile project team or a potential agile project coordinator should check which resources are required and actually available over the course of the project. To do this, they can use a simple →P.4.1 static resource plan or, for complex and large agile projects, a →P.4.2 time-resolved resource plan. If the agile project team is known, additional agile resource planning can be carried out directly on the taskboard or Kanban board.

Markers (e.g. colored sticky dots) can indicate who is currently working on the upcoming tasks. The number of available markers per person can be assigned depending on the person's availability. The limitation ensures a realistic assignment of tasks and prevents unnecessary bottlenecks during project work.

P.5 Cost planning

Task:	**P.5.1 Create static cost plan** (plan-based as well as agile)
Project lifecycle:	Initialization | Definition | Planning | Control | Closing
Team size: / **Team distribution:**	
Precondition:	P.4 Resource planning
Subsequent tasks:	M Monitoring and Control
Summary:	A static cost plan is a simple list of costs planned in the project based on the resources required.

Aim:

The aim is to determine the costs that will be incurred in the project based on the resources required. With a simple static cost plan, there is no scheduling of costs.

Instruction:

The tasks planned for working on the project deliverables are usually planned as work packages and activities in plan-based projects and often as user stories and tasks in agile projects. The resources required for each task are determined as part of → P.4 resource planning. The costs associated with the use of resources are now determined and added up. Typical cost types can be taken from the following figure:

Personnel costs	Material costs	Infrastructure and service costs		Capital and other costs
▪ salaries ▪ wages ▪ social costs ▪ training costs	▪ raw materials and supplies ▪ tools ▪ warehousing ▪ packaging	▪ rooms ▪ rents, leases ▪ energy ▪ IT, telephone, internet ▪ postage, courier	▪ travel ▪ consulting ▪ hospitality ▪ licenses/ commissions	▪ depreciation and amortization ▪ capital commitment ▪ risk costs ▪ taxes and fees ▪ insurance

Structure:

In small and simple medium-sized projects, it is sufficient to identify and check the costs using a list or table, see the example below. However, an earned value analysis for project monitoring and control cannot be carried out on the basis of a static cost plan.

	A	B	C	D	E
1	**ID**	**work package (or user story)**	**resources**	**cost type**	**cost**
2	1.2.1	design system architecture	10 staff hours	salaries	1,500 US$
3	1.2.1	design system architecture	1 licence ArchiDesigner XT)	licences	3,000 US$
4	1.2.2	...	...	...	...

Task:	**P.5.2 Create time-resolved cost plan**	plan-based as well as agile
Project lifecycle:	Initialization / Definition / **Planning** / Control / Closing	
Team size:		**Team distribution:**
Precondition:	P.4.2 Create time-resolved resource plan	
Subsequent tasks:	M Monitoring and control	
Summary:	A time-resolved cost plan lists the planned costs per time and as a total cost line. It is a prerequisite for the subsequent application of the earned value analysis for project monitoring and control..	

Aim:

The aim is to plan the costs arising for the project.

Instruction:

Based on the resource plan, the associated costs are determined for all resources and aggregated per time unit. The total cost line indicates when a certain cost level is reached in the project.

Structure:

The figure on the right shows a possible structure. As with the ➔ P.4.2 resource plan, attention must be paid to appropriate time scaling.

Resource plan (required resources in hours [h] or quantity [#])

			CW 4	CW 5	CW 6	CW 7	CW 8	CW 9	CW 10	CW 11	
Activity 1:	Requirements engineer [h]		30	30							
Activity 2:	SW developer [h]				120	120	120	120	120		
Activity 3:	SW developer [h]							30	30		
	Test engineer [h]							60	60		
	Test system (prototype) [#]							1	1		
Activity 4:	Requirements engineer [h]									30	
	Technical writer [h]									60	
Demand (total)	Requirements engineer [h]		30	30						30	
	SW developer [h]				120	120	120	150	150		
	Test engineer [h]							60	60		
	Technical writer [h]									60	
	Test system (prototype) [#]							1	1		
Cost calculation											
Type of resource		Cost rate									
Requirements engineer [1,000 US$]		100 US$/h	3	3						3	
SW developer [1,000 US$]		100 US$/h			12	12	12	15	15		
Test engineer [1,000 US$]		100 US$/h							6		6
Technical writer [1,000 US$]		100 US$/h								6	
Test system (prototype) [1,000 US$]		5,000 US$/#						5	5		
Cost per period [1,000 US$ per week]			**3**	**3**	**12**	**12**	**12**	**26**	**26**	**9**	
Total costs [1,000 US$]			**3**	**6**	**18**	**30**	**42**	**68**	**94**	**103**	

w. = week → time

Task:	**P.5.3 Create cost plan in agile projects**	plan-based / as well as / agile
Project lifecycle:	Initialization / Definition / Planning / Control / Closing	
Team size:		**Team distribution:**
Precondition:	P.3 Scheduling	
Subsequent tasks:	M Monitoring and control	
Summary:	Clients usually expect projects to adhere to a cost budget, which must be planned and approved in advance. The basis for this is a cost plan.	

Aim:
The aim is to plan project costs in agile projects.

Instruction:
Basically, → P.5.1 static cost plans or → P.5.2 time-resolved cost plans can also be created in agile projects. In plan-based projects, the scope of the project is usually firmly defined by requirements and specifications. If there are difficulties during implementation, time and costs scale accordingly. In many agile projects, on the other hand, the duration (fixed sprints in Scrum) and costs (stable project teams) are fixed and the project scope is scaled, resulting in an inverted magic triangle, see figure below. However, not all agile projects work so consistently. For many, time and costs also scale in the event of conflict.

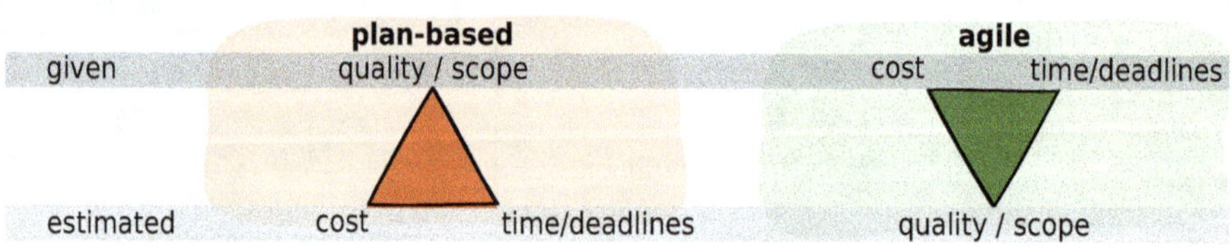

An important concept in agile cost planning is the agile fixed price. With this, the client and the project team first agree on the rough scope of the project. Only individual, well-defined and predictable parts are planned in detail. This is followed by a feasibility phase that ends at a checkpoint. Only at this point the actual fixed price is agreed. If the project partners are dissatisfied with the result of the feasibility phase, they can terminate the collaboration. The client receives the previous results in return for a corresponding fee. It is important that the partners agree on how the scope of the project (including changes) will be managed from the checkpoint onwards. Change budgets in the form of story points can be agreed for this purpose.

Task:	P.5.4 Plan (cost-based) contingency and management reserve	plan-based as well as agile
Project lifecycle:	Initialization | Definition | **Planning** | Control | Closing	
Team size:		Team distribution:
Precondition:	P.4 Resource planning O.3 Risk management	
Subsequent tasks:	M Monitoring and control	
Summary:	Reserves (e.g. cost buffers) can protect the project against cost overruns in the event of emerging risks. However, cost buffers must not be set arbitrarily, but must be planned and monitored transparently depending on the project risks.	

Aim:
The aim is to plan reserves (e.g. cost buffers) to safeguard the budget of the project in the event of risks occurring.

Instruction:
A distinction is usually made between two different types of buffer:

- The contingency reserve protects the project against known risks.
- The management reserve protects the project against unknown risks.

To determine the **contingency reserve**, the identified risks are analyzed with regard to their potential to make the project more expensive. A cost buffer is planned depending on this potential. Risk control measures that are planned in any case are not included in the contingency reserve, as they are not subject to uncertainties but are implemented regardless of whether a risk occurs.

The **management reserve** represents a buffer for cost overruns due to unforeseen risks. As the risks are therefore unknown, the size of the management reserve is based on experience from past projects and the company's risk awareness.

Contingency and management reserve should be reported separately. If a risk occurs and additional costs result, these additional costs are subtracted from the respective buffer. The reporting of contingency and management reserve requires an awareness of possible risks and their consequences on the part of the client of the project.

M Monitoring and control

M.1 Iterative detail and scenario planning

Task:	**M.1.1 Detail tasks**	plan based as well as agile
Project lifecycle:	Initialization / Definition / Planning / **Control** / Closing	
Team size:		**Team distribution:**
Precondition:	P Planning	
Subsequent tasks:	M.2 Determine project progress	
Summary:	During the monitoring and control phase, further detailing of the planned tasks and milestones may be necessary. Efforts actually incurred during implementation for the subsequent progress analysis must be recorded.	

Aim:
The aim is to update and detail the planned tasks, if necessary, and to record the actual efforts booked for the tasks.

Instruction:
As part of → P.3 scheduling, work packages for plan-based projects are divided into activities. During the completion of the activities, changes may occur that must be recorded and taken into account in the plans. In this case, the activities must be updated and the schedules, resource plans and cost plans revised accordingly.

In addition to updating the activities, the efforts involved in working on the activities should also be recorded. The actual efforts recorded are required for controlling the project with the help of the earned value analysis and for a later post-calculation and analysis of the project.

In addition, the recording of efforts serves to secure experience as part of knowledge management. If the actual effort involved in activities is recorded and documented in such a way that it can be easily found in future projects, this can significantly improve effort estimation.

Structure:
Many software packages for project management allow the effort to be recorded directly by the employees using respective software or web-based applications. However, effort can also be recorded without complex software, for example in shared lists in a spreadsheet.

Task:	M.1.2 Update action items	plan-based as well as agile
Project lifecycle:	Initialization | Definition | Planning | **Control** | Closing	
Team size:		Team distribution: 
Precondition:	P Planning	
Subsequent tasks:	M.2 Determine project progress	
Summary:	Lists of action items help to monitor open action points and document decisions in a traceable and sustainable manner. They are often kept in the form of lists in word processing programs, spreadsheets or web-based applications.	

Aim:

The aim is to efficiently and sustainably document open action items and decisions made during project execution.

Instruction:

Foreseeable tasks are planned and managed in projects in the form of work packages, activities, user stories or tasks. In addition, there are many small tasks and activities to be tracked. In addition, decisions are made that should be documented in such a way that both the decision itself and the reasons that led to the respective decision are comprehensible.

Without appropriate documentation of action points and decisions, there is a risk that these will be forgotten or repeatedly questioned, leading to inefficient projects. However, creating extensive logs for documentation is time-consuming and therefore unpopular with many project participants. Lists of action items are an efficient option for documentation.

Structure:

The typical structure of an action item list is outlined in the figure below. In addition to the action item or decision, the person responsible, the originally planned and actually expected fulfillment date, the status and the priority can be documented.

#	action item	description	respon-sible	approved by	prio	started	due	expected	status	comments
1	update project brief	updating the project brief to release the planning phase	project manager	program manager	1	Aug. 10th 2025	Aug. 31st 2025	Aug. 31st 2025	work in progress	-
2	finalize contract with supplier	a contract for work and services is to be established with company XY for the development of the ball bearings	supplier manager	project manager	2	Aug. 17th 2025	Aug. 31st 2025	Sep. 18th 2025	rests	delay due to plant vacations at the supplier

Task:	**M.1.3 Conduct daily Scrum**		plan-based as well as agile
Project lifecycle:	Initialization | Definition | Planning | **Control** | Closing		
Team size:		**Team distribution:**	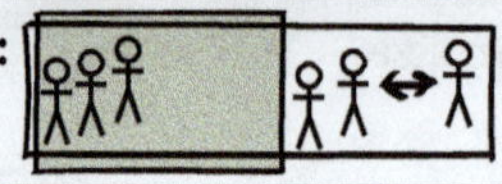
Precondition:	P Planning		
Subsequent tasks:	M.2 Determine project progress		
Summary:	During the sprints of agile projects according to Scrum, short meetings take place every day, so-called daily Scrums. These meetings are used to discuss what has been achieved since the last meeting, what needs to be done next and what obstacles might prevent implementation.		

Aim:

The aim is to synchronize the work in agile projects according to Scrum, including status monitoring, identifying obstacles, initiating control measures and planning the next steps.

Instruction and structure:

In Scrum projects, the daily Scrum is organized by the developers themselves. The Scrum master can optionally take part. The daily Scrum takes place at the same time every day, often in the morning. The figure below shows a typical agenda. The daily scrum is carried out standing directly at the taskboard and is the only recurring project meeting during the sprint execution phase. The duration must not exceed 15 minutes. Problems that the team cannot solve themselves are recorded by the Scrum master in the so-called → M.1.5 impediment backlog and taken along for resolution.

Agenda Daily Scrum (max. 15 min.)
- What has been achieved since the last daily Scrum?
- What should be achieved by the next daily Scrum?
- What obstacles are there?
- What support is needed to become faster/better?

Remark:

Short daily meetings can also be useful in plan-based and agile projects that do not work according to Scrum. Another common name is a stand-up meeting. Such meetings create a sense of urgency, solution orientation and tackling the problem. It is a signal that the project is to progress and that obstacles are to be solved immediately. It is important to reduce the number of participants to really relevant and active team members. The nature of active stand-up meetings differs significantly from purely informative (passive) meetings.

Task:	M.1.4 Update sprint backlog	plan-based as well as agile
Project lifecycle:	Initialization / Definition / Planning / **Control** / Closing	
Team size:		Team distribution:
Precondition:	P.3.5 Create sprint backlog	
Subsequent tasks:	M.2 Determine project progress	
Summary:	The sprint backlog contains the user stories to be implemented in the current sprint and the tasks to be completed. It is created and maintained by the team of developers.	

Aim:

The aim is to update the sprint backlog while the project is being managed. The sprint backlog is updated at the start of a new sprint.

Instruction:

The sprint backlog is created at the start of a sprint in a sprint planning event, see also the figure on the right. In the first part of the sprint planning, the product owner presents the user stories and works out the goal for the sprint with the developers. The developers then define the scope of the user stories to be implemented.

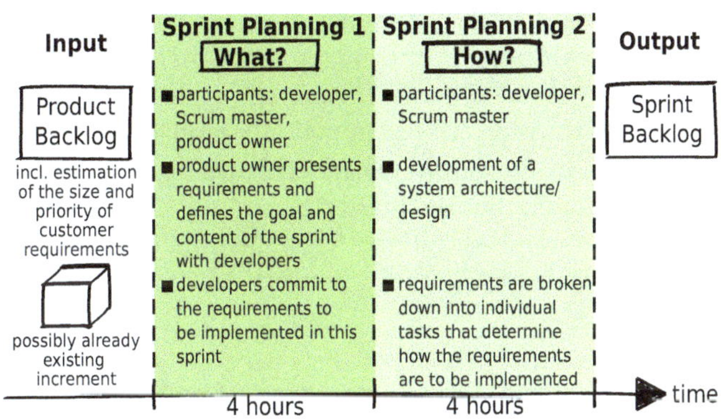

In the second part of the sprint planning, the user stories are then broken down into individual tasks to be completed; it is determined how the implementation should take place.

The sprint backlog contains the user stories that the developers have firmly committed to the product owner for the current sprint. Changes, for example due to difficulties during implementation, should not be made without due consideration. If changes are necessary, the product owner should be involved. Minor changes that do not affect the sprint goal or the agreed user stories, but their design, can and should be made independently by the developers and recorded in the sprint backlog.

Task:	**M.1.5 Maintain impediment backlog** — plan-based as well as agile
Project lifecycle:	Initialization / Definition / Planning / **Control** / Closing
Team size:	**Team distribution:**
Precondition:	P Planning
Subsequent tasks:	M.2 Determine project progress
Summary:	In the Impediment Backlog, action items are monitored for the elimination of obstacles during project implementation. The Impediment Backlog is managed by the Scrum master and updated in the daily Scrum.

Aim:
The aim is to systematically identify obstacles that could disrupt the course of the project and to manage and monitor their elimination.

Instruction:
The impediment backlog is usually managed by the Scrum master and contains obstacles that the developers cannot remove themselves. It has a similar character as the → M.1.2 list of action items. The impediments can be quantified and prioritized in terms of the risk to the project. New impediments are addressed by the team in the daily Scrum, documented by the Scrum master and then attempted to be resolved. In one of the next daily Scrums, the Scrum master can report to the developers on the completion or possible compromises. Alternatively, the impediment backlog can be filled by the team using a list (web-based or on a network drive) as required and viewed and edited by the Scrum Master.

Structure:
The following figure shows an example of an impediment backlog. The structure of the list should be adapted to the project environment and the way the team works. Alternatively, the impediment backlog can also be displayed as a taskboard.

	A	B	C	D	E	F	G
1	**ID**	**impediment**	**affected user story**	**date of occurrence**	**presumed impact**	**prio**	**status/ comment**
2	I1	For the completion of S410_U, the missing deliveries from RagStor are required.	S410_U	Oct. 3rd 2025	implementation of S410_U only in the next sprint	2 - medium	done/RagStor will deliver prototypes at short notice
3	I2	Tony was removed from the project without replacement; there is a lack of expert knowledge for SQL databases.	S375_P	Oct. 4th 2025	significant delay of the entire project	1 - high	in progress/Tony is to provide at least advisory support

Task:	**M.1.6 Update business case**		plan-based as well as agile
Project lifecycle:	Initialization | Definition | Planning | **Control** | Closing		
Team size:		**Team distribution:**	
Precondition:	P Planning		
Subsequent tasks:	M.2 Determine project progress		
Summary:	A business case justifies the existence of a project and contains a quantitative comparison of income and the necessary expenses. During the course of the project, new findings may emerge that influence the business case and thus the justification for the project. These findings must be incorporated into the business case and evaluated.		

Aim:
The aim is to regularly check that the business case is up to date and to revise it accordingly.

Instruction:
During the original → I.1.6 preparation of the business case, extensive data was collected, processed and analyzed. This led to the decision that the project should be launched. This justification for the project must be reviewed regularly. Cost increases, changes in customer expectations, new competitors on the market and other influencing factors affect the business case.

The data originally collected must therefore be checked to ensure it is up to date and revised if necessary. This data can come from market observations, stakeholder analyses or directly from the project, for example in the form of detailed risk analyses.

Structure:
The structure of the business case does not change when it is updated.

Consequences:
If new findings emerge, it is necessary to react to them. If the justification for the project is confirmed, it can be continued as planned. If the justification fades, it must be checked whether the project can be changed so that it is still a success. If this is not the case, the project should be discontinued.

M.2 Determine project progress

Task:	**M.2.1 Estimate progress**		plan-based as well as agile
Project lifecycle:	Initialization | Definition | Planning | Control | Closing		
Team size:		**Team distribution:**	
Precondition:	P Planning		
Subsequent tasks:	M.3 Analyze project progress		
Summary:	The percent complete PC (also: progress or degree of progress) is the ratio of the work completed to the planned total work on a selected reporting date. If the actual work performed cannot be determined in any other way, the degree of completion of a task is determined by estimation .		

Aim:
The aim is to determine the degree of completion of a task or the entire project, i.e. the ratio of actual work completed to total work planned on the selected reporting date.

Instruction:
When determining the percentage of completion by estimation, the project manager asks the person responsible for a task in the project about the degree of completion of the task. The percent complete PC is often given as a percentage of the total work planned and therefore increases from 0 percent at the beginning of the task to 100 percent at the end of the task. Estimating the degree of completion is a simple and quick method, but provides subjective and sometimes overly optimistic values. There are several reasons for this

- The 90 percent syndrome means that the degree of completion is often overestimated. Instead of an estimated 90 percent, the task is only 30 to 60 percent complete. This may be due to the fact that the remaining tasks required to solve the problem appear simple when the solution is known and possible problems are suppressed or overlooked.
- In addition to unconscious miscalculation, the degree of completion can also be deliberately misstated, for example in the assumption that the current project delay can be made up for.

The estimation of the percentage of completion should therefore only be used for tasks that are not critical or for which alternative methods are not possible.

The project manager calculates the overall percent complete of the project from the estimated values of the individual tasks by weighting the individual estimated values with the planned total costs of the individual tasks and adding them up.

Task:	M.2.2 Measure proportionalities	plan-based as well as agile
Project lifecycle:	Initialization | Definition | Planning | **Control** | Closing	
Team size:		Team distribution:
Precondition:	P Planning	
Subsequent tasks:	M.3 Analyze project progress	
Summary:	The percent complete (also: progress or degree of progress) of the project, i.e. the ratio of a completed task or project to the planned total task or project on a reporting date, can be determined indirectly via proportionality relationships. This is helpful when the degree of completion is difficult to measure directly.	

Aim:
The aim is to determine the degree of completion of a task or the entire project, i.e. the ratio of the work actually completed to the planned total work on the selected reporting date.

Instruction:
The percent complete can often only be determined using proportionality relationships. A distinction is made between quantity, time and secondary proportionalities:

- With the quantity based proportionality, the quantity of work completed on the reporting date is set in relation to the total quantity. To do this, the total quantity must be made up of several partial quantities that are as equal as possible.
  Example: A total of 100 requirements for a product have been collected for the specifications. When preparing the specifications, one employee has already worked on 70 requirements. The percent complete is therefore 70 percent.

- In the case of time proportionality, the degree of completion is determined as the ratio of the time already worked to the estimated total working time of the task. This should always be viewed critically if the total working time was estimated incorrectly. However, in combination with the →M.2.3 determination of the estimate to complete, the assessment of progress using time proportionality can provide good results.

- Secondary proportionality does not measure the degree of completion of the actual task, but other quantities proportional to the task. Example: In the "Create process analysis" work package, the percent complete is determined by measuring the number of expert interviews with process participants. The number is set in relation to the total number of planned interviews.

Task:	M.2.3 Determine estimate to complete	plan-based — as well as — agile
Project lifecycle:	Initialization · Definition · Planning · Control · Closing	
Team size:		Team distribution:
Precondition:	P Planning	
Subsequent tasks:	M.3 Analyze project progress	
Summary:	When determining the percent complete, the actual work completed is set in relation to the planned total work. The latter was estimated during the planning phase. Due to favorable or unfavorable circumstances, the required total effort may change during project execution; it should therefore be updated in due course by means of a residual effort estimate.	

Aim:
The aim is to determine the degree of completion of a task or the entire project, i.e. the ratio of the work actually completed to the planned total work on the selected reporting date.

Instruction:
All → M.2.2 proportionality methods have a common weakness: If the total quantity (quantity of tasks to be performed, quantity of estimated time or scope of the assumed secondary quantity) changes, the basis of the ratio calculation and thus the calculated percent complete is incorrect.
Therefore, when applying the proportionality method, a residual effort estimate should always be made at the same time. Whenever the actual work performed is queried (in the form of quantities, times or secondary quantities), the person working on the task is asked to provide an estimate of the residual (open) effort which is required to complete the task. The degree of completion is then determined from the ratio of the work actually completed to the sum of the work actually completed and the outstanding work. The results of the residual effort estimate can be used not only to determine the percent complete, but also to update the project plans.
Example: An employee is to prepare the requirements specification with an estimated 100 requirements. After 70 requirements have been elaborated, she initially reports a degree of completion of 70 percent for the work package. To be on the safe side, the project manager carries out a residual effort estimate together with the employee, which shows that a total of 140 requirements will probably have to be prepared. As a result, the progress according to the current situation is only 50 percent.

Task:	**M.2.4 Determine weighted milestones**	plan-based as well as agile
Project lifecycle:	Initialization / Definition / Planning / Control / Closing	
Team size:		**Team distribution:**
Precondition:	P Planning	
Subsequent tasks:	M.3 Analyze project progress	
Summary:	In the weighted milestone method (also known as the status step method, tasks are broken down into several sub-task (i.e. status steps or milestones). Each sub-task is linked to a deliverable (plan-based: milestone, agile: definition of done). If this deliverable is completed, a certain percent complete is achieved.	

Aim:
The aim is to determine the degree of completion of a task or the entire project, i.e. the ratio of the work actually completed to the planned total work on the selected reporting date.

Instruction:
In this method, tasks are subdivided into several sub-tasks (i.e. weighted milestones or status steps). The subdivision should be chosen in such a way that a status step can be completed as regularly as possible. A status step is completed when a pre-defined deliverable is achieved. This deliverable is linked to a certain degree of completion of the task. For example, the task "create software architecture" can be divided into the following status steps:

status steps	percent complete PC	planned value PV
requirements analyzed	10 %	1,000 US$
system architecture defined	30 %	3,000 US$
detail architecture created	70 %	7,000 US$
detail architecture approved	90 %	9,000 US$
documentation of architecture completed	100 %	10,000 US$
		BAC = 10,000 US$

percent complete linked to the status steps of the task

optionally, each status step can also be linked to the planned value PV

If the responsible employee reports to the project manager during the control phase of the project that the detailed architecture has been created, the project manager can book a degree of completion of 70 percent (and planned costs of 7,000 US$). By separating the definition of the status steps in the planning phase and the application in the control phase, the status step method is relatively objective. The accuracy can be adjusted by using more or fewer status steps.

Task:	**M.2.5 Percent-start/ percent-finish method**	plan-based | as well as | agile	
Project lifecycle:	Initialization | Definition | Planning | Control | Closing		
Team size:		**Team distribution:**	
Precondition:	P Planning		
Subsequent tasks:	M.3 Analyze project progress		
Summary:	The percent-start/percent-finish method is a technocratic, objective method that only roughly determines the degree of completion of a task on the basis of clear rules. A previously defined percentage of completion is booked as progress at the start of work on a task and at the end of work.		

Aim:

The aim is to determine the degree of completion of a task or the entire project, i.e. the ratio of the work actually completed to the planned total work on the selected reporting date.

Instruction:

With the percent-start/percent-finish method, a predefined percent complete of X percent is recorded as soon as a task is started. The remaining Y (= 100 % - X) percent is booked as soon as the task has been completed. Typical values for X and Y are 0/100 percent, 50/50 percent and 20/80 percent. In principle, however, any combination is possible. With the percent-start/percent-finish method, the planned costs PV can be determined approximately in addition to the percent complete PC. The figure below explains how to calculate the percent complete PC and the planned value PV for the 0/100 and the 50/50 percent-start/percent-finish method.

At the actual start of the task, the percent complete PC (degree of completion) is set to X percent. At the actual end, the remaining Y percent is then booked. At the planned start of the task, the planned value of the task is set to X percent of the total costs BAC of the task. At the planned end of the task (regardless of the actual end), the planned value is set to the planned total costs BAC.

The method is relatively inaccurate, but objective and easy to use. It is particularly suitable for small, non-critical tasks and in projects with a large number of tasks.

	0/100 method		50/50 method	
	PC	PV	PC	PV
from the actual start of the task	0 %		50 %	
from the actual end of the task	100 %		100 %	
from the planned start of the task		0		0.5xBAC
from the planned end of the task		BAC		BAC

Task:	M.2.6 Monitor milestones	plan-based / as well as / agile
Project lifecycle:	Initialization / Definition / Planning / **Control** / Closing	
Team size:		Team distribution:
Precondition:	P Planning	
Subsequent tasks:	M.3 Analyze project progress	
Summary:	In many projects, especially plan-based projects, there are milestones that represent important events in the course of the project. Adherence to the milestones must be monitored throughout the project life cycle and countermeasures taken as early as possible in the event of deviations.	

Aim:
The aim is to determine the degree of completion of a task or the entire project, i.e. the ratio of the work actually completed to the planned total work on the selected reporting date.

Instruction:
Milestones are (project) events with particular significance. Examples are
- the achievement of an important subtarget or
- important deadlines, such as trade fairs, shipments or customer commitments, for which a (partial) result of the project is required. .

In the definition or planning phase of the project, milestones were defined, the adherence to which must be monitored now in the control and monitoring phase. A common tool for this is the → M.3.3 milestone trend analysis.

The prerequisite for their application is that the currently planned milestone dates are also checked on each reporting date on which the project progress is monitored. It must be examined whether the milestones can realistically be achieved. The project manager should therefore document which date is planned for each milestone and which date is realistic based on the current status of the project.

The more milestones are planned within a project, the earlier deviations can be detected. Milestone deadlines that are not met are always an alarm signal for a project delay. However, too many milestones increase the monitoring and analysis effort and can therefore disrupt the course of the project.

Task:	**M.2.7 Verify deliverables**		plan-based as well as agile
Project lifecycle:	Initialization / Definition / Planning / **Control** / Closing		
Team size:		**Team distribution:**	
Precondition:	P Planning		
Subsequent tasks:	M.3 Analyze project progress		
Summary:	As the project progresses, more and more requirements regarding the project deliverables are implemented and the project approaches its completion. The implementation of the specified requirements is checked through verification.		

Aim:
The aim is to inspect the implementation of the specified requirements for the project deliverables by verification.

Instruction:
Verification is the unambiguous proof (usually by measurement) that a certain requirement has been fulfilled. Verification answers the question of whether it has been implemented correctly (in the sense of the requirement).
In plan-based projects, each entry of the requirements specifications is often checked through verification. In agile projects, verification is carried out by checking whether the requirements have been implemented in accordance with the definition of done. In some industries, a documented 100 % verification of the specified requirements is mandatory. A test case must then be defined, carried out and documented for each requirement.
Verification can be carried out on an ongoing basis. In this case, the requirement is always verified immediately after it has been implemented. Alternatively, the control phase of the project can be divided into an implementation phase and a verification phase. In this case, verification only follows the completion of the implementation of all requirements.
If verification is carried out on an ongoing basis, there is a risk that interactions with requirements implemented at a later date will not be recognized. Assume that the implementation of a requirement has been successfully verified. After the implementation of further requirements, the implementation of the already successfully verified requirement no longer works. This is due to interactions between the respective implementations.
Conversely, the strict separation of implementation and verification can prolong the project, as errors in the implementation of individual requirements are only recognized at a late stage.

Task:	M.2.8 Validate deliverables (plan-based as well as agile)
Project lifecycle:	Initialization / Definition / Planning / **Control** / Closing
Team size:	(1 – 4+ persons) — Team distribution: (co-located ↔ distributed)
Precondition:	P Planning
Subsequent tasks:	M.3 Analyze project progress
Summary:	As the project progresses, more and more requirements for the project object are implemented and the project object approaches its final version. The implementation of the customer requirements is checked in the form of validation.

Aim:
The aim is to inspect the implementation of the customer requirements for the project deliverables by validation.

Instruction:
Validation is the proof that the customer's requirements have been met. Validation answers the question of whether the right thing has been developed for the customer.

In plan-based projects, the customer requirements are checked through validation. In agile projects based on Scrum, validation can be carried out by the customer themselves, for example as part of the sprint review.

In some industries, a documented 100 percent check of customer requirements through validation is mandatory. A test case must then be defined, carried out and documented for each requirement.

Validation can only take place once the project deliverables have progressed to the point where the customer's requirements can be tested. Use cases can be defined for this purpose, for example. These can be reproduced during validation using the project deliverables for validation.

As with verification, it is also important to ensure that validation does not take place too early in the project life cycle. If work is still being carried out as part of the project, customer requirements that have already been positively validated may be changed. These must then be re-validated. Conversely, early validation can provide early feedback on whether the project deliverables will meet the customer requirements.

Task:	**M.2.9 Measure key performance indicators**	plan-based as well as agile
Project lifecycle:	Initialization | Definition | Planning | **Control** | Closing	
Team size:		**Team distribution:**
Precondition:	P Planning	
Subsequent tasks:	M.3 Analyze project progress	
Summary:	Key performance indicators can help to record and subsequently evaluate the performance of the project. Measures for project control can be derived from this.	

Aim:
The aim is to determine project-relevant key performance indicators for later analysis of the progress and control of the project.

Instruction:
Typical key performance indicators that can be recorded within a project are

- resource requirements (staff hours per week),
- earned value (value of the work completed),
- percent complete (degree of progress, progress),
- actual costs spent,
- planned value on the reporting date,
- ...

typical indicators in plan-based projects

- velocity (completed story points per sprint),
- cycle time (duration of a requirement from specification through implementation and testing to final release)
- amount of open story points or user stories (shown on the burndown chart)
- ...

typical indicators in agile projects

The way in which key performance indicators are measured and subsequently analyzed should be chosen with care. Every recording, data archiving and analysis requires effort. Therefore, only those key performance indicators should be collected that are actually used to analyze and control the project. The frequency of recording should be adapted to the rate of change of the key performance indicators and the frequency with which control measures are initiated. The recording of key performance indicators can lead to resistance from employees. Key performance indicators make performance transparent. At the same time, they condense performance into a single value. Key performance indicators can provide important information about team performance. However, they usually say nothing about the causes of poor performance.

Task:	M.2.10 Measure as a tool for retrospective control	plan-based as well as agile
Project lifecycle:	Initialization / Definition / Planning / **Control** / Closing	
Team size:		Team distribution:
Precondition:	P Planning	
Subsequent tasks:	M.3 Analyze project progress	
Summary:	Measuring the actual effort and costs invested per task and comparing them with the planned effort and costs per task are important components of knowledge management and controlling. When a task is completed, the effort and costs actually invested must therefore be recorded, compared with the planned data and saved as experience for future projects. Any deviations from the plan must be analyzed and deducted from the available project buffer (contingency or management reserve).	

Aim:
The aim is to compare the actual effort of tasks with the planned tasks. The results of the comparison must be analyzed and saved for future projects.

Instruction:
When a task is completed, the effort invested in it or the costs incurred are recorded and saved so that the information can be used for future projects. These can then draw on the empirical values for comparable tasks and improve the effort estimation. Tasks must be documented in sufficient detail so that they can be compared at a later date.

In addition, the actual expenses are compared with the planned efforts. If there are deviations, it must be investigated whether these are due to foreseen risks or unexpected ones. If foreseen risks occur, the planned contingency reserve must be reduced by the difference between the planned effort and the actual effort.

If unexpected risks arise for which no buffer has been planned as part of the contingency reserve, the management reserve must be reduced by the difference between the planned effort and the actual expense.

Task:	M.2.11 Update burndown chart	plan-based as well as agile
Project lifecycle:	Initialization / Definition / Planning / **Control** / Closing	
Team size:		Team distribution:
Precondition:	P.3.8 Prepare burndown chart	
Subsequent tasks:	M.3 Analyze project progress	
Summary:	The burndown chart is used to determine and visualize the progress of the project. It must be updated before it can be analyzed in order to initiate measures for project control.	

Aim:
The aim is to update the burndown chart for later analysis of the project.

Instruction:
The diagram created in → P.3.8 Prepare burndown chart must be updated regularly in order to be analyzed subsequently. For most teams, the update takes place as part of the → M.1.3 Daily Scrum.

In order to visualize the progress achieved on the burndown chart, the open work shown there should be quantified in such a way that a daily progress is achieved. If the open work is counted using open tasks, for example, the tasks should be dimensioned in such a way that a task can be completed between two daily Scrums. Otherwise, the horizontal curve shown in the figure below will occur. Such a progression does not allow any statement to be made about the progress of work. It also irritates important stakeholders such as the product owner, as it gives the impression that the team is not making progress.

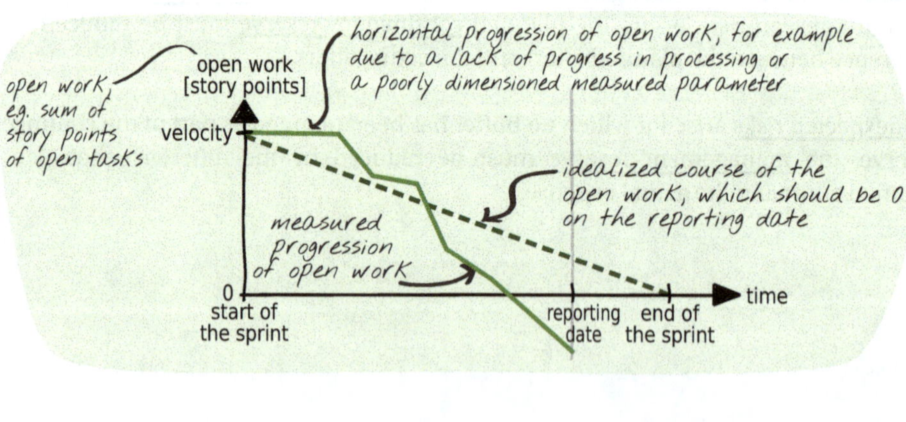

Task:	M.2.12 Measure Kanban metrics	plan-based as well as agile
Project lifecycle:	Initialization / Definition / Planning / **Control** / Closing	
Team size:		Team distribution:
Precondition:	P.3.7 Prepare Kanban board	
Subsequent tasks:	M.3 Analyze project progress	
Summary:	Optimizing the flow of work in Kanban projects requires the flow to be measured. There are various metrics for this, which must be defined and filled with measurements.	

Aim:
The aim is to measure the flow of work in projects according to Kanban in order to subsequently optimize it.

Instruction:
Typical metrics in projects according to Kanban are

- the lead time of a task through all processing stages or
- the lead time per stage,
- the throughput, i.e. the number of completed tasks per time unit,
- the average waiting time between two stages and
- the cumulative flow, i.e. the cumulative number of tasks in the individual stages.

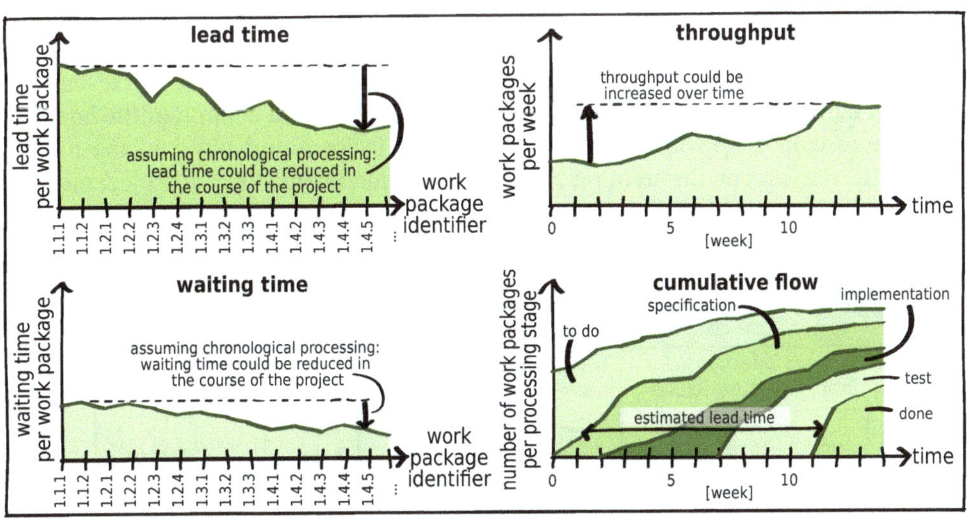

Task:	**M.2.13 Update communication board**	plan-based as well as agile	
Project lifecycle:	Initialization / Definition / Planning / **Control** / Closing		
Team size:		**Team distribution:**	
Precondition:	- -		
Subsequent tasks:	M.3 Analyze project progress		
Summary:	The communication board is an interactive (physical or electronic/virtual) board on which current tasks, questions or problems are documented and comments, solutions and decisions are collected.		

Aim:

The aim is efficient and interactive communication and documentation of tasks, questions, problems and their solutions.

Instruction and structure:

The communication board (also known as a voting board or decision board) can be created and used in all project phases. It should either be placed in a central location accessible to all team members or implemented as an electronic board. The structure is shown in the figure below. In contrast to the → M.1.2 list of action items, the communication board encourages participation and the willingness to take responsibility through its transparent presentation. It can be a central management tool for daily stand-up meetings or serve as a repository of ideas between regular project meetings.

Employees who encounter problems, have questions or want to raise ideas for discussion write their contributions on cards and pin them in the left-hand column of the board. If an employee comes up with an answer, they write it down and pin it in the middle column. The decision on the solution to the problem documented in the project meeting is documented in the right-hand column. Decisions can be transferred to an official record for archiving purposes. Alternatively, the communication board can be photographed regularly or kept electronically.

M.3 Analyze project progress

Task:	M.3.1 Analyze earned value	plan-based as well as agile
Project lifecycle:	Initialization / Definition / Planning / **Control** / Closing	
Team size:		Team distribution:
Precondition:	M.2 Determine project progress	
Subsequent tasks:	M.4 Manage progress	
Summary:	Earned value analysis is an integrative method of determining and analyzing progress. It includes the work completed and its value, the planned costs and the actual costs. It provides information on the time and cost-related status of the project and allows forecasts to be made.	

Aim:
The aim is to analyze the progress of the project in terms of time and costs.

Instruction:
The project manager knows the planned total costs BAC (budget at completion) and regularly determines the actual costs spent AC, the planned value PV and the degree of completion PC (percent complete). From this, the earned value EV, which indicates the planned value of the work actually performed on the reporting date, can be calculated: EV = PC x BAC. The actual costs AC are determined by means of cost monitoring. The planned value PV follow from the ➔ P.5.2 time-resolved cost plan and the percent complete PC from the ➔ M.2 progress determination. The figure at the top right shows an example of the progression of costs up to a key date. These values can now be used to calculate and analyze the cost performance index CPI (Cost Performance Index) and the schedule performance index SPI (Schedule Performance Index), see figure on the right.

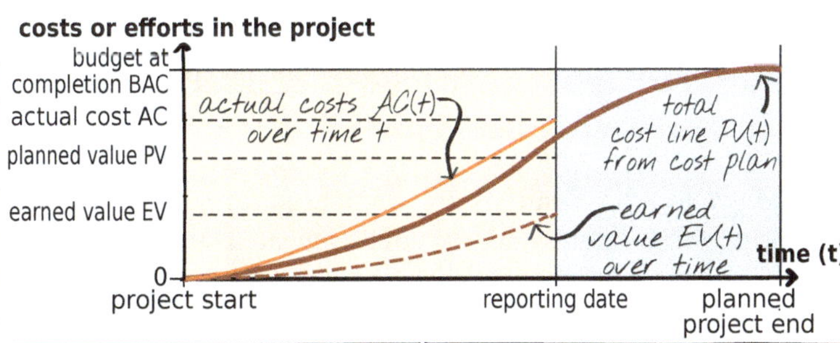

Cost analysis		
Cost performance index $CPI = \frac{EV}{AC}$		
The following statements can be derived:		
The project is on budget as planned.	EV = AC	CPI = 1
The project costs more than planned.	EV < AC	CPI < 1
The project costs less than planned.	EV > AC	CPI > 1

Schedule analysis		
Schedule performance index $SPI = \frac{EV}{PV}$		
The following statements can be derived:		
The project is on schedule as planned.	EV = AC	SPI = 1
The project is slower than planned.	EV < AC	SPI < 1
The project is faster than planned.	EV > AC	SPI > 1

Task:	M.3.2 Create forecasts	plan-based as well as agile
Project lifecycle:	Initialization / Definition / Planning / **Control** / Closing	
Team size:	⚲ … ⚲⚲⚲⚲	Team distribution: ⚲⚲⚲ … ⚲⚲↔⚲
Precondition:	M.2 Determine project progress	
Subsequent tasks:	M.4 Manage progress	
Summary:	Earned value analysis is an integrative method of determining and analyzing progress. It includes the work completed and its value, the planned costs and the actual costs. It provides information on the time and cost-related status of the project and allows forecasts to be made.	

Aim:
The aim is to calculate forecasts on the further course of the project using earned value analysis.

Instruction:
The earned value analysis allows the calculation of forecasts EAC (estimate at completion) over the further course of the project EAC_c until the end of the project. The calculation is based on the schedule performance index SPI and the cost performance index CPI, which are explained in ➔ M.3.1 earned value analysis. A distinction is made between linear and additive forecasts.
The linear forecast assumes that past developments will continue proportionally into the future. For example, if personnel costs have increased in the past due to wage increases, they will continue to do so in the future. The forecast is calculated using the following formulas:

$$EAC_C = \frac{BAC}{CPI} \qquad EAC_t = \frac{\text{planned project duration}}{SPI}$$

The additive forecast assumes that the reason for a deviation from the plan in the past was of a one-time nature and will not be repeated. The calculation is based on the actual costs AC and the earned value EV:

$$EAC_c = BAC + AC - EV$$

The forecast formulas presented here simplify the further course of the project and can only serve as an initial indication. If the actual course of the project deviates from the planned one, the schedule and cost plan should always be updated.

Task:	M.3.3 Analyze milestone trend	plan-based as well as agile	
Project lifecycle:	Initialization | Definition | Planning | **Control** | Closing		
Team size:		Team distribution:	
Precondition:	M.2 Determine project progress		
Subsequent tasks:	M.4 Manage progress		
Summary:	Milestone trend analysis is used to assess the actual progression of milestones over time. Milestone trend analysis is often used in progress reports for management, as it focuses on important points in time in the project. For this very reason, however, it is only suitable to a limited extent as the sole analysis tool in larger projects.		

Aim:
The aim is to analyze important milestones in the project.

Instruction:
If the schedule is reduced to important milestones, their observation over the course of the project can be used to analyze progress. In the milestone trend analysis, the milestone data for all reporting dates, i.e. all dates on which the project status is determined, are recorded and entered in a diagram. An example is shown in the diagram below.

The reporting dates are plotted on the horizontal axis and the milestone dates planned for the reporting date are plotted on the vertical axis. If the progress of a milestone moves upwards, there is a risk of project delay. The delay often begins at an early milestone and then spreads to the other milestones.

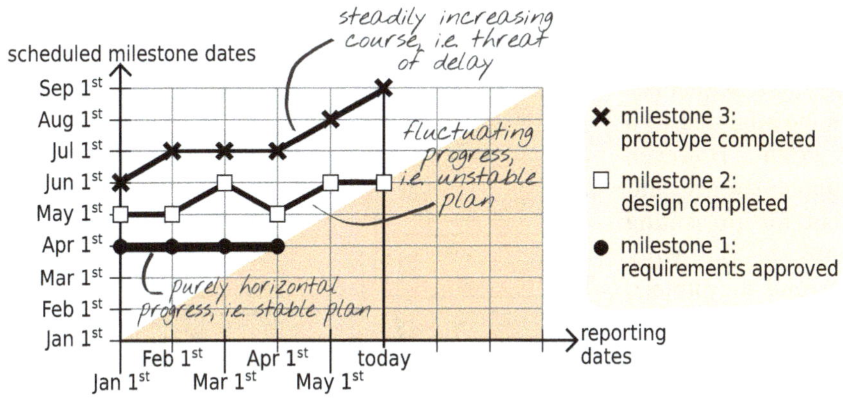

Task:	M.3.4 Analyze work breakdown structure	plan-based as well as agile
Project lifecycle:	Initialization | Definition | Planning | **Control** | Closing	
Team size:		Team distribution:
Precondition:	M.2 Determine project progress	
Subsequent tasks:	M.4 Manage progress	
Summary:	The work breakdown structure contains all of the project's work packages. An extended representation of the work breakdown structure allows the status of the work packages to be visualized and used for further analysis.	

Aim:
The aim is to analyze the status of a project's work packages.

Instruction:
The work breakdown structure created as a hierarchical diagram (see example below) or as a tabular list is supplemented with data describing the project status. This data depends on the respective project and the tools used there. Examples of such data are

- traffic lights for work packages and projects to quickly assess the status: It is important that the traffic light colors are clearly defined, for example green for "everything is going according to plan", yellow for "deviations from plan with promising counter-measures to achieve the plan" and red for "deviations from plan without such countermeasures " and
- key figures of the earned value analysis, for example percent complete (PC), planned value (PV), earned value (EV), actual cost (AC) and others.
- ...

As the work breakdown structure is not a schedule, time sequences can only be analyzed to a limited extent. However, as a supplementary and clear tool for presenting the status of the work packages and the project, the extended presentation of the work breakdown structure is well suited.

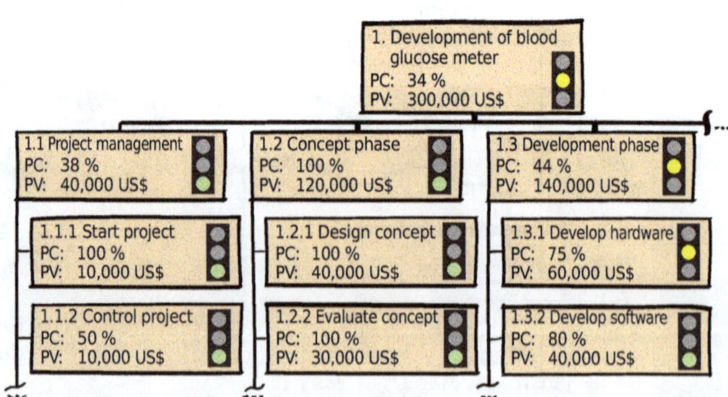

Task:	M.3.5 Analyze Gantt chart	plan-based as well as agile
Project lifecycle:	Initialization · Definition · Planning · **Control** · Closing	
Team size:		Team distribution:
Precondition:	M.2 Determine project progress	
Subsequent tasks:	M.4 Manage progress	
Summary:	The Gantt chart contains all project activities. A comparison of the currently planned project progress with the initially created baseline plan can be helpful for analyzing the progress of the project over time.	

Aim:

The aim is to analyze the progress of the project over time using the Gantt chart.

Instruction:

In the Gantt chart, not only the currently planned course of the project can be visualized in the form of bars along a time axis, but also the originally planned course of the project, which is often called the baseline plan. The figure below shows an example of the basic plan with black bars. The actual progress of processing the tasks up to the current reporting date is visualized with grey bars. The expected future progress based on outstanding work and a → M.2.3 residual effort estimate is illustrated with white bars. In addition, the percent complete PC (progress or degree of completion) is shown as a numerical value in percent.

Common project management software allows you to save and visualize several base plans, which can then be used for comparison with the actual progress. If the expected future course deviates from the originally planned course, → M.4 control measures must be initiated.

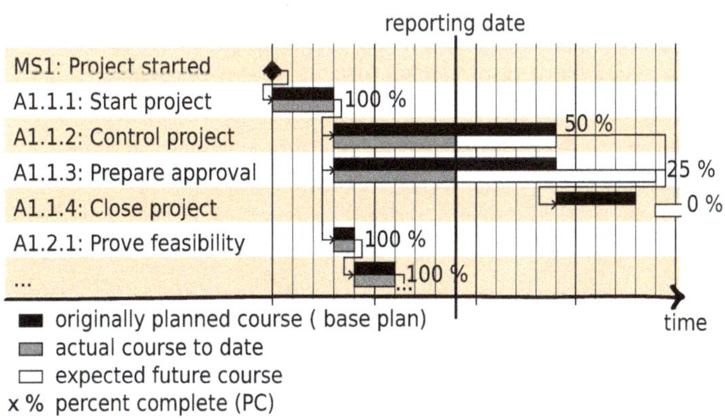

Task:	M.3.6 Analyze status of objectives	plan-based as well as agile
Project lifecycle:	Initialization / Definition / Planning / Control / Closing	
Team size:		Team distribution:
Precondition:	M.2 Determine project progress	
Subsequent tasks:	M.4 Manage progress	
Summary:	A simple method for reflecting on the status of the objectives set and deriving the necessary measures for project control is to review the status of the objectives. Objectives that are assessed as being at risk must be analyzed and measures initiated.	

Aim:
The aim is to analyze the project's achievement of objectives by means of a review and the subsequent identification of measures to ensure that all objectives will be achieved.

Instruction:
The project manager has the project team vote at regular intervals on whether the team members believe that a certain objective (or requirement or milestone, etc.) will be achieved according to plan. The vote can take place on a whiteboard as part of a regular meeting, for example, as illustrated in the figure on the right.

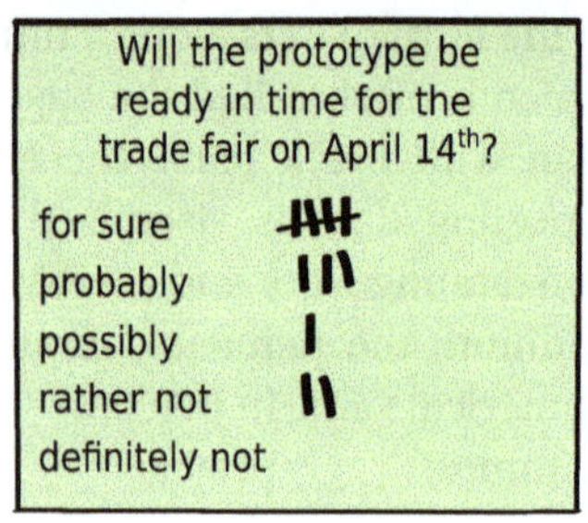

If there are doubts about the fulfillment of the objective or the implementation of the requirement or the achievement of the milestone, the reasons for this can be investigated and discussed. Measures can then be developed to ensure achievement.

In contrast to purely quantitative methods for analyzing project progress, this form of coordination deliberately incorporates soft, subjective factors in order to clarify issues on this basis. This form of analysis can be used very well in the context of daily Scrums or regular project meetings at the Kanban board to capture the mood of the team.

In principle, it can also be useful in plan-based projects, but is not widespread there.

Task:	**M.3.7 Analyze taskboard**	plan-based as well as agile
Project lifecycle:	Initialization / Definition / Planning / Control / Closing	
Team size:		**Team distribution:**
Precondition:	M.2 Determine project progress	
Subsequent tasks:	M.4 Manage progress	
Summary:	The taskboard is a planning, monitoring and control tool in agile projects according to Scrum. As part of the progress analysis, the taskboard helps to identify deficits in the implementation and initiate measures for project control.	

Aim:
The aim is to analyze the progress of the project using the taskboard.

Instruction:
In agile projects according to Scrum, the taskboard visualizes the sprint backlog, i.e. the tasks pending (to do) in the current sprint, see the figure on the right.

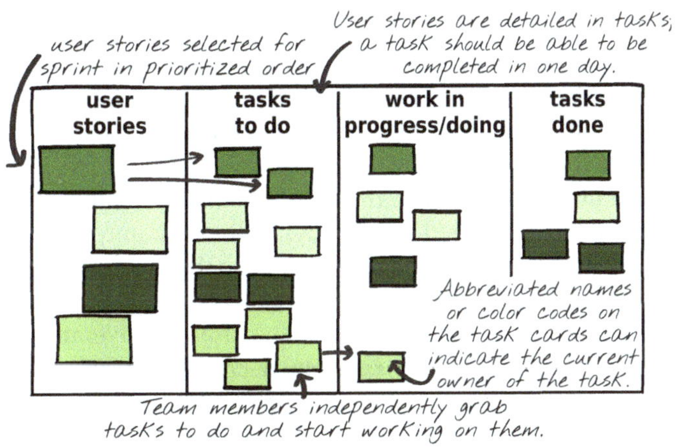

The developers know their velocity, i.e. the number of story points that can be processed within a sprint.

As each task should be provided with an estimate of its size in story points, the taskboard can be used to quickly determine how many story points still need to be completed. By comparing this with the remaining time available in the current sprint, it is possible to check whether the open tasks can still be completed on time.

The taskboard can also be used to regularly check whether the tasks marked as completed really correspond to the defined definition of done, i.e. whether all agreed activities within the task (e.g. code review, test, documentation, etc.) have really been completed. In the event of deviations, the team initiates appropriate control measures.

Task:	M.3.8 Analyze burndown chart	plan-based as well as agile
Project lifecycle:	Initialization / Definition / Planning / Control / Closing	
Team size:		Team distribution:
Precondition:	M.2 Determine project progress	
Subsequent tasks:	M.4 Manage progress	
Summary:	The burndown chart visualizes the amount of open work in story points, number of user stories, number of tasks or another predefined unit.	

Aim:
The aim is to analyze the progress of the project using the burndown chart.

Instruction:
Open work is measured and regularly presented in a unit defined by the project team (usually story points). In agile projects according to Scrum, the recording takes place as part of the daily Scrum. The open tasks or user stories that have not yet been completed are determined and the associated story points are added up. This open number of story points is then plotted over time (days since the start of the sprint) in a diagram. An example can be seen in the figure below.
Instead of the story points, the number of tasks or other variables representing the open work can also be added up directly. However, if only the tasks are counted, it should be noted that these are not necessarily of equal size and the representation can therefore be misleading. The burndown chart can also be used in plan-based projects, for example to show the number of open defects in the project deliverables to be created.
Ideally, the amount of open work decreases linearly until the end of the sprint. In practice, the process is usually not linear. It takes a few days after the start of the sprint before the first tasks can be completed. The curve therefore initially runs horizontally, without this necessarily indicating a project delay.

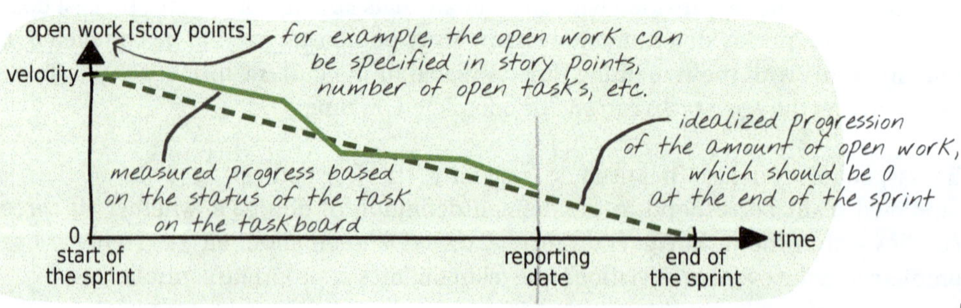

Task:	M.3.9 Conduct sprint review	plan-based as well as agile
Project lifecycle:	Initialization / Definition / Planning / Control / Closing	
Team size:		Team distribution:
Precondition:	M.2 Determine project progress	
Subsequent tasks:	M.4 Manage progress	
Summary:	At the end of the implementation phase of a sprint, a new product increment is created from all completed tasks and made available for review or use. This product increment is analyzed as part of the sprint review.	

Aim:
The aim is to analyze the progress of the project by evaluating the product increment at the end of a sprint in projects based on Scrum.

Instruction:
A sprint review is usually conducted as illustrated in the figure below. The Scrum master invites the developers, the product owner and other stakeholders determined by the product owner, e.g. customers. They take part in the sprint review in whole or in part. The product owner presents the agreed goal of the sprint and the team explains what has actually been achieved. The current increment is discussed and evaluated.

The product backlog is checked and, if necessary, supplemented, revised and reprioritized. At the end, the next steps are discussed. The duration of the sprint review is approximately half a day for a 30-day sprint.

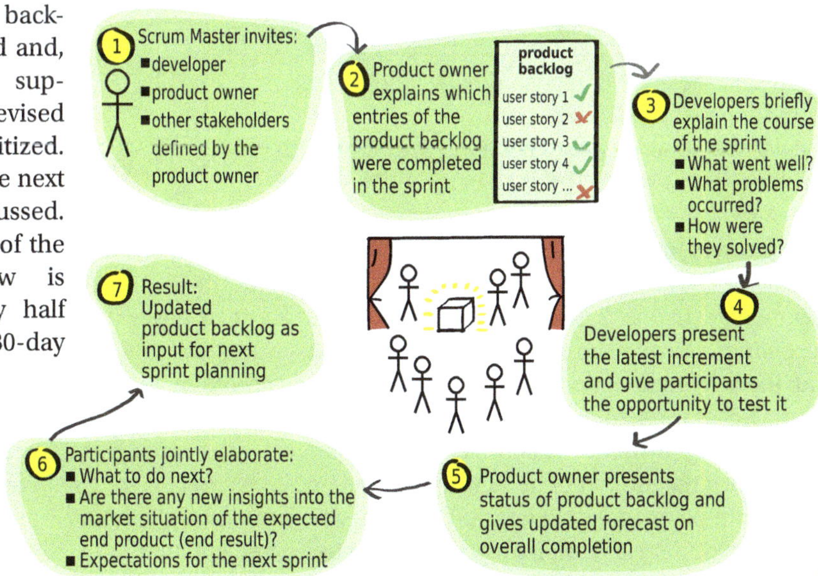

Task:	M.3.10 Conduct sprint retrospective	plan-based as well as agile
Project lifecycle:	Initialization / Definition / Planning / Control / Closing	
Team size:		Team distribution:
Precondition:	M.2 Determine project progress	
Subsequent tasks:	M.4 Manage progress	
Summary:	At the end of the implementation phase of a sprint, a new product increment is created from all completed tasks and made available for review or use. This product increment is analyzed as part of the sprint review.	

Aim:

The sprint retrospective is used to analyze team-internal collaboration in agile projects according to Scrum and to define measures to improve collaboration and increase effectiveness and efficiency.

Instruction:

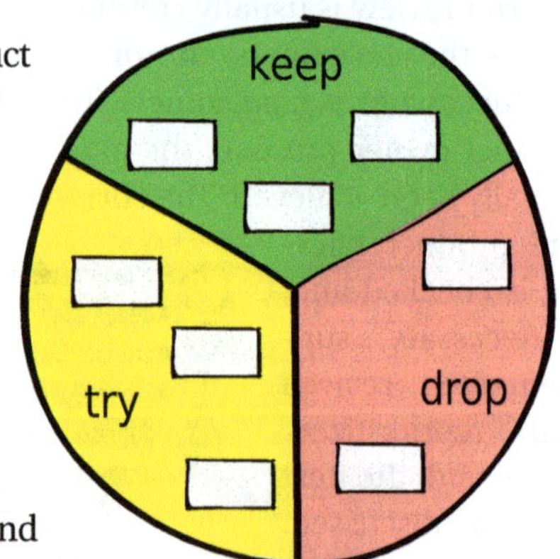

While the sprint review focuses on the product increment and the fulfillment of customer requirements, the sprint retrospective examines the team's internal collaboration and the interfaces to other areas of the organization. In the end, this also benefits the customer, as the team becomes faster, the results achieve a higher quality and overall motivation increases.

Usually, the Scrum master invites the Scrum team to the sprint retrospective. The duration should be chosen so that there is sufficient time for a thorough analysis and the development of measures for improvement. For a sprint duration of 30 days, a retrospective lasting around 3 hours is commonly scheduled. At the beginning of the retrospective, positive and negative aspects of the ending sprint can be collected in a team brainstorming session. These are supplemented with ideas for future improvements and categorized according to the keep-drop-try scheme illustrated in the figure above. Of course, other methods can also be used instead of the keep-drop-try method. The collected and categorized ideas are then evaluated together as a team and different points of view are discussed. In the next step, concrete measures for change are planned and prioritized. Responsible persons are appointed for the measures rated as important and implementation is discussed and initiated.

Task:	**M.3.11 Analyze Kanban board** (plan-based as well as agile)
Project lifecycle:	Initialization / Definition / Planning / **Control** / Closing
Team size:	**Team distribution:**
Precondition:	M.2 Determine project progress
Subsequent tasks:	M.4 Manage progress
Summary:	The Kanban board is a planning, monitoring and control tool in agile projects based on Kanban. When analyzing progress, the Kanban board helps to identify deficits in implementation and initiate measures for project control.

Aim:
The aim is to analyze the progress of the project using the Kanban board.

Instruction:
The Kanban board visualizes the stages that a task must pass through from the start to the end of the project, see figure below. Important Kanban practices, such as visualizing the tasks to be completed, limiting tasks that have been started and measuring and optimizing the flow of tasks through the project, can be implemented directly on the Kanban board. If too many work packages are stuck at one stage for a long period of time, this indicates a bottleneck. The blockage must be resolved. The reason for this may be that employees want to work on too many work packages in parallel, thereby limiting the progress of the work packages.
The Kanban board can also be used to regularly check whether the work packages marked as completed really correspond to the specified definition of done, i.e. whether all agreed tasks (e.g. code review, test, documentation, etc.) have really been accomplished. Another aspect that can be checked on the Kanban board is the number of work packages waiting to be transferred to the next stage. These work packages correspond to expensive buffers or temporary storage of unfinished work. Work packages on hold should be reduced as much as possible. In the event of deviations, the team initiates appropriate measures.

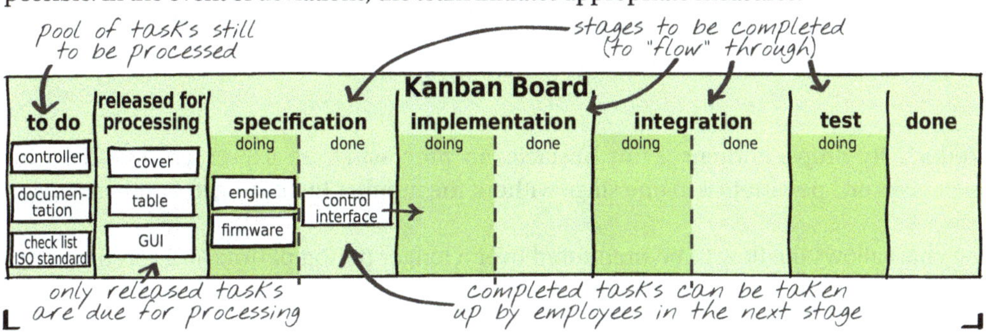

Task:	M.3.12 Analyze cumulative flow		plan-based as well as agile
Project lifecycle:	Initialization | Definition | Planning | Control | Closing		
Team size:		Team distribution:	
Precondition:	M.2 Determine project progress		
Subsequent tasks:	M.4 Manage progress		
Summary:	The analysis of the cumulative flow is based on the recurring recording of the status of the work packages on the Kanban board. It allows the development of the flow to be analyzed over a longer period of time.		

Aim:
The aim is to analyze the progress of the project based on the flow of work over a longer period of time.

Instruction:
The Kanban board visualizes the stages that a task must pass through from the start to the end of the project. The work packages processed there can be totaled for each stage (hence the name: cumulative flow). The Kanban board thus reflects the status on a specific reporting date.

When analyzing the cumulative flow, the number of tasks currently being processed in the individual stages is recorded regularly, for example as part of the daily meeting at the Kanban board. This number per stage is noted and plotted in a chart. The illustration on the right shows an example of such a chart. The diagram thus shows the flow through the individual stages over time.

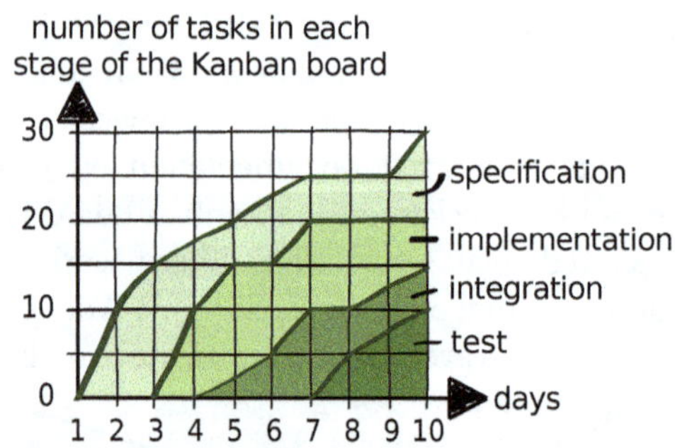

Productivity drops, bottlenecks or obstacles in processing are expressed by a drop in open tasks or a persistence in one stage without the number increasing in another stage.

The chart allows the flow to be monitored over a longer period of time and can therefore help to analyze the progress of the project.

Task:	M.3.13 Analyze buffer consumption — plan-based as well as agile
Project lifecycle:	Initialization / Definition / Planning / **Control** / Closing
Team size:	Team distribution:
Precondition:	M.2 Determine project progress
Subsequent tasks:	M.4 Manage progress
Summary:	Contingency and management reserves are planned in order to cover the time and cost-related effects of known and unknown risks when they occur. The current residual buffers must be monitored and analyzed as part of the progress analysis.

Aim:
The aim is to analyze existing residual buffers in the contingency and management reserve.

Instruction:
Contingency and management reserves were planned depending on potential project risks. Both types of buffer can be viewed as time and cost buffers. If a known risk occurs during project execution, the contingency reserve is reduced by the amount of the additional effort. In the event of an unforeseen risk, the management reserve is reduced accordingly.

The remaining buffer can be plotted along a time axis, as illustrated in the example below. If the remaining buffer is above a reference line (green area in the figure below), less buffer is used than planned. The project can therefore be completed earlier or less expensively than planned taking the buffer into account.

If the residual buffer is below this reference line, more buffer is consumed than planned. Despite the planned buffers, the project will probably be completed later or more expensively than planned taking the buffers into account.

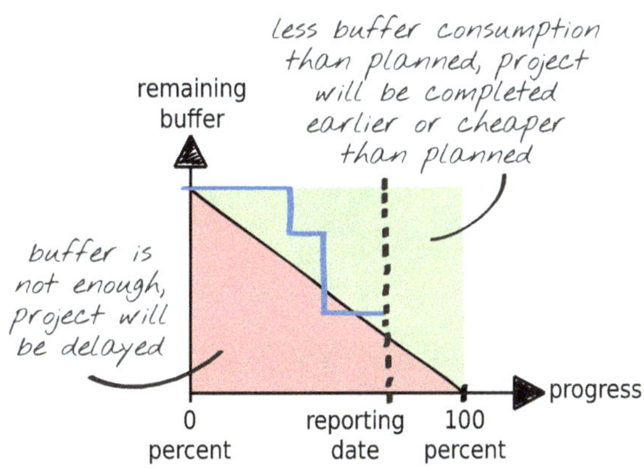

Task:	M.3.14 Conduct A3 problem analysis — plan-based as well as agile
Project lifecycle:	Initialization / Definition / Planning / **Control** / Closing
Team size:	Team distribution:
Precondition:	M.2 Determine project progress
Subsequent tasks:	M.4 Manage progress
Summary:	Problem analysis using the A3 report is a method of structured problem solving. The idea is to solve a problem using an A3-sized template. The problem is analyzed, root causes identified and measures derived to solve it.

Aim:
The aim is the structured analysis of a problem and the derivation and verification of measures to solve it.

Instruction:
The A3 report is a template form that helps with structured problem solving. The procedure follows the principle of continuous improvement and the Japanese Kaizen (change for the better). The following illustration shows an example of an A3 report. It is important that the identified solution measures are also implemented and checked for effectiveness in the subsequent → M.4 management phase.

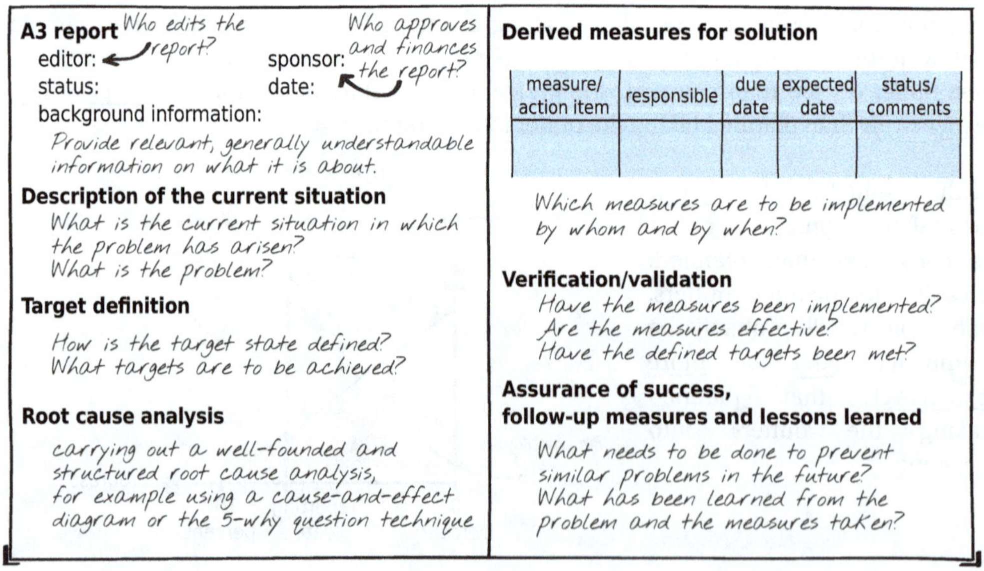

Task:	M.3.15 Analyze cost trend	plan-based as well as agile
Project lifecycle:	Initialization / Definition / Planning / **Control** / Closing	
Team size:	1 – 4+	Team distribution: co-located – distributed
Precondition:	M.2 Determine project progress	
Subsequent tasks:	M.4 Manage progress	
Summary:	The cost trend analysis examines the actual development of costs over time. It is often used in progress reports for management, as it provides a clear overview of the cost trend. The costs can be shown for work packages, sub-projects or for the entire project.	

Aim:
The aim is to analyze the development of important cost items in the project over time.

Instruction:
The first step is to determine which cost items are to be planned, regularly monitored and reported. For example, costs can be broken down by project phase, sub-project or work package.

The reporting dates are plotted on the horizontal axis and the cost items according to the plan at the time of the respective reporting dates are plotted on the vertical axis.

If the course of a cost item moves upwards, there is a risk of cost increases. In the example on the left, the call for tenders ends up being slightly cheaper than originally planned on January 1st, while the costs for the training concepts increase significantly and are even more expensive than the planned costs for IT procurement on the reporting date.

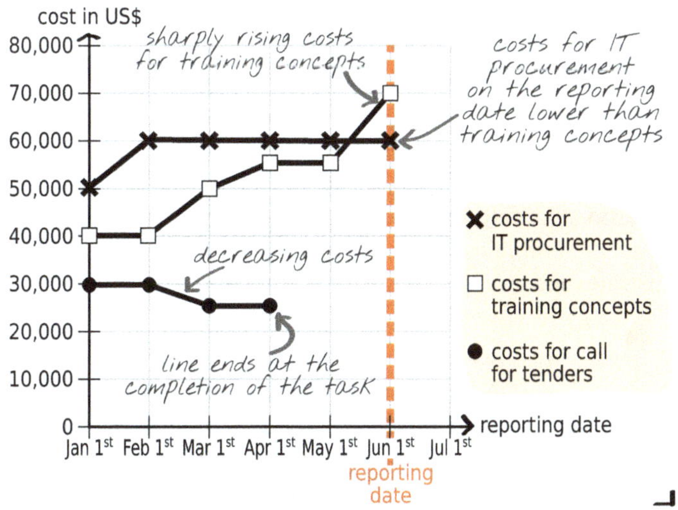

M.4 Manage progress

Task:	M.4.1 Plan scenarios	plan-based as well as agile
Project lifecycle:	Initialization / Definition / Planning / Control / Closing	
Team size:		Team distribution:
Precondition:	M.3 Analyze project progress	
Subsequent tasks:	depending on project M.1 Iterative detail/scenario planning C Closing project	
Summary:	Scenarios can be planned and optimized in order to examine the effect of potential control measures and select the best one based on this.	

Aim:
The aim is to identify the best control measure(s) through scenario planning.

Instruction:
Scenarios are forward-looking views of possible future project developments. They are usually created step by step as follows:

- Determine the current status, see →M.2 determine project progress.
- Analyze the causes of the deviation from the plan, see →M.3 analyze project progress.
- Identify and weight factors that influence the further course of the project.
- Define assumptions as to how the factors will develop in the future and what impact they will have on the course of the project.
- Bundling various assumptions into different scenarios, for example a pessimistic, a realistic and an optimistic scenario.

Scenario planning can be extended as required. Project managers should always be aware of possible scenarios of probable outcomes. The following figure illustrates scenario planning in a simple way.

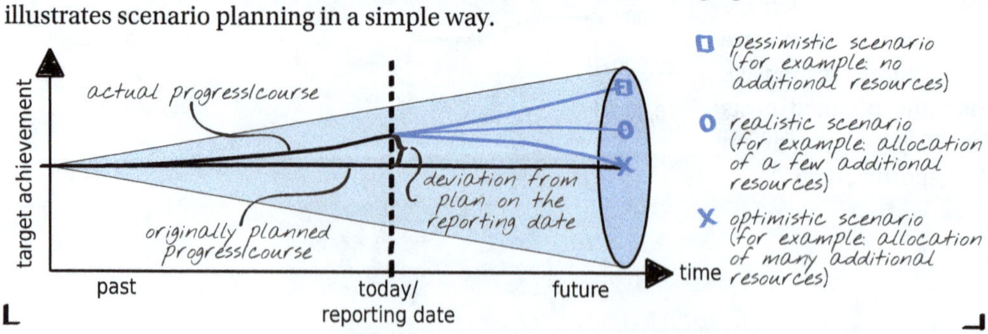

Task:	M.4.2 **Change resources** — plan-based as well as agile
Project lifecycle:	Initialization · Definition · Planning · **Control** · Closing
Team size:	 **Team distribution:** 
Precondition:	M.3 Analyze project progress
Subsequent tasks:	depending on project M.1 Iterative detail/scenario planning C Closing project
Summary:	If the analysis of the project shows that the actual progress deviates from the original plan, control measures must be initiated. One way to reduce these deviations is to change the quantity or quality of the resources allocated.

Aim:
The aim is to reduce deviations from the project plan by changing the quantity or quality of allocated resources.

Instruction:
If the project is slower than planned, additional resources can be requested to take on tasks and thus make faster progress. However, additional resources also increase the overall project costs, meaning that the possible adherence to the deadline target is achieved at the sacrifice of not meeting the cost target.

It should be noted that the necessary training of additional employees costs money and time: Employees already working on the project have to train the new employees at the expense of their actual tasks. As a result, there is a short-term risk of additional project delays. Overall, the costs for new employees therefore increase over-proportionally. The positive effects of the management measure only pay off in the medium to long term. For this reason, this instrument of project control should not be used shortly before the end of the project.

A change in available resources can also be achieved within the project without additional new employees, for example by

- a redistribution of employees towards a bottleneck or a relief of employees from tasks that have nothing to do with the bottleneck,
- overtime or suspension of vacation, or
- purchasing external services or outsourcing tasks.

Task:	M.4.3 Reduce effort (plan-based as well as agile)
Project lifecycle:	Initialization / Definition / Planning / **Control** / Closing
Team size:	Team distribution:
Precondition:	M.3 Analyze project progress
Subsequent tasks:	depending on project M.1 Iterative detail/scenario planning C Closing project
Summary:	If the analysis of the project shows that the actual progress deviates from the original plan, control measures must be initiated. One way to minimize these deviations is to reduce the effort required to implement the tasks.

Aim:
The aim is to minimize deviations from the project plan by reducing the effort required to implement the tasks.

Instruction:
In principle, effort optimization should be an integral part of every project planning and implementation. However, at the latest when time-related or cost-related deviations between the actual course of the project and the original plan have been identified, it must be examined how the part of the project deliverables that has not yet been processed can be realized with the least possible effort.

Methods such as value analysis or general creativity techniques can be used for this purpose. Established solutions should be challenged. It should be checked whether comparable results can be achieved in other, faster and cheaper ways.

Purchasing external expertise can also reduce your own workload and thus speed up or reduce the cost of the project.

The reduction in effort can also make sense in combination with a minor change to the project scope: Performance requirements are often set without thinking sufficiently in advance about the effort involved. Sometimes small compromises in the fulfillment of the performance requirements lead to significantly reduced effort.

Task:	**M.4.4 Increase productivity**		plan-based as well as agile
Project lifecycle:	Initialization | Definition | Planning | **Control** | Closing		
Team size:		**Team distribution:**	
Precondition:	M.3 Analyze project progress		
Subsequent tasks:	depending on project M.1 Iterative detail/scenario planning C Closing project		
Summary:	If the analysis of the project shows that the actual progress deviates from the original plan, control measures must be initiated. One way to minimize these deviations is to identify and implement measures that increase productivity.		

Aim:
The aim is to reduce deviations from the project plan by increasing productivity.

Instruction:
Good project managers constantly challenge working methods and tasks and check whether they contribute to the success of the project. Lean management concepts help to reduce any form of waste or avoid it altogether.

Examples of measures to increase productivity are

- training measures for employees to enable them to perform their tasks more efficiently, for example by learning to use software programs better and faster,
- the use of better tools that enable tasks to be completed faster or with better quality, for example through machines with higher throughput or software programs with more relevant functionality, and
- maintaining or increasing employee commitment, for example by creating working conditions that meet employees' needs, by valuing their work or by offering financial or other incentives.

A typical mistake is to try to increase productivity through measures that extend working hours in the short term (overtime, weekend work), but which in the long term lead to overload, declining commitment and a reduction in productivity.

Task:	M.4.5 Improve process quality and collaboration	plan-based as well as agile
Project lifecycle:	Initialization / Definition / Planning / **Control** / Closing	
Team size:		Team distribution:
Precondition:	M.3 Analyze project progress	
Subsequent tasks:	depending on project M.1 Iterative detail/scenario planning C Closing project	
Summary:	If the analysis of the project shows that the actual progress deviates from the original plan, control measures must be initiated. One way to reduce these deviations is to improve process quality and collaboration.	

Aim:
The aim is to reduce deviations from the project plan by improving process quality and collaboration.

Instruction:
Poor internal processes in the project and at the interfaces to external partners can impair the efficiency and effectiveness of the project. Project managers should therefore regularly check whether there are structures and processes that hinder the success of the project.

Symptoms of poor process quality and collaboration problems include conflicts between several employees or between employees and external partners. Also, if the speed of work is slower than expected or slows down during the project, poor process quality or disrupted collaboration may be the root cause.

Opportunities for improvement include, for example:

- Check communication interfaces and optimize them; escalation and reporting channels may be longer than necessary or meeting cycles too short or too long
- Replacing missing or incorrect key performance indicators for project management with suitable key performance indicators
- Early detection, resolution and permanent avoidance of conflicts

Task:	M.4.6 Change project scope	plan-based as well as agile
Project lifecycle:	Initialization | Definition | Planning | **Control** | Closing	
Team size:		Team distribution: 
Precondition:	M.3 Analyze project progress	
Subsequent tasks:	depending on project M.1 Iterative detail/scenario planning C Closing project	
Summary:	With Scrum, controlling by adjusting the scope is an integral part of the project. This can also be used in other projects, for example to achieve cost or time targets or to consider changing customer requirements.	

Aim:
The aim is to control the project by changing its scope.

Instruction:
Projects can be controlled by changing their scope. If additional requirements are added to the project, more or new customer needs can be taken into account. This usually makes the project more expensive or takes longer. Conversely, if the scope is reduced, the project will probably be cheaper or can be completed earlier.

The agile manifesto emphasizes the value of change and places it above strict adherence to a plan. This is why changing the scope in agile projects is usually an integral part of the process. In plan-based projects, on the other hand, the scope is documented in the requirements and functional specifications and is initially regarded as firmly agreed. In both agile and plan-based projects, the client, as the most important stakeholder, must be satisfied at the end. If the achievement of the deadline or cost target is at risk and this is more important to the client than the agreed functionality, this can be reduced in favor of achieving the deadline or cost target. However, the scope should always be changed in agreement with the client.

The changes can be made in various ways, e.g.

- stepwise deployment of the project deliverables in stages,
- quality reductions,
- changes to the scope so that the implementation effort is reduced, or
- reduction of the scope with the elimination of entire work packages.

Task:	M.4.7 Prioritize	plan-based as well as agile
Project lifecycle:	Initialization / Definition / Planning / Control / Closing	
Team size:		Team distribution:
Precondition:	M.3 Analyze project progress	
Subsequent tasks:	depending on project M.1 Iterative detail/scenario planning C Closing project	
Summary:	In the control phase, there are regularly many tasks, such as creating documents, processing tests or solving errors in the system. Not all tasks are equally important for the success of the project. An unfocused approach can jeopardize this and the achievement of the really important project goals. Therefore, the tasks at hand must be carefully prioritized.	

Aim:
The aim is to control the project by prioritizing upcoming tasks.

Instruction:
During project execution, depending on the respective project, there are a variety of tasks, for example

- the preparation of documents (specifications, design documents, declarations of conformity to standards and directives, etc.),
- the organization of events and important meetings,
- the completion of tests as part of the verification and validation of the project deliverables, or
- the solution of errors in the system.

Not all tasks are equally important for achieving the overall project goals and thus for ensuring the success of the project. Solving errors in particular takes up a lot of time in the control phase of technical research and development projects. If the project team does not focus on the tasks that are really critical to success, it runs the risk of getting bogged down and not finishing within the set deadline and the allocated budget.
All tasks should therefore be prioritized on an ongoing basis and this prioritization should be communicated and followed within the team. This also and especially applies to solving errors. Perfectionists often demand that all errors be solved, which is rarely

Task:	M.4.8 Implement measures of the sprint retrospective	plan-based as well as agile
Project lifecycle:	Initialization | Definition | Planning | **Control** | Closing	
Team size:		Team distribution:
Precondition:	M.3 Analyze project progress	
Subsequent tasks:	depending on project M.1 Iterative detail/scenario planning C Closing project	
Summary:	The sprint retrospective is used to analyze team-internal collaboration in agile projects according to Scrum and to define measures to improve collaboration. During project control, these measures must be implemented and checked for effectiveness.	

Aim:

The aim is to manage the project by implementing the measures of the sprint retrospective.

Instruction:

The sprint retrospective analyzes the team's collaboration and the interfaces to other areas of the organization. The keep-drop-try scheme presented in → M.3.10 sprint retrospective helps to strengthen structures and processes that are worth maintaining, to identify structures and processes that urgently need to be changed and to identify ideas that have potential for improvement.
Examples include

- use of different software tools to enable better software engineering, programming, troubleshooting, or documentation
- changed communication interfaces to important stakeholders
- concretization of the definition of done after previous misunderstandings
- solving resource bottlenecks in the team

In addition, the Scrum master records measures in the → M.1.5 impediment backlog during the sprint. All measures in the sprint retrospective and the impediment backlog can only be effective if they are implemented. This must be ensured as part of project control. In addition, the effectiveness of the measures should be checked to ensure that the problems are actually solved and experience is gathered for future effective measures.

C Closing

C.1 Deploy and handover

Task:	C.1.1 Handover project deliverables	plan-based – as well as – agile
Project lifecycle:	Initialization / Definition / Planning / Control / **Closing**	
Team size:		Team distribution:
Precondition:	M Monitor and control project	
Subsequent tasks:	C.2 Analyze project	
Summary:	The completed project deliverables must be handed over to and accepted by the client.	

Aim:
The aim is for the project deliverables to be accepted and approved by the project client.

Instruction:
The completed project deliverables, including all components, the associated documentation and any other agreed items, must be handed over by the project to the client. The acceptance of the project deliverables can be divided into two parts:

- handover of the project deliverables from the project to the client and, if necessary, careful inspection of the deliverables by the client
- acceptance of the project deliverables by the client

This acceptance is typically accompanied by **legal consequences** for the client:

- liability periods for defects usually begin now
- reversal of the burden of proof to the client
- transfer of risk to the client
- due date of (outstanding) payments

Due to the great importance of these legal consequences, the handover and acceptance should be recorded carefully. A note should be made in the minutes,

- what is being handed over (precise description of the project deliverables),
- how it will be handed over (free of defects or with defects) and
- what rework is to be carried out and by when.

The minutes are signed by the client and the project manager and included in the project documentation. In the event of existing defects, acceptance can be granted conditionally. Existing defects should be clearly documented and any resulting rework should be specified with a deadline. Acceptance criteria for the final acceptance should also be defined.

C.2 Analyze project

Task:	C.2.1 Conduct post-calculation	plan-based as well as agile
Project lifecycle:	Initialization / Definition / Planning / Control / **Closing**	
Team size:		Team distribution:
Precondition:	M Monitor and control project	
Subsequent tasks:	C.2 Analyze project	
Summary:	Post-calculation is used to compare the originally estimated costs with the actual costs incurred. This allows the originally planned profitability targets of the project to be checked.	

Aim:
The aim is to review the profitability targets with the help of a post-calculation.

Instruction:
Post-calculation involves comparing the originally estimated costs and expenses with those actually incurred. Any deviations are analyzed. Depending on the agreement with the client, the recalculation can lead to further payment claims against the client or reimbursements by the contractor.
The following illustration shows an example of a post-calculation. Potential causes of deviations between planned and actual costs are listed below:

- Uncertainties at the time of cost estimation
- Changes in general conditions such as raw material prices
- Change requests due to changed customer requirements
- Changes due to the impossibility of implementing a selected solution

The post-calculation should also be considered in the → C.2.3 lessons learned.

work package	planned BAC [US$]	approved change request (CR)	new BAC [US$]	AC [US$]	deviation [US$]	reason
1.3.2 implement database	20,000	CR100053: addition of a further address line in the data model	24,000	26,000	+2,000	late modification of the data model was more complex than estimated
1.3.3 implement printer interface	10,000	-	-	10,000	-	-
1.3.4 design user interface	30,000	CR100054: added requirement that interface can also be used on mobile devices	45,000	40,000	-5,000	purchase of a software library led to simpler implementation

Task:	C.2.2 Evaluate project	plan-based as well as agile
Project lifecycle:	Initialization / Definition / Planning / Control / **Closing**	
Team size:		Team distribution:
Precondition:	M Monitor and control project	
Subsequent tasks:	C.2 Analyze project	
Summary:	A systematic project evaluation helps to check the appreciation of the project by the important stakeholders. The evaluation can be conducted as part of a (sprint) retrospective, a workshop, an informal survey or a questionnaire.	

Aim:
The aim is to check the appreciation of the project by relevant stakeholders.

Instruction:
The appreciation of the most important stakeholders determines whether the project is perceived as a success or a failure. This appreciation can be determined in different ways, for example through

- a retrospective,
- an analysis workshop,
- informal interviews
- structured interviews, or
- questionnaires.

The most important stakeholders usually include the client of the project, important customers and users, suppliers and involved departments of the company. The company's own project team can also provide valuable information. An example questionnaire is shown in the illustration on the right. It is important that the questions are asked in such a way that the answers allow a meaningful analysis.

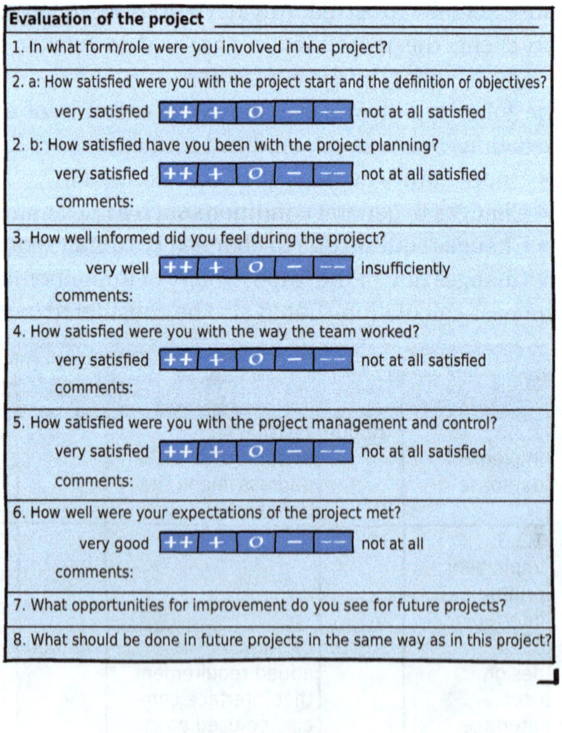
Evaluation of the project ______

1. In what form/role were you involved in the project?

2. a: How satisfied were you with the project start and the definition of objectives?
very satisfied ++ + o – –– not at all satisfied

2. b: How satisfied have you been with the project planning?
very satisfied ++ + o – –– not at all satisfied
comments:

3. How well informed did you feel during the project?
very well ++ + o – –– insufficiently
comments:

4. How satisfied were you with the way the team worked?
very satisfied ++ + o – –– not at all satisfied
comments:

5. How satisfied were you with the project management and control?
very satisfied ++ + o – –– not at all satisfied
comments:

6. How well were your expectations of the project met?
very good ++ + o – –– not at all
comments:

7. What opportunities for improvement do you see for future projects?

8. What should be done in future projects in the same way as in this project?

Task:	**C.2.3 Analyze lessons learned**		plan-based as well as agile
Project lifecycle:	Initialization | Definition | Planning | Control | **Closing**		
Team size:		**Team distribution:**	
Precondition:	M Monitor and control project		
Subsequent tasks:	C.3 Dissolve project organization		
Summary:	Identifying and analyzing lessons learned are part of knowledge management in projects. At the end of the project, the knowledge gained over the course of the project must be secured, evaluated, consolidated and made available for future projects.		

Aim:
The aim is to provide relevant experience for future projects.

Instruction:
Due to the unique nature of projects, there is a high risk that positive and negative experiences will be lost at the end of the project and not used for future projects. Successful project-oriented companies therefore establish a culture of learning and knowledge transfer. When recording lessons learned, knowledge gained in the project, such as

- unforeseen risks and successful countermeasures,
- effort estimates and actual efforts of completed tasks,
- dealing with stakeholders,
- successful or failed project management tools, ...

is determined.
Successful knowledge management only works if certain requirements are met: The experience must be

- secured,
- evaluated in terms of its contribution to the success of the project,
- presented in a condensed and compact form (nobody reads long project reports), and
- easy to find and actually used for new projects.

To secure the lessons learned, it is recommended that the project manager and the team regularly document their experiences during the course of the project, which can then be consolidated at the end of the project. The preparation can later be published on the intranet (wiki, templates, checklists, tables with experiences from the effort estimation), disseminated as part of regular training sessions for colleagues, or incorporated into the further development of the company's own process model.

C.3 Dissolve project organization

Task:	C.3.1 **Recognize achievements**		plan-based as well as agile	
Project lifecycle:	Initialization	Definition	Planning	Control / Closing
Team size:		**Team distribution:**		
Precondition:	M Monitor and control project			
Subsequent tasks:	C.3 Dissolve project organization			
Summary:	Before the project is dissolved, the achievements of the team members must be recognized and communicated to the disciplinary supervisors.			

Aim:
The aim is to recognize and communicate the achievements of a project's team members.

Instruction:
Project staff make a significant contribution to the success of a project by

- taking responsibility for tasks,
- working on tasks,
- proactively recognizing and counteracting risks, and
- promoting the customer's appreciation of the project.

In order to accomplish these contributions, project team members sometimes have to suspend line tasks or tasks from other projects and justify this to their disciplinary supervisors. However, these disciplinary supervisors will usually evaluate the employees and make further career development dependent on their performance.

It is therefore an important task of the project manager to ensure that the achievements made in the project are recognized and also communicated to the respective disciplinary supervisors of the employees. Otherwise, there is a risk that employees will prefer line tasks or tasks from other projects because they consider these to be more career-boosting.

Task:	**C.3.2 Create final project report** — plan-based / as well as / agile
Project lifecycle:	Initialization / Definition / Planning / Control / **Closing**
Team size:	
Team distribution:	
Precondition:	M Monitor and control project
Subsequent tasks:	C.3 Dissolve project organization
Summary:	The final project report provides a summary description of the project. It is the last progress report or status report of the project before its dissolution.

Aim:
The aim is to produce a final project report as a means of stakeholder communication.

Instruction:
The final project report summarizes the final project status for relevant stakeholders before the project is closed. These stakeholders usually include

- the client,
- the members of the project team, and
- other stakeholders who were closely involved in the project.

The final project report can be structured similarly to the other progress or status reports of the project. However, the content focuses on the result and the lessons learned:

- Overall assessment of the project
- Outline of the course of the project including particular challenges and their solution
- Analysis of the project deliverables and their achievement
- Follow-up projects, outlook, sustainability aspects

The final project report should be written to meet the needs of the target group and be as short and concise as possible. In the hectic day-to-day business of projects and companies, nobody will read pages of text in which the relevant information disappears in an unstructured way.

Like the progress or status reports, the final project report is approved by the project client.

Task:	**C.3.3 Release resources** — plan-based as well as agile
Project lifecycle:	Initialization | Definition | Planning | Control | Closing
Team size:	
Team distribution:	
Precondition:	M Monitor and control project
Subsequent tasks:	C.3 Dissolve project organization
Summary:	If there are no more tasks for individual team members or the entire team in the project, they are released for other project or line tasks.

Aim:
The aim is to release project team members for other projects and tasks.

Instruction:
Once the project is completed and no open tasks remain, project team members must be released for other tasks in projects or the line. Despite this rational and obviously necessary measure, implementation is not always conflict-free, as emotions are associated with the project. Various aspects come together and should be taken into account when dissolving the team:

- For a long time, the project was "home" for many team members who are now finding it difficult to let go of the collaboration and tasks they have grown fond of.
- Other team members are no longer interested in the remaining work required to complete the project, as they are already working on exciting new projects in which they see a longer-term perspective.
- Still other team members do not yet have a follow-up project. The completion of the current project means uncertainty for them.

Project managers must not ignore these emotions, otherwise there is a risk of conflict. The completion of the project object can be delayed

- in order to keep the cherished project alive or, conversely,
- because required team members no longer prioritize the project and are already involved in new projects.

The release of team members should therefore be planned in advance. Together with the resource managers, the project manager must help to ensure that employees find follow-up projects and, conversely, that all team members originally allocated and still required for project completion are available as planned.

Task:	**C.3.4 Celebrate project completion**	plan-based as well as agile	
Project lifecycle:	Initialization | Definition | Planning | Control | Closing		
Team size:		**Team distribution:**	
Precondition:	M Monitor and control project		
Subsequent tasks:	- -		
Summary:	The end of the project must be celebrated appropriately. This shows appreciation for the work that has been done and also marks the end of the project.		

Aim:
The aim is to conclude the project with an adequate celebration.

Instruction:
The project tasks have been completed, the project deliverables have been handed over to the client, all remaining work has been finished and the project knowledge has been analyzed and stored.

Now it is time to honor the achievement in the form of an appropriate celebration and at the same time mark the completion of the project. The scope of the celebration should match the project. A small organizational project can be concluded with a meal shared by the project team. A production planning project in which a new production line has been set up will probably celebrate its completion with the client, suppliers and other stakeholders on a larger scale.

It is important to think about a worthy conclusion to the project at an early stage, to announce the date in good time and to plan the budget for the celebration.

It is also important that the celebration also suits the relevant stakeholders. The project team must not get the impression that they have done the work, but that they have been ignored when it comes to recognizing the results.

A poorly planned or inappropriate project closing ceremony can have the opposite effect and give the impression of a lack of appreciation. In the long term, this leads to a drop in motivation and frustration.

9 Glossary

The glossary summarizes the key definitions you have learned in this book.

Activity
An activity is a temporal process element that represents a self-contained assignment. A work package can be converted into one or more activities. While a work package defines what is to be completed, activities define the (temporal) sequences of the assignments. Activities are not part of the work breakdown structure.

Actual cost
The actual cost is the cost actually spent up to a specific reporting date.

Agile project management
Agile project management is an overarching term for agile process models such as Scrum and Kanban. Agile is to be understood in accordance with the values of the manifesto for agile software development.

Agile release train (ART)
The agile release train (ART) consists of several agile teams. These iteratively develop value stream-oriented solutions with other stakeholders, deliver them and operate them if necessary.

Authority
An authority is a permission and power to perform certain actions and make decisions.

Budget at completion
The budget at completion BAC is the total cost of the project at the scheduled completion date, according to the cost plan.

Burndown chart
The burndown chart is a diagram on which open remaining efforts are displayed as a function of time. These remaining efforts can be specified in staff hours, the number of open requirements, story points or other units. By a planned deadline, these remaining efforts should have been "burned down" to zero.

Business case
A business case is an economic justification for a project and contains a quantitative comparison of the expected income and the corresponding expenses for a project.

Capacity
The capacity indicates how many resources of a certain type (people or material resources) or qualification (e.g. requirements engineers, software developers, integrators) are available per period of time.

Claim
A claim is a contractual demand that can be made on the other party of the contract. This requires that the claim is already defined in the contract, in an amendment agreement, or an approved change order.

Claim management
The task of claim management is to monitor deviations and changes to contracts. These are evaluated with respect to their consequences for the project and, if necessary, enforced against the contractual partner in the form of additional claims.

Communication
Communication is the process of passing information from a sending to a receiving person.

Competence
A competence is the proven ability to apply knowledge and skills appropriately to the situation.

Configuration
A configuration describes the functional and physical characteristics of a product, a service or an object in general.

Configuration management
Configuration management ensures the transparency, traceability and consistency of all functional and physical characteristics of an object throughout its entire lifetime.

Configuration unit
A configuration unit fulfills one or more of the characteristics defined in the configuration.

Conflict
A conflict is a disagreement between two or more parties with different interests.

Conflict management
Conflict management refers to the process of conflict identification, analysis, resolution and sustainable conflict avoidance.

Contingency reserve
The contingency reserve is a buffer for efforts and costs from identified, i.e. known, risks. The amount of the contingency reserve is determined from the assessment of the risks. If a risk occurs, the associated efforts and costs can be taken from the contingency reserve.

Contract
Contracts govern the relationship between two or more parties. They define what is to be provided by whom to whom and which rules apply. Contracts are the basis of legal transactions.

Contract management
Contract management is one of the tasks of project management. It serves to prepare, conclude and process contracts.

Crisis
A crisis is a seemingly hopeless conflict that can lead to a project being stopped or even canceled completely.

Critical chain
The critical chain is the sequence of tasks that represent the actual bottleneck, taking into account the resources actually available in the project.

Critical chain project management
Critical chain project management is based on the theory of constraints. The planning of work packages is designed to optimize the management of bottleneck resources and avoid waste through harmful multitasking.

Critical path
The critical path is the sequence of activities and milestones in which there are the fewest (usually no) floats. The critical path therefore determines the duration of the overall project.

Daily Scrum
The daily Scrum is the short daily meeting of the developers. It is always held at the same time as a regular meeting. The daily Scrum lasts a maximum of 15 minutes. There is a brief discussion about what has been achieved since the last meeting, what is to be achieved until the next meeting and what obstacles there are on the way or who can help to solve the obstacles.

Data
Data are characters (numbers, letters, etc.) that describe a certain issue and are administered on a data carrier (paper, hard disk, etc.).

Default in performance
Default in performance arise from violations of obligations by contracting parties. Defaults can be, for example, incomplete deliveries, delays or poor performance.

Definition of done
The definition of done is the formal description of the finished implementation of a product backlog entry. When the definition of done is reached, this implementation supplements the previous increment.

Developer
The developers consist of interdisciplinary experts. Alongside the Scrum master and the product owner, they form the Scrum team.

Disciplinary supervision
Disciplinary supervision relates to compliance by all employees with the standards of conduct and behavior applicable in the company when performing their duties. In the event of negative or positive deviations from these standards, the disciplinary supervisor can discipline or reward employees. The tasks of the disciplinary supervisor include approving vacation, granting salary increases and assigning employees to specific tasks and projects.

Dissonant leadership
Dissonant leadership arouses negative feelings, promotes stress, builds up pressure and discourages employees in the long term. However, employees can be encouraged to show greater commitment in the short term.

Document
A document contains information that is summarized in the form of text, images or video and audio recordings in a self-contained form. Documents are usually stored in paper form or electronically as a file or website.

Documentation
The term "documentation" refers to the complete life cycle of documents, from their creation and revision to their archiving and destruction.

Duration
The duration or lead time is the pure time required to complete a work package or activity, taking into account the resources available.

Earned value
The earned value EV is the (planned) value of the (actual) work performed on a chosen date. It is calculated as the product of the percent complete PC and the budget at completion BAC.

Effort
Effort refers to the need for or consumption of time, money or human and material resources to complete a task.

Emotional intelligence
Emotional intelligence is the ability to correctly perceive, understand and influence one's own and other people's feelings.

Epic
An epic is a large, still vague user story. The requirement can only be roughly outlined and its size (complexity) cannot be reliably estimated.

Escalation
An escalation is the structured delegation of a problem or decision to the next higher hierarchy level if it cannot be solved at the current hierarchy level.

Estimate at completion
The estimate at completion EAC is the forecast of the project progress at the planned completion date.

Estimation meeting
In the estimation meeting, the developers familiarize themselves with the current status of the product backlog. They estimate the size of the entries in the product backlog. The product owner is available to answer questions regarding the entries.

Extrinsic motivation
With extrinsic motivation, the incentive to do something comes from the outside. The incentives can be positive, such as bonuses and rewards, or negative, such as salary cuts, not being considered for promotions and blocking vacations.

Flow chart
The flow chart defines the logical sequence of tasks in the project.

Free float
The free float is the maximum time of an activity that this activity can be delayed from its early position without delaying the subsequent activity from its early position.

Function
A person's function describes the purpose and substantive tasks associated with their position in the company. Job descriptions should contain responsibilities and authorizations.

Functional supervision
In the case of functional supervision, the supervisors are authorized to specify the manner in which tasks are to be performed. They decide on the modalities of carrying out a certain task.

Hybrid project management
Hybrid project management refers to the combined use of methods, roles, processes and phases of different standards or process models.

Impediment backlog
In the impediment backlog, the Scrum master notes all impediments that the team encounters while working. It is the Scrum master's responsibility to take care of these obstacles.

Increment
An increment is a usable and tested (intermediate) result of a product. Increments build on each other in such a way that they represent a step towards achieving the product goal.

Independent float
The independent float is the time span by which an activity or milestone can be postponed if the predecessor is in the late position and the successor is in the early position.

Information
An information is an intangible good that has a value for the recipient (of the information).

Intrinsic motivation
With intrinsic motivation, the incentive for a certain behavior comes from within us and from the activity itself. We enjoy doing something.

Iron triangle
The iron triangle of project management embodies the traditional view of projects, which are characterized by the service or quality to be delivered (first corner), the necessary costs (second corner) and the time required (third corner).

Kanban
Kanban is known from production control as a method for optimizing value chains. In project management, Kanban is a process model that visualizes and optimizes project workflows. The central instrument is the Kanban board; this is supplemented by additional practices such as visualizing the work flow, limiting the amount of work started and continuous improvement.

Kanban board

The Kanban board is the central visualization and control tool for Kanban. Unlike the task board, which displays the status of individual tasks to be performed, the Kanban board visualizes the workflow through the individual processing stages. For a technical requirement, this flow can be visualized by the stages requirement, design, implementation, integration, test, documentation and completed. Each stage has two columns on the Kanban board, one for "to do" and one for "in progress".

Knowledge

Knowledge is based on information and data and enables a person to solve problems.

Lead time

The duration or lead time is the pure time required to complete a work package or activity, taking into account the resources available.

Leadership

Leadership is the direct or indirect guidance and influencing of others to achieve a goal.

Leadership concepts

Leadership concepts are general procedures and measures for achieving objectives.

Leadership style

Leadership styles are fundamental behavioral patterns of the person who takes on the leadership task. Leadership styles express the attitude towards employees.

Lean project management

Lean project management applies the principles of lean management to project management. The core idea of lean management is to focus on activities that create value for the customer and to avoid all waste that does not contribute to value creation.

Management

Management refers either to the control and organization of something, e.g. a business or an enterprise, or it refers to the people who are responsible for an organization.

Management reserve

The management reserve is a buffer for efforts and costs from unknown risks. Its amount is determined from the experience of past projects and the company's risk policy. If an unexpected risk occurs, the resulting efforts and costs are compensated from the management reserve.

Material resources

Material resources include materials (e.g. for the construction of prototypes), consumables, operating resources, infrastructure (offices, laboratories), equipment and energy.

Milestone

Milestones are events of particular importance. They define important points in time in the schedule at which certain deliverables must be available.

Minimum viable product

A minimum viable product is a product that offers the minimum number of features to be usable and valuable for the customer.

Mission

The mission of an organization describes the way to implement the vision. It defines how the vision is to be achieved (daily actions), which offers (products, services, etc.) characterize the organization and which values are important in the process.

Modern project management

Modern project management refers to situationally appropriate project management that facilitates efficient and effective project implementation. It can include both plan-based (traditional, predictive) and agile components.

Motivation
Motivation is a state that drives people to engage in target-oriented behavior. It is often understood as a force that explains why we start, continue or end a certain behavior at a certain point in time.

Network diagram
Network diagrams represent the schedule graphically or in the form of tables with the help of activities, milestones and relationships.

Opportunities
Risks and opportunities are potential, unplanned events that are characterized by a probability of occurrence and an impact on the project objectives. Risks counteract the project objectives, i.e. the project suffers damage. Opportunities support the objectives.

Organizational chart
An organizational chart visualizes the organizational hierarchy, structures organizational units and depicts relationships between them.

Parkinson's law
Work expands so as to fill the time available for its completion.

Percent complete
The percent complete PC is the ratio of work already completed to the total workload on a chosen date.

Phase
A phase is a time-limited section that differs in content from other sections.

Phase model
Phase models are part of a process model and structure the project in phases. Phases can be run through sequentially (one after the other), partially in parallel, completely in parallel or recurrently (multiple times).

Phase plan
Phase plans are rough project plans that structure the project into phases. For each phase, tasks to be completed, required resources, costs and deadlines are roughly defined. Phasing is usually based on a phase model. Its phases are used as the starting point for phase planning.

Plan
A plan is a (usually simplified) representation of an expected (future) target state. The plan can contain (work) steps to achieve this target state, premises and other information to be considered.

Planned value
The planned value PV are the planned costs at a chosen date. At the end of the project, the planned value corresponds to the budget at completion.

Planning poker
Planning poker is an estimation method similar to the broadband Delphi method, in which the participants place their estimate face down on the table using playing cards with the estimated value on them. Only then are the cards turned over and the estimates discussed. The estimated values on the playing cards are roughly based on the Fibonacci numbers and are often: 0, 1, 2, 3, 5, 8, 13, 20, 40, 100 and ?, whereby the question mark indicates that the estimator cannot provide an estimated value.

Portfolio
A (project) portfolio comprises a series of projects and/or programs that are not necessarily related to one another. Rather, the projects and programs are grouped together for administrative purposes, for joint controlling or for cross-project resource management.

Process
A process is a directed sequence of activities and events that converts input variables (status, documents ...) into output variables (status, documents ...).

Process elements
Process elements of a schedule describe activities, milestones and relationships.

Process model
A process model combines methods and elements, processes and phases for a standardized project process. It consists of structural, procedural and functional building blocks.

Product backlog
The product backlog is the only source from which the Scrum team derives its work. It contains the overall product objectives and other entries in the form of prioritized requirements to achieve the product objectives. The product owner is responsible for the creation and continuous maintenance of the product backlog.

Product owner
The product owner is responsible for the commercial success of the product to be developed and tries to maximize this success. He or she represents the relevant customer requirements from a business perspective and summarizes these in a so-called product backlog (i.e. a catalog of requirements). The individual user needs are prioritized in this product backlog and made available to the developers.

Program
A program consists of a series of closely linked projects that pursue a common, overarching strategic goal.

Progress report
Progress reports are report forms that summarize the project status on a specific reporting date.

Project
A project is usually a unique endeavor that can be distinguished from other tasks. It has limited time, financial, personnel and material resources. Projects pursue defined goals and have a project-specific organization.

Project file
The project file refers to the set of all project-related documents.

Project governance
The project governance refers to the management framework in which decisions are made. It comprises rules, processes and structures for the organizational structuring of the project and for ensuring the ability to act.

Project lifecycle
The project lifecycle describes the course of a project from the very start of the project to its end.

Project management
Project management refers to the entirety of tasks, methods and means from the areas of definition, planning, control, closure and management for the successful completion of projects.

Project management success
Project management success refers to the recognition of the project results by the relevant interested parties (= stakeholders).

Project organization
The project organization is a group of people and the related infrastructure for which agreements are made regarding authority, relationships and responsibilities in alignment with the business and functional processes.

Project phase
Project phases structure the project into separate time periods and thus define the rough course of the project.

Project plan
The project plan includes all plans that exist in a project.

Project portfolio
A project portfolio comprises a series of projects and/or programs that are not necessarily related to one another. Rather, the projects and programs are grouped together for administrative purposes, for joint controlling or for cross-project resource management.

Project structure
The project structure refers to the entirety of tasks, work packages and activities that together represent the project. Structuring can be based on different aspects, such as time, function or object-related. In principle, the project structure comprises both the process structure and the organizational structure of a project, whereby only the organizational structure is defined in the work breakdown structure.

Quality
Quality is the degree to which an object fulfills certain requirements.

Quality gate
Quality gates are milestones with a review character. A quality gate defines in advance which quality criteria must be met in order to be allowed to enter the next project phase. The quality criteria are checked by a defined committee.

Quality management
Quality management comprises the definition of quality policy, the derivation of quality objectives, their planning and control as well as the continuous improvement of quality.

Relationship
Relationships link activities and milestones. They establish a quantifiable and logical relationship between the elements of the schedule.

Release plan
The release plan is an overview plan that shows which user stories are to be implemented in which sprint. It therefore takes into account the priority of the user stories, their estimated size, the velocity of the team and the strategic and economic constraints of the product.

Report
Reports in the sense of progress reports are report forms that summarize the project status on a specific reporting date.

Requirement
A requirement is the documented representation of a condition or capability needed to solve a problem or achieve an objective, or a condition or capability of a system to fulfill a contract, standard, specification or other document.

Requirements specification
With the requirements specification, the project documents how the user needs are to be implemented.

Resonant leadership
Resonant leadership arouses positive feelings, generates enthusiasm and passion and thus strengthens motivation and commitment to the project.

Responsibility
A responsibility is an obligation associated with a specific task to ensure that the right things are done to fulfill the task or achieve a goal and to take ownership of it.

Resources
Resources are human, financial and material resources in quantifiable quantities that are required for the successful completion of the project.

Risk
Risks and opportunities are potential, unplanned events that are characterized by a probability of occurrence and an impact on the project objectives. Risks counteract the project objectives, i.e. the project suffers damage. Opportunities support the objectives.

Risk buffer
A risk buffer, also known as a safety buffer, is a temporal, financial or qualitative buffer for unplanned situations. This buffer is used to set aside additional time, money or resources for difficult project phases or work packages.

Risk index
The risk index is the product of the probability of occurrence and damage category. It can be calculated for each individual risk. The sum of the risk indices of all individual risks is the overall risk index of the project.

Role
A role describes a position in the project that is associated with a certain responsibility and the corresponding authority. Usually, a role is tied to certain expectations from other stakeholders.

Root element
The root element refers to the top level of the work breakdown structure, i.e. the entire project task.

Safety buffer
A safety buffer, also known as a risk reserve, is a temporal, financial or qualitative buffer for unplanned situations. This buffer is used to reserve additional time, additional money or additional resources for challenging project phases or work packages.

Scenario
Scenarios are models of the future; they are intended to help illustrate the potential future course of a project. They go further than the pure forecast values of the earned value analysis. Soft factors and additional estimates can also be incorporated into the development of a scenario.

Schedule
The schedule completes the flow chart with specific deadlines by taking into account realistic lead and waiting times.

Scrum
Scrum is a well-known representative of agile process models. The project deliverables are completed incrementally in fixed time periods (sprints).

Scrum master
The Scrum master is an expert in the planning and implementation of projects with Scrum. They support the whole team in working efficiently and promote Scrum to people outside the team. The Scrum master is also a coach who supports the product owner and the developers with all questions relating to the process model. The Scrum master also solves problems that the team cannot solve itself.

Scrum team
The Scrum team consists of the developers, the product owner and the Scrum master.

Sprint
A sprint is a fixed period of time in which the requirements for achieving the sprint goal are implemented and an increment is created. In practice, typical sprints usually last between 14 and 30 days.

Sprint backlog
The sprint backlog contains the sprint goal of the current sprint, the entries (user stories) selected from the product backlog for implementation and the tasks derived from them to achieve the sprint goal.

Sprint planning

The sprint planning is intended to answer two questions: 1) What is the goal of the sprint and which entries in the product backlog (user stories) should be implemented to achieve the goal? 2) Which tasks have to be completed to implement the selected entries, i.e. how is the implementation to be carried out?

Sprint retrospective

In the sprint retrospective, cooperation within the team and the interfaces to other areas of the organization are analyzed and opportunities for improvement are planned with regard to increasing quality and effectiveness. The sprint retrospective takes place at the end of each sprint after the sprint review.

Sprint review

In the sprint review, the increment is presented and evaluated by the Scrum team. Participants are the Scrum team and other stakeholders who may be invited by the Product Owner, for example customers. The sprint review is a working meeting that is subject to the timebox principle.

Stakeholder

Stakeholders are persons or groups of persons who are interested in the project, who are involved in the project, who can influence the project or who are or who feel affected by the project.

Story point

Story points are a measure of the size of a user story in Scrum. The size reflects the complexity of the requirements documented in the user story and is independent of the effort required. The latter also depends on the skill of the person working on it; this is not taken into account when estimating the story points.

Strategy

A strategy represents a fundamental, long-term behavior of an organizational unit towards its environment to achieve long-term goals.

Student syndrome

Student syndrome refers to the observation that many people tend to complete pending tasks at the last minute.

Sub-project

A sub-project is an organizational sub-unit of the project that is separate from other units in terms of its work packages or duties. It is made up of several work packages. A sub-project manager is responsible for it. Sub-projects can refer to other work breakdown structures and be detailed there.

Sub-task

A sub-task is obviously a part of a task and can itself be subdivided into further sub-tasks and/or work packages. A sub-task can be used for structuring in the work breakdown structure. However, it can neither represent the overall project nor be at the lowest level of the work breakdown structure, which is reserved for work packages.

Tailoring

Tailoring refers to the individual adaptation of a standardized process model to a specific project.

Task

A task is a work assignment under defined conditions and using certain resources. Depending on the scope and hierarchical level, a task can be an entire program, a project, a sub-project, a work package, an activity or an even smaller unit. Tasks can be subdivided into sub-tasks.

Taskboard

The taskboard is used to visualize the sprint backlog in Scrum. The user stories of the current sprint are broken down into tasks and usually ordered according to their processing status in columns “to do”, “in progress” and “done”.